FRONTIERS IN INVERTEBRATE PHYSIOLOGY

A Collection of Reviews

Volume 1

Non-Bilaterian Phyla

Frontiers in Invertebrate Physiology: A Collection of Reviews, 3-volume set
Hardback ISBN: 978-1-77491-398-7
Paperback ISBN: 978-1-77491-399-4

Volume 1: Non-Bilaterian Phyla
Hardback ISBN: 978-1-77491-400-7
Ebook ISBN: 978-1-00340-331-9
Paperback ISBN: 978-1-77491-401-4

Volume 2: Crustacea
Hardback ISBN: 978-1-77491-402-1
Ebook ISBN: 978-1-00340-501-6
Paperback ISBN: 978-1-77491-403-8

Volume 3: Annelida and Echinodermata
Hardback ISBN: 978-1-77491-404-5
Ebook ISBN: 978-1-00340-518-4
Paperback ISBN: 978-1-77491-405-2

Cover photo credits:
Left: Sea Sponges on coral reef of Maldives.

Second from left: *Trichoplax adhaerens.* Photo by B. Schierwater. https://creativecommons.org/licenses/by/4.0/deed.en

Third from left: Staghorn coral (*Acropora cervicornis*). Photo by Albert Kok. Public domain.

Right: The yellow tube sponge, *Aplysina fistularis*, the purple vase sponge, *Niphates digitalis*, the red encrusting sponge, *Spiratrella coccinea*, and the gray rope sponge, *Callyspongia* sp. Caribbean Sea, Cayman Islands. Public domain.

FRONTIERS IN INVERTEBRATE PHYSIOLOGY

A Collection of Reviews

Volume 1

Non-Bilaterian Phyla

Edited by

Saber Saleuddin, PhD

Sally P. Leys, PhD

Robert D. Roer, PhD

Iain C. Wilkie, PhD

APPLE ACADEMIC PRESS

First edition published 2024

Apple Academic Press Inc.
1265 Goldenrod Circle, NE,
Palm Bay, FL 32905 USA

760 Laurentian Drive, Unit 19,
Burlington, ON L7N 0A4, CANADA

CRC Press
2385 NW Executive Center Drive,
Suite 320, Boca Raton FL 33431

4 Park Square, Milton Park,
Abingdon, Oxon, OX14 4RN UK

Apple Academic Press exclusively co-publishes with CRC Press, an imprint of Taylor & Francis Group, LLC

Library and Archives Canada Cataloguing in Publication

Title: Frontiers in invertebrate physiology: A Collection of Reviews / edited by Saber Saleuddin, PhD, Sally P. Leys, PhD, Robert D. Roer, PhD, Iain C. Wilkie, PhD.
Names: Saleuddin, Saber, editor. | Leys, Sally P., editor. | Roer, Robert D. (Robert David), 1952-editor. | Wilkie, Iain C., editor.
Description: First edition. | Includes bibliographical references and indexes. | Contents: Volume 1. Non-bilaterian phyla.
Identifiers: Canadiana (print) 20230551483 | Canadiana (ebook) 20230551521 | ISBN 9781774913987 (set ; hardcover) | ISBN 9781774913994 (set ; softcover) | ISBN 9781774914007 (v. 1 ; hardcover) | ISBN 9781774914014 (v. 1 ; softcover) | ISBN 9781774913987 (set ; ebook) | ISBN 9781003403319 (v. 1 ; ebook)
Subjects: LCSH: Invertebrates—Physiology.
Classification: LCC QL364 .F76 2024 | DDC 592—dc23

Library of Congress Cataloging-in-Publication Data

CIP data on file with US Library of Congress

ISBN: 978-1-77491-400-7 (hbk)
ISBN: 978-1-77491-401-4 (pbk)
ISBN: 978-1-00340-331-9 (ebk)

Dedication

The editors dedicate this series to our mentors—
Vera Fretter (Saleuddin), George Mackie (Leys),
Karl Wilbur (Roer), and Norman Millott (Wilkie)—for their
knowledge and encouragement in guiding us to become
invertebrate physiologists.

About the Editors

Saber Saleuddin, PhD
University Professor Emeritus, Department of Biology, York University in Toronto, Ontario, Canada

Saber Saleuddin, PhD, is a University Professor Emeritus in the Department of Biology, York University in Toronto, Ontario, Canada. His research on biomineralization in mollusks started at the University of Alberta in Edmonton, Canada, and continued at Duke University, Durham, North Carolina, in the laboratory of Karl Wilbur. Though offered a teaching position at Duke University, he accepted a faculty position at York University, Canada, where he taught for nearly 40 years. His outstanding contributions in teaching, research, and administration were recognized by York University in awarding him a university professorship. He has published extensively in international journals and has co-edited seven books on molluscan physiology, including several published by Apple Academic Press. He served as co-editor of the *Canadian Journal of Zoology* for 18 years and was president of the Canadian Society of Zoologists, from whom he received a Distinguished Service Medal. He was Chair of the Department of Biology, acting Dean of Research, and Chair of the Senate of York University. He received his early education in Bangladesh and his doctorate from the University of Reading in the UK.

Sally P. Leys, PhD
Professor, Department of Biological Sciences at the University of Alberta, Edmonton, Canada

Sally P. Leys, PhD, is a Professor in the Department of Biological Sciences at the University of Alberta, in Edmonton, Canada. She held a Commander C. Bellairs Postdoctoral Fellowship from McGill University, Canada, for postdoctoral research in Barbados (1997) and an NSERC Postdoctoral Fellowship from 1998–2000 with which she worked both at the University Aix Marseille, France, and the University of Queensland, Australia. She won an Natural Sciences and Engineering Research Council of Canada (NSERC) Women's University Research Award in 2000 and was Assistant Professor at the University of Victoria, British Columbia. In 2002 Dr. Leys became a Canada Research Chair Tier II in Evolutionary and Developmental Biology at the University of Alberta. Professor Leys' research interests broadly concern understanding the origin of multicellularity in metazoans and, more specifically, the cellular and molecular characteristics of representative groups of non-bilaterian animals (sponges, ctenophores, placozoans, and cnidarians). Most of her work centers on studies on sponges, and she is a world expert on many aspects of sponge biology, but in particular on the physiology, ecology, and cell biology of glass sponges. Dr. Leys has authored over 100 journal articles, book chapters, and advisory reports. She has supervised many graduate students. She is presently a member of the Council of Ocean Networks Canada and was the President of the Canadian Society of Zoologists from 2020-2021. She obtained her BSc from the University of British Columbia (1990) and PhD from the University of Victoria under George Mackie in 1996, for which she received the Canadian Society of Zoologists Cameron Award in 1997.

Robert D. Roer, PhD
Professor, Emeritus of Biology and Marine Biology,
University of North Carolina Wilmington, North Carolina, USA

Robert D. Roer, PhD, is a Professor Emeritus of Biology and Marine Biology at the University of North Carolina, Wilmington North Caroline, USA. He joined the faculty at UNCW in 1979 and was promoted to the rank of professor in 1990. Dr. Roer served as Dean of the Graduate School and Research and Chief Research Officer at UNCW from 2002 to 2012. He was the recipient of UNCW's first Graduate Mentor Award and of the Chancellor's Teaching Excellence Award. He retired from the faculty in 2016. Dr. Roer is a member of the Society for Integrative and Comparative Biology (SICB), the Society of Sigma Xi, The Crustacean Society, and Phi Kappa Phi. He served as Treasurer of SICB and President of the NC Conference of Graduate Schools. He was on the executive board of the Conference of Southern Graduate Schools. Dr. Roer conducted research principally in the fields of biomineralization and ion transport and is the author of numerous journal articles and book chapters. His research was funded by grants from NSF, NOAA, NASA, NC Sea Grant, and other state and national agencies. He received his ScB in Aquatic Biology from Brown University in 1974 and his PhD in Zoology from Duke University in 1979.

Iain C. Wilkie, PhD
Affiliate Researcher, University of Glasgow, Glasgow, Scotland

Iain C. Wilkie, PhD, is an Affiliate Researcher in the School of Biodiversity, One Health and Veterinary Medicine at the University of Glasgow Scotland. He held a faculty position at Glasgow Caledonian University for over 30 years. He retired from there as Head of Department of Biological and Biomedical Sciences in 2008 in order to pursue his research interests, which are echinoderm physiology and functional morphology, invertebrate matrix biology, and the physiology and mechanics of animal detachment mechanisms. He has published over 90 papers and is on the editorial board of *Zoomorphology* and is Chief Editor of *The Glasgow Naturalist*. In 2006 he was awarded an honorary doctorate in Biodiversity and Evolutionary Biology by the University of Milan, Italy. He was recently in receipt of a Leverhulme Emeritus Fellowship. Dr. Wilkie earned a BSc and PhD in Zoology from the University of Glasgow, Scotland.

Contents

Contributors

D. Allemand
Centre Scientifique de Monaco, Department of Marine Biology, Monaco

Kristina Bayer
Marine Symbioses Unit, Department of Marine Ecology, GEOMAR Helmholtz Center for Ocean Research Kiel, Düsternbrooker Weg 20, Kiel, Germany

Kathrin Busch
Marine Symbioses Unit, Department of Marine Ecology, GEOMAR Helmholtz Center for Ocean Research Kiel, Düsternbrooker Weg 20, Kiel, Germany

N. Caminiti-Segonds
Centre Scientifique de Monaco, Department of Marine Biology, Monaco

L. Capasso
Centre Scientifique de Monaco, Department of Marine Biology, Monaco

Ching-Fong Chang
Center of Excellence for the Oceans, National Taiwan Ocean University; Department of Aquaculture, Taiwan National Taiwan Ocean University, Keelung, Taiwan

Wassim Elkhatib
Department of Biology, University of Toronto Mississauga, Mississauga, Canada

Emma J. K. Esposito
University of Alberta, Edmonton, AB, Canada

P. Ganot
Centre Scientifique de Monaco, Department of Marine Biology, Monaco

Julia Gauberg
Department of Biology, University of Toronto Mississauga, Mississauga, Canada

Amanda S. Kahn
Moss Landing Marine Laboratories and San Jose State University, California, USA

Vasliki Koutsouveli
RD3 Marine Symbioses, GEOMAR Helmholtz Center for Ocean Research Kiel, Düsternbrooker Weg 20, Kiel, Germany

Emilio Lanna
Federal University of Bahia, Biology Institute. Barão de Jeremoabo Street, s/n. Campus de Ondina, Salvador, BA, Brazil

Sally P. Leys
Department of Biological Sciences, University of Alberta, Edmonton, AB, Canada

María López-Acosta
Department of Oceanology, Instituto de Investigaciones Marinas (IIM-CSIC), Vigo, Pontevedra, Spain

Manuel Maldonado
Department of Marine Ecology, Center for Advanced Studies of Blanes (CEAB-CSIC), Blanes, Girona, Spain

Angela M. Marulanda-Gómez
Marine Symbioses Unit, Department of Marine Ecology, GEOMAR Helmholtz Center for Ocean Research Kiel, Düsternbrooker Weg 20, Kiel, Germany

Teresa M. Morganti
Max Planck Institute for Marine Microbiology; HGF MPG Joint Research Group for Deep-Sea Ecology and Technology, Celsiusstr. 1, Bremen, Germany

Lucía Pita
Marine Symbioses Unit, Department of Marine Ecology, GEOMAR Helmholtz Center for Ocean Research Kiel, Düsternbrooker Weg 20, Kiel, Germany; Department of Marine Biology and Oceanography, Institut de Ciències del Mar, ICM-CSIC, Passeig de la Barceloneta 37-49, Barcelona, Spain

Ana Riesgo
Department of Biodiversity and Evolutionary Biology, Museo Nacional de Ciencias Naturales (CSIC), Calle José Gutiérrez Abascal 2, Madrid, Spain; Department of Life Sciences, Natural History Museum, London, UK

Richard A. Satterlie
Department of Biology and Marine Biology, University of North Carolina Wilmington, Wilmington, NC

Adriano Senatore
Department of Biology, University of Toronto Mississauga, Mississauga, Canada

Shinya Shikina
Institute of Marine Environment and Ecology, National Taiwan Ocean University;
Center of Excellence for the Oceans, National Taiwan Ocean University, Keelung, Taiwan

Anhadvir Singh
Department of Biology, University of Toronto Mississauga, Mississauga, Canada

E. Tambutté
Centre Scientifique de Monaco, Department of Marine Biology, Monaco

S. Tambutté
Centre Scientifique de Monaco, Department of Marine Biology, Monaco

N. Techer
Centre Scientifique de Monaco, Department of Marine Biology, Monaco

A. A. Venn
Centre Scientifique de Monaco, Department of Marine Biology, Monaco

D. Zoccola
Centre Scientifique de Monaco, Department of Marine Biology, Monaco

Abbreviations

ABTP	Ashbridges Bay Treatment Plant
ACC	amorphous calcium carbonate
ACC-H2O	hydrated amorphous calcium carbonate
AKDF	AMPA-Kainate-Delta-Phi
AMPs	antimicrobial peptides
ANR	assimilatory nitrate reduction
AOA	ammonia-oxidizing archaea
AOB	ammonia-oxidizing bacteria
APO	assimilatory phosphite oxidation
ASICs	acid-sensing ion channels
ATD	amino-terminal domain
ATP	adenosine triphosphate
BAP	bacterial alkaline phosphatase
CICR	Ca^{2+}-induced Ca^{2+} release
COGs	cluster orthologous groups
CPA	crystallization by particle attachment
CRISPR	clustered regularly interspaced short palindromic repeats
CRY	cryptochrome
DAG	diacylglycerol
DIC	dissolved inorganic carbon
DNRA	dissimilatory nitrate reduction to ammonium
DOC	dissolved organic carbon
DOM	dissolved organic matter
DON	dissolved organic nitrogen
DOP	dissolved organic P
DPO	dissimilatory phosphite oxidation
ECM	extracellular calcifying medium
EMP	earth microbiome project
ENaC	epithelial sodium channel
ER	endoplasmic reticulum
ESD	environmentally induced sex determination
FAD	flavin adenine dinucleotide
FISH	fluorescence *in situ* hybridization

GABA	gamma-aminobutyric acid
GAD	glutamic acid decarboxylase
GE	growing edge
GIRK	G protein activated inward rectifying K^+
GPCRs	G-protein coupled receptors
HIFα	hypoxia-inducible transcription factor
HMA	high microbial abundance
HMG	high mobility group
HPLC	high-performance liquid chromatography
HPO42	hydrogen phosphate
IP3	inositol triphosphate
IRFs	interferon regulatory factors
IRs	ionotropic receptors
ISH	*in situ* hybridization
LBD	ligand binding domain
LC-MS	liquid chromatography-mass spectrometry
LH	luteinizing hormone
LMA	low microbial abundance
LP	lipoprotein
LPS	lipopolysaccharide
LRR	leucine-rich repeat
LSP	lateral skeleton preparative
MALDI	matrix-assisted laser desorption/ionization
MALDI-IMS	matrix-assisted laser desorption-ionization imaging mass spectrometry
MAMPs	microbe-associated molecular patterns
MAP	mitogen-activated protein
MAPK	mitogen-activated protein kinase
MNN	motor nerve net
MRTF	myocardin-related transcription factor
NED	non-embryonic development
NF-Kb	nuclear factor kappa B
NH3	ammonia
NH4+	ammonium
NO	nitric oxide
NOB	nitrite-oxidizing bacteria
NOD	nucleotide-binding and oligomerization domain
NOS	nitric oxide synthase
NR	nerve ring

NRPS	nonribosomal peptide synthetase
NRs	nuclear receptors
OA	ocean acidification
OM	organic matrix
PAM	peptidylglycine monooxygenase
PHD	prolyl-hydroxylase domain
PKA	protein kinase A
PKC	protein kinase C
PKS	polyketide synthetase
PMCA	plasma membrane calcium ATPase
PMCs	primary mesenchyme cells
PNG	peptidoglycan
POC	particulate organic carbon
Poly-P	polyphosphate
PRRs	pattern recognition receptors
RA	retinoic acid
RGS	regulators of G protein signaling
ROVs	remotely operated vehicles
RTK	receptor tyrosine kinase
SDH	sorbitol dehydrogenase
SEM	scanning electron microscopy
SR	scavenger receptor
SRCRs	scavenger receptor cysteine-rich
SSC	sponge-specific clusters
TEM	transmission electron microscopy
TLRs	toll-like receptors
TRP	transient receptor potential
VHL	von Hippel Lindau

Acknowledgments

The editors want to thank our co-authors, colleagues, and friends for contributing their incredible knowledge, time and talents to this endeavor. All have worked hard to make this a current and important volume on issues in the physiology of the invertebrates. You have our sincerest gratitude.

Preface

The earliest invertebrate fossils are some 300 million years older than the first recognized chordates. This early appearance was followed by the extraordinary diversification exhibited by the fossil forms of the Cambrian (550 million years ago). The varied forms have been accompanied by an extreme proliferation in numbers. Invertebrates exhibit the largest number of species and the largest number of individuals. They occupy virtually every conceivable ecological niche. They are economically important in food chains, they recycle organic waste, are crucial pollinators of plants and they are sources of food. They are also medically important as parasites causing important diseases in both humans and livestock. The contributions of invertebrate studies in biology have been significant. Studies on *Drosophila* gave rise to how development is controlled at the molecular level. The structure and anatomy of molluscan neurons were crucial in our understanding of the action potential of the process of learning. Recent genomic studies using *Caenorhabditis* show that progeny remember hostile bacteria. These examples show the importance of the remarkable diversity of structures exhibited in specific invertebrates which offer insights into processes whose importance in our survival we may not even appreciate at this time but may be of crucial importance in the future.

Saber Saleuddin's association with the Apple Academic Press goes back to 2017 when he coedited with Spencer Mukai a series on the *Physiology of Molluscs: A Collection of Selected Reviews* (2-volume set), and then in 2020, he coedited a series with Angela Lange and Ian Orchard on *Advances in Invertebrate Endocrinology* (2-volume set). The latter series had no papers on crustaceans and helminths, and the present series with Sally Leys, Robert D. Roer, and Iain C. Wilkie now attempts to address this deficiency. This new series, although not covering all taxa, has papers that represent important taxa and we are grateful to the current authors for their contributions. The papers in this series from prominent invertebrate physiologists are timely and appropriate in the context of significant advancement in the phase of post-genomic information. Many of the papers are at the cutting edge of research using invertebrates and have opened our understanding of evolutionary and regulatory pathways and increased the depth of our understanding.

Volume 1 contains chapters representing non-bilaterians (sponges, cnidarians, placozoans). The focus on sponge biology has recently been on symbiosis, nutrient uptake, and sensory biology. The focus on cnidarians has primarily been on biomineralization, the nervous system, and development while the contribution of placozoans examines neuropeptide biology involved in feeding. We have added a previously published paper on neuropeptides in cnidarians.

Volume 2 has chapters on crustacean physiology and contains diverse physiological topics ranging from molting, respiration, water balance, biomineralization, bioreceptors, and temperature regulation to the land adaptation of terrestrial crustaceans. Each paper is comprehensive, and all add new knowledge to crustacean biology.

Volume 3 contains papers on echinoderms and annelids. Temporary adhesion and regeneration are two important areas in echinoderm biology and the review of juxtaligamental cells, which may regulate the mechanical properties of connective tissue, is an important new addition; a previously published paper on echinoderm neuropeptides is included. Annelid physiology is represented by three new reviews covering the neurobiology of locomotion in leeches, regeneration, and reproduction as well as a paper previously published that examines the neuro-endocrine-immune response in annelids.

Editing a series in invertebrate physiology is a daunting task because of the enormous range of topics involved. The chapters in these volumes provide readers with up-to-date coverage of topical issues, and for this, we are indebted to the authors. We are also grateful to reviewers for their critical appraisals which made our work easier. The papers in this series will advance the knowledge of invertebrate physiology and serve as an important resource for advanced undergraduate students, graduate students, and researchers/ professionals.

We are grateful to Sandra Sickels, Ashish Kumar, and especially Rakesh Kumar of AAP for their support, encouragement, and patience and we appreciate the role of AAP in scientific publications.

—Saber Saleuddin, PhD
Sally Leys, PhD
Robert D. Roer, PhD
Iain C. Wilkie, PhD

CHAPTER 1

Physiology of Reproduction in Porifera

EMILIO LANNA,[1] ANA RIESGO,[2,3] VASLIKI KOUTSOUVELI,[4] and SALLY P. LEYS[5]

[1]*Federal University of Bahia, Biology Institute, Campus de Ondina. Salvador, BA, Brazil*

[2]*Department of Biodiversity and Evolutionary Biology, Museo Nacional de Ciencias Naturales (CSIC), Madrid, Spain*

[3]*Department of Life Sciences, Natural History Museum, London, UK*

[4]*RD3 Marine Symbioses, GEOMAR Helmholtz Centre for Ocean Research Kiel, Kiel, Germany*

[5]*Department of Biological Sciences, University of Alberta, Edmonton, Canada*

ABSTRACT

Sponges (Porifera) are one of the most ancient animals present on the planet. They are aquatic, filter-feeding sessile metazoans that rely on asexual and sexual reproduction. These animals have a long history on Earth and had plenty of time to develop different reproductive strategies. Here, we review different aspects of the physiology of reproduction in Porifera. This chapter is divided into six sections. In the first section, we present general features of sponge reproduction, such as factors that trigger the onset of their reproduction, as well as the periodicity of their reproductive cycles. The molecular basis of the hormonal control of gametogenesis is presented although sponges have no endocrine system. The second section deals with gametogenesis,

Frontiers in Invertebrate Physiology: A Collection of Reviews, Volume 1: Non-Bilaterian Phyla.
Saber Saleuddin, Sally P. Leys, Robert D. Roer, and Iain C. Wilkie, (Eds.)

including how sex and the germline are determined and maintained in this group, how oocytes and spermatozoa are formed and nourished, and how they behave once released. The third section reviews different topics about the reproductive mode. Here, we discuss the dichotomy in reproductive mode: oviparity vs. viviparity, the spatial distribution of the reproductive elements in the sponge tissue, the effect of symbiosis in reproduction (and vice-versa), and energetic trade-offs during reproduction. The fourth section describes fertilization, and we cover the factors controlling the spawning events and how the sperm are attracted and recognized by the egg. The diversity of developmental modes, the molecular control of sponge embryonic development, and the maternal-embryo relationship are discussed in the fifth section. Finally, in the sixth section, the types of asexual reproduction, factors influencing budding, gemmulation, hibernation, and gemmule development are described. Knowledge about the physiology of reproduction of sponges is still fragmentary and based on studies in very few species. Consequently, there are many generalizations that need further investigation. However, evidence-based on morphological, experimental, and molecular data demonstrates that their physiology is not very different from that of other metazoans.

1.1 INTRODUCTION

1.1.1 WHAT IS A SPONGE?

Sponges (phylum Porifera) are largely immobile, aquatic animals (Bergquist, 1978; Leys & Hill, 2012; Leys & Riesgo, 2012). Their body plan is adapted to filter water to extract food, exchange gas, and expel wastes (in most cases, except in carnivorous sponges [Cladorhizidae, Poecilosclerida, Demospongiae]). Within the range of physiological activities they carry out, perhaps the best studied are feeding and reproduction. The body of most sponges, except for some carnivorous sponges, is organized around the aquiferous system, a series of epithelial-lined canals that are supported by a collagenous mesohyl. The mesohyl is an extracellular space that houses a diversity of cell types as well as a few or many microbial symbionts (Figure 1.1(A)). In general, the canals forming the aquiferous system begin at openings in the outer epithelium called ostia and lead as epithelial-lined tubes through to choanocyte chambers. Here choanocytes, cells with a flagellum surrounded by a collar of microvilli, create a unidirectional water current from the ostia to the osculum. Filtered water leaves the chambers to excurrent canals, which merge to form larger canals, finally opening to a large atrium or directly to the osculum, where the water will leave the sponge in a jet-like manner (e.g., Korotkova, 1981; Simpson, 1984; Da Hora et al., 2018). The description above relates to a leuconoid type of aquiferous system found in three of the four classes of sponges. However, in the Calcarea, four different types of aquiferous systems (asconoid, syconoid, sylleibid, and solenoid) are found, varying in the extensiveness of the canal system. Sponges also vary significantly in the thickness of the mesohyl (generally referred to as thick or thin-walled sponges (Pavans de Ceccatty, 1974)), their microbial symbiotic content (high or low microbial abundance (e.g., Vacelet & Donadey, 1977; Taylor et al., 2007; Webster & Taylor, 2012; Thacker & Freeman, 2012)), their general body shape (encrusting, massive, etc. (Boury-Esnault & Rutzler, 1997)) and, of course, spicule nature and complement. Through the aquiferous system, sponges do not only feed, but they absorb or release gametes and other reproductive elements such as embryos/larvae. Thus, the type of aquiferous system and the mode of reproduction established in the different classes and species are interconnected.

The phylum Porifera is monophyletic. There is currently a debate as to whether Ctenophora or Porifera is the sister group to all metazoans (Figure 1.1(B) and (C); Dunn et al., 2008; Pick et al., 2010; Pisani et al., 2015; Feuda et al., 2017; Whelan et al., 2017; Laumer et al., 2019). Although phylogenomic studies have argued both outcomes (*see* Li et al., 2021 for a review), whether Porifera or Ctenophora is the sister group to all metazoans does not explain what the ancestor was like, and both hypotheses offer a rich collection of avenues to explore the origin of animal characters including muscle, nervous systems, enzyme-secreting gastral layers, and mechanisms of reproduction. Currently, there are approximately 9,500 known extant species of sponges (De Voogd et al., 2022). They are found since the early Cambrian era (ca. 535 Mya, (Antcliffe, 2015)) at least, although molecular clock analyses (Gold et al., 2016; Plese et al., 2021) and intriguing fossils (Turner, 2021) place the origin of sponges a little earlier in the Proterozoic. This diversity of species is distributed into four monophyletic classes (Figure 1.1(D)): Demospongiae, the most speciose, has sponges with and without mineralized skeletons. Some lineages of the group have their body supported by spongin fibers only. However, in the lineages that possess spicules, these structures are made of silica (although some can secrete an aragonite base in addition to producing siliceous spicules) and their choanoderm and pinacoderm are cellular; Hexactinellida, known as glass-sponges are found usually in deep or cold water and comprise sponges with siliceous spicules, but their tissues, including choanoderm and pinacoderm, are syncytial. The class Homoscleromorpha is the least speciose of the four and comprises marine sponges with siliceous (or lacking) spicules. The choanoderm and pinacoderm are cellular and are conspicuously underlined by basement membranes. And finally, the fourth class of Porifera is Calcarea. Although their choanoderm and pinacoderm are also cellular, their skeleton is made of calcium carbonate. In addition to these general characteristics, different aspects of the reproduction of sponges are correlated with their phylogeny, namely the cell types that originate the gametes, and the type of larva, among others.

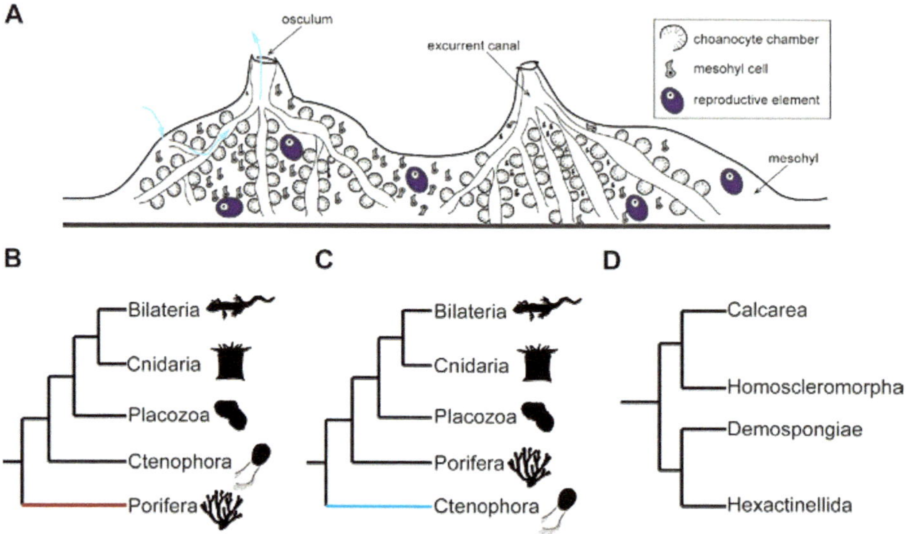

FIGURE 1.1 Porifera body plan and phylogeny. (A) Schematic drawing of the sponge body plan showing the mesohyl and the components of the aquiferous system; (B) and (C) competing hypothesis of the sister groups of metazoans: (B) Porifera and (C) Ctenophora; (D) phylogeny of the classes of Porifera.

Source: Silhouette images from phylopic.org.

Sponges are sessile organisms living on the benthos; the vast majority of sponges are marine, but a group of demosponges – Spongillida – invaded the freshwater. Sponges cannot move more than at a rate of cell crawling, although recent studies suggest some sort of slow locomotion for some species (Maldonado & Uriz, 1999; Morganti et al., 2021). Therefore, one of their strategies to deal with competitors and predators is to produce a plethora of secondary metabolites, many of which are actually produced by prokaryotic symbionts living in association with the sponge. These metabolites have been intensely studied for several decades (*see* Busch et al., 2022). They have a different chemical composition and serve a wide array of functions for humans: anti-tumor, anti-inflammatory, anti-malarial, anti-mycobacterial, and anti-fungal properties; yet most of these metabolites are still not being used in the treatment of several diseases because synthesizing them in the laboratory is very costly and inefficient (Koopmans et al., 2009). Consequently, understanding the reproduction and development of these animals is pivotal to preparing protocols to produce larger biomasses of the sponges and/or management plans to extract them from nature.

Finally, sponges are fundamental organisms ecologically, being key species in the communities where they are found. They play a plethora of roles in the benthic realm, from being food for spongivorous fishes and turtles to serving as a refuge for several invertebrates and vertebrates for survival and reproduction, and consolidating the substrate (Archer et al., 2020). They are also important in the benthopelagic coupling of food webs, as they filter large quantities of water all the time (Skelton & Strand, 2013; McMurray et al., 2014; Kahn et al., 2015). They are also important in the biogeochemical cycle of different chemical products (e.g., N, P, Si) and can form reefs (animal forests) in different parts of the world (*see* Maldonado et al., 2022, this volume). Different species can also be important for different steps of ecological succession in a given habitat. Their distribution, biomass, and growth, among other characteristics, will impact the reproduction of the organisms and vice versa. An important characteristic of the sponges is the diversity of associations they have with other organisms (Wulff, 2006). From minute microbes living in their mesohyl to the refuge of marine turtles, all sponges are living in association with other organisms. Depending on the density of microbes living in association with the sponges, they can be classified as organisms having low microbial abundance (the LMA sponges) and others that have high microbial abundance (HMA) in their mesohyl (Gloeckner et al., 2014).

The reproduction of sponges is therefore crucial to understand the evolution of animals, providing a framework for obtaining their metabolites for socioeconomic use by humans, and better understanding and predicting their ecological roles. In the next sections, we describe different physiological aspects of their reproduction. First, we present the general features of sponge reproduction. Next, we discuss gametogenesis, reproductive mode, fertilization, and embryogenesis. Finally, we describe different aspects of their asexual reproduction, a striking characteristic of the group.

1.2 GENERAL FEATURES OF SEXUAL REPRODUCTION

Reproduction has to be tightly controlled to be successful. If a single individual starts its reproductive process out of phase with the rest of the population, it wastes energy that could have been allocated to other physiological processes. Death could be a consequence of this mismatch. At the same time, the organism needs to control the amount of energy it allocates to its offspring (i.e., fecundity) and how often it reproduces. Therefore, different species are generally adapted to control their reproductive timing, periodicity, and frequency of reproduction over time (Stearns, 1976).

1.2.1 WHAT TRIGGERS THE ONSET OF REPRODUCTION IN PORIFERA?

Sponge gametes are produced throughout the body: eggs and sperm are not concentrated in a single structure (e.g., ovary or testis; see below) although they can be associated with choanocyte chambers. Sponges have some types of hormonal control (see below), and as with many other marine invertebrates, the onset of reproduction in sponges is triggered by environmental cues. Among these cues, temperature has been recognized as one of the main factors controlling the reproduction of sponges (e.g., Elvin, 1976; Fell, 1976; Simpson, 1980; Kaye & Reiswig, 1991; Ereskovsky, 2000; Maldonado & Riesgo, 2008a; Riesgo & Maldonado, 2008; Lanna et al., 2015, 2018, 2021). From polar to tropical regions, species appear to be adapted to reproduce within a determinate range of temperatures. Some polar demosponges reproduce when the water temperature decreases (Ereskovsky, 2000), but usually, the period of reproduction is linked to the increase in the temperature of the water. Interestingly, different species of sponges living in sympatry, consequently under the influence of the same thermal regime, either in a temperate region of the Mediterranean Sea (Riesgo & Maldonado, 2008) or in a tropical beach of the Atlantic Ocean (Lanna et al., 2018), reproduce at different periods of the year. Therefore, the way in which temperature influences physiological responses in organisms varies from species to species. It was suggested that a minimum temperature of the water should be reached for sponges to begin developing gametes (e.g., Lanna et al., 2021). However, after this minimum temperature has been achieved in the environment, the sponge will start to produce more gametes with increasing temperature. Therefore, in addition to determining when sponges become reproductive, temperature also seems to influence the amount of energy the sponge invests in the next generation (Lanna et al., 2015, 2021; Calazans & Lanna, 2019). It should be noted that in the deep sea, where the temperature is constant, the fallout of plankton from the surface – vertical particle flux – is thought to influence the onset of reproduction in at least some species (Witte, 1996).

While temperature seems to control the number of reproductive elements being produced by the sponge, photoperiod usually controls the period when the sponge reproduces. It was already suggested that in regions where the temperature variation throughout the year is low, the production of gametes should rely on other stimuli that have more intense changes, such as photoperiod (Calazans & Lanna, 2019). This environmental factor was already implicated in the reproductive activity of different demosponges and some calcareous sponges (e.g., Calazans & Lanna, 2019). However, as adult sponges do not have any clear light-sensitive organs, the exact way in which the sponge senses the changes in the day's length is not fully understood. Genes involved in setting circadian rhythms are one option, and they have been recently found to control larval release in the demosponge *Amphimedon queenslandica*. Two potentially circadian genes, *AqPARa* and *AqCry2*, show oscillating expression patterns correlating with larval release in *A. queenslandica* (Jindrich et al., 2017). Another possible mechanism for detecting light comes from the symbionts living in association with the sponge. In *Dysidea janiae*, for instance, the photoperiod was indirectly related to the development of embryos. This species lives in obligatory symbiosis with macroalgae and more light would mean more photosynthetic products for the sponge. In this case, it was suggested that the sponge could dedicate more energy to other vital functions, while the photosynthetic products of the algae could help nourish the embryos (Cajado & Lanna, 2021).

Other environmental factors that could contribute to the onset and maintenance of the reproduction include physicochemical modifications in the water (e.g., tides, pH, salinity) and biological factors, mainly related to the availability of food (which can come in different types: bacterioplankton, phytoplankton, or dissolved organic matter (DOM)). *Dysidea avara*, for instance, reproduces during the summer in the Mediterranean Sea (Mariani et al., 2005). It is known that this species feeds on different sources of plankton, however, during the summer, the main food source is bacteria present in the plankton (Ribes et al., 1999; Mariani et al., 2005). In the tropics, the calcareous sponge *Paraleucilla magna* reproduces continuously all year round, and the density of embryos in the mesohyl is correlated with the dynamics of the bacterioplankton in the region (Lanna et al., 2015). It is known that sponges, especially the HMA ones, are able to feed on DOM (Reiswig, 1981; Yahel et al., 2003; De Goeij et al., 2008). Although no direct observation that DOM is related to the onset and maintenance of the reproduction of sponges has been made so far, the reproductive periods of tropical demosponges living in coastal regions are related to the local rainfall (Lanna et al., 2018, 2021; Cajado & Lanna, 2021). In this region, the salinity is slightly modified by the rainfall, and it is expected that the runoff of coastal DOM will impact the reproduction of the sponges. DOM was also implicated in the reproductive period of deep-sea sponges (Witte, 1996). In the Norwegian fjords, some *Geodia* species reproduce after the phytoplankton pulse at the surface waters. At the end of the phytoplankton bloom, the organic snow sinks into the deep waters, feeding the sponges, and triggering the onset of reproduction (Witte, 1996; Spetland et al., 2007; Koutsouveli et al., 2020a). Other exogenous factors could also influence the reproductive period of sponges (e.g., wind, hydrodynamism, etc.). Although experimental investigations are still needed to confirm the role of these different factors in the control of the reproductive period and the fecundity rate of the sponges, it is clear now that each species has different environmental triggers that cause the onset of reproductive activity, and which determine how long reproduction lasts.

In addition to the exogenous cues, the endogenous condition of the sponges also influences when, how often, and how much the organism will invest in reproduction. The size of the sponge is one of the main characteristics that influence reproduction. Apparently, many sponge species have a minimum size at which they can start their reproductive activity, and the fecundity of the species seems to be related to the size of the sponge (usually, the larger the sponge, the more gametes, embryos, and larvae it produces). An interesting contrast can be drawn between three species of calcareous sponges that have the potential to invade new areas (i.e., non-indigenous species). Individuals of *P. magna*, *Sycettusa hastifera*, and *Heteropia glomerosa* start reproducing when they are still small. However, the size of the individual is positively related to the fecundity of *S. hastifera*, but this characteristic was not observed in the other two species (Lanna et al., 2007; Lanna & Klautau, 2015, 2018; Calazans & Lanna, 2019). Similarly, the size of the sponge was also shown to be positively related to the increase of fecundity rates in four out of five *Geodia* spp., with the exception of *G. macandrewii* (Koutsouveli et al., 2020a). Interestingly, in the branched *H. glomerosa* the number of tubes comprising the sponge is related to the fecundity of the sponge (Calazans & Lanna, 2019). This species starts its growth as a single tube with an apical osculum and additional tube branches are added as the sponge is growing. More tubes might mean that the water circulation and respiration increase, providing a better environment for gametogenesis. Additionally, it could be that the sponge relies on a strategy to first produce new branches before starting to produce the gametes (Calazans & Lanna, 2019). Another endogenous characteristic that seems to regulate reproduction is found in some freshwater species. In these habitats, the sponge needs to go through a diapause stage (forming a desiccation-resistant gemmule, see below) before starting to carry out its gametogenesis (Gilbert, 1974). Other endogenous characteristics are very likely to influence reproductive activity but are still poorly investigated. One of them is the fluctuation in the production of secondary metabolites, which usually varies over the year (*see* Section 1.4.4).

1.2.2 PERIODICITY OF REPRODUCTION

The reproduction will interfere with the homeostasis of the organism, as the energetic budget needs to be shifted to gamete production and embryonic and larval nourishment. Therefore, the organism needs to balance its reproductive activity with other physiological activities. In this sense, the sponges have different strategies depending on the phylogeny and/or the ecology of the group. Some species tend to invest large amounts of energy in a single reproductive event that is highly coordinated in the population. This high investment often results in the death of the organism. This life history strategy in which the organism invests all of its resources in reproduction, subsequently dying, is known as semelparity. Semelparous sponges are usually observed in the class Calcarea, in which the species tend to produce many embryos and larvae in a single period of the year followed by the disappearance of individuals in the population (Lanna et al., 2007). Some tetillid species seem to have this strategy, too. *Tetilla japonica* reproduces every two years and dies after the second year (Watanabe, 1978). *Tetilla* is a gonochoristic (separate sexes) oviparous species that become reproductive after one year and must invest all its energy in gamete production.

Even in species in which the populations have continuous reproductive activity, it is suggested that the somatic maintenance and survival of the individual are severely affected by reproduction (Lanna & Klautau, 2018; Calazans & Lanna, 2019). Semelparous sponges are likely to be found also in the other classes, especially in opportunistic species living in very unstable regions. For instance, the reproductive individuals of the

Mediterranean demosponge *Crella elegans* massively receded or even disappeared after larval release (Perez-Porro et al., 2012). On the other end of the spectrum, most demosponges, hexactinellids, and homoscleromorph species have several reproductive cycles during their lifespan (i.e., iteroparity). The individuals of the demosponge *Geodia cydonium*, for instance, spawn their gametes during the summer. After that, they have a resting period and start to produce a new batch of gametes during the spring of the next year (Mercurio et al., 2007). *Xestospongia muta*, also known as the redwood of the reef, can reach 100 years old in the Caribbean reefs (McMurray et al., 2008). There, they apparently reproduce every year in the warmer months (mid-April to late-May) (Neely & Butler, 2020), and therefore reproduction does not seem to affect the survival of this giant sponge. Broadly, one can infer that demosponges tend to be at the iteroparous end of the continuous, while calcareous sponges are at the semelparous one. But interestingly, the fecundity (i.e., the quantity of reproductive elements) of demosponges is usually lower than in Calcarea. Lanna & Klautau (2018) suggested that this difference in the reproductive investment (and subsequent survival or mortality of the organism) might be evolutionarily constrained in these two classes of Porifera. However, the exact mechanism by which these characteristics are controlled in the sponge is still to be understood.

1.2.3 DO SPONGES HAVE HORMONAL CONTROL OF THEIR REPRODUCTION?

The driving forces regulating the reproductive season in most sponge species studied so far are abiotic factors in their surroundings (e.g., temperature, light, tide, and food). However, these factors should further trigger some endogenous signals of the sponge which in turn control the processes of gametogenesis, spawning, and fertilization. Although sponge spawning is not usually as spectacular as coral spawning, there are some examples of quite beautiful events of sponge gamete release in gonochoristic species, in which the oocytes and sperm are finely synchronized (e.g., Reiswig, 1976; Hoppe & Reichert, 1987; Ritson-Williams et al., 2005; Röthig & Voolstra, 2016; Neely & Butler, 2020). The cues for the synchronization of reproduction and spawning are still unknown, but some molecules have been proposed to be responsible, such as mycosporine-like amino acids (MAAS) (Bandaranayake et al., 1997).

Sex steroids are among the most widespread triggers of reproduction in vertebrates and invertebrates (Jones et al., 2017), although the steroid signaling pathways have only been extensively studied in a few marine invertebrates (Köhler et al., 2007). In corals, the Gonadotropin-Releasing Hormone (GnRH) dramatically increases during the spawning period, stimulating the activation of steroidogenesis and the increase in testosterone and oestradiol. The coral GnRH also has been found to stimulate the release of the reproductive hormone LH (luteinizing hormone) in fish, indicating that hormonal control plays an important role both in gamete development and mass release in corals (Twan et al., 2005, 2006). Sponges have orthologous genes responsible for cystine-knot structures, which are part of several secreted proteins and glycoprotein hormone families, and features of glycoprotein hormones (follicle-stimulating hormone, LH, thyroid-stimulating hormone) (Roch & Sherwood, 2014). However, even though glycoprotein hormones are responsible for regulating reproduction in animals, their role in hormonal control in sponge reproduction has not yet been explored. Regarding steroids, sponges possess the most diverse and biogenetically unprecedented array of unconventional steroids in the entire animal kingdom (Aiello et al., 1999), but only their cytotoxic properties as well as their role in maintaining cellular structures and function as nuclear receptor ligands have been studied (Fiorucci et al., 2012). One of the roles of steroid hormones, especially endogenous estrogens such as 17β-estradiol (E2), is triggering the synthesis of vitellogenin, which is the main precursor of the egg-yolk proteins (vitellins) forming the yolk stored in the oocytes. Transcriptomic data revealed the overexpression of several genes that are engaged in the pathway of estrogen production (e.g., 3-keto-steroid reductase, estradiol 17-beta-dehydrogenase 8, very-long-chain 3-oxoacyl-CoA reductase, 17-beta-hydroxysteroid dehydrogenase, estradiol 17-beta-dehydrogenase), testosterone biosynthesis pathway (testosterone 17-beta-dehydrogenase 3), as well as homologs of vitellogenin proteins (Riesgo et al., 2014b), during female gametogenesis in *Geodia* species (Koutsouveli et al., 2020b). In addition, the expression of genes involved in the synthesis of male-related hormones (*lutropin-choriogonadotropin hormone receptor*) has also been found in sponges (Koutsouveli et al., 2020b). Therefore, it is highly possible that already in Porifera, some kind of chemical communication with hormones (similar to those regulating the reproduction in other animals) could control the synchronization of gametogenesis and the simultaneous spawning within a population, and more extensive studies are necessary to further understand this topic.

1.3 GAMETOGENESIS

Sexual reproduction has appeared only once in the common ancestor of eukaryotes (Hörandl, 2009; Birdsell & Wills, 2003; Speijer et al., 2015). The advantages of sexual reproduction, i.e., the potential elimination of deleterious mutations, higher variety of genotypes, and greater overall fitness of a population (Williams, 1975; Barton & Charlesworth, 1998), seem to surmount its two-fold cost (Lehtonen et al., 2012; Gibson et al., 2017), rendering this mode as the main way to reproduce in all animals. Especially in long-lived sessile organisms, such as sponges, sexual reproduction represents an ecological advantage, since the resulting genetically diverse populations of the same species may exploit different niches or resources within the same habitat, reducing competition among individuals and allowing their coexistence (Song et al., 2012). Meiosis and genetic recombination are the very essences of sexual reproduction, and they are conserved across the animal kingdom, as well as in unicellular eukaryotes and fungi (Carr et al., 2010; Halary et al., 2011; Hofstatter & Lahr, 2019). Both gametic cells are generated by meiosis in sponges and anisogamy of female and male gametes is observed, with much smaller male (~ 2–4 μm) than female (~ 40–100 μm) gametes. This mode of anisogamy extends through all of Metazoa (Kirk, 2006; Lehtonen et al., 2016).

The main stages of oogenesis: (i) germ cell formation; (ii) meiotic arrest; and (iii) maturation, growth, and vitellogenesis) (Telfer & McLaughlin, 2007; Woodruff et al., 2018) and spermatogenesis: (a) germ cell formation; (b) mitosis; (c) two meiotic cycles; (d) chromatin condensation; and (e) flagellum formation and capacitation) (Auger, 2018) occurring in all animals, have been previously observed by both cytological (e.g., Tuzet et al., 1970; Kaye & Reiswig, 1991; Riesgo & Maldonado, 2009a, b; Lanna & Klautau, 2010) and molecular studies (Leininger et al., 2014; Fierro-Constaín et al., 2017; Koutsouveli et al., 2020b) in Porifera. Overall, the processes of gametogenesis and embryogenesis in sponges are morphologically and molecularly similar to those in other metazoans (e.g., Riesgo & Maldonado, 2009a, b; Ereskovsky et al., 2013; Lanna & Klautau, 2010, 2016; Koutsouveli et al., 2018, 2020b). The most essential difference for sponges compared to other animals is that they lack gonads (i.e., the reproductive organs in which the gametes are generated). In this phylum, gametes are either spread in the mesohyl (female gametes) or in most cases are developed within cysts (development of the spermatocytes) (Figures 1.2(A) and 1.3(A)). Additionally, a striking difference relates to the mechanisms for germline determination. One aspect that is still poorly known is the mechanism sponges use for sex specification, which has been suspected to be both molecularly and environmentally controlled.

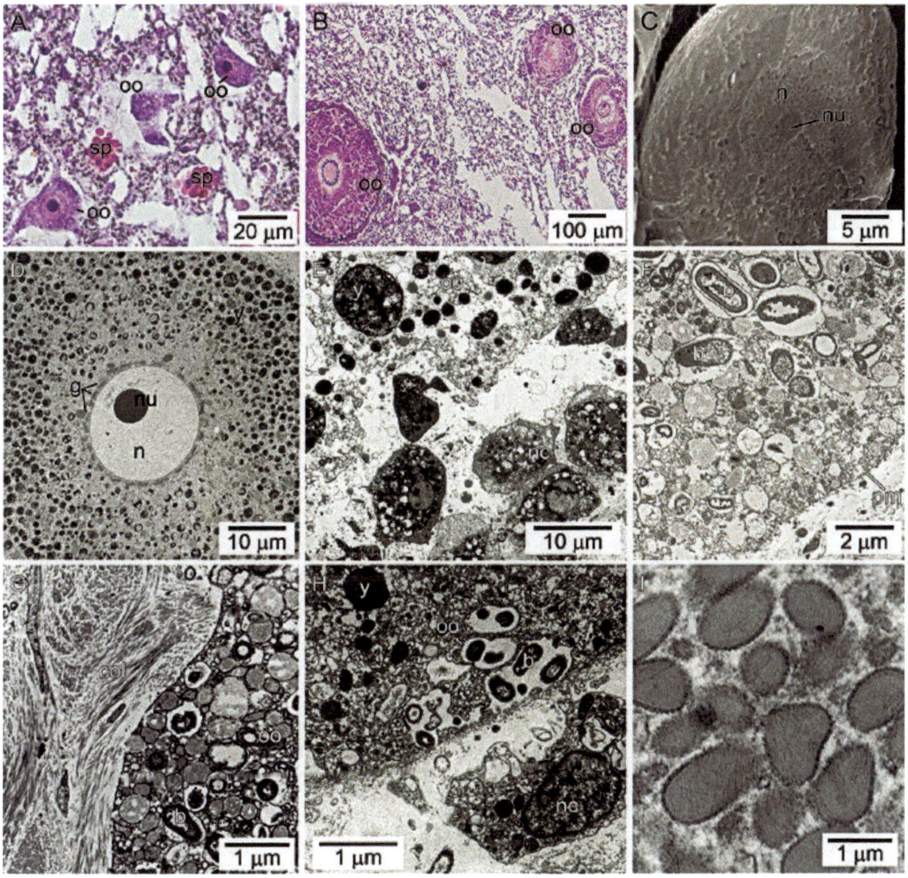

FIGURE 1.2 Oogenesis of Porifera. (A) Distribution of oocytes in the mesohyl of *Geodia barreti*. (B) Oocytes of *Stylocordila borealis*. (C) Oocyte of *Paraleucilla magna*. (D) Ooplasm of the oocyte of *Raspaciona aculeata*. Note the presence of several Golgi apparatuses and a clear ooplasm next to the nucleus. (E) The peripheral region of the oocyte of *Ircinia fasciculata* shows the presence of nurse cells and some heterogeneous yolk in the ooplasm. (F) and (G) Periphery of the oocyte of *Geodia macandrewi* showing several bacteria inside vesicles. In (G), a large amount of collagen is observed surrounding the oocyte. (H) Nurse cell in contact with the oocyte of *Borojevia brasiliensis*. (I) Detail of yolk inclusions of *B. brasiliensis*. ((A) and (B): Light microscopy; (C): scanning electron microscopy; (D)–(F): Transmission electron microscopy).

Legends: b: bacteria; col: collagen fibrils; g: Golgi apparatus; n: nucleus; nc: nurse cells; nu: nucleolus; and y: yolk inclusion.

1.3.1 SEX DETERMINATION

Sex occupies a central role in the evolution of eukaryotes. It may have appeared as a repair mechanism to avoid damaging single DNA strands (Bernstein et al., 2013; Hörandl & Dave, 2018) but since then, sex has had profound implications for animal biology and ecology. The evolution of separate sexes has occurred several times independently both in plants and animals (Bachtrog et al., 2014), including sponges (Riesgo et al., 2014a), suggesting that there is an evolutionary cost associated with hermaphroditism. Given the many independent acquisitions, there is a bewildering number of molecular mechanisms underpinning sex determination in animals (Bachtrog et al., 2014; Beukeboom & Perrin, 2014), from entirely genetic (GSD) to environmentally induced sex determination (ESD). Within GSD, there is also a panoply of genetic triggers (Bachtrog et al., 2014), the most commonly understood being the expression of the sex-determining locus *Sry* (Kashimada & Koopman, 2010). *Sry* belongs to the SOX gene family (SoxA), which is developmentally important in all animals (Fortunato et al., 2012) and has been identified in several sponges, their expression is tightly linked to oogenesis (Larroux et al., 2006; Fortunato et al., 2012; Koutsouveli et al., 2020b). A XX/XY sex-determination system in cnidarians has been suggested on the basis of association studies between allele frequencies in SNP data and phenotypic sex (Pratlong et al., 2017). A few candidate loci have been proposed as sex-determining genes in invertebrates, including *DMRT* (Picard et al., 2015) and *FEM/TRA* (Haag et al., 2002). Gonochoric corals have several orthologs of *DMRT* genes, with female-specific (Traylor-Knowles et al., 2015; Chen et al., 2016) and male-specific expression (Traylor-Knowles et al., 2015). *DMRT* genes also occur in hermaphroditic corals with expression patterns coinciding with the timing of sexual differentiation (Miller et al., 2003). But the picture is not homogeneous across Cnidaria. While in some corals it seems that sex is determined genetically (Traylor-Knowles et al., 2015; Chen et al., 2016; Siebert & Juliano, 2017), in a common hydrozoan, the temperature is a major factor controlling the sex ratio. In *Clytia hemisphaerica*, low temperatures may lead to the loss of female germline cells, with drastic consequences for population survival (Carré & Carré, 2000). In sponges, both *FEM* and *DMRT* genes are present, but although *FEM* is prevalent across the four orders of sponges, *DMRT* is exclusively present in Homoscleromorpha (Riesgo et al., 2014b).

SOX genes are a particularly interesting family of genes with a complex involvement in sex determination across animals (Kiefer, 2007). The male sex-determining factor *Sry* was the first SOX gene identified, but it took more than three decades for researchers to understand the mechanisms of action of this Y-linked gene, which was involved in testis development. SOX genes are part of the high mobility group (HMG) superfamily of proteins, which divides into two clades, one with specific DNA binding properties and one without (Kiefer, 2007). SOX are animal-specific transcription factors (both activators and repressors) that are known from sponges to vertebrates (Jager et al., 2008), with roles in development of several tissues and organs (Kiefer, 2007; Jager et al., 2008; Fortunato et al., 2012), and a fundamental role in sex determination in several animal groups (Koopman, 2005; Jager et al., 2008; Manolakou et al., 2006).

Sponges can use many strategies to sexually reproduce, from gonochorism to successive or contemporaneous hermaphroditism. Although the sexual strategy was thought to bear some phylogenetic history (Sarà, 1993), it is now well-established that sexuality is a plastic strategy (Riesgo et al., 2014a) that may or may not be genetically determined, but that is definitely influenced by physiology and environment (Simpson, 1984). While most sponge species exhibit the same sexual strategy over their entire life and distribution range, there are examples of sponge species that can reproduce both by gonochorism and hermaphroditism (Gilbert & Simpson, 1976; Fell & Jacob, 1979; Ereskovsky, 2000). Although those species' sexuality might be influenced by sex ratio or environmental conditions, there are also interesting studies pointing to genetic control of sexuality in sponges. In several freshwater sponges, the male or female sex is maintained throughout life, and the sex of all the asexually produced offspring is always the same, coinciding also with that of the parent (Mukai, 1990), which strongly suggests genetic sex determination. However, the possibility of environmentally mediated sex determination in sponges cannot be ruled out. In marine sponges, *FEM* expression can be modulated by temperature (Perovic-Ottstadt et al., 2004), and also in freshwater sponges, early reports of sponge reproduction suggest environmental control could underpin sex reversal observations (Van de Vyver & Willenz, 1975) and sex ratios skewed from 1:1 (Simpson, 1984).

1.3.2 GERMLINE DETERMINATION

Sponges are known for having incredible regeneration capacities, from rapid wound healing to whole-body regeneration after complete dissociation of the body into individual cells (Wilson, 1907; Lavrov & Kosevich, 2014;

Eerkes-Medrano et al., 2015). The molecular machinery deployed during sponge regeneration is remarkably similar to that in other metazoans (Kenny et al., 2018), but one of the key aspects of such amazing capacities might be the high number of totipotent (or pluripotent) cells present in the sponge body. There are two types of cells with different degrees of potency: archaeocytes, which are considered the true totipotent cells but still are able to perform "somatic" functions, like delivery of nutrients, and which are capable of differentiating into all other cells of sponges; and choanocytes, which filter out the bacteria from the water and which have certain pluripotency as they are capable of transdifferentiation into some of the other sponge cell types (Simpson, 1984; Funayama, 2013; Fierro-Constaín et al., 2017). Among other cell types, both archaeocytes and choanocytes can differentiate into female and male gametes (Simpson, 1984; Fierro-Constaín et al., 2017).

While the germ cell line in most bilaterians (e.g., Ecdysozoa and Chordata) is found exclusively in the gonads and its role is to generate the gametes (Extavour & Akam, 2003; Extavour, 2007; Juliano & Wessel, 2010; Richardson & Lehmann, 2010), sponges instead have primordial stem cells that are responsible for both generating somatic stem cells and germ cells (Voronina et al., 2008; Funayama et al., 2010; Juliano et al., 2010; Funayama, 2013). Before the onset of gametogenesis, these primordial totipotent and pluripotent cells (archaeocytes and choanocytes, respectively) are transformed into germ cells with transdifferentiation and differentiation processes with the ultimate goal of generating the gametes (Lévi, 1956; Tuzet et al., 1970; Simpson, 1984; Riesgo et al., 2007a, b; Maldonado & Riesgo, 2008b; Ereskovsky, 2010; Lanna & Klautau, 2010, 2016).

Although the complete stem cell gene toolkit (Ewen-Campen et al., 2010) in sponges is not fully unraveled yet, several studies have identified key components expressed in totipotent cells and gametes (Funayama et al., 2010; Leininger et al., 2014; Riesgo et al., 2014b; Fierro-Constaín et al., 2017; Koutsouveli et al., 2020b). Many of these genes have been found expressed in stem cells of sponges (archaeocytes and choanocytes) (Leininger et al., 2014) and are overexpressed during female and male gametogenesis (Koutsouveli et al., 2020b) supporting the statement of common regulation of stem and germline in sponges. However, there is no evidence yet of the direct relationship between archaeocytes or choanocytes expressing germline markers and those later becoming gametes. Indeed, some of these genes (*piwi*, *nanos*, and *vasa*) are part of the totipotency toolkit (Juliano & Wessel, 2010), and therefore it is still unclear whether they are more involved in maintaining the totipotency of archaeocytes and choanocytes or determining the germline fate of these cells (Riesgo & Solana, 2021).

1.3.3 OOGENESIS: CELL NUTRITION AND DEVELOPMENT

As described in Section 1.2.1, the reproductive season and reproductive strategy of most species are determined by the available nutrients in the surrounding, as the whole process requires high amounts of energy. Nutrients are not only important for the reproductive adults, but during gametogenesis and embryogenesis, the oocytes and embryos also accumulate nutrients, in the phase called vitellogenesis. Because of the lecithotrophic nature of sponge larvae, the accumulated yolk in gametes and embryos serves as an energetic reserve for further development during their drifting period in the water column until their benthic settlement and transformation to an adult sponge (Maldonado & Bergquist, 2002). Consequently, the amount of nutrients accumulated, and also the type of yolk, predetermines to a high extent, the survival rate and the dispersal time of the larvae. Usually, oocytes can deposit the yolk scattered in the ooplasm (Figure 1.2(B) and (C)) or more externally (i.e., next to the oolemma) (Figure 1.2(D)).

The yolk can have a lipid, protein, or glycogen origin and is formed either in homogenous droplets or in heterogeneous platelets after a mixture of different types of nutrients (Figure 1.2(D), (G), and (I); e.g., Fell, 1974; Simpson, 1984; Maldonado & Riesgo, 2008b). Within the maternal tissue, the gametic cells/embryos phagocytize nutrients directly from the mesohyl, through pseudopodia, or accessory cells, also called nurse cells, synthesize the yolk, and provide it to the gametic cells/embryos, a process called heterosynthesis (Figure 1.2(E) and (I); e.g., Fell, 1969, 1974; Simpson, 1984; Maldonado & Riesgo, 2008b; Koutsouveli et al., 2018). The nurse cells are found in close proximity (Figure 1.2(E)), in direct contact (Figure 1.2(H)), or entirely phagocytized and digested by the gametic cells/embryos for nutrient translocation. Another potential source of energy to produce the yolk is the bacteria. In several species, these microbes are phagocytized by the oocyte and digested in vacuoles inside the ooplasm (Figure 1.2(F) and (H)). Gametic cells also produce the nutrients themselves via the Golgi apparatus and endoplasmic reticulum (ER) a process called autosynthesis (e.g., Fell, 1974; Simpson, 1984; Maldonado & Riesgo, 2008b).

Among the nutrients, lipids in particular provide the propagule with high energy content for active swimming (Lee, 1991; Heras et al., 2000;

Viladrich et al., 2016; Maier et al., 2020), they are metabolized faster than proteins (Clarke, 1980), and they also serve for buoyancy (e.g., Clarke, 1980; Phleger, 1998; Pond & Tarling, 2011). The lipid yolk seems to be predominant or exclusively present in some cold-water or deep-sea species (Riesgo et al., 2015; Koutsouveli et al., 2020a), compared to their temperate counterparts (Sciscioli et al., 1994; Riesgo et al., 2015). Due to their higher energy and faster metabolization, lipids could possibly enhance the quick environmental changes in the more extreme habitats that sometimes are poor in nutrients (Giese, 1966; Clarke, 1980). Among the lipids, triacylglycerides, are one of the main forms of energy storage in sponges (together with sterols) (Parzanini et al., 2018) and seem to be the lipid category most involved in yolk formation in sponges (Koutsouveli et al., 2021, 2022), as in most lecithotrophic animals.

1.3.4 SPERMATOGENESIS – TYPES OF SPERM CELLS AND ECOLOGY

Spermatogenesis is a much shorter process than oogenesis in general in sponges. In most species, it usually lasts for two weeks to one month (Fell, 1974; Reiswig, 1983; Boury-Esnault & Jamieson, 1999; Riesgo & Maldonado, 2008; Koutsouveli et al., 2020a); although the presence of spermatic cysts can be detected for several months (e.g., Courregès & Fell, 1989; Boury-Esnault & Vacelet, 1994; Vasconcellos et al., 2019). The spermatogenic process is conserved across animals, and its cellular milestones can be easily recognized in sponges (Figure 1.3(B), (C), and (F); e.g., Fell, 1974; Reiswig, 1983; Riesgo et al., 2007; Riesgo & Maldonado, 2009b; Vasconcellos et al., 2019). The sperm morphology of sponges was regarded as "primitive" for many years because most sponges were described as having sperm cells with round heads, no acrosome, and a midpiece with few mitochondria (Figure 1.3(D) and (G); Fell, 1974; Reiswig, 1983). But in 1984, Baccetti & collaborators detected, for the first time, the presence of an acrosome in the sperm of the homoscleromorph sponge *Oscarella lobularis*. Since then, acrosomes (or acrosomal vesicles) have been observed in other species of Homoscleromorpha (*Corticium candelabrum*, Riesgo et al., 2007a), in the demosponge orders Suberitida (*Suberites massa* – Diaz & Connes, 1980), Haplosclerida (*Petrosia ficiformis* – Maldonado & Riesgo, 2009), and Poecilosclerida (*Lycopodina occidentalis*, *Crambe crambe*, *Tedania ignis*, *Crellomima imparidens* and *Hymedesmia irregularis* – Tripepi et al., 1984; Riesgo et al., 2007b; Riesgo & Maldonado, 2009b; Vasconcellos et al., 2019;

Ereskovsky & Tokina, 2022); and in Calcarea (*Sycon calcaravis* – Nakamura et al. (1998), *P. magna* – Lanna & Klautau (2010)). In addition to the acrosome hallmark of eumetazoan complexity, sponge sperm is also not round in many species, being long and slender in freshwater species (Paulus, 1989) and most poecilosclerids (Figure 1.3(E) and (G); Riesgo et al., 2007b; Riesgo & Maldonado, 2009b; Ereskovsky, 2010; Vasconcellos et al., 2019; Ereskovsky & Tokina, 2022). Round sperm has been observed in many other sponge species, from calcareous sponges (Riesgo et al., 2018) to demosponges (e.g., Fell, 1974; Reiswig, 1983; Ereskovsky, 2010; Koutsouveli et al., 2020a). Surprisingly, a sophisticated V-shaped spermatozoon is observed in the poecilosclerid *Crambe crambe* (Figure 1.3(F) and (G)).

In general, the morphology of the sperm is tightly correlated with the mode of fertilization. Round sperm is generally associated with mass spawning and external fertilization in the water, drifting with the current until reaching a female oocyte or a female sponge that would take it in by filtering (Riesgo et al., 2007a). In the case of long, slender modified sperm, this shape can be related to internal fertilization, as the sperm will need to burrow into the female sponge body until reaching the oocytes (Riesgo et al., 2007b; Riesgo & Maldonado, 2009b). However, the fertilization process in sponges is far from being well understood. For most oviparous species, fertilization occurs in the water, but in viviparous sponges the fertilization is internal. Although rarely observed (Gatenby, 1920; Gaino et al., 1987; Riesgo et al., 2007a), fertilization is mediated by choanocytes in some sponges, which take up the sperm and transport it to the oocyte (*see* Section 1.5.2).

1.4 REPRODUCTIVE MODE

1.4.1 OVIPARITY VS. VIVIPARITY

Sponges usually release their sperm through the osculum. However, while some sponges release oocytes and fertilization will occur in the environment (oviparity), other species brood their embryos until the larvae ripen in the choanosome and finally are released (viviparity). Release of larvae can be through the osculum, or they can break through the sponge's tissues (as observed in some carnivorous sponges of the family Cladorhizidae; Chu & Reiswig, 2014). According to Riesgo et al. (2014a), the ancestral reproductive condition of the sponges is viviparity. In Hexactinellida, Homoscleromorpha, and Calcarea all species studied so far are viviparous. In Demospongiae, viviparity is also the ancestral condition, but oviparity

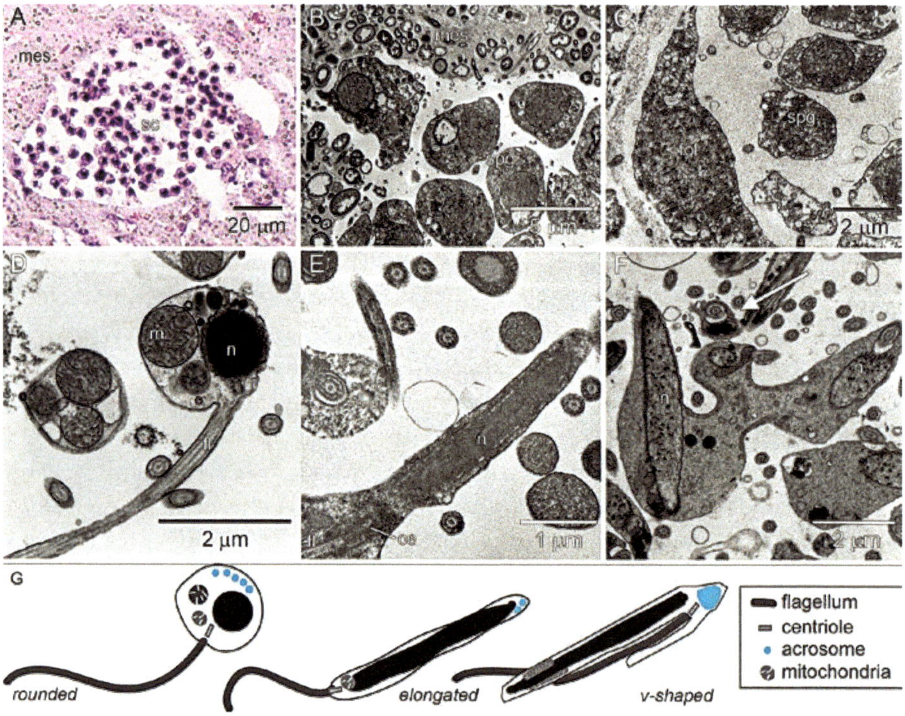

FIGURE 1.3 Spermatogenesis in Porifera. (A) Spermatic cyst in the choanosome of *Craniella cranium*. (B) Spermatogonia inside the spermatic cyst of *Geodia hentscheli*. (C) Follicle cells surrounding the spermatocytes of *Tedania ignis*. (D) The rounded-shaped spermatozoon of *Petrosia ficiformis*. (E) The elongated spermatozoon of *Tedania ignis*. (F) Spermatid of *Crambe crambe*. The arrow points to a cross-section of a mature spermatozoon. (G) Schematic drawings of the main different types of spermatozoa found in Porifera. The acrosome (or pre-acrosomal vesicles) is present in some species. ((A): Light microscopy; (B)–(F): Transmission electron microscopy).

Legends: ce: centriole; fol: follicle; m: mitochondria; mes: mesohyl; n: nucleus; sc: spermatic cyst; spg: spermatogonia; and spc: spermatocytes.

evolved independently in different demosponge lineages. Interestingly, within some clades that are oviparous, some lineages have reversed the reproductive condition to viviparity (for example in Poecilosclerida) (Riesgo et al., 2014a). The reason for this evolutionary transition in the Porifera is still unknown, but we can provide a hypothesis based on the physiological and morphological features of sponges. It is known that embryos require good ventilation (i.e., oxygenation) to develop. Most hexactinellids, calcareous sponges, and homoscleromorphs have a thin body with a well-developed aquiferous system (i.e., the proportion of canals and chambers to mesohyl in the choanosome of the sponge is relatively high). Therefore, the embryos are usually located close to the circulating water and have a good amount of oxygen to develop during brooding time. This is also the case of encrusting demosponges (like most Poecilosclerida, which are also viviparous), where the surface of the sponge and the aquiferous system comprise a large area/volume of the animal. On the other hand, oviparous sponges usually have a thick mesohyl that is less easily oxygenated (i.e., the proportion of aquiferous system/mesohyl is low; Riesgo et al., 2014a). This characteristic probably gives rise to regions in the choanosome, where an anoxic microbiota will be important for the functioning of the sponge body (Hoffmann et al., 2005), and therefore embryos would not have their oxygen requirements met.

1.4.2 DISTRIBUTION OF THE REPRODUCTIVE ELEMENTS IN THE SPONGE TISSUE

It is widely accepted that sponges lack regionalization in their bodies. Therefore, gametes are usually spread uniformly through all parts of the sponge. Usually, the reproductive elements are found closer to choanocyte chambers and canals (i.e., in the choanosome). As discussed, this distribution might be related to the oxygenation provided by the water flow within the canal system. Moreover, this location also facilitates the spawning process, as the gametes and embryos would need only to cross a thin barrier to reach the aquiferous system and be released through the osculum. However, some species do have a regionalized distribution of embryos and larvae. Some demosponges seem to be adapted to protect their embryos against predation and the negative effects of hydrodynamics forces by restricting them to basal portions of the organism (Reiswig, 1973; Fell, 1976; Fromont, 1994). In the demosponge *A. queenslandica*, embryos are restricted to a region of the sponge known as brood chambers (Leys & Degnan, 2002). This characteristic was already observed in other Haplosclerida (Fromont, 1994) and in the bath sponge *Spongia officinalis* (Baldacconi et al., 2007). These brood chambers are a region of the sponge where many embryos (20–150) are found. It is usually delimited by a thin follicle formed by pinacocyte-like cells (Baldacconi et al., 2007). In *A. queenslandica* the embryos in a given chamber are not necessarily in the same developmental stage (Leys & Degnan, 2002), suggesting that the chamber is a more 'fixed' structure in the body of this iteroparous species that reproduce year-round. In some species of the subclass Calcinea, the reproductive activity disrupts the organization of the sponge. In the genus *Ascandra*, several tubes of the organism become 'nests' where embryos develop. In *Leucetta giribeti* the embryos are also clumped in a region of the mesohyl (Borojevic, 1969; Riesgo et al., 2018). Consequently, these changes in the general morphology of the sponge with the disruption of the aquiferous system will have a profound impact on the physiology of the organism. The regionalization of the body with some modules of the aquiferous system becoming brooding units could therefore be an important adaptation to keep the organism alive through the reproductive period in iteroparous sponges.

1.4.3 SYMBIOSIS AND REPRODUCTION

Sponges are well known for their wide variety of interactions with micro- and macro-organisms. The presence and influence of the microbiota on the physiology of Porifera have been widely recognized in the last decades (Erwin et al., 2015). The sponge microbiota is usually associated with different nutrient cycle pathways (such as N, S, and P) within the host, but also provides some autotrophic nutrition to the holobiont, especially in HMA sponges (Erwin et al., 2015). Therefore, it is expected that these associated organisms will play an important role in the reproduction and other aspects of sponges' physiology. Consequently, at least part of the microbiota of the sponges is transferred from generation to generation (vertical transmission) in most oviparous and viviparous sponges (Maldonado, 2007). Usually, in oviparous species, the bacteria are captured by the oocytes still in the mesohyl and are kept in membrane-bound compartments in the ooplasm (Maldonado, 2007; Koutsouveli et al., 2020a, b). In turn, viviparous sponges usually transfer the microbiota during embryogenesis and symbionts enter the larval cavity through the cleavage furrows or other openings of the embryo (such as the pore in calcareous stomoblastulae, see below) (Maldonado, 2007; Lanna & Klautau, 2012). These observations are in line with the fact that the microbiota of an individual is more similar to that of other individuals of the same species than with neighboring sponges or the surrounding environment (e.g., Erwin et al., 2015). However, recently this view was challenged, suggesting that vertical transmission is restricted to a microbial core, while the bulk of the symbionts is later acquired by the adult through filtration (Björk et al., 2019). Interestingly, sibling larvae inherit different sets of microbes from the same parent, probably to better fit the environmental requirements of the larvae settling in different locations (Björk et al., 2019). In addition to the different roles in metabolic activities within the sponge, microbes also serve as a food source for oocytes, embryos, and larvae (Maldonado, 2007; Riesgo & Maldonado, 2009a; Lanna & Klautau, 2012). As discussed in Section 1.2.1, variation in the density of bacterioplankton (i.e., exogenous bacteria) being consumed by the sponge can trigger reproductive activity in both HMA and LMA sponges, despite the fact that the associated microbiota is stable throughout the year (Erwin et al., 2015). In some cases, microbes living in the mesohyl are ingested by cells (either oocytes or embryonic cells) and are digested within phagosomes in the cytoplasm. In some species, e.g., *Paraleucilla magna* and *Sycettusa hastifera* (Calcarea), and *Axinella damicornis* and *Phorbas areolatus* (Demospongiae), the digested bacteria are used as a substrate to produce the yolk observed in the ooplasm and in the embryonic cells (e.g., Riesgo & Maldonado, 2009b; Lanna & Klautau, 2010, 2016; Koutsouveli et al., 2018). In addition to phagocytosis of the microbes, it is possible that the microbiota influences oocyte production in other ways. Analysis of lipid-biomarkers indicates a transfer of biomass of some sulfur-reducing and other anoxic bacteria to the sponge cells in *Geodia barretti* (Hoffmann et al., 2005). Similarly, it is possible that macroalgae living in obligatory association with some species of sponges also contribute to the reproduction of the sponge, such as the observation reported previously that the photoperiod in a tropical beach influences the development of embryos in the demosponge *Dysidea janiae* (Cajado & Lanna, 2021). There, the authors suggested that photoperiod would influence the photosynthetic rates of the algae and some products of the photosynthesis could, then, be used as an energetic source to produce embryos.

1.4.4 DOES REPRODUCTION LEAD TO ENERGETIC TRADE-OFFS?

During reproduction, a large number of choanocytes are transformed into reproductive cells. In the freshwater sponge *Racekiela ryderi*, 32% of its volume was occupied by spermatic cysts (Courregès & Fell, 1989), which likely reduced its pumping ability. In the demosponges *Svenzea zeai* and *C. crambe* the pumping rate was found to decrease during their reproductive cycles, and a possible explanation was the reduction of the number of choanocyte chambers and the disorganization of the canal system (Turon et al., 1999; López-Legentil & Turon, 2010). This could affect the nutrient uptake and the regular physiological processes of the sponge during reproduction. In parallel, the formation of the gametes is a highly energetic process as has been clearly shown in other marine invertebrates (e.g., Doughty & Shine, 1997; Coma et al., 1998; Harii et al., 2002).

The production of secondary metabolites is a physiological activity that is costly for the sponge (Leong & Pawlik, 2011; Ivanisevic et al., 2011a, b). In theory, reproduction should interfere with the concentration of these metabolites, generating a trade-off between the toxicity of the sponge and its reproductive activity. Leong & Pawlik (2011) observed that some Caribbean sponges which produce large amounts of metabolites to defend the species against predation, do not produce fewer propagules than those species that are "undefended." The Mediterranean demosponge *Dysidea avara* produces a potent metabolite known as avarol, which has several biomedical applications. The concentration of avarol in the tissue of the sponges fluctuates during the year, but during the period when the population is reproducing (spring to early summer), the concentration of this metabolite was higher than in non-brooding periods of the species (de Caralt et al., 2018). This is also the case of *Oscarella tuberculata* (Homoscleromorpha) in the Mediterranean Sea. In this homoscleromorph, the presence of lysophospholipids was associated with the presence of embryos and larvae in the species (Ivanisevic et al., 2011a). On the other hand, another species of the same genus (*Oscarella balibaloi*) as well as the demosponge *C. crambe*, both in the Mediterranean Sea, do show a trade-off between the production of their secondary metabolites and reproductive activity (Turon et al., 1996; Ivanisevic et al., 2011b). A possible explanation for the absence of trade-offs in the aforementioned species is that the peaks of production of metabolites and reproductive propagules occur in the summer when more food is available for the species (Leong & Pawlik, 2011). Therefore, it seems that the expected trade-offs among growth, secondary metabolite production, and reproductive output are less obvious than expected in the theory and that each species of sponges seems to have taken different evolutionary trajectories to allocate their resources among these three costly metabolic activities.

The investment in sexual and asexual reproduction is another potential source for energetic trade-offs in sponges. Usually, either the sponge invests in sexual reproduction (i.e., gametes and embryos) or in asexual reproduction (i.e., budding and gemmules) (Ereskovsky et al., 2017). Korotkova (1988) hypothesized that the incompatibility of sexual and somatic morphogenesis (i.e., asexual reproduction) should be observed in sponges because these forms of development will involve the same cell types (mainly the totipotent archaeocytes). In the haplosclerid *Haliclona* (*Halichoclona*) *fulva* budding occurs mainly during the winter, while the production of eggs is restricted to the summer months of the year (Ereskovsky et al., 2017). Freshwater demosponges (Spongillida) also showed a similar strategy. Gugel (2001) investigated the life cycle of five species of Spongillida along the river Rhine, in Germany. The author found that while *Ephydatia fluviatilis* relied mostly on its larvae (not gemmules) to colonize new substrates, other Spongillida species usually use passively transported gemmules. The number of larvae produced by the former species was higher than in the gemmulating species, and in fact, some of the species did not produce any larvae during the study but were usually gemmulating. These results suggest that in some populations of Spongillida, individuals produce either gemmules or larvae. From the studies of Mukai however (Mukai, 1990), we know that other populations of the same species can produce both gemmules in the winter months and larvae in the summer months.

Fragmentation is another way in which sponges can reproduce asexually and is usually observed in branching sponges (Wulff, 1991). In the Caribbean reefs, two sympatric species of the demosponge *Callyspongia* demonstrate a trade-off between fragmentation and sexual reproduction (Leong & Pawlik, 2010). These two species have similar abundance in the reefs; however, while *C. vaginalis* relies more on asexual reproduction by budding (with little investment in the production of larvae), *C. armigera* produces around 13 times the number of larvae and seldom reproduces by fragmentation (Leong & Pawlik, 2010). On the other hand, Maldonado & Uriz (1999) observed a synergy between fragmentation and sexual reproduction in the demosponge *Scopalina lophyropoda*. They found that fragments (generated by wave action or foraging by fishes or echinoderms) contained brooding chambers full of embryos. They expected that these fragments while reattaching to the substrate, would reabsorb the embryos to have more energy for the whole-body regeneration of the sponge. However, the embryos were not only kept intact in the tissue of the regenerating sponge but after 4–8 days of the fragmentation larvae started to be released (Maldonado & Uriz, 1999).

1.5 FERTILIZATION

1.5.1 CONTROL OF SPAWNING

Sponges usually release massive amounts of sperm into the water and male gametes swim and eventually encounter the eggs. While some species are sperm casters (i.e., sperm is released in the water, enters another sponge through the canal system, and fertilization occurs inside the parental sponges, resulting in a viviparous development mode), others release both male and female gametes into the water (i.e., broadcasters/oviparity). Independently of the reproductive mode, some control of the spawning is expected, as the release of spermatozoa without the existence of oocytes (either inside other sponges or in the water) is a waste of energy for the organism. Gamete release in sponges that spawn can be associated with both exogenous and endogenous controls.

As discussed earlier, water temperature is one of the main factors controlling the reproductive timing in Porifera. This and other environmental factors will synchronize/regulate the production and consequent maturation of gametes. In oviparous species, as expected, large amounts of sperm and oocytes are produced within the parental sponge tissue during the reproductive season. In the demosponge, *Neofibularia nolitangere* up to 50% and 35% of the tissue of the female and male sponges, respectively, are occupied by gametes (Hoppe & Reichert, 1987). Lanna et al. (2018) observed that more gametes are produced by oviparous species when compared to viviparous ones. This observation is probably related to the fact that in oviparous species, gametes need to be released in synchrony and in large amounts to find each other and carry out fertilization. Therefore, while the reproductive period of the species can be controlled by annual cycles, the onset of the spawning is likely related to environmental cues within narrower periods, such as diel, tidal, and/or lunar cycles.

The effect of the moon phase on the spawning of corals is well known, such that it is even possible to know ahead when colonies in the Great Barrier Reef, for instance, will release their gametes. In sponges, the effect of the moon phases on the spawning of gametes is still fragmentary. Lévi & Lévi (1976) observed that the gamete release of *Chondrosia reniformis* at the Mediterranean Sea occurs during the new moon. Reiswig (1983) suggested that Caribbean sponges released their gametes in synchrony, mainly biased by the moon phase. Recently, Neely & Butler (2020) observed that the Caribbean populations of the giant coral reef sponge *Xestospongia muta* mainly spawn around the first quarter moon, but a second pulse also occurs just before the third quarter moon. Other species of the same genus found in Indonesia (*X. testudinaria*) release their gametes close to the full moon, and *X. bergquistia* spawns 1–7 days after the new moon or 6–7 days after the full moon in Australia. However, sponges of the orders Agelasida and Biemnida spawn from 3–5 days after the full moon. The influence of the moon can be related to the illumination during the night, by affecting the tides, or the synergy of these two factors. There is another factor controlling the timing of spawning, known as the *time giver* (zeitgeber), related to the amount of time of darkness. In *X. muta*, the spawning events occur mainly in the morning, probably triggered by the beginning of the night in the day before. This is also the case with other species of the genus *Xestospongia* found in the Pacific. However, in other oviparous species (such as *Agelas* sp., *Neofibularia nolitangere*, *Aplysina archeri*, and *Geodia neptuni*), spawning occurs in the afternoon (Reiswig, 1976; Hoppe & Reichert, 1987; Sarano, 1991; Fromont & Bergquist, 1994; Röthig & Voolstra, 2016). It is still unclear whether the period of the day when the spawning occurs is phylogenetically constrained. Moreover, the molecular mechanism used to detect these subtle changes in illumination during the night is still contentious.

The role of pheromones cannot be ruled out in controlling the spawning event in broadcasting sponges. In the Caribbean population of *N. nolitangere*, gamete mass release usually started with females randomly located in the reef. It was suggested that females undergoing oogenesis might release some pheromones into the water to promote the maturation of male individuals (Hoppe & Reichert, 1987). This supposition has gained support recently, as the pheromone *attractin* was found in the transcriptome of sponges belonging to all classes (Riesgo et al., 2014b).

Besides the gamete release in oviparous sponges, it is expected that the larval release of the viviparous species should also be somehow controlled. It is assumed that most viviparous sponges will release their larvae early in the morning (at dawn). In a series of controlled experiments, Amano (1986, 1988) demonstrated that independent of the light regime (at least in the laboratory), two species of Demospongiae from Japan release their larvae in the first hours of the day. However, while in *Halichondria* (*Halichondria*) *panicea* the main stimulus is the beginning of the night (i.e., the time past between the onset of darkness and the dawn), in *Callyspongia* (*Callyspongia*) *nuda* the beginning of the light 24 h before the larval release is the main stimulus of this event. There are also several observations in the field indicating that viviparous sponges release larvae in the first hours of the day (*see* Neely & Butler, 2020). However, in tropical sponges that are found along the coast of Brazil, anecdotal observations indicate that both demosponges and calcareous sponges do not have a preferred timing for the release of larvae.

How does the sponge transmit these external stimuli, especially the ones related to light, to the spawning behavior? In corals, it is suggested that circadian clock genes (related to cryptochrome (CRY) genes) control the expression of genes that will affect the physiology of the organism. CRY genes are known from a number of species of sponges including *Amphimedon queenslandica*, *Suberites domuncula*, and *Aphrocallistes vastus* (Müller et al., 2010; Rivera et al., 2012). In *A. queenslandica* the gene for cryptochrome 2 (*Cry2*) was found to be associated with phototactic behavior of the larva. Jindrich et al. (2017) found the presence of genes related to the circadian clock in other metazoans in the genome of *A. queenslandica*, and that some of these genes (orthologs of *cryptochrome*) do oscillate depending on the illumination. In the case of viviparous sponges, is it the parental sponge or the larvae in the mesohyl that receives the photic stimulus? Amano (1988) suggested that the adult sponge lacks photoreceptors, so it is unlikely that it is responsible for receiving this cue. However, the tissue of haplosclerid species investigated by him was thin and transparent, which is not the case in most sponges. Embryos (as shown below) are covered by a thin follicle and light would need to reach the photoreceptors present in the larvae (Leys et al., 2002; Rivera et al., 2012). In this case, after receiving the light stimulus, the larvae could start to move inside the follicle, break this epithelium, and migrate through the mesohyl to reach the excurrent canals. This movement might take 24 h after a dark regime, resulting in the larval release on the next day (Amano, 1988). The importance of the microbiota living in association with the sponges, especially microsymbionts that photosynthesize, cannot be excluded. However, so far, no study has checked the influence of these symbionts on the spawning of the sponges.

Sperm is released from the spermatic cysts into the aquiferous system. Spawning of the sperm usually occurs in clouds, which, in massive sponges can look like an impressive stream of smoke leaving the osculum. In the oviparous species however, eggs can either be spawned singly or in small groups (as in *Polymastia boletiformis* [Polymastiida], *Suberites ficus* [Suberitida], *Tethya aurantium* [Tethyida], and *Raspailia ramosa* [Axinellida]) or in massive discharges (as in *Axinella damicornis* [Axinellida], *N. nolitangere*, and *X. muta*). Eggs can either be negatively (*N. nolitangere* and *X. muta*) or positively buoyant. Fertilization is a poorly known event of the reproduction of sponges. In oviparous species, it is well-documented for *Tetilla japonica* (Spirophorida) and *Petrosia ficiformis* (Haplosclerida) (Watanabe, 1978; Maldonado & Riesgo, 2009). In viviparous species, the sperm needs to locate the eggs that are generally scattered in the mesohyl and to cross the barriers between the canal system and the oocytes. This is probably related to some sort of chemotaxis.

1.5.2 SPERM-OOCYTE ATTRACTION AND RECOGNITION

Gametes have to find and recognize each other to carry out fertilization. In most animals, the attraction of the sperm to the oocytes is carried out by chemical cues. In Porifera, this issue is still poorly understood. As mentioned above, some pheromones might be involved in the spawning process of the spermatozoa. Koutsouveli et al. (2020b) detected the presence of different genes related to the maturation and the capacitation of the sperm in *Geodia* spp. They proposed that the male would filter female chemoattractants, consequently capacitating the sperm still within the spermatic cysts. In oviparous species, both gametes are released in the water column. Maldonado & Riesgo (2008b) reported that the Caribbean *Chondrilla nucula* [Chondrosiida] releases sperm three days after the female spawning, and fertilization (in laboratory conditions) can generate embryos. In other three species of oviparous demosponges (*Chondrosia reniformis* [Chondrosiida], *Petrosia ficiformis* [Haplosclerida], and *Aplysina aerophoba* [Verongiida]), sperm collected from male sponges was put in contact with oocytes collected from female ones. *In vitro*, the spermatozoa surrounded the eggs and started the fertilization process (Lévi & Lévi, 1976; Maldonado, 2009; Maldonado & Riesgo, 2009). This attraction is probably related to some chemoattractants as even after the formation of the fertilization envelope, sperm are still observed surrounding the embryo (Maldonado, 2009). After the penetration of the sperm cell into the eggs of *Tetilla* spp. and *P. ficiformis*, a non-cellular fertilization envelope develops surrounding the zygote. This membrane is formed quickly (ca. 1 min) and, in *Tetilla* spp., is made of fibrous material expelled by peripheral vacuoles in the cytoplasm of the eggs (Watanabe, 1978; Watanabe & Masuda, 1990). How the sperm is recognized by the oocyte is another poorly understood process. In most animals, the acrosome plays an important role in this recognition process. The presence of sperm cells with pro-acrosomal vesicles has been repeatedly observed (*see* Section 1.3.4), which is an indication that this structure, as in other animals, might also be important in the fertilization of sponges.

While external fertilization occurs in some lineages of demosponges, all other lineages of Porifera are sperm casters. It has been argued that even in some alleged oviparous sponges the sperm enters the body of the oocyte-bearing animal, fertilizes the oocytes and, instead of oocytes being spawned, the zygote is released through the aquiferous system (Sidri et al., 2005; Maldonado & Riesgo, 2008b). Mass spawning events were never observed for viviparous sponges, probably because the sperm is released in small quantities given that it does not need to encounter the oocytes diluted in the water. In these sponges, fertilization is usually associated with an indirect process. In Calcarea, at least, the sperm is engulfed by a choanocyte, and put in a vesicle where it changes its shape, losing its flagellum and becoming more condensed (electron dense). This structure is now called the spermiocyst. The choanocyte changes its shape, leaves the choanoderm, and acquires an amoeboid shape, becoming the carrier cell. Next, the carrier cell ingresses in the mesohyl and moves toward the oocyte. When the carrier cell approximates the oocyte, their plasma membranes fuse, and the spermiocyst is inserted in the cytoplasm of the oocyte (Gatenby, 1920; Gallissian, 1980, 1988; Gallissian & Vacelet, 1990, 1992; Nakamura et al., 1998; Watanabe & Okada, 1998). After the entrance of the spermiocyst, the oocyte resumes the meiotic process and releases the polar body (Watanabe & Okada, 1998). While this event is relatively well documented in Calcarea, outside Calcarea it has only been observed in some homoscleromorphs (*O. lobularis* and *C. candelabrum*) (Tuzet & Paris, 1964; Riesgo et al., 2007a), and therefore the existence of carrier-cells in Demospongiae was already questioned (Reiswig et al., 1976). However, it is known that the sperm is captured by choanocytes in *Hippospongia communis* (Tuzet & Pavans de Ceccatty, 1958) or enters the mesohyl and is captured by amoebocytes next to an oocyte in *Cliona viridis* and *Haliclona* (*Haliclona*) *simulans* (Tuzet, 1930, 1932). In any case, after being captured, the further processes of the spermatozoa and the movements of the sperm-carrying-amoebocytes are very similar to that described for calcareous sponges. This indirect fertilization process, which is mediated by somatic cells of the "mother," was considered related to the absence of acrosome in some species of demosponges (*see* Kaye & Reiswig, 1991); however, *S. calcaravis* forms a complex of carrier-cells to fertilize its eggs and has a conspicuous acrosome (Nakamura et al., 1998; Watanabe & Okada, 1998). As mentioned above, the type of spermatozoa varies in Porifera. In many demosponges, instead of a round sperm cell, the spermatozoa are elongated (Reiswig, 1983; Boury-Esnault & Jamieson, 1999; Vasconcellos et al., 2019; Ereskovsky & Tokina, 2022) and a highly modified V-shaped sperm cell was observed in *C. crambe* (Riesgo & Maldonado, 2009b). The V-shaped morphology of the spermatozoon of *C. crambe* can be associated with the passage of this gamete through the choanoderm and to move in the gelatinous mesohyl (Riesgo & Maldonado, 2009b). This shape would favor the rotation of the whole cell, consequently, facilitating this movement in the very viscous environment where the oocyte is located. Elongated sperm cells (as observed in other Poecilosclerida) might also pass through the choanoderm and move actively toward the oocyte (Vasconcellos et al., 2019). Carnivorous sponges (Poecilosclerida) are viviparous but lack an active aquiferous system. In this case, the sperm is packed within a spermatophore-like structure that after being released from the surface of the sponge remains buoyant until being captured by an oocyte-bearing partner by unusual forceps-shaped spicules that are in its outer coat. There, the spermatophore is digested, releasing the elongated sperm cells which burrow into the mesohyl, actively moving to reach the oocytes (Riesgo et al., 2007b).

Independent of whether the sponge is oviparous or viviparous, it seems that fertilization is an efficient process in Porifera. Approximately 70% of the oocytes that are released in the haplosclerids *Xestospongia bergquistae* and *Petrosia ficiformis* are fertilized and start their external development (Fromont & Bergquist, 1994; Maldonado & Riesgo, 2009). In viviparous species, fertilization is probably higher than that (> 90%), as the degeneration and reabsorption of eggs are not usually observed (Riesgo et al., 2007a; Lanna et al., 2018, 2021).

1.6 EMBRYO DEVELOPMENT

1.6.1 DIVERSITY OF DEVELOPMENTAL MODES

The vast majority of sponges have an indirect type of development, i.e., they produce free-living (either swimming or crawling) larvae. The exception is found in the order Spirophorida, which has direct development in different species of *Tetilla* (Watanabe, 1957, 1978; Watanabe & Masuda, 1990). In *T. japonica* and *Tetilla serica*, fertilization occurs externally, and development occurs outside of the parental sponge. The zygotes attach to the substrate and go through successive cleavages while attached so that when the morula stage is reached, the embryos are still inside the fertilization membrane. In some cases, cleavages occur before attachment, and the rounded embryo attaches to the substrate by the migration of cells located at the periphery of the embryo. In either case, those cells form the basopinacoderm and the external pinacoderm of the sponge. The inner cells form the initial leuconoid aquiferous system. As such, a larva is never formed in this genus (Watanabe & Masuda, 1990).

There are eight larval types in Porifera, most of them being characteristic and exclusive of some clades (Figure 1.4; Maldonado & Bergquist, 2002; Leys & Ereskovsky, 2006). These larval types are defined not only by the general morphology of the larvae, but also by their respective embryonic development (i.e., cleavage types, and mechanisms of morphogenesis). The class Hexactinellida has a *trichimella* larva (Figure 1.4(A)), which has been documented best in *Oopsacas minuta* (Boury-Esnault et al., 1999; Leys et al., 2006, 2016); other larval types might exist, but they have not so far been documented. In this group, archaeocyte congeries (groups) develop into sperm, while others become oocytes. Fertilization, although never documented, is internal and embryogenesis takes place in the choanosome of the parental sponge (viviparous). Cleavage is total and equal until the formation of coeloblastula. These cells then divide to form an inner and an external layer, a process called delamination (Boury-Esnault et al., 1999). The cells of the external layer connect to each other through plugged cytoplasmic bridges. Some of these cells appear to form the ciliated cells of the larva, and the remaining cells appear to fuse to form an external syncytium. The cells of the inner layer, which are initially larger than the outer cells, also fuse to form the inner region of the embryo. Some of the internal cells differentiate in sclerocytes (which secretes the stauractines spicules of the larvae), and other cells appear to form choanoblasts which give rise to collar-flagella units (eunucleate choanocytes) which form tiny spherical chambers in the larva (Leys et al., 2006). The rest remains as one large syncytium, containing yolk and lipids, which, as a seemingly amorphous cobwebbed tissue surrounds and connects with the choanoblasts and sclerocytes. In the end, all tissues are connected in one syncytium, often using the characteristic plugged junctions of hexactinellids sponges, which are in the cytoplasmic bridges connecting different regions of syncytia. The resulting larva is elongated with a clear anteroposterior polarity: a rounded anterior pole, a central ciliated zone, and a conical posterior pole. The main tissue of the larvae is the trabecular syncytium, which forms the inner cell mass and the external ciliated epithelium. Choanocyte chambers, formed by choanoblasts with several collar-flagella units as extensions (as in the adult sponge), are usually observed among the inner cell mass (Figure 1.5(A); Leys et al., 2006).

The class Homoscleromorpha has a *cinctoblastula* larva, best described from the genera *Oscarella*, *Plakina* (Boury-Esnault et al., 2003; Ereskovsky et al., 2010) and *Corticium* (de Caralt et al., 2007; Maldonado & Riesgo, 2008a). Embryogenesis occurs entirely in the choanosome of the parental sponge (viviparous development). Cleavages are total and initially equal, and asynchronous. However, soon cleavage becomes mostly chaotic, developing a stereoblastula with similar-sized cells. Later, the cells start to migrate to the periphery of the embryo, in a centrifugal migration, creating a large cavity within the embryo. The larval epithelium starts to differentiate by cells forming apical flagella and producing a basement membrane. At the same time, desmosome-like junctions are formed at the apical portion of the cells, tightening up the epithelium (Boury-Esnault et al., 2003). The larvae are entirely ciliated, with cilia having the same size all along the larval body, but the anterior-posterior polarity is seen in a larger anterior pole and tapered

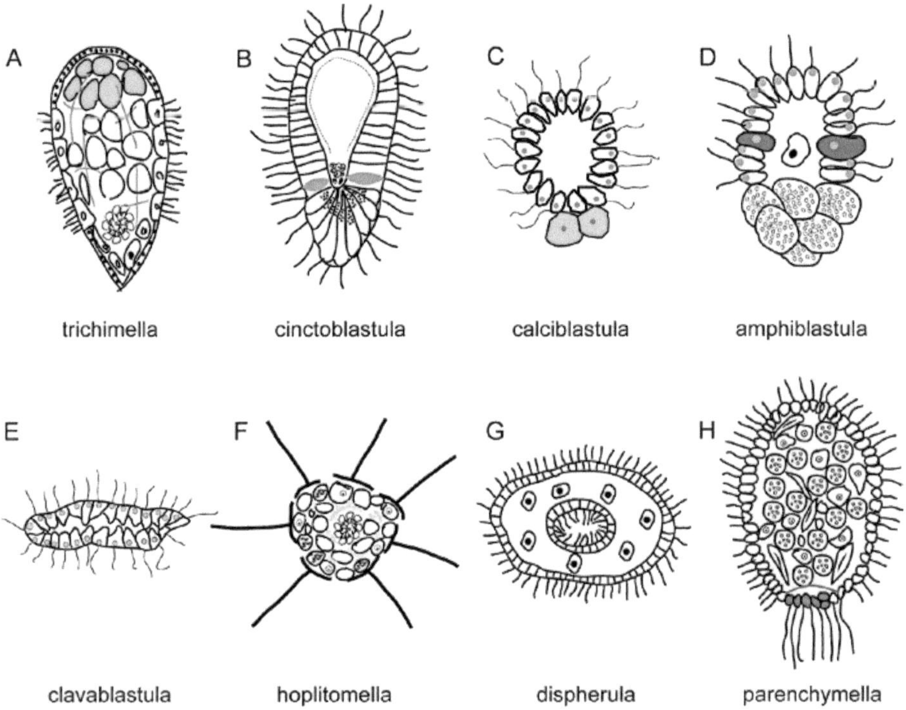

FIGURE 1.4 Schematic drawings showing the larval diversity in Porifera. (A) Trichimella of the class Hexactinellida; (B) Cinctoblastula of the class Homoscleromorpha; (C) Calciblastula of the subclass Calcinea (Calcarea); (D) Amphiblastula of the subclass Calcaronea (Calcarea); (E) Clavablastula of the orders Chondrillida and Verongida (Demospongiae); (F) Hoplitomella of the family Alectonidae (Demospongiae); (G) Dispherula of the order Halisarcida (Demospongiae); and (H) Parenchymella type 'iii' of the order Haplosclerida [there are other variations on the morphology of parenchymella and this type of larvae is widespread in Demospongiae, see text].

posterior pole (which is also usually darker than the anterior). The larva can be divided into three regions (anterio-lateral, posterio-lateral, and the posterior regions). Each region comprises ciliated cells with different nuclear positions and cytoplasm content (Figure 1.4(B); Boury-Esnault et al., 2003; Ereskovsky et al., 2010; de Caralt et al., 2007; Maldonado & Riesgo, 2008a).

The class Calcarea has two different types of development: the *calciblastula* in the subclass Calcinea, and the *amphiblastula* in the subclass Calcaronea, both larvae develop inside the parental sponge (viviparous) (Figure 1.4(C)). In Calcinea, the zygote passes through successive total and equal cleavages, forming a hollow blastula with mostly similar cells, except for a small group of larger blastomeres located in the future posterior pole of the embryo. The epithelium becomes entirely ciliated, except for those large blastomeres (Ereskovsky, 2010; Riesgo et al., 2018). In Calcaronea development of the amphiblastula also involves total cleavages. However, after the third round of cleavages, macromeres, and micromeres are formed. The division rate appears faster in micromeres, and cleavage is spiral-like resulting in a cup-shaped hollow embryo, with an opening among the macromeres. The micromeres produce a cilium facing into the embryonic cavity. Four cells located in each quadrant at the equator of the embryo do not have cilia and become cross-cells. In this stage, the embryo is called stomoblastula. Later, the ciliated cells of the stomoblastula pass through the pore among the macromeres, inverting the epithelium (i.e., it turns inside-out) to expose the cilia of the ciliated cells. During this process, some maternal cells immigrate into the larval cavity, which is later sealed with the closure of the pore among the macromeres (Figures 1.4(C) and 1.5(B); Dubscq & Tuzet, 1937; Gallissian, 1988; Gallissian & Vacelet, 1992; Lanna & Klautau, 2012).

The class Demospongiae has the highest diversity of larval types: *clavablastula*, *hoplitomella*, *dispherula*, and *parenchymella* (Figure 1.4; Maldonado & Bergquist, 2002). The *clavablastula* is an artificial name that encompasses different larvae that develop from blastula-like embryos and then become entirely ciliated and hollow (Maldonado & Bergquist, 2002; Wörheide et al., 2012). Consequently, some variation is expected in this developmental type. The formation of the larva seen in *Chondrilla australiensis* is an example of this type of development but should not be generalized to all "*clavablastula*" type larvae. In *C. australiensis*, fertilization takes place externally or can occur still inside the parental sponge, but the zygote is quickly released. The cleavages occur externally, being total and equal, leading to a blastula with a small cavity surrounded by a single-layered epithelium. Some maternal cells can ingress in the larval cavity. All cells of the larva are ciliated and appear to have the same cytoplasmic constitution, but have different shapes, being flattened at the posterior pole and cuboidal elsewhere (Figure 1.4(E); Usher & Ereskovsky, 2005). The *hoplitomella* developmental type is restricted to sponges of Thoosidae (Thoosina, Tetractinellida) (Vacelet, 1999; Wörheide et al., 2012). In this group, fertilization is internal, but the zygote is soon released, and development occurs externally (oviparous). Cleavage is total, equal, and chaotic, generating a stereoblastula. There is an unusual amount of collagen among the blastomeres and in the space between the cells and the follicle covering the embryo. Larval sclerocytes first differentiate and start to secrete some long spicules that are thought to retard sinking by the larva. In *Alectona millari*, some granular cells and choanocyte chambers develop inside the inner cell mass. In addition to the long spicules, other small and plate-like ones are also secreted and cover the larva, generating an armored structure. The larva has no anteroposterior polarity and no cilia. The larva has three layers: the external, with the plate-like spicules and pinacocytes; an intermediary, with a dense accumulation of archaeocytes, bacteriocytes, and granular cells; and the inner layer, with loosely non-identified cell types and the choanocyte chambers (Figure 1.4(F); Vacelet, 1999). The *dispherula* type of larval development is characteristic of the Halisarcidae (Gonobobleva & Ereskovsky, 2004; Wörheide et al., 2012). Fertilization and embryogenesis occur internally (viviparous). Cleavages are total and equal. The blastomeres form a single layer covering the internal cavity until reaching the blastula stage. It seems that some species of Halisarcidae are poecilogonous. In *Halisarca dujardini*, some embryos develop into larvae from this coeloblastula stage, forming a single-layered ciliated epithelium. Other embryos have an internal mass of cells that might have entered into the cavity after migrating from the epithelium. Finally, a third type of embryogenesis can also be observed in which the embryo folds the epithelium (invagination) into its cavity, pushing inside the embryo a group of cells that will form a spherical structure, which eventually loses contact with the external epithelium. The larva is fully ciliated, but anteroposterior polarity can be observed as the posterior pole is less densely ciliated (Figure 1.4(G); Gonobobleva & Ereskovsky, 2004).

The best-known larva of Demospongiae is the *parenchymella* (Figure 1.4(H)). This type of development usually occurs internally (viviparous), and a solid ciliated larva is formed. Cleavages are total and initially equal, but soon become chaotic (often due to heavy amounts of yolk in the blastomeres). Cleavages lead to a solid stereoblastula (Figure 1.5(C) and (D)). In some lineages, the blastomeres of the stereoblastula differentiate early into collencytes, sclerocytes, archaeocytes, and choanocytes. However, in most embryos, smaller cells (micromeres) migrate towards the external portion of the embryo, forming an external layer, where they differentiate in ciliated cells. In addition to the ciliated cells, some globular cells and flask-shaped cells might be observed in the external layer of the sponge. Some parenchymellae have a subepithelial region, where roundish cells are observed. In the inner region, different cell types (e.g., sclerocytes, archaeocytes, collencytes), spicules, intra, and intercellular symbionts, and collagen fibrils are observed. There are at least three different morphological variations of the parenchymella depending on the larval form and pattern of ciliation (Maldonado & Bergquist, 2002). The parenchymella type (i) was described for some species of Poecilosclerida, Suberitida, and Dictyonellidae. The larvae of this type are elongated (with a tapering posterior pole) and entirely and uniformly ciliated or, in some species, have the posterior end naked (Figure 1.5(E)). The parenchymella type (ii) was observed in some Spongillida (freshwater sponges). In this group, the larva is pyramidal and has a wide and rounded anterior pole. A large secondary cavity is present next to the anterior pole and, in many species, choanocyte chambers are present (Maldonado & Bergquist, 2002). Finally, the parenchymella type (iii) is recognized in some species of Haplosclerida and Dictyoceratida (Figure 1.4). It is characterized by being bullet-shaped and entirely ciliated, except for the anterior end, which is slightly flatter and lacks cilia. The posterior pole is frequently surrounded by a ring formed by pigmented cells that have longer cilia. To form this ring of pigmented cells, still, during the cleavages, some small blastomeres become pigmented and the embryo acquires a salt-and-pepper morphology. Later, these pigmented cells migrate towards the posterior pole to form, first the "spot," and then the "ring" (which is a photosensitive structure of the larva).

1.6.2 MOLECULAR CONTROL OF EMBRYONIC DEVELOPMENT IN SPONGES

Sponges possess homologs of many eumetazoan developmental genes in their genomes. Among these genes, several families of transcription factors (e.g., *Pax*, *Sox*, *Fox*, and *T-box*), components of the signaling pathways (e.g., *Wnt*, *TGF-beta*, receptor tyrosine kinase (RTK), *Hedgehog*, and *Notch*/*Delta*), and cellular adhesion molecules (CAM, e.g., *integrin*s and *cadherins*) are present, although some of pathways and/or species have significant absences (reviewed in Adamska et al., 2011; Adamska, 2016; Riesgo et al., 2014b). Despite the vast diversity of developmental types, knowledge about the molecular control of embryonic development is restricted to a few species, mainly the haplosclerid *Amphimedon queenslandica* (parenchymella type (iii), Figure 1.4(H)) and the calcaronean *Sycon ciliatum* (amphiblastula, Figure 1.4(D)). The control of the cell division and early organization of the embryos was reviewed recently (Degnan et al., 2015).

How the inversion process seen in *S. ciliatum* and other calcaroneans is controlled, is still unknown. However, the epithelium turns inside-out and exposes the cilia that were originally pointing inwards in the embryo to the outside and anterior portion of the amphiblastula (Degnan et al., 2015). The organization of the embryo of *A. queenslandica*, however, seems to be more complex. After cleavage, cell movement sorts the different types of blastomeres to form an inner and an outer layer. The outer layer is enriched with ciliated cells, sclerocytes, and pigment cells. The inner layer has large granular cells that barely touch each other and are surrounded by extracellular matrix (ECM). The cell sorting is probably related to the different adhesive properties of the cells. Cells in both layers keep moving to be arranged in the anteroposterior axis of the embryo. Then, just before the 'cloud' stage, the embryo passes through a compaction, and cell movement ceases. During the 'spot' stage, pigment cells accumulate in the posterior region of the embryo being later spaced by some non-pigmented micromeres in the center of the spot. The outer layer of the embryo is formed only by ciliated micromeres (except at the anterior and the posterior ends where cilia are not present). The next stage ('ring') is similar to the larva. Flask cells become evident close to the anterior region and are embedded in the ciliated epithelium. In the inner region, several cell types, including sclerocytes, are present. At the posterior region, the pigment cells combine with the long-ciliated ring cells to form a photosensory organ. Interestingly, grafting pigment cells in different stages to host embryos induce the formation of ectopic rings in the new location, indicating that the fate of these cells is determined early in the development (reviewed in Degnan et al., 2015).

Although the adult sponges lack a clear polarity, most larvae of Porifera have an anteroposterior axis (Manuel, 2009). This polarity seems to be acquired during embryonic development and is likely related to the expression of *Wnt* ligands. In *A. queenslandica*, some cells start to express *Wnt*-ligands before the 'spot' stage. When the embryo reaches the 'ring' stage, the *Wnt*-expressing cells are situated around the ring, at the posterior region (Adamska et al., 2007). In *S. ciliatum*, the expression of *Wnt*-ligands occurs mainly at the granular cells of the stomoblastula (i.e., previous to the inversion process). When the amphiblastula is formed, homologues of this gene are expressed also at the posterior pole of the larva (Leininger et al., 2014). After metamorphosis, in both *A. queenslandica* and *S. ciliatum* the osculum region also expresses *Wnt*. In the Spongillida sponge *E. muelleri*, the treatment of sponges hatching from gemmules with drugs that mimics or inhibit the *Wnt* signaling pathway interferes in the development of the aquiferous system (Windsor & Leys, 2010; Windsor-Reid et al., 2018). Moreover, *Wnt* homologs are also related to the establishment of ostia, canals, and oscula in *O. lobularis* (Homoscleromorpha, Lapebie et al., 2009). Together, this information supports the idea that the *Wnt* pathway is related to the establishment of the polarity of sponges, if not the anteroposterior, at least the aquiferous system polarity. However, Windsor-Reid et al. (2018) observed that different

Wnts are expressed in individual cells of the hatchling Spongillida sponges, rather than in a specific region. Additionally, glass sponges seem to lack *Wnt* whatsoever (Schenkelaars et al., 2017). Schenkelaars et al. (2017) suggested that *Wnt* pathway might work in different ways depending on the sponge class and, instead of marking a specific polarity (as occurs in eumetazoans), these genes might be orchestrating different functions of the cells.

1.6.3 MATERNAL-EMBRYO RELATIONSHIP IN VIVIPAROUS SPONGES

In most viviparous sponges, regardless of the type of development, a follicle formed by maternal cells encapsulates the developing embryo (Figures 1.5(B) and (D)). This follicle can vary from a single layer of squamous cells to multiple layers of amoeboid cells surrounding the embryo. It is usually formed during oogenesis when nurse cells are providing nutrition to the oocytes, and it becomes more specialized as embryogenesis progresses. Usually, the follicle cells are separated from the embryo by a thin layer of collagen. The follicle layer can help protect the embryo, but they are mostly considered to have a nutritive role. Evidence for the nutritive role can be seen by the increase in the size of the embryos inside the follicle or the increase in the number of macromeres, the migration of maternal cells into the embryonic cavity followed by their digestion by the embryo cells, and the presence of the same type of inclusions in the cells of the follicle and in the embryo (Lanna & Klautau, 2012; Ostrovsky et al., 2015). The role of the follicle in defining the polarity of the embryo has also been suggested and is something that needs further investigation (reviewed by Degnan et al., 2015).

1.7 ASEXUAL REPRODUCTION AND DEVELOPMENT

1.7.1 TYPES OF ASEXUAL REPRODUCTION

Sponges are well known for their capacity for regeneration, and this is closely tied to asexual reproduction. This characteristic is likely related to the maintenance of stem cells during their whole life cycle (*see* Section 1.3.2). There are four types of asexual reproduction in Porifera: fragmentation, non-embryonic development (NED, also called whole-body regeneration or somatic embryogenesis), budding, and gemmulation. In theory, all sponges are capable of reproducing by fragmentation and NED. Fragmentation can be caused by external forces (e.g., wave action, incomplete predation, and substrate partitioning; Wulff, 1991) or endogenously driven. In the latter

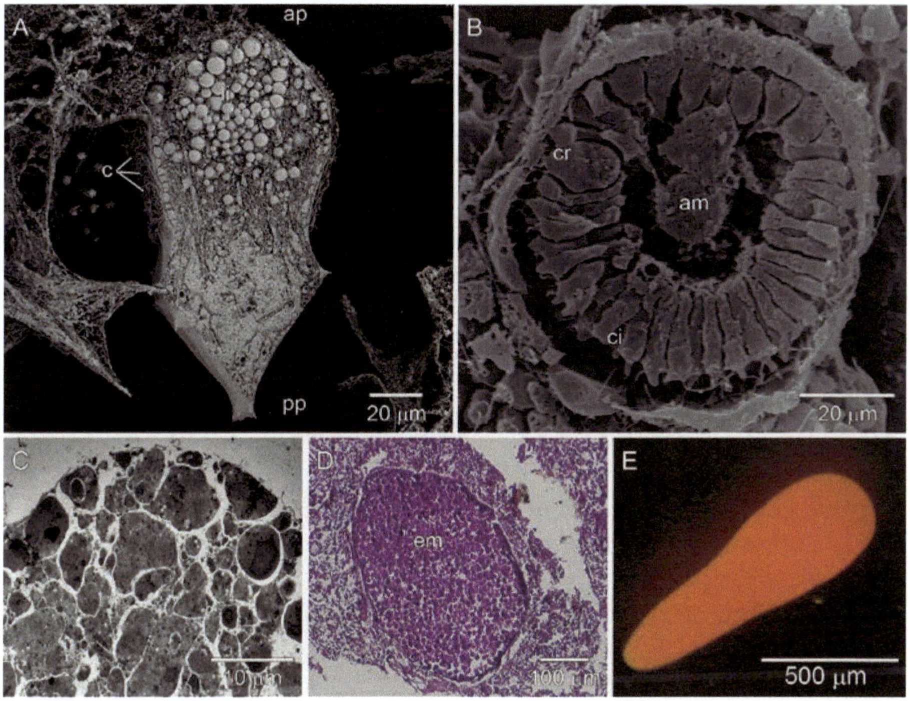

FIGURE 1.5 Some embryos and larvae of Porifera. (A) Longitudinal fracture of the trichimella of *Oopsacas minuta*. (B) Transversal fracture of the amphiblastula of *Paraleucilla magna*. (C) The embryo of *Kirkpatrickia variolosa* shows the result of chaotic cleavage. (D) The late embryo of *Stylocordila borealis* in the choanosome. (E) Free swimming parenchymella type 'ii' of *Lissondendorix colombiensis*. ((A) and (B): scanning electron microscopy; (C): Transmission electron microscopy; (D): Light microscopy; (E): live image).

Legends: am: amoeboid cells; ap: anterior pole; c: cilia; ci: ciliated cells; cr: cross cells; em: embryo; and pp: posterior pole.

case, parts of the sponge pinch off leaving two separate masses. This is seen in a range of encrusting and spherical sponges but is often commented on in species of *Tethya* where tendrils are extended and left attached to the substrate to develop into new individuals (Cardone et al., 2010; Singh & Thankur, 2018; Shaffer et al., 2021). NED consists of the reconstruction of a functional sponge after the dissociation of the sponge into individual cells, first described by Wilson (1907). In this type of asexual reproduction, clumps of cells reorganize to form a sphere of amoeboid cells surrounded by pinacocyte-like cells called an 'aggregate' (Galtsoff, 1925). These cells can be derived from adults or even larvae (Borojevic & Lévi, 1964, 1965). The factors that lead to adhesion of the cells have been shown to be often spherical proteoglycans, which although they share attributes with hyaluronan-type adhesion found in other animals, are sponge specific, and have been called 'spongicans' (Fernandez-Busquets & Burger, 2003). This process has been developed for maintaining sponges in a semi-embryonic state in culture, and these small sponges are called 'primmorphs' (Custodio et al., 1998; Le Pennec et al., 2003). In continuing development, aggregates or primmorphs attach to the substrate, at which point the inner cells transdifferentiate and move to form the aquiferous system of the sponge. A single osculum opens in the apical region of the sponge and a new individual will resume its functions. Although it is suggested that NED can occur in any sponge, it seems that ecological and morphological constraints, such as the type of aquiferous system of the sponge and cell complement, interfere with the capability of reproduction through NED (Eerkes-Medrano et al., 2015; Lanna & Klautau, 2019).

Budding and gemmulation are more coordinated processes, compared to the other asexual reproduction types. Budding is observed mainly in demosponges, especially in Tethyida and Axinellida, but it is also found in some homoscleromorphs (Boury-Esnault, 1970; Corriero et al., 1996; Gaino et al., 2006; Rocher et al., 2020). There are, roughly, two types of budding, one that consists in the accumulation of cells in the mesohyl (as in *Tethya* spp.) and another that consists of swellings of the surface of the sponge (as in *Axinella* and *Oscarella*). In the first type, budding consists of the accumulation of groups of cells (mostly archaeocytes and cells with inclusions) inside the sponge, being later covered by a pinacoderm and sprouting out from the surface of an adult to develop in a new individual (Gaino et al., 2006). The organizing region of the budding can be directed either by mesohyl cells or by the cortex of the sponge. In either instance, the sponge deposits some collagen in this region which seems to attract archaeocytes and cells with inclusions from the mesohyl. Usually, the cortical region of the sponge reorganizes its spicules to prepare a spicule tract where the cells migrate to form a bulged structure at the tips of the filaments (the 'bud') (Gaino et al., 2006). At the tip of the filaments, the bud is covered by a layer of pinacoderm and, in some species, spicules. This pinacoderm encases extracellular matrix and amoeboid cells, mostly archaeocytes and different cells with inclusions. The size and the number of buds, as well as the types of cells inside them, vary from species to species. Some species can even have functional choanocyte chambers within the bud (Simpson, 1984). In *Oscarella lobularis* and *O. tuberculata*, budding occurs as swellings at the surface of the sponge. In this case, finger-like protrusions appear all along the body of the sponge. These protrusions contain all the tissues of the sponge: exo- and endopinacoderm, mesohyl, and choanoderm. Subsequently, the protrusions swell due to the volume increase in the central cavity and the tightening of the basal portion to create a small stalk connecting the bud and the parental sponge. The presence of the aquiferous system and protruding spicules have been interpreted to suggest that buds could float adrift in the currents after being separated from the parental sponge (Ereskovsky & Tokina, 2007).

Depending on the species, budding can be either constant in the population or occurs seasonally (Simpson, 1984; Corriero et al., 1996). In *Tethya* spp., the ability to bud seems to be defined genetically, as sympatric species experiencing the same environment, produce buds in different periods of the year. The water temperature is related to this phenomenon, usually, more buds are developed as a response to decreasing temperatures (Corriero et al., 1998). The deposition of sediments or the growth of epibionts on the cortex of the sponge can also influence the production of buds (Gaino et al., 2006). In controlled conditions, artificial fragmentation of sponges can also lead to the production of buds (Gaino et al., 2006; Rocher et al., 2020).

Finally, gemmules are found in freshwater sponges (Spongillida) and some marine demosponges (e.g., *Chalinula loosanoffi*, *Suberites domuncula;* Connes et al., 1978; Simpson, 1984). In this type of asexual reproduction, totipotent cells accumulate in a given region of the mesohyl (often close to the substrate) and pass through successive divisions. Concomitantly, these cells accumulate complex yolk inclusions, while other cells secrete a thick collagenous envelope surrounding the archaeocytes. Depending on the species, the collagenous envelope also contains species-specific spicule types (gemmoscleres) embedded in it. The envelope completely surrounds the gemmule, with a slightly thinner region at one point developed into a small opening, the micropyle, through which the gemmule senses the environment (Simpson, 1984). The production of gemmules is part of the life cycle of different lineages of freshwater sponges (although it is absent in some families of the order Spongillida). Although gemmules do not disperse themselves, they are transported by the wind or by animals, e.g., birds, mammals, and fish either by foraging on the sponges or by an entanglement of the gemmoscleres in the feathers or fur (Van Leeuwen et al., 2017). Their main function in the life cycle of the sponge, however, is assumed to be as protection against harsh conditions. The gemmule coat allows some gas exchange (Rasmont, 1962; Loomis et al., 2009), but is resistant to desiccation and freezing (Barbeau et al., 1989; Ungemach et al., 1997). Consequently, gemmules can be considered resting bodies (Simpson, 1984).

The mechanisms by which both buds and gemmules are triggered to begin new development are still unresolved. Buds separate from the parent sponge and attach to the substrate and start spreading their cells to start the development of the juvenile. Gemmules typically begin development when temperatures increase after the winter dormant period. Cells in the gemmule differentiate by nuclear division into two and four nuclei per cell as they move toward the micropyle. The first cells to exit the micropyle, form the pinacoderm and once that is formed, the remaining cells exit, differentiate into archaeocytes, and from those into the range of cell types that make up the aquiferous system (Höhr, 1977).

1.7.2 GEMMULATION, HIBERNATION, AND GEMMULE HATCHING

The gemmulation process is much more ordered than the other asexual reproduction processes. It was observed that, at least in *Ephydatia fluviatilis*, the sponge must have a minimum size before starting gemmulation (Rasmont, 1963). This is likely related to a minimum threshold of archaeocytes that are going to be used during the process. The initial phase is marked by a chemotactic movement of cells (archaeocytes, lophocytes, and spongocytes) toward a specific region of the mesohyl. The factors contributing to the aggregation point are still unknown, but apparently, several initial aggregation points are formed and some of them will overcome the others, recruiting more cells. These characteristic spaces the aggregation points regularly. In experimental conditions, the onset of the aggregation can be stimulated by the treatment with methyl xanthine or theophylline (Rasmont, 1974). The cells grow before starting to move. Once at the aggregation point, these cells will accumulate the yolk platelets (containing lipids and proteins) after being in contact with trophocytes (reviewed in Simpson, 1984). In the meantime, the coat (a rigid structure that covers the mass of cells) begins to be secreted. Two epithelia surround the aggregate: the internal epithelium comprises flattened archaeocytes with some platelets, and the external one has spongocytes. These epithelia form initially at one pole of the aggregate, expanding their limits to cover it entirely. The area in which both extremes of the epithelia meet, passes through a special morphogenesis, marking where the micropyle will be formed. As the spongocytes arrange themselves to form the external epithelium, microsclerocytes move to carry their gemmoscleres in formation (aided by the 'companion cells') into the space between both epithelia. This space is filled with collagen secreted by the spongocytes. Later, the cells of the internal epithelium disappear. In some species, vacuoles of the spongocytes are reinforced by collagen fibrils, stabilizing "holes" within the coat, which acquire an alveolar structure that will be important for the fluctuation of the gemmule. At the external portion of the gemmule, a thin extra layer of collagen is finally secreted by the spongocyte, forming the husk of the gemmule. The development of the micropyle membrane is oriented in a swirl of the spongocyte epithelium and the internal one. In this region, there are no alveolar layers or gemmoscleres. The tissue that forms the micropyle is continuous with both epithelia and the swirl of the spongocytes is probably the main driver of the difference in the orientation of the collagen fibrils in this region (Langenbruch, 1981). The structure of the micropyle varies from species to species. At the end of coat formation, the yolky archaeocytes in the aggregate become binucleate, differentiating in thesocytes. The thesocytes are the single cells that are found in the inner cell mass of the gemmule (Langenbruch, 1981; Simpson, 1984). The resulting gemmules are very resilient and can survive in challenging environments, as the thesocytes can enter in quiescent mode.

The gemmulation process is usually linked to changes in the environment, either freezing or drying conditions of the water body. The way in which the sponge senses that it is time to start producing the gemmules is still unknown. The best hypothesis still in use was proposed in 1935 by Pourbaix: the older parts of the sponge suffer nutritional and respiration difficulties, triggering the development of the aggregation points. However, there are some factors that seem to influence the size and amount of gemmules produced by a sponge. Food source and the volume of the sponge are important factors controlling the production of gemmules. Depending on the quality of the food, even small sponges are able to produce more gemmules than larger sponges that are not well-fed (Rasmont, 1963). The length of time during which the gemmules are kept at 3°C also influences the amount and the size of gemmules produced in the next cycle by the sponge that hatched from those gemmules, suggesting that the current overwintering conditions will influence the gemmulation process of the next winter. Additionally, sponges sharing the same habitat start producing gemmules at different periods of the year. Together, these factors indicate that an endogenous control of the gemmulation is present. Light and temperature can also influence the size and the amount of gemmules produced by the sponge (Rasmont, 1970). Additionally, low silica concentration dissolved in the water, in experimental conditions, leads to the production of more gemmules with half of the size of the regular gemmules of *E. fluviatilis* (Peetermans-Pé et al., 1975). In these low silica conditions, the adult sponges produce less spicules. The authors suggested that this condition could allow the formation of more aggregation points in the mesohyl and/or provide more sclerocytes that could be recruited to form the gemmules. In these gemmules, the gemmule coat lacks spicules, and although the presence and morphology of thesocytes are the same, they are found in half of the original quantity.

Gemmules can be kept in freezing conditions for months and will still hatch to form new juveniles (Barbeau et al., 1989; Leys et al., 2019). However, the quiescence capacity of the gemmules varies from species to species. An interesting scenario appears when comparing two species of *Ephydatia.* While in *E. mulleri* the gemmules need to pass through a period of cold before being able to hatch (i.e., they pass through a diapause independently of the environment where they are found), *E. fluviatilis* produces gemmules that can hatch as soon as they find better environmental conditions (Rasmont, 1954). Consequently, some authors have proposed that there are two types of gemmules: those that go through diapause and those that only stay quiescent for a while. According to Rasmont (1954), *Spongilla lacustris* has both types of gemmules. He found that green gemmules (harboring symbiont algae) could hatch and form new individuals as soon as they are exposed to favorable conditions, but brown gemmules (which were larger and lacked symbiotic algae) needed to go through a period of diapause (Rasmont, 1954). The factors that control entry into or exit from quiescence/diapause of the thesocytes are also poorly understood. However, Loomis et al. (2009) suggest that when the thesocytes accumulate yolk platelets, glycogen is converted to sorbitol, increasing the osmotic pressure of the cells. When the environment becomes harsh (either by freezing or desiccation), metabolism becomes depressed. They suggest that *sorbitol dehydrogenase* (*SDH*) genes are downregulated, and the osmotic pressure remains high during the quiescent period, inhibiting cell division and reducing the metabolism of the cells. Release from exposure to harsh conditions (either cold or desiccation) can activate the *SDH* pathway leading to a decrease in the osmolarity of the cells, allowing them to resume their division and metabolic activity.

Finally, the cells leave the gemmule through the micropyle and start to attach to the substrate. Basopinacocytes attach the gemmule to the substrate and pinacocytes form a second epithelial layer over the basopinacocytes. An intermediary layer of thesocytes fills the space between these two first layers. The thesocytes differentiate into archaeocytes, sclerocytes, and choanocytes. The sclerocytes secrete the first spicules which are transported by a special type of cell, positioned, and anchored to the substrate to erect the external epithelial layer like a tent (Borojevic, 1971; Nakayama et al., 2015). The archaeocytes also differentiate into choanocytes, each dividing twice to form the incipient choanocyte chambers which appear to increase in size and number of choanocytes by immigration of new archaeocytes dividing in the mesohyl adjacent to the chamber (Kahn & Leys, 2016). Very early on, chambers open to lacunae, while incipient incurrent canals bring water in from the first few ostial openings in the surface of the sponge (Weissenfels, 1981). With time, the lacunae merge into canals and, as flow increases through this seemingly disorganized network, eventually the lacunae merge into larger canals. It appears that an excurrent opening develops almost immediately, and looks like a large ostium because water does exit the sponge even before a clearly defined and raised osculum develops (Leys, personal observation).

The osculum is an important player in the organization of this primordial aquiferous system. Experiments ablating or transposing oscula result in the disorganization of the morphology of the sponge (Mergner, 1964, 1966; Windsor & Leys, 2010). The gene expression profile from hatching to juvenile formation was investigated in *E. mulleri* (Kenny et al., 2020). The pattern of expression of the genes related to this process is similar to the shifts occurring in the development of other animals. First, genes related to the breakdown of energetic reserves (e.g., *arachidonate* pathway to a breakdown of glycogen), then genes involved in cell motility and the production of the extracellular matrix are activated. In the late stages, genes related to structural maintenance (e.g., type IV collagen, *contactin*, *laminin*, etc.), homeostasis, and the immune system are upregulated. *Wnt*, *TGF-beta*, and *Hedgehog* genes are expressed mainly when hatching and during the formation of the spicules, choanocyte chambers, and the aquiferous system.

1.8 CONCLUSIONS

The phylum Porifera is an ancient lineage of metazoans that evolved different reproductive strategies along its history on the planet. Knowledge about the physiology of reproduction of Porifera is still fragmentary and is based on studies in very few species. Consequently, there are many generalizations that may not apply broadly to the phylum, and new investigations may lead to different findings that can contradict or point to different ideas about the physiology of sponges. However, evidence based on morphological, experimental, and molecular data demonstrates that the reproductive physiology of sponges shares many characteristics with other metazoans. Based on this review, we highlight some open questions that can drive the next steps in the physiology of the reproduction of sponges.

- We suggest that in the next years, it would be useful to obtain more genomic and transcriptomic data to enhance our understanding of how many aspects of the reproductive cycle, sexuality, gametogenesis, and embryo and asexual development are controlled.
- Controlled experiments to verify the effect of exogenous factors influencing the onset of gametogenesis should be carried out.
- How reproduction affects the filtration rate in sponges should be examined.
- Is there a germ cell lineage in Porifera? This is an ongoing question, but if there is, then what determines this lineage during the life cycle of a sponge?
- How is spawning controlled in sponges? RNAseq studies could help to identify the molecular machinery involved in this process.
- How does fertilization occur in several groups of Porifera? Is the acrosome involved in this process?
- Much more work is needed on larval structure and behavior across a diversity of sponges.
- Studies on the molecular developmental biology of diverse larval types are very much needed.
- What is the molecular machinery controlling the formation of gemmules and similarly, what molecular pathways are are involved during the hatching of sponges from gemmules?
- Finally, how does the development of sponges from gemmules compare to the development of larvae from an embryo and the development of a juvenile sponge from a metamorphosing larva?

These are just some examples of future directions that would greatly improve our knowledge about the challenging, but very exciting field of the reproductive physiology of Porifera.

ACKNOWLEDGMENTS

We thank Dr. Manuel Maldonado (CEAB-CSIC) for providing images of embryos. Funding is acknowledged from FAPESB and CNPq (EL), the Spanish Ministry of Science and Innovation (PID2019-105769GB-I00)

(AR), and NSERC 2016-05446 (SPL). All authors contributed equally to all the steps of this work.

KEYWORDS

- **fertilization**
- **gametogenesis**
- **larval types**
- **molecular data**
- **Porifera**
- **sponges**

REFERENCES

Adamska, M., (2016). Sponges as models to study emergence of complex animals. *Curr. Opin. Gen. Dev., 39*, 21–28.

Adamska, M., Degnan, B. M., Green, K., & Zwafink, C., (2011). What sponges can tell us about the evolution of developmental processes. *Zoology, 114*, 1–10.

Adamska, M., Degnan, S. M., Green, K. M., Adamski, M., Craigie, A., Larroux, C., & Degnan, M. B., (2007). Wnt and TGF-β expression in the sponge *Amphimedon queenslandica* and the origin of metazoan embryonic patterning. *PLoS One, 2*, e1031.

Aiello, A., Fattorusso, E., & Menna, M., (1999). Steroids from sponges: Recent reports. *Steroids, 64*(10), 687.

Amano, S., (1986). Larval release in response to light signal by the intertidal sponge *Halichondria panicea. Biol. Bull., 171*, 371–378.

Amano, S., (1988). Morning release of larvae controlled by the light in an intertidal sponge, *Callyspongia ramosa. Biol. Bull., 175*, 181–185.

Antcliffe, J., (2015). The oldest compelling evidence for sponges is still early Cambrian in age – reply to Love and Summons (2015). *Palaeontology, 58*, 1–3.

Archer, S. K., Kahn, A. S., Thiess, M., Law, L., Leys, S. P., Johannessen, S., C., Layman, C. A., et al., (2020). Foundation species abundance influences food web topology on glass sponge reefs. *Front. Mar. Sci., 7*, 549478.

Auger, J., (2018). Spermatogenic cells – structure. *Encycl. Reprod., 1*, 53–60.

Baccetti, B., Gaino, E., & Sarà, M., (1986). A sponge with acrosome: *Oscarella lobularis. J. Ultrastruct. Mol. Struct. Res., 94*, 195–198.

Bachtrog, D., Mank, J. E., Peichel, C. L., Kirkpatrick, M., Otto, S. P., Ashman, T. L., Hahn, M. W., et al., (2014). Sex determination: Why so many ways of doing it? *PLoS Biol., 12*(7), e1001899.

Baldacconi, R., Nonnis-Marzano, C., Gaino, E., & Corriero, G., (2007). Sexual reproduction, larval development and release in *Spongia officinalis* L. (Porifera, Demospongiae) from the Apulian coast. *Mar. Biol., 152*, 969–979.

Bandaranayake, W. M., Bourne, D. J., & Sim, G. R., (1997). Chemical composition during maturing and spawning of the sponge *Dysidea herbacea* (Porifera: Demospongiae*). Comp. Biochem. Physiol. Part B: Biochem. Mol. Biol., 118*(4), 851–859.

Barbeau, M. A., Reiswig, H. M., & Rath, C. L., (1989). Hatching of freshwater sponge gemmules after low temperature exposure: *Ephydatia müelleri* (Porifera: Spongillidae). *J. Thermal Biol., 14*, 225–231.

Barton, N. H., & Charlesworth, B., (1998). Why sex and recombination? *Science, 281*(5385), 1986–1990.

Bergquist, P. R., (1978). *Sponges*. University of California, Berkeley and Los Angeles.

Bernstein, C., Prasad, A. R., Nfonsam, V., & Bernstein, H., (2013). DNA damage, DNA repair and cancer. In: Chen, C., (ed.), *New Research Directions in DNA Repair* (pp. 413–465). IntechOpen.

Beukeboom, L. W., & Perrin, N., (2014). *The Evolution of Sex Determination*. Oxford University Press, USA.

Birdsell, J., & Wills, C., (2003). The evolutionary origin and maintenance of sexual recombination: A review of contemporary models. *Evol. Biol., 33*, 27–138.

Björk, J. R., Díez-Vives, C., Astudillo-García, C., Archie, E. A., & Montoya, M. J., (2019). Vertical transmission of sponge microbiota is inconsistent and unfaithful. *Nat. Ecol. Evol., 3*(8), 1172–1183.

Borojevic, R., & Lévi, C., (1964). Étude au microscope électronique des cellules d'éponge *Ophlitaspongia seriata* (Grant) au cours de la réorganisation après dissociation. *Z. Zellforsch. Mikrosk. Anat., 64*, 708–725.

Borojevic, R., & Lévi, C., (1965). Morphogenese experimentale d'une éponge a partir de cellules de la larve dissociee. *Z. Zellforsch. Mikrosk. Anat., 68*, 57–69.

Borojevic, R., (1969). Étude du développement et de la différenciation cellulaire d'éponges calcaires calcinéennes (genres *Clathrina* et *Ascandra*). *Ann. Embryol. Morphogen., 2*, 15–36.

Borojevic, R., (1971). Le comportement des cellules d'éponges lors de processus morphogénétiques. *Année Biol., 10*, 533–545.

Boury-Esnault, N., & Jamieson, B. G. M., (1999). Porifera. In: Adiyodi, K. G., & Adiyodi, R. G., (eds.), *Reproductive Biology of Invertebrates; IX. Progress in Male Gamete Ultrastructure and Phylogeny* (pp. 1–20). John Wiley and Sons.

Boury-Esnault, N., & Rutzler, K., (1997). Thesaurus of sponge morphology. *Smithsonian Contribution to Zoology, 596*, 1–55.

Boury-Esnault, N., & Vacelet, J., (1994). Preliminary studies on the organization and development of a hexactinellid sponge from a Mediterranean cave, *Oopsacas minuta*. In: van Soest, R.W.M., van Kempen, Th.M.G. & Braekman, J.-C. (eds.), *Sponges in Time and Space* (pp. 407–415). Balkema: Rotterdam.

Boury-Esnault, N., (1970). Un phénoméne de bourgeonnement externe chez l'éponge *Axinella damicornis* (Esper.). *Cah. Biol. Mar., 11*, 491–496.

Boury-Esnault, N., Efremova, S., Bézac, C., & Vacelet, J., (1999). Reproduction of a hexactinellid sponge: First description of gastrulation by cellular delamination in the *Porifera. Invertebr. Reprod. Dev., 35*, 187–201.

Boury-Esnault, N., Ereskovsky, A. V., Bézac, C., & Tokina, D. B., (2003). Larval development in homoscleromorpha (Porifera, Demospongiae). *Invert. Biol., 122*, 187–202.

Busch, K., et al. (2024). Sponge symbiosis: microbes make an essential part of what it means to be a sponge. *Frontiers in Invertebrate Physiology*. Saleuddin, S., Leys, S.P., Roer, R.D., and Wilkie, I. C. Apple Academic Press. Vol. 1 Non-Bilaterian Phyla.

Cajado, B., & Lanna, E., (2021). How can environmental variables affect the sexual reproduction of a tropical symbiotic sponge? *Mar. Biodiversity, 51*, 23.

Calazans, V. P. S. B., & Lanna, E., (2019). Influence of endogenous and exogenous factors in the reproductive output of a cryptogenic calcareous sponge. *Mar. Biodiv., 49*, 2837–2850.

Cardone, F., Gaino, E., & Corriero, G., (2010). The budding process in *Tethya citrina* Sarà & Melone (Porifera, Demospongiae) and the incidence of post-buds in sponge population maintenance. *J. Exp. Mar. Biol. Ecol., 389*, 93–100.

Carr, M., Leadbeater, B. S., & Baldauf, S. L., (2010). Conserved meiotic genes point to sex in the choanoflagellates. *J. Eukaryot. Microbiol., 57*(1), 56–62.

Carré, D., & Carré, C., (2000). Origin of germ cells, sex determination, and sex inversion in medusae of the genus *Clytia* (Hydrozoa, Leptomedusae): The influence of temperature. *J. Exp. Zool., 287*(3), 233–242.

Chen, C. J., Shikina, S., Chen, W. J., Chung, Y. J., Chiu, Y. L., Bertrand, J. A., Lee, Y. H., & Chang, F. C., (2016). A novel female-specific and sexual reproduction-associated Dmrt gene discovered in the stony coral, *Euphyllia ancora*. *Biol. Reprod., 94*(2), 40–41.

Chu, J. W. F., & Reiswig, M. H., (2014). Mechanisms of propagule release in the carnivorous sponge *Asbestopluma occidentalis*. *Invertebr. Biol., 133*, 109–120.

Clarke, A., (1980). A reappraisal of the concept of metabolic cold adaptation in polar marine invertebrates. *Biol. J. Linn. Soc., 14*(1), 77–92.

Coma, R., Ribes, M., J-Gili, M., & Zabala, M., (1998). An energetic approach to the study of life-history traits of two modular colonial benthic invertebrates. *Mar. Ecol. Prog. Ser., 162*, 89–103.

Connes, R., Carriere, D., & Paris, J., (1978). Etude du développement des gemmules chez la démosponge marine: *Suberites domuncula* (Olivi) Nardo. *Ann. Sci. Nat. Zool. Biol. Anim., 20*, 357–385.

Corriero, G., Liaci-Scalera, L., Marzano, C. N., & Gaino, E., (1998). Reproductive strategies of *Mycale contarenii* (Porifera: Demospongiae). *Mar. Biol., 131*, 319–327.

Corriero, G., Sarà, M., & Vaccaro, P., (1996). Sexual and asexual reproduction in two species of *Tethya* (Porifera: Demospongiae) from a Mediterranean coastal lagoon. *Mar. Biol., 126*, 175–181.

Courrèges, V. C., & Fell, E. P., (1989). Sexual and asexual reproduction by the freshwater sponge *Anheteromeyenia ryderi*, with emphasis on spermatogenic activity. *Trans. Am. Microsc. Soc., 108*, 127–138.

Custodio, M. R., Prokic, I., Steffen, R., Koziol, C., Borojevic, R., Brümmer, F., Nickel, M., & Müller, W. E. G., (1998). Primmorphs generated from dissociated cells of the sponge *Suberites domuncula*: A model system for studies of cell proliferation and cell death. *Mech. Ageing Dev., 105*, 45–59.

Da Hora, J., Cavalcanti, F. F., & Lanna, E., (2018). Anatomy and ultrastructure of the tropical sponge *Cladocroce caelum* (Haplosclerida, Demospongiae). *J. Morphol., 279*, 1872–1886.

De Caralt, S., González, J., Turon, X., & Uriz, J. M., (2018). Reproductive strategies of two common sympatric Mediterranean sponges: *Dysidea avara* (Dictyoceratida) and *Phorbas tenacior* (Poecilosclerida). *PeerJ., 6*, e5458.

De Caralt, S., Uriz, M. J., Ereskovsky, A. V., & Wijffels, H. R., (2007). Embryo development of *Corticium candelabrum* (Demospongiae: Homosclerophorida). *Invertebr. Biol., 126*, 211–219.

De Goeij, J. M., Moodley, L., Houtekamer, M., Carballeira, N. M., & Duyl, F. C. V., (2008). Tracing ^{13}C-enriched dissolved and particulate organic carbon in the bacteria-containing coral reef sponge *Halisarca caerulea*: Evidence for DOM feeding. *Limnol. Oceanography, 53*(4), 1376–1386.

De Voogd, N. J., Alvarez, B., Boury-Esnault, N., Carballo, J. L., Cárdenas, P., Díaz, M. C., Dohrmann, M., et al., (2022). *World Porifera Database*. Accessed on May 3, 2023 at https:marinespecies.org/porifera. doi: 10.14284/359.

Degnan, B. M., Adamska, M., Richards, G. S., Larroux, C., Leininger, S., Bergum, B., Calcino, A., et al., (2015). *Porifera*. In: Wanninger, A., (ed.), *Evolutionary Developmental Biology of Invertebrates 1: Introduction, Non-Bilateria, Acoelomorpha, Xenoturbellida, Chaetognatha* (pp. 65–106). Springer-Verlag.

Diaz, J. P., & Connes, R., (1980). Étude ultrastructurale de la spermatogénèse d'une démosponge. *Biol. Cell., 38*, 225–230.

Doughty, P., & Shine, R., (1997). Detecting life history trade-offs: Measuring energy stores in "capital" breeders reveals costs of reproduction. *Oecologia., 110*(4), 508–513.

Duboscq, O., & Tuzet, O., (1937). L'ovogenèse, la fécondation et les premiers stades du développement des éponges calcaires. *Arch. Zool. Exp. Gén., 79*, 157–316.

Dunn, C. W., Hejnol, A., Matus, D. Q., Pang, K., Browne, W. E., Smith, S. A., et al., (2008). Broad phylogenomic sampling improves resolution of the animal tree of life. *Nature, 452*, 745–749.

Eerkes-Medrano, D., Feehan, C. J., & Leys, P. S., (2015). Sponge cell aggregation: Checkpoints in development indicate a high level of organismal complexity. *Invertebr. Biol., 134*, 1–18.

Elvin, D. W., (1976). Seasonal growth and reproduction of an intertidal sponge, *Haliclona permollis* (Bowerbank). *Biol. Bull., 151*, 108–125.

Ereskovsky, A. V., & Tokina, B. D., (2007). Asexual reproduction in homoscleromorph sponges (Porifera; Homoscleromorpha). *Mar. Biol., 151*, 425–434.

Ereskovsky, A. V., (2000). Reproduction cycles and strategies of the cold-water sponges *Halisarca dujardini* (Demospongiae, Halisarcida), *Myxilla incrustans* and *Iophon piceus* (Demospongiae, Poecilosclerida) from the White Sea. *Biol. Bull., 198*, 77–87.

Ereskovsky, A. V., (2010). *The Comparative Embryology of Sponges*. Springer.

Ereskovsky, A. V., Geronimo, A., & Pérez, T., (2017). Asexual and puzzling sexual reproduction of the Mediterranean sponge *Haliclona fulva* (Demospongiae): Life cycle and cytological structures. *Invertebr. Biol., 136*, 403–421.

Ereskovsky, A. V., Renard, E., & Borchiellini, C., (2013). Cellular and molecular processes leading to embryo formation in sponges: Evidences for high conservation of processes throughout animal evolution. *Dev. Genes. Evol., 223*(1, 2), 5–22.

Ereskovsky, A., & Tokina, D., D., (2022). Ultrastructural research of spermiogenesis in two sponges, *Crellomima imparidens* and *Hymedesmia irregularis* (Demospongiae): New evidence of sperms with acrosome in sponges. *J. Morphol., 283*, 333–345.

Ereskovsky, A., Konyukov, P., & Tokina, D., (2010). Morphogenesis accompanying larval metamorphosis in *Plakina trilopha* (Porifera, Homoscleromorpha). *Zoomorphology., 129*, 21–31.

Erwin, P. M., Coma, R., Lopez-Sendino, P., Serrano, E., & Ribes, M., (2015). Stable symbionts across the HMA-LMA dichotomy: Low seasonal and inter-annual variation in sponge-associated bacteria from taxonomically diverse hosts. *FEMS Microbiol. Ecol., 91*(10), fiv115.

Ewen-Campen, B., Schwager, E. E., & Extavour, G. C., (2010). The molecular machinery of germ line specification. *Mol. Reprod. Dev., 77*(1), 3–18.

Extavour, C. G., & Akam, M., (2003). Mechanisms of germ cell specification across the metazoans: Epigenesis and preformation. *Development, 130*(24), 5869–5884.

Extavour, C. G., (2007). Evolution of the bilaterian germ line: Lineage origin and modulation of specification mechanisms. *Integr. Comp. Biol., 47*(5), 770–785.

Fell, P. E., & Jacob, F. W., (1979). Reproduction and development of *Halichondria* sp. in the Mystic estuary, Connecticut. *Biol. Bull., 155*, 62–75.

Fell, P. E., (1969). The involvement of nurse cells in oogenesis and embryonic development in the marine sponge, *Haliclona ecbasis*. *J. Morphol., 127*(2), 133–149.

Fell, P. E., (1974). *Porifera*. In: Giese, A., & Pearse, J. S., (eds.), *Reproduction of Marine Invertebrates* (pp. 51–132). New York: Academic Press.

Fell, P. E., (1976). The reproduction of *Haliclona loosanoffi* and its apparent relationship to water temperature. *Biol. Bull., 150*, 200–210.

Fernandez-Busquets, X., & Burger, M. M., (2003). Circular proteoglycans from sponges: First members of the spongican family. *Cell. Mol. Life Sci., 60*(1), 88–112.

Feuda, R., Dohrmann, M., Pett, W., Philippe, H., Rota-Stabelli, O., Lartillot, N., Wörheide, G., & Pisani, D., (2017). Improved modeling of compositional heterogeneity supports sponges as sister to all other animals. *Curr. Biol., 27*(24), 3864–3870.e3864.

Fierro-Constaín, L., Schenkelaars, Q., Gazave, E., Haguenauer, A., Rocher, C., Ereskovsky, A., Borchiellini, C., & Renard, E., (2017). The conservation of the germline multipotency program, from sponges to vertebrates: A stepping stone to understanding the somatic and germline origins. *Genome Biol. Evol., 9*, 474–488.

Fiorucci, S., Distrutti, E., Bifulco, G., D'Auria, M. V., & Zampella, A., (2012). Marine sponge steroids as nuclear receptor ligands. *Trends Pharmacol. Sci., 33*(11), 591–601.

Fortunato, S., Adamski, M., Bergum, B., Guder, C., Jordal, S., Leininger, S., Zwafink, C., Rapp, H. T., & Adamska, M., (2012). Genome-wide analysis of the sox family in the calcareous sponge *Sycon ciliatum*: Multiple genes with unique expression patterns. *EvoDevo, 3*, 14.

Fromont, J., & Bergquist, R. P., (1994). Reproductive biology of three sponge species of the genus *Xestospongia* (Porifera: Demospongiae: Petrosida) from the great barrier reef. *Coral Reefs, 13*, 119–126.

Fromont, J., (1994). Reproductive development and timing of tropical sponges (Order Haplosclerida) from the Great Barrier Reef, Australia. *Coral Reefs., 13*(2), 127–133.

Funayama, N., (2013). The stem cell system in demosponges: Suggested involvement of two types of cells: Archeocytes (active stem cells) and choanocytes (food-entrapping flagellated cells). *Dev. Genes Evol., 223*, 23–38.

Funayama, N., Nakatsukasa, M., Mohri, K., Masuda, Y., & Agata, K., (2010). *Piwi* expression in archeocytes and choanocytes in demosponges: Insights into the stem cell system in demosponges. *Evol. Dev., 12*, 275–287.

Gaino, E., Burlando, B., & Buffa, P., (1987). Ultrastructural study of oogenesis and fertilization in *Sycon ciliatum* (Porifera: Calcispongiae). *Int. J. Invertebr. Reprod. Dev., 11*, 73–82.

Gaino, E., Scalera, L. L., Sciscioli, M., & Corriero, G., (2006). Investigation of the budding process in *Tethya citrina* and *Tethya aurantium* (Porifera, Demospongiae). *Zoomorphology., 125*, 87–97.

Gallissian, M. F., & Vacelet, J., (1990). Fertilization and nutrition of the oocyte in the calcified sponge *Petrobiona massiliana*. In: Rützler, K., (ed.), *New Perspectives in Sponge Biology* (pp. 175–181). Washington, Smithsonian Institution Press.
Gallissian, M. F., & Vacelet, J., (1992). Ultrastructure of the oocyte and embryo of the calcified sponge, *Petrobiona massiliana* (Porifera, Calcarea). *Zoomorphology., 112*, 133–141.
Gallissian, M. F., (1980). Étude ultrastructurale de la fécondation chez *Grantia compressa* F. *Int. J. Invertebr. Reprod., 2*, 321–329.
Gallissian, M. F., (1988). Étude ultrastructurale de l'ovogenèse et de la fécondation chez *Leucilla endoumensis* (Spongiaire, Calcarea). *C. R. Hebd. Séances Acad. Sci., 306*(III), 245–252.
Galtsoff, P. S., (1925). Regeneration after dissociation (an experimental study on sponges) I. Behavior of dissociated cells of *Microciona prolifera* under normal and altered conditions. *J. Exp. Zool., 42*(1), 183–221.
Gatenby, J. B., (1920). The germ-cells, fertilization, and early development of *Grantia* (*Sycon*) *compressa*. Linn. *J. Zool., 36*, 261–297.
Gibson, A. K., Delph, L. F., & Lively, M. C., (2017). The two-fold cost of sex: Experimental evidence from a natural system. *Evol. Lett., 1*(1), 6–15.
Giese, A. C., (1966). Lipids in the economy of marine invertebrates. *Physiol. Rev., 46*(2), 244–298.
Gilbert, J. J., & Simpson, L. T., (1976). Sex reversal in a freshwater sponge. *J. Exp. Zool., 195*, 145–151.
Gilbert, J. J., (1974). Field experiments on sexuality in the freshwater sponge *Spongilla lacustris*. The control of oocyte production and the fate of unfertilized oocytes. *J. Exp. Zool., 188*, 165–178.
Gloeckner, V., Wehrl, M., Moitinho-Silva, L., Gernert, C., Schupp, P., Pawlik, J. R., Lindquist, N. L., et al., (2014). The HMA-LMA dichotomy revisited: An electron microscopical survey of 56 sponge species. *Biol. Bull., 227*(1),78–88.
Gold, D. A., Grabenstatter, J., De Mendoza, A., Riesgo, A., Ruiz-Trillo, I., & Summons, E. R., (2016). Sterol and genomic analyses validate the sponge biomarker hypothesis. *Proc. Nat. Acad. Sci., 113*, 2684–2689.
Gonobobleva, E., & Ereskovsky, V. A., (2004). Polymorphism in free-swimming larvae of *Halisarca dujardini* (Demospongiae, Halisarcida). *Boll. Mus. Ist. Biol. Univ. Genova., 68*, 349–356.
Gugel, J., (2001). Life cycles and ecological interactions of freshwater sponges (Porifera, Spongillidae) in the River Rhine in Germany. *Limnologica, 31*, 185–198.
Haag, E. S., Wang, S., & Kimble, J., J., (2002). Rapid coevolution of the nematode sex-determining genes fem-3 and tra-2. *Curr. Biol., 12*(23), 2035–2041.
Halary, S., Malik, S., Lildhar, L., Slamovits, C. H., Hijri, M., & Corradi, N., (2011). Conserved meiotic machinery in *Glomus* spp., a putatively ancient asexual fungal lineage. *Genome Biol. Evol., 3*, 950–958.
Harii, S., Kayanne, H., Takigawa, H. T., Hayashibara, T. H., & Yamamoto, M., (2002). Larval survivorship, competency periods and settlement of two brooding corals, *Heliopora coerulea* and *Pocillopora damicornis*. *Mar. Biol., 141*, 39–46.
Heras, H., Pollero, R. J., Gonzalez-Baró, M. R., & Pollero, J. R., (2000). Lipid and fatty acid composition and energy partitioning during embryo development in the shrimp *Macrobrachium borellii*. *Lipids, 35*(6), 645–651.

Hoffmann, F., Larsen, O., Thiel, V., Rapp, H. T., Pape, T., Michaelis, W., & Reitner, J., (2005). An anaerobic world in sponges. *Geomicrobiol. J., 22*, 1–10.

Höhr, D., (1977). Differenzierungsvorgänge in der keimenden gemmula von *Ephydatia fluviatilis*. *Roux Arch. Dev. Biol., 182*, 329–346.

Hoppe, W. F., & Reichert, M. J. M., (1987). Predictable annual mass release of gametes by the coral reef sponge *Neofibularia nolitangere* (Porifera: Demospongiae). *Mar. Biol., 94*, 277–285.

Hörandl, E. A., (2009). A combinational theory for maintenance of sex. *Heredity, 103*(6), 445–457.

Hörandl, E., & Dave, S., (2018). How oxygen gave rise to eukaryotic sex. *Proc. R. Soc. B., 285*, 20172706.

Ivanisevic, J., Perez, T., Ereskovsky, A. V., Barnathan, G., & Thomas, O. P., (2011a). Lysophospholipids in the Mediterranean sponge *Oscarella tuberculata*: Seasonal variability and putative biological role. *J. Chem. Ecol., 37*, 537–545.

Ivanisevic, J., Thomas, O. P., Pedel, L., Pénez, N., Ereskovsky, A. V., Culioli, G. R., & Pérez, T., (2011b). Biochemical trade-offs: Evidence for ecologically linked secondary metabolism of the sponge *Oscarella balibaloi*. *PLOS One, 6*, e28059.

Jager, M., Quéinnec, E., Chiori, R., Le Guyader, H., & Manuel, M., (2008). Insights into the early evolution of SOX genes from expression analyses in a ctenophore. *J. Exp. Zool. B: Mol. Dev. Evol., 310*(8), 650–667.

Jindrich, K., Roper, K. E., Lemon, S., Degnan, B. M., Reitzel, A. M., & Degnan, M. S., (2017). Origin of the animal circadian clock: Diurnal and light-entrained gene expression in the sponge *Amphimedon queenslandica*. *Front. Mar. Sci., 4*, 327.

Jones, B. L., Walker, C., Azizi, B., Tolbert, L., Williams, L. D., & Snell, W. T., (2017). Conservation of estrogen receptor function in invertebrate reproduction. *BMC Evol. Biol., 17*, 65.

Juliano, C. E., Swartz, S. Z., & Wessel, M. G., (2010). A conserved germline multipotency program. *Development, 137*(24), 4113–4126.

Juliano, C., & Wessel, G., (2010). Versatile germline genes. *Science, 329*(5992), 640–641.

Kahn, A. S., & Leys, P. S., (2016). The role of cell replacement in benthic–pelagic coupling by suspension feeders. *R. Soc. Open Sci., 3*(11), 160484.

Kahn, A. S., Yahel, G., Chu, J. W. F., Tunnicliffe, V., & Leys, P. S., (2015). Benthic grazing and carbon sequestration by deep-water glass sponge reefs. *Limnol. Oceanograph., 60*(1), 78–88.

Kashimada, K., & Koopman, P., (2010). Sry: The master switch in mammalian sex determination. *Development, 137*(23), 3921–3930.

Kaye, H., & Reiswig, M. H., (1991). Sexual reproduction in four Caribbean commercial sponges. I. Reproductive cycles and spermatogenesis. *Invertebr. Reprod. Dev., 19*, 1–11.

Kenny, N. J., De Goeij, J. M., De Bakker, D. M., Whalen, C. G., Berezikov, E., & Riesgo, A., (2018). Towards the identification of ancestrally shared regenerative mechanisms across the Metazoa: A transcriptomic case study in the demosponge *Halisarca caerulea*. *Mar. Genomics., 37*, 135–147.

Kenny, N. J., Francis, W. R., Rivera-Vicens, R. E., Juravel, K., De Mendoza, A., Diez-Vives, C., Lister, R., et al., (2020). Tracing animal genomic evolution with the chromosomal-level assembly of the freshwater sponge *Ephydatia muelleri*. *Nat. Commun., 11*, 3676.

Kiefer, J. C., (2007). Back to basics: Sox genes. *Dev. Dyn., 236*, 2356–2366.

Kirk, D. L., (2006). Oogamy: Inventing the sexes. *Curr. Biol., 16*(24), R1028–1030.

Köhler, H. R., Kloas, W., Schirling, M., Lutz, I., Reye, A. L., Langen, J. S., Triebskorn, R., Nagel, R., & Schönfelder, G., (2007). Sex steroid receptor evolution and signaling in aquatic invertebrates. *Ecotoxicology, 16*(1), 131–143.

Koopman, P., (2005). Sex determination: A tale of two sox genes. *Trends Genet., 21*(7), 367–370.

Koopmans, M., Martens, D., & Wijffels, H. R., (2009). Towards commercial production of sponge medicines. *Mar. Drugs., 7*(4), 787–802.

Korotkova, G. P., (1988). Integrative mechanisms and morphogenesis (to problem of ontogenesis evolution). *Zhur. Obshey Biol., 69*, 464–475.

Korotkova, G., (1981). General characteristics of sponge organization. In: Korotkova, G., (ed.), *Morphogenesis in Sponges*. Leningrad: Leningrad University Press.

Koutsouveli, V., Balgoma, D., Checa, A., Hedeland, M., Riesgo, A., & Cárdenas, P., (2022). Oogenesis and lipid metabolism in the deep-sea sponge *Phakellia ventilabrum* (Linnaeus, 1767). *Sci. Rep. 12*(1):6317.

Koutsouveli, V., Cárdenas, P., Conejero, M., Rapp, H. T., & Riesgo, A., (2020a). Reproductive biology of *Geodia* species (Porifera, Tetractinellida) from Boreo-Arctic North-Atlantic deep-sea sponge grounds. *Front. Mar. Sci., 7*, 595267.

Koutsouveli, V., Cárdenas, P., Santodomingo, N., Marina, A., Morato, E., Rapp, H. T., & Riesgo, A., (2020b). The molecular machinery of gametogenesis in *Geodia* demosponges (Porifera), evolutionary origins of a conserved toolkit across animals. *Mol. Biol. Evol., 37*, 3485–3506.

Koutsouveli, V., Taboada, S., Moles, J., Cristobo, J., Ríos, P., Bertran, A., Solà, J., Avila, C., & Riesgo, A., (2018). Insights into the reproduction of some Antarctic dendroceratid, poecilosclerid, and haplosclerid demosponges. *PLOS One, 13*, e0192267.

Langenbruch, P. F., (1981). Zur Enstehung der Gemmulae bei *Ephydatia fluviatilis* L. (Porifera). *Zoomorpholy, 97*, 263–284.

Lanna, E., & Klautau, M., (2010). Oogenesis and spermatogenesis in *Paraleucilla magna* (Porifera, Calcarea). *Zoomorphology, 129*, 249–261.

Lanna, E., & Klautau, M., (2012). Embryogenesis and larval ultrastructure in *Paraleucilla magna* (Calcarea, Calcaronea), with remarks on the epilarval trophocyte epithelium ("placental membrane"). *Zoomorphology, 131*, 277–292.

Lanna, E., & Klautau, M., (2016). Some aspects of the oogenesis of three species of clathrinid sponges (Calcarea, Porifera). *J. Mar. Biol. Ass. U. K., 96*, 529–539.

Lanna, E., & Klautau, M., (2018). Life history and reproductive dynamics of the cryptogenic calcareous sponge *Sycettusa hastifera* (Porifera, Calcarea) living in tropical rocky shores. *J. Mar. Biol. Ass. U. K., 98,* 505–514.

Lanna, E., & Klautau, M., (2019). The choanoderm of *Sycettusa hastifera* (Calcarea, Porifera) is able to generate new individuals. *Invertebr. Biol., 138*, e12262.

Lanna, E., Cajado, B., Santos-da-Silva, C., Da Hora, J., Porto, U., & Vasconcellos, V., (2018). Is the Orton's rule still valid? Tropical sponge fecundity, rather than periodicity, is modulated by temperature and other proximal cues. *Hydrobiologia, 815*, 187–205.

Lanna, E., Monteiro, L. C., & Klautau, M., (2007). Life cycle of *Paraleucilla magna* Klautau, Monteiro and Borojevic, 2004 (Porifera, Calcarea). In: Custódio, M. R., Lôbo-Hajdu, G., Hajdu, E., & Muricy, G., (eds.), *Porifera Research – Biodiversity, Innovation and Sustainability.* (pp. 413–418). Museu Nacional – Série Livros.

Lanna, E., Paranhos, R., Paiva, P. C., & Klautau, M., (2015). Environmental effects on the reproduction and fecundity of the introduced calcareous sponge *Paraleucilla magna* in Rio de Janeiro, Brazil. *Mar. Ecol., 36*, 1075–1087.

Lanna, E., Santos, D., Carvalho, V., Vilela, B., Santos, G. G., Skinner, L. F., Segal, B., Carraro, J. L., & Pinheiro, U., (2021). Different environmental drivers influence the reproduction of a tropical sponge (*Tedania ignis*) along a latitudinal gradient. *Mar. Ecol. Prog. Ser., 667*, 25–41.

Lapebie, P., Gazave, E., Ereskovsky, A., Derelle, R., Bézac, C., Renard, E., Houliston, E., & Borchiellini, C., (2009). WNT/β-catenin signaling and epithelial patterning in the homoscleromorph sponge *Oscarella*. *Plos One, 4*, e5823.

Larroux, C., Fahey, B., Liubicich, D., Hinman, V. F., Gauthier, M., Gongora, M., Green, K., et al., (2006). Developmental expression of transcription factor genes in a demosponge: Insights into the origin of metazoan multicellularity. *Evol. Dev., 8*, 150–173.

Laumer, C. E., Fernández, R., Lemer, S., Combosch, D., Kocot, K. M., Riesgo, A., Andrade, S. C., et al., (2019). Revisiting metazoan phylogeny with genomic sampling of all phyla. *Proc. Roy. Soc. B., 286*, 20190831.

Lavrov, A. I., & Kosevich, A. I., (2014). Sponge cell reaggregation: Mechanisms and dynamics of the process. *Russ. J. of Dev. Biol., 45*(4), 205–223.

Le Pennec, G., Perovic, S., Ammar, M., Grebenjuk, V., Steffen, R., Brummer, F., & Muller, W., (2003). Cultivation of primmorphs from the marine sponge *Suberites domuncula*: Morphogenetic potential of silicon and iron. *Biotechnology, 100*, 93–108.

Lee, R. F., (1991). Lipoproteins from the hemolymph and ovaries of marine invertebrates. In: Houlihan, D. F., Livingstone, D. R., & Lee, R. F., (eds.), *Advances in Comparative and Environmental Physiology* (pp. 187–207). Springer Berlin Heidelberg.

Lehtonen, J., Jennions, M. D., & Kokko, H., (2012). The many costs of sex. *Trends Ecol. Evol., 27*(3), 172–178.

Lehtonen, J., Kokko, H., & Parker, A. G., (2016). What do isogamous organisms teach us about sex and the two sexes? *Philos. Trans. R. Soc. Lond. B Biol. Sci., 371*(1706), 20150532.

Leininger, S., Adamski, M., Bergum, B., Guder, C., Liu, J., Laplante, M., Bråte, J., et al., (2014). Developmental gene expression provides clues to relationships between sponge and eumetazoan body plans. *Nat. Commun., 5*, 3905.

Leong, W., & Pawlik, R. J., (2010). Fragments or propagules? Reproductive tradeoffs among *Callyspongia* spp. from Florida coral reefs. *Oikos, 119*, 1417–1422.

Leong, W., & Pawlik, R. J., (2011). Comparison of reproductive patterns among 7 Caribbean sponge species does not reveal a resource trade-off with chemical defenses. *J. Exp. Mar. Biol. Ecol., 401*, 80–84.

Lévi, C., & Lévi, P., (1976). Embryogénèse de *Chondrosia reniformis* (Nardo), démosponge ovipare, et transmission des bactéries symbiotiques. *Ann. Sci. Nat. Zool. Biol. Anim., 18*(4), 367–380.

Lévi, C., (1956). Étude des *Halisarca* de Roscoff. Embryologie et systématique des démosponges. *Arch. Zool. Exp. Gén., 93*, 1–181.

Leys, S. P., & Degnan, M. B., (2002). Embryogenesis and metamorphosis in a haplosclerid demosponge: Gastrulation and transdifferentiation of larval ciliated cells to choanocytes. *Invertebr. Biol., 121*, 171–189.

Leys, S. P., & Ereskovsky, V. A., (2006). Embryogenesis and larval differentiation in sponges. *Can. J. Zool., 84*, 262–287.

Leys, S. P., & Hill, A., (2012). The physiology and molecular biology of sponge tissues. *Adv. Mar. Biol., 62*, 1–56.
Leys, S. P., & Riesgo, A., (2012). Epithelia, an evolutionary novelty of metazoans. *J. Exp. Zool. B: Mol. Dev. Evol., 318B*, 438–447.
Leys, S. P., Cheung, E., & Boury-Esnault, N., (2006). Embryogenesis in the glass sponge *Oopsacas minuta*: Formation of syncytia by fusion of blastomeres. *Integr. Comp. Biol., 46*, 104–117.
Leys, S. P., Cronin, T. W., Degnan, B. M., & Marshall, N. J., (2002). Spectral sensitivity in a sponge larva. *J. Comp. Physiol. A., 188*, 199–202.
Leys, S. P., et al. (2024). The physiology of sponge behaviour. *Frontiers in Invertebrate Physiology.* Salleudin, S., Leys, S.P., Roer, R.D., and Wilkie, I.C., Apple Press. Vol. 1 Non-Bilaterian Phyla.
Leys, S. P., Zaman, A. K., & Boury-Esnault, N., (2016). Three-dimensional fate mapping of larval tissues through metamorphosis in the glass sponge *Oopsacas minuta. Invertebr. Biol., 135*, 259–272.
Leys, S., Grombacher, L., & Hill, A., (2019). Hatching and Freezing Gemmules from the Freshwater Sponge *Ephydatia muelleri.* Protocols.io. doi.org/10.17504/protocols.io.863hzgn.
Li, Y., Shen, X. X., Evans, B., Dunn, C. W., & Rokas, A., (2021). Rooting the animal tree of life. *Mol. Biol. Evol., 38*(10), 4322–4333.
Loomis, S., Bettridge, A., & Branchini, R. B., (2009). The effects of elevated osmotic concentration on control of germination in the gemmules of freshwater sponges *Eunapius fragilis* and *Anheteromeyania ryderi. Physiol. Biochem. Zool., 82*, 388–395.
López-Legentil, S., & Turon, X., (2010). Pumping water or producing larvae? Oscula occlusion during the reproductive period of the sponge *Svenzea zeai. Zool. Stud., 50*, 395.
Maier, S. R., Bannister, R. J., Van, O. D., & Kutti, T., (2020). Seasonal controls on the diet, metabolic activity, tissue reserves and growth of the cold-water coral *Lophelia pertusa. Coral Reefs., 39*(1), 173–187.
Maldonado, M., & Bergquist, R. P., (2002). Phylum Porifera. In: Young, C. M., Sewell, M. A., & Rice, M. E., (eds.), *Atlas of Marine Invertebrate Larvae* (pp. 21–50). Academic Press.
Maldonado, M., & Riesgo, A., (2008a). Reproductive output in a Mediterranean population of the homosclerophorid *Corticium candelabrum* (Porifera, Demospongiae), with notes on the ultrastructure and behavior of the larva. *Mar. Ecol., 29*, 298–316.
Maldonado, M., & Riesgo, A., (2008b). Reproduction in Porifera: A synoptic overview. *Treb. Soc. Catalana Biol., 59*, 29–49.
Maldonado, M., & Riesgo, A., (2009). Gametogenesis, embryogenesis, and larval features of the oviparous sponge *Petrosia ficiformis* (Haplosclerida, Demospongiae). *Mar. Biol., 156*, 2181–2197.
Maldonado, M., & Uriz, M. J., (1999a). An experimental approach to the ecological significance of microhabitat-scale movement in an encrusting sponge. *Mar. Ecol. Prog. Ser., 185*, 239–255.
Maldonado, M., & Uriz, M. J., (1999b). Sexual propagation by sponge fragments. *Nature, 398*, 476.
Maldonado, M., (2007). Intergenerational transmission of symbiotic bacteria in oviparous and viviparous demosponges, with emphasis on intracytoplasmically-compartmented bacterial types. *J. Mar. Biol. Assoc. U.K., 87*, 1701–1713.

Maldonado, M., (2009). Embryonic development of verongid demosponges supports the independent acquisition of spongin skeletons as an alternative to the siliceous skeleton of sponges. *Biol. J. Linnean Soc., 97*, 427–447.

Manolakou, P., Lavranos, G., & Angelopoulou, R., (2006). Molecular patterns of sex determination in the animal kingdom: A comparative study of the biology of reproduction. *Reprod. Biol. Endocrinol., 4*(1), 1–23.

Manuel, M., (2009). Early evolution of symmetry and polarity in metazoan body plans. *C. R. Biol., 332*, 184–209.

Mariani, S., Uriz, M. J., & Turon, X., (2005). The dynamics of sponge larvae assemblages from northwestern Mediterranean nearshore bottoms. *J. Plankton Res., 27*, 249–262.

McMurray, S. E., Pawlik, J. R., & Finelli, M. C., (2014). Trait-mediated ecosystem impacts: How morphology and size affect pumping rates of the Caribbean giant barrel sponge. *Aquatic Biol., 23*(1), 1–13.

McMurray, S., Blum, J., & Pawlik, J., (2008). Redwood of the reef: Growth and age of the giant barrel sponge *Xestospongia muta* in the Florida Keys. *Mar. Biol., 155*, 159–171.

Mercurio, M., Corriero, G., & Gaino, E., (2007). A 3-year investigation of sexual reproduction in *Geodia cydonium* (Jameson 1811) (Porifera, Demospongiae) from a semi-enclosed Mediterranean bay. *Mar. Biol., 151*, 1491–1500.

Mergner, H., (1964). Uber die inducktion neuer oscularrohre bei *Ephydatia fluviatilis*. Roux's Arch. *Dev. Biol., 155*, 9–128.

Mergner, H., (1966). Zum nachweis der artspezifitat des induktionsstoffes bei oscularrohrneubildungen von spongilliden. *Vehr. Deut. Z., 30*, 522–564.

Miller, S. W., Hayward, D. C., Bunch, T. A., Miller, D. J., Ball, E. E., Bardwell, V. J., Zarkower, D., & Brower, L. D., (2003). A DM domain protein from a coral, *Acropora millepora*, homologous to proteins important for sex determination. *Evol. Dev., 5*(3), 251–258.

Morganti, T. M., Purser, A., Rapp, H. T., German, C. R., Jakuba, M. V., Hehemann, L., Blendl, J., et al., (2021). *In situ* observation of sponge trails suggests common sponge locomotion in the deep central Arctic. *Curr. Biol., 31*(8), R368–R370.

Mukai, H., (1990). *Further Studies on Growth and Sex Differentiation in Four Species of Freshwater Sponges* (Vol. 39, pp. 41–56). Sci. Rep. Faculty of Education, Gunma University.

Müller, W. E., Schröder, H. C., Markl, J. S., Grebenjuk, V. A., Korzhev, M., Steffen, R., & Wang, X., (2013). Cryptochrome in sponges: A key molecule linking photoreception with phototransduction. *J. Histochem. Cytochem., 61*(11), 814–832.

Nakamura, Y., Okada, K., & Watanabe, Y., (1998). The ultrastructure of spermatozoon and its structural change in the choanocytes of *Sycon calcaravis* Hôzawa. *Sponge Sciences: Multidisciplinary Perspectives; Proceedings of the International Conference on Sponge Science* (pp. 179–191). Springer-Verlag.

Nakayama, S., Arima, K., Kawai, K., Mohri, K., Inui, C., Sugano, W., Koba, H., et al., (2015). Dynamic transport and cementation of skeletal elements build up the pole-and-beam structured skeleton of sponges. *Curr. Biol., 25*, P2549–2554.

Neely, K. L., & Butler, B. C., (2020). Seasonal, lunar, and diel patterns in spawning by the giant barrel sponge, *Xestospongia muta*. *Coral Reefs, 39*, 1511–1515.

Ostrovsky, A. N., Lidgard, S., Gordon, D. P., Schwaha, T., Genikhovich, G., & Ereskovsky, V. A., (2015). Matrotrophy and placentation in invertebrates: A new paradigm. *Biol. Rev., 91*, 673–711.

Parzanini, C., Parrish, C., Hamel, J., & Mercier, A., (2018). Functional diversity and nutritional content in a deep-sea faunal assemblage through total lipid, lipid class, and fatty acid analyses. *PLoS One, 13*(11), e0207395.

Paulus, W., (1989). Ultrastructural investigation of spermatogenesis in *Spongilla lacustris* and *Ephydatia fluviatilis* (Porifera, Spongillidae). *Zoomorphology, 109*(3), 123–130.

Pavans de Ceccatty, M. (1974). Coordination in sponges. The foundations of integration. *Amer. Zool., 14*, 895–903.

Peeterrnans-Pé, C. J., De Vos, L., & Rasmont, R., (1975). Reproduction asexuée de l'Éponge siliceuse *Ephydatia fluviatilis* L. dans un milieu forterment apauv rien silice. *Vie Milieu., 25*, 187–196.

Perez-Porro, A. R., Gonzalez, J., & Uriz, J. M., (2012). Reproductive traits explain contrasting ecological features in sponges: The sympatric poecilosclerids *Hemimycale columella* and *Crella elegans* as examples. *Hydrobiologia, 687*, 315–330.

Perovic-Ottstadt, S., Cetkovic, H., Gamulin, V., Schröder, H. C., Kropf, K., Moss, C., Korzhev, M., et al., (2004). Molecular markers for germ cell differentiation in the demosponge *Suberites domuncula. Int. J. Dev. Biol., 48*, 293–305.

Phleger, C. F., (1998). Buoyancy in marine fishes: Direct and indirect role of lipids. *Am. Zool., 38*(2), 321–330.

Picard, M. A. L., Cosseau, C., Mouahid, G., Duval, D., Grunau, C., Toulza, E., Allienne, J. F., & Boissier, J., (2015). The roles of Dmrt (double sex/male-abnormal-3 related transcription factor) genes in sex determination and differentiation mechanisms: Ubiquity and diversity across the animal kingdom. *Comp. Rend. Biol., 338*(7), 451–462.

Pick, K. S., Philippe, H., Schreiber, F., Erpenbeck, D., Jackson, D. J., Wrede, P., Wiens, M., et al., (2010). Improved phylogenomic taxon sampling noticeably affects nonbilaterian relationships. *Mol. Biol. Evol., 27*, 1983–1987.

Pisani, D., Pett, W., Dohrmann, M., Feuda, R., Rota-Stabelli, O., Philippe, H., Lartillot, N., & Wörheide, G., (2015). Genomic data do not support comb jellies as the sister group to all other animals. *Proc. Natl. Acad. Sci. U. S. A., 112*, 15402–15407.

Plese, B., Kenny, N. J., Rossi, M. E., Cárdenas, P., Schuster, A., Taboada, S., Koutsouveli, V., & Riesgo, A., (2021). Mitochondrial evolution in the Demospongiae (Porifera): Phylogeny, divergence time, and genome biology. *Mol. Phyl. Evol., 155*, 107011.

Pond, D. W., & Tarling, A. G., (2011). Phase transitions of wax esters adjust buoyancy in diapausing *Calanoides acutus. Limnol. Oceanogr., 56*(4), 1310–1318.

Pourbaix, N., (1935). Formation histochiminique des gemmules d'éponges. *Ann. Soc. Roy. Zool. Belg., 66*, 33–37.

Pratlong, M., Haguenauer, A., Chenesseau, S., Brener, K., Mitta, G., Toulza, E., Bonabaud, M., et al., (2017). Evidence for a genetic sex determination in Cnidaria, the Mediterranean red coral (*Corallium rubrum*). *Roy. Soc. Open Sci., 4*(3), p.160880.

Rasmont, R., (1954). La diapause chez les spongillides. *Bull. Acad. Roy. Belg., 40*, 288–304.

Rasmont, R., (1962). The physiology of gemmulation in fresh-water sponges. *Symp. Soc. Study Dev. Growth., 20*, 3–25.

Rasmont, R., (1963). Le rôle de la taille et de la nutrition dans le determinisme de la gemmulation chez les spongillides. *Dev. Biol., 8*, 243–271.

Rasmont, R., (1970). Some new aspects of the physiology of fresh-water sponges. In: Fry, W. G., (ed.), *The Biology of the Porifera* (Vol. 25, pp. 415–422). Academic Press, London. Sym. Zool. Soc. Lond.,.

Rasmont, R., (1974). Stimulation of cell aggregation by theophylline in the asexual reproduction of fresh-water sponges (*Ephydatia fluviatilis*). *Experientia, 30*, 792–794.

Reiswig, H. M., (1973). Population dynamics of three Jamaican demospongiae. *Bull. Mar. Sci., 23*, 191–226.

Reiswig, H. M., (1976). Natural gamete release and oviparity in Caribbean Demospongiae. In: Harrison, F., (ed.), *Aspects of Sponge Biology* (pp. 99–112). New York: Academic Press.

Reiswig, H. M., (1981). Partial carbon and energy budgets of the bacteriosponge *Verongia fistularis* (Porifera: Demospongiae) in Barbados. *Mar. Ecol., 2*(4), 273–293.

Reiswig, H. M., (1983). *Porifera*. In: Adiyodi, K. G., & Adiyodi, R. G., (eds.), *Reproductive biology of Invertebrates, Spermatogenesis and Sperm Function* (Vol. II, pp. 1–21). John Wiley & Sons.

Reiswig, H. M., Chen, W. T., Fell, P. E., & Simpson, L. T., (1976). Is the "carrier-cell" system of fertilization in sponges supported by convincing evidence? In: Harrison, F., (ed.), *Aspects of Sponge Biology* (pp. 25–27).

Ribes, M., Coma, R., & Gili, & J, M., (1999). Natural diet and grazing rate of the temperate sponge *Dysidea avara* (Demospongiae, Dendroceratida) throughout an annual cycle. *Mar. Ecol. Prog. Ser., 176*, 179–190.

Richardson, B. E., & Lehmann, R., (2010). Mechanisms guiding primordial germ cell migration: Strategies from different organisms. *Nat. Rev. Mol. Cell. Biol., 11*(1), 37–49.

Riesgo, A., & Maldonado, M., (2008). Differences in reproductive timing among sponges sharing habitat and thermal regime. *Invertebr. Biol., 127*, 357–367.

Riesgo, A., & Maldonado, M., (2009a). Ultrastructure of oogenesis of two oviparous demosponges: *Axinella damicornis* and *Raspaciona aculeata* (Porifera). *Tissue Cell, 41*, 51–65.

Riesgo, A., & Maldonado, M., (2009b). An unexpectedly sophisticated, V-shaped spermatozoon in Demospongiae (Porifera): Reproductive and evolutionary implications. *Biol. J. Linnean Soc., 97*, 413–426.

Riesgo, A., & Solana, J., (2021). Chapter 4: Evolution of the animal germline: Insights from animal lineages with remarkable regenerating capabilities. In: Leys, S. P., & Hejnol, A., (eds.), *Origin and Evolution of Metazoan Cell Types* (pp. 47–74). CRC Press.

Riesgo, A., Cavalcanti, F. F., Kenny, N. J., Ríos, P., Cristobo, J., & Lanna, E., (2018). Integrative systematics of clathrinid sponges: Morphological, reproductive and phylogenetic characterization of a new species of *Leucetta* from Antarctica (Porifera, Calcarea, Calcinea) with notes on the occurrence of flagellated sperm. *Invertebr. Syst., 32*, 827–841.

Riesgo, A., Farrar, N., Windsor, P.J., Giribet, G. & Leys, S.P. (2014b). The analysis of eight transcriptomes from all poriferan classes reveals surprising genetic complexity in sponges. *Mol. Biol. Evol. 31,* 1102–1120.

Riesgo, A., Maldonado, M., & Durfort, M., (2007a). Dynamics of gametogenesis, embryogenesis, and larval release in a Mediterranean homosclerophorid demosponge. *Mar. Freshw. Res., 58*, 398–417.

Riesgo, A., Novo, M., Sharma, P. P., Peterson, M., Maldonado, M., & Giribet, G., (2014a). Inferring the ancestral sexuality and reproductive condition in sponges (Porifera). *Zool. Scr., 43*, 101–117.

Riesgo, A., Taboada, S., Sánchez-Vila, L., Solà, J., Bertran, A., & Avila, C., (2015). Some like it fat: Comparative ultrastructure of the embryo in two demosponges of the genus *Mycale* (Order Poecilosclerida) from Antarctica and the Caribbean. *PLoS One, 10*, e0118805.

Riesgo, A., Taylor, C., & Leys, S. P., (2007b). Reproduction in a carnivorous sponge: The significance of the absence of an aquiferous system to the sponge body plan. *Evol. Dev., 9*, 618–631.

Ritson-Williams, R., Becerro, M. A., & Paul, J. V., (2005). Spawning of the giant barrel sponge *Xestospongia muta* in Belize. *Coral Reefs, 24*, 160–160.

Rivera, A. S., Ozturk, N., Fahey, B., Plachetzki, D. C., Degnan, B. M., Sancar, A., & Oakley, H. T., (2012). Blue-light-receptive cryptochrome is expressed in a sponge eye lacking neurons and opsin. *J. Exp. Biol., 215*, 1278–1286.

Roch, G. J., & Sherwood, M. G., (2014). Glycoprotein hormones and their receptors emerged at the origin of metazoans. *Genome Biol. Evol., 6*(6), 1466–1479.

Rocher, C., Vernale, A., Fierro-Constaín, L., Séjourné, N., Chenesseau, S., Marschal, C., Le Goff, E., et al., (2020). The Buds of Oscarella Lobularis (Porifera): A New Convenient Model for Sponge Cell and Developmental Biology. bioRxiv 2020.06.23.167296.

Röthig, T., & Voolstra, R. C., (2016). *Xestospongia testudinaria* nighttime mass spawning observation in Indonesia. *Galaxea, J. Coral Reef Studies, 18*(1), 1–2.

Sarà, M., (1993). - Porifera. In: Adiyodi, K. G., & Adiyodi, R. G., (eds.), *Reproductive Biology of Invertebrates* (Vol. V, pp. 1–29). John Wiley and Sons, New York.

Sarano, F., (1991). Synchronized spawning in Indonesian sponges. *Coral Reefs., 10*(3), 166–166.

Schenkelaars, Q., Pratlong, M., Kodjabachian, L., Fierro-Constain, L., Vacelet, J., Le Bivic, R. E. A., & Borchiellini, C., (2017). Animal multicellularity and polarity without Wnt signaling. *Sci. Rep., 7*, 15383.

Sciscioli, M., Lepore, E., Gherardi, M., & Scalera, L. L., (1994). Transfer of symbiotic bacteria in the mature oocyte of *Geodia cydonium* (Porifera, Demospongiae): An ultrastructural study. *Cah. Biol. Mar., 35*, 471–478.

Shaffer, M. R., Davy, S. K., Maldonado, M., & Bell, J. J., (2020). Seasonally driven sexual and asexual reproduction in temperate *Tethya* species. *Biol. Bull., 238*(2), 89–105.

Sidri, M., Milanese, M., & Brümmer, F., (2005). First observation on egg release in the oviparous sponge *Chondrilla nucula* (Mediterranean Sea). *Invertebr. Biol., 124*, 91–97.

Siebert, S., & Juliano, E. C., (2017). Sex, polyps, and medusae: Determination and maintenance of sex in cnidarians. *Mol. Reprod. Dev., 84*, 105–119.

Simpson, T., (1980). Reproductive processes in sponges: A critical evaluation of current data and views. *Int. J. Invertebr. Reprod., 2*, 251–269.

Simpson, T., (1984). *The Cell Biology of Sponges*. Springer, New York.

Singh, A., & Thakur, L. N., (2018). Influence of spatial competitor on asexual reproduction of the marine sponge *Cinachyrella* cf. *cavernosa* (Porifera, Demospongiae). *Hydrobiologia, 809*, 247–263.

Skelton, J., & Strand, M., (2013). Trophic ecology of a freshwater sponge (*Spongilla lacustris*) revealed by stable isotope analysis. *Hydrobiologia, 709*(1), 227–235.

Song, Y., Scheu, S., & Drossel, B., (2012). The ecological advantage of sexual reproduction in multicellular long-lived organisms. *J. Evol. Biol., 25*(3), 556–565.

Speijer, D., Lukeš, J., & Eliáš, M., (2015). Sex is a ubiquitous, ancient, and inherent attribute of eukaryotic life. *Proc. Natl. Acad. Sci. U S A., 112*(29), 8827–8834.

Spetland, F., Rapp, H. T., Hoffmann, F., & Tendal, S. O., (2007). Sexual reproduction of *Geodia barreti* Bowerbank, 1858 (Porifera, Astrophorida) in two Scandinavian fjords. In: Custódio, M. R., Lôbo-Hajdu, G., Hajdu, E., & Muricy, G., (eds.), *Porifera Research – Biodiversity, Innovation and Sustainability* (pp. 613–620). Museu Nacional - Série Livros.

Stearns, S. C., (1976). Life-history tactics: A review of the ideas. *Quart. Rev. Biol., 51*, 3–47.
Taylor, M. W., Radax, R., Steger, D., & Wagner, M., (2007). Sponge-associated microorganisms: Evolution, ecology, and biotechnological potential. *Microbiol. Mol. Biol. Rev., 71*(2), 295–347.
Telfer, E. E., & McLaughlin, M., (2007). Natural history of the mammalian oocyte. *Reprod. Biomed. Online, 15*(3), 288–295.
Thacker, R. W., & Freeman, J. C., (2012). Sponge–microbe symbioses: Recent advances and new directions. *Adv. Mar. Biol., 62*, 57–111.
Traylor-Knowles, N. G., Kane, E. G., Sombatsaphay, V., Finnerty, J. R., & Reitzel, M. A., (2015). Sex-specific and developmental expression of Dmrt genes in the starlet sea anemone, *Nematostella vectensis*. *EvoDevo., 6*(1), 1–6.
Tripepi, S., Longo, O., & La Camera, R., (1984). A new pattern of spermiogenesis in the sponge *Crambe crambe*: Preliminary observations. In: Csanády, A., Röhlich, P., & Szabó, D., (eds.), *Proceedings of the Eighth European Congress on Electron Microscopy* (pp. 2073–2074). Budapest: Program Committee.
Turner, E. C., (2021). Possible poriferan body fossils in early Neoproterozoic microbial reefs. *Nature, 596*, 87–91.
Turon, X., Becerro, M. A., & Uriz, J. M., (1996). Seasonal patterns of toxicity in benthic invertebrates: The encrusting sponge *Crambe crambe* (Poecilosclerida). *Oikos, 75*, 33–40.
Turon, X., Uriz, M. J., & Willenz, P., (1999). Cuticular linings and remodelisation processes in *Crambe crambe* (Demospongiae: Poecilosclerida). *Mem. Queensl. Mus., 44*, 617–625.
Tuzet, O., & Paris, J., (1964). La spermatogenèse, l'ovogenèse, la fécondation et les premiers stades de développement chez *Octavella galangaui*. *Vie Milieu., 15*, 309–327.
Tuzet, O., & Pavans De, C. M., (1958). La spermatogenèse, l'ovogenèse, la fécondation et les premiers stades du développement d'*Hippospongia communis* Lmk (*H. equina* O.S.). *Bull. Biol., Fr. Belgique., 42*, 331–348.
Tuzet, O., (1930). Sur la fécondation de l'éponge siliceuse *Cliona viridis* Schmidt. *C. R, Séances Acad. Sci. Paris., 191*, 1095–1097.
Tuzet, O., (1932). Recherches sur l'histologie des Eponges: *Reniera elegans* et *R. simulans*. *Arc. Zool. Exp. Gen., 74*, 169–192.
Tuzet, O., Garrone, R., & Pavans De, C. M., (1970). Observations ultrastructurales sur la spermatogènese chez la démosponge *Aplysilla rosea* Schulze (Dendroceratida): Une métaplasie exemplaire. *Ann. Sci. Nat. Zool., 12*, 27–50.
Twan, W. H., Hwang, J. S., Lee, Y. H., Jeng, S. R., Yueh, W. S., Tung, Y. H., Wu, H. F., et al., (2006). The presence and ancestral role of gonadotropin-releasing hormone in the reproduction of scleractinian coral, *Euphyllia ancora*. *Endocrinology, 147*, 397–406.
Twan, W. H., Wu, H. F., Hwang, J. S., Lee, Y. H., & Chang, C. F., (2005). Corals have already evolved the vertebrate-type hormone system in the sexual reproduction. *Fish. Physiol. Biochem., 31,* 111–115.
Ungemach, L. F., Souza, K., Fell, P. E., & Loomis, S. H., (1997). Possession and loss of cold tolerance by sponge gemmules: A comparative study. *Invertebr. Biol., 116*(1), 1–5.
Usher, K. M., & Ereskovskyz, V. A., (2005). Larval development, ultrastructure and metamorphosis in *Chondrilla australiensis* Carter, 1873. (Demospongiae, Chondrosida, Chondrillidae). *Invertebr. Reprod. Dev., 47*, 51–62.
Vacelet, J., & Donadey, C., (1977). Electron microscope study of the association between some sponges and bacteria. *J. Exp. Mar. Biol. Ecol., 30*, 301–314.

Vacelet, J., (1999). Planktonic armoured propagules of the excavating sponge *Alectona* (Porifera: Demospongiae) are larvae: Evidence from *Alectona wallichii* and *A. mesatlantica* sp. nov. *Mem. Queensland Mus., 44*, 627–642.

Van de Vyver, G., & Willenz, P., (1975). An experimental study of the life-cycle of the fresh-water sponge *Ephydatia fluvitilis* in its natural surroundings. *Roux's Arch. Dev. Biol., 177*(1), 41–52.

van Leeuwen (2017). Great cormorants reveal overlooked secondary dispersal of plants and invertebrates by piscivorous waterbirds. *Biol. Lett., 13*, 20170406.

Vasconcellos, V., Willenz, P., Ereskovsky, A., & Lanna, E., (2019). Comparative ultrastructure of the spermatogenesis of three species of *Poecilosclerida* (Porifera, Demospongiae). *Zoomorphology, 138*, 1–12.

Viladrich, N., Bramanti, L., Tsounis, G., Chocarro, B., Martínez-Quitana, A., Ambroso, S., Madurell, T., & Rossi, S., (2016). Variation in lipid and free fatty acid content during spawning in two temperate octocorals with different reproductive strategies: Surface versus internal brooder. *Coral Reefs, 35*, 1033–1045.

Voronina, E., Lopez, M., Juliano, C. E., Gustafson, E., Song, J. L., Extavour, C., George, S., et al., (2008). Vasa protein expression is restricted to the small micromeres of the sea urchin, but is inducible in other lineages early in development. *Dev. Biol., 314*, 276–286.

Watanabe, Y., & Masuda, Y., (1990). Structure of fiber bundles in the egg of *Tetilla japonica* and their possible function in development. In: Rützler, K., (ed.), *New Perspectives in Sponge Biology*. Smithsonian Institution Press, Washington, DC.

Watanabe, Y., & Okada, K., (1998). The involvement of two carrier cells in fertilization and the ultrastructure of the spermiocyst in *Sycon calcaravis*. *Sponge Sciences: Multidisciplinary Perspectives: Proc. Int. Conf. on Sponge Science* (pp. 193–202). Springer-Verlag.

Watanabe, Y., (1957). Development of *Tethya serica* Lebwohl, a tetraxonian sponge. I. Observations on external changes. *Nat. Sci. Rep. Ochanomizu Univ., 8*, 97–104.

Watanabe, Y., (1978). The development of two species of *Tetilla* (Demosponge). *Nat. Sci. Rep. Ochanomizu Univ., 29*, 71–106.

Webster, N. S., & Taylor, W. M., (2012). Marine sponges and their microbial symbionts: Love and other relationships. *Environ. Microbiol., 14*, 335–346.

Weissenfels, N., (1981). Bau und Funktion des Sübwasserschwamms *Ephydatia fluviatilits* L. (*Porifera*) VIII. Die entstehung und entwicklung der kragengeisselkammern und ihre verbindung mit dem ausfuhrenden kanalsystem. *Zoomorphology, 98*, 35–45.

Whelan, N. V., Kocot, K. M., Moroz, T. P., Mukherjee, K., Williams, P., Paulay, G., Moroz, L. L., & Halanych, M. K., (2017). Ctenophore relationships and their placement as the sister group to all other animals. *Nat. Ecol. Evol., 1*(11), 1737–1746.

Williams, G. C., (1975). *Sex and Evolution*. Princeton: Princeton University Press.

Wilson, H. V., (1907). On some phenomena of coalescence and regeneration in sponges. *J. Exp. Zool., 5*, 245–258.

Windsor, P. J., & Leys, P. S., (2010). Wnt signaling and induction in the sponge aquiferous system: Evidence for an ancient origin of the organizer. *Evol. Dev., 12*, 484–493.

Windsor, R. P. J., Matveev, E., McClymont, A., Posfai, D., Hill, A. L., & Leys, P. S., (2018). Wnt signaling and polarity in freshwater sponges. *BMC Evol. Biol., 18*, 12.

Witte, U., (1996). Seasonal reproduction in deep-sea sponges – triggered by vertical particle flux? *Mar. Biol., 124*, 571–581.

Woodruff, T. K., Lee, H. C., Lee, J., & Edmonds, M., (2018). Oocyte maturation during folliculogenesis. In: *Encyclopedia of Reproduction* (2nd edn., Vol. 3, pp. 172–175). Elsevier.

Wörheide, G., Dohrmann, M., Erpenbeck, D., Larroux, C., Maldonado, M., & Voigt, O., (2012). Deep phylogeny and evolution of sponges (Phylum Porifera). *Adv. Mar. Biol., 61*, 1–78.

Wulff, J. L., (1991). Asexual fragmentation, genotype success, and population dynamics of erect branching sponges. *J. Exp. Mar. Biol. Ecol., 149*, 227–247.

Wulff, J. L., (2006). Ecological interactions of marine sponges. *Can. J. Zool., 84*, 146–166.

Yahel, G., Sharp, J. H., Marie, D., Hase, C., & Genin, A., (2003). *In situ* feeding and element removal in the symbiont-bearing sponge *Theonella swinhoei*: Bulk DOC is the major source for carbon. *Limnol. Oceanograph., 48*(1), 141–149.

CHAPTER 2

The Physiology of Sponge Behavior

SALLY P. LEYS,[1] EMMA J. K. ESPOSITO,[1] and AMANDA S. KAHN[2]

[1]*University of Alberta, Edmonton, Alberta, Canada*

[2]*Moss Landing Marine Laboratories and San Jose State University, California, USA*

ABSTRACT

Sponges (Porifera) are benthic suspension feeders that are of importance to aquatic system functioning because of their vast filtration capacity and their role as habitat formers. Increasingly the role of sponge behavior is also being recognized because it changes how sponges filter and thereby their wellbeing and ecological roles. Sponges lack neurons, gap junctions or synapses; rapid behavior is only known in glass sponges which can propagate electrical signals along syncytial tissues to arrest the feeding current. Glass sponges are syncytial; all others are cellular and yet all sponges share similar slow contractions of their aquiferous system. Contractions are rhythmic and in response to specific mechanical and chemical stimuli. The timing of rhythmic contractions depends upon the size of the sponge and type of aquiferous system, ranging from as 'fast' as 40 minutes for submillimeter sized sponges, to an hour for millimeter sized, half a day for fist-sized sponges and several days for larger sponges. Deep sea glass sponges take from a day to nearly a week to contract and expand. For cellular sponges the tissues involved are endopinacoderm of subdermal space, canals, ostia, and osculum. Evidence suggests the osculum is the common sensory organ but responses differ by species and habitat, with some that are accustomed to constant current responding to slow water, and others accustomed to still or slow moving water responding to current, usually by constricting oscula

Frontiers in Invertebrate Physiology: A Collection of Reviews, Volume 1: Non-Bilaterian Phyla.
Saber Saleuddin, Sally P. Leys, Robert D. Roer, and Iain C. Wilkie, (Eds.)

and canals. Many species are sensitive to sudden changes in the temperature, turbidity, and the chemical environment. Signaling molecules may include glutamate, with modulation by nitric oxide (NO), GABA, ATP, and cAMP which may prevent uncoordinated responses. Specifics of signaling and location of receptors are still unknown. Other behaviors by sponges include crawling/translocation, larval swimming in response to light and orientation, and settlement. Despite over a century of work, sponge behavior is only beginning to be understood beyond our bilaterian lens and areas that merit new or renewed investigation are highlighted.

2.1 INTRODUCTION

A fundamental feature of animals is the ability to move in a coordinated manner. This usually requires the presence of ion channels as well as the ability to receive signals from the environment, generate a cellular signal, and transmit it to an effector (Hille, 1992). While single celled protists display remarkable behaviors, including predation, chemotaxis, and phototaxis (Brunet & Arendt, 2016; Kirchman, 2018; Brunet et al., 2019), in most animals behavior is coordinated by nerves, and the ability to integrate signals and responses varies among and between lineages. Some metazoans, however, have no neurons and yet they carry out a range of behaviors using mechanisms we are only just starting to understand. These include animals that have clearly lost or have highly reduced nervous tissues such as myxozoans and dicyemids (Gruhl & Okamura, 2015; Lu et al., 2019; Zverkov et al., 2019). Myxozoans are parasitic cnidarians and dicyemids are parasites that live in the kidneys of cephalopods; both have either lost neurons entirely or have neurons that lack the characteristic features of a neuron such as synapse and axon. There are also free-living phyla that lack neurons, like placozoans (Schuchert, 1993; Syed & Schierwater, 2002; Miller & Ball, 2005; Ryan & Chiodin, 2015; Varoqueaux et al., 2018; Moroz et al., 2021b) and sponges (Jones, 1962; Nickel, 2010; Leys, 2015). Because nervous systems are widely assumed to be required for coordinated behaviors, Gruhl & Okamura (2015, p. 165) state that in myxozoans their absence is 'hard to accept when there is evidence of concerted movement.' The same is very true for sponges.

Sponges range widely in form from delicate encrusting films to large and stalked or barrel-shaped; all affect their surrounding environment by the continual filtration of large volumes of water and by forming habitats for a multitude of other creatures (Maldonado et al., 2012). They are also all active and responsive to their environment, and yet have no neurons (Nickel, 2010; Leys, 2015; Leys et al., 2019).

Sponges, together with ctenophores, placozoans, and cnidarians, are non-bilaterians, meaning they lack bilateral symmetry. Like the other non-bilaterian phyla, sponges have many distinct body characteristics; however, they are most often described by traits that bilaterians have but that they lack, such as polarity and differentiated muscle and neuronal tissues (Dunn et al., 2015). Like all animals sponges have epithelia (reviewed in Renard et al., 2021), which have been demonstrated to seal as effectively as frog epithelia (Adams, 2010; Adams et al., 2010), and which carry out a range of protective and secretory functions. Epithelia are contractile (Bagby, 1966; Nickel et al., 2011; Colgren & Nichols, 2021), but there is no rapidly contracting muscle tissue in a sponge. While tissues are embryologically derived in sponges, it is difficult to align mechanisms of embryogenesis with those known from bilaterians (Brien, 1967; Borojevic, 1970; Leys, 2004; Ereskovsky & Dondua, 2006; Leys, 2007; Nakanishi et al., 2014). From what is currently understood, sponge embryonic 'layers' do not give rise to muscle, connective tissues, or neurons.

Biologists are puzzled by the absence of neurons because, even without them, sponges can display a range of slow contractile behaviors. Many cringe in response to damage (Aristotle in Parker, 1910), some have rhythmic contractions (Nickel, 2004), most sneeze in response to sediments (Ludeman et al., 2014), and some can also crawl very slowly, possibly to create a space for offspring (Bond & Harris, 1988; Maldonado & Uriz, 1999; Morganti et al., 2021a). These are all behaviors that happen without neurons or muscle and which we discuss in this review. Some of these behaviors could be described as reflex responses and may be localized to small regions of the body, but how others that involve the whole body and which have pacemaker-like frequency are coordinated and spread, remains unknown.

Another well-known feature that distinguishes sponges from other metazoans is their use of choanocytes to feed. Choanocytes are flagellated collar cells that line cup-shaped chambers that make up the bulk of the sponge's body – the choanosome. The sinusoidal beat of the choanocyte flagellum draws water through a mesh created by a collar of microvilli. The current created by choanocytes draws water surrounding the sponge through a multitude of ~20 micron-wide openings in the animal's surface, called ostia, into well-defined canals that lead to the chambers (Figure 2.1) (Weissenfels, 1989). At the chambers and elsewhere along the canal system, food and oxygen are extracted and wastes excreted. Used water is collected into a single excurrent canal and vented out of a raised chimney-like structure called the osculum, that functions to make sure the waste water is not ingested a second time (Bidder, 1923). That process alone generates behavior because if a sponge senses it is ingesting poor water it can constrict incurrent openings (ostia) or canals, or it constricts the excurrent vent (osculum), which reduces the intake of water; if a sponge senses it is ingesting a lot of sediment or harsh chemicals, it can contract its whole body within minutes (Parker, 1919; Elliott & Leys, 2007; Nickel, 2010; Ludeman et al., 2014). Some sponges also contract regularly, either to increase flushing and oxygenation of their tissues, or to get rid of wastes (Weissenfels, 1990; Kornder et al., 2022). But how does a sponge do this without neurons, without easily-identifiable sensory organs connected to neurons, and without typical coordinating pathways? Of the four classes of sponges, only one group – Hexactinellida, or glass sponges – is able to stop the flagella pumps by using electrical signals. Glass sponges have syncytial tissues that allow slow, 5 s-long action potentials to travel unimpeded through the whole body of the sponge, arresting the feeding current (Leys & Mackie, 1997), and while this signaling system functions effectively as an 'alternative' to a nervous system (Leys et al., 1999), as far we know, syncytial tissues and electrical conduction are unique to the class Hexactinellida.

Over the years, observations and experimental data have illustrated that sponge tissues and behaviors are challenging for biologists to understand, and much of our thinking about their structure and function (morphology/histology and behavior) is coloured by our bilaterian bias (Dunn et al., 2015). Despite the weight of evidence showing that sponges have tissues (Nichols et al., 2006; Leys et al., 2009; Leys & Riesgo, 2011; Nichols et al., 2012; reviewed in Renard et al., 2021), that regions of their bodies function as organs (Windsor & Leys, 2010; Ludeman et al., 2014; Windsor, 2014), and that they sense and respond to their environment (reviewed in Nickel, 2010; Leys, 2015), there is still a tendency for writers to refer to sponges as "simple."

As there is a heavy emphasis on the presence or absence of a nervous system in this group, especially with the recent phylogenetic data which suggest that Porifera may have branched off from other metazoans after animals with a nervous system – ctenophores – already existed (Ryan et al., 2013; Moroz, 2015; Whelan et al., 2015; Moroz & Kohn, 2016), we begin with an overview of the phylogeny of non-bilaterian animals. Next we give a brief overview of the current understanding of how nervous systems arose. We then focus most of our attention on the physiology of coordination of

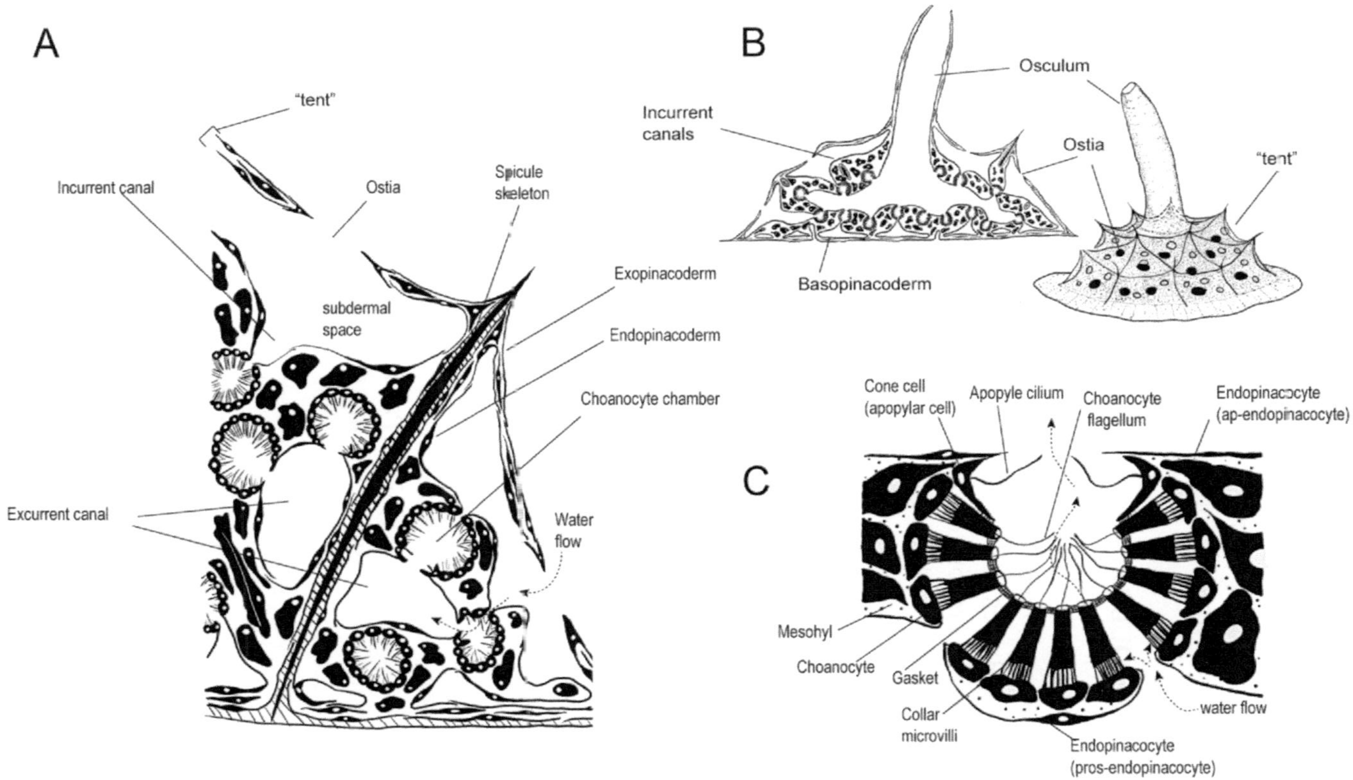

FIGURE 2.1 Basic anatomy of a sponge (redrawn from Weissenfels, 1989). (A) Cross section of body wall showing openings (ostia) through which water enters to a subdermal space lined by endopinacoderm. (B) Overview of the position of ostia, oscula, and the internal tissue containing choanocyte chambers. (C) A single choanocyte chamber of the freshwater sponge *Ephydatia fluviatilis* showing choanocytes lying on a extracellular mesohyl bounded by endopinacocytes, and bound at the upper region of their collars by a glycocalyx (extracellular) mesh that forms a gasket separating the chamber into two parts. Downstream of the gasket water exits through the apopyle which is formed by L-shaped or 'cone' cells which have a single cilium whose forward stroke is out of the chamber.

activities in sponges, looking at the types of behavior found, the tissue pathways of coordination, and the present understanding of the physiological basis for coordination. We have a separate section on locomotion in sponges, and on larval behavior, as well as a section on feeding arrests by electrically propagated events in glass sponges (Hexactinellida). We end with a short section that addresses future directions and questions that merit further study.

2.2 PHYLOGENY: COULD SPONGES HAVE LOST A NERVOUS SYSTEM?

Until the 1990s sponges were unquestioned as sister to the remaining metazoans (Hyman, 1940; Brusca et al., 2016). That view was largely based on the absence of clear germ layers formed during embryogenesis, unlike in ctenophores and cnidarians (Byrum & Martindale, 2004; Leys, 2004). The absence of germ layers was interpreted to imply an absence of tissues (Ereskovsky & Dondua, 2006). But physiology also played a role in this view. Early work that explored contractions in sponges concluded that sponges lack cells that allow rapid coordinated behavior (reviewed in Parker, 1919; Jones, 1962; Pavans de Ceccatty, 1974a). A key piece of evidence was the inability of electrical stimuli to trigger contractions in sponges and the absence of propagated action potentials in cellular sponges (Prosser et al., 1962). The fact that contractions were still present in solutions where potassium concentrations should have depolarized the cells was interpreted as evidence that sponges lack the ability to propagate any action potential (Prosser et al., 1962). In contrast, nerve nets known from cnidarians and ctenophores, which allowed these animals to rapidly coordinate behavior, were considered hierarchically as evolutionarily advanced, and it was therefore assumed that sponges were the earliest branching non-bilaterians because they lacked systems such as these (Pavans de Ceccatty, 1974a, b). Max Pavans de Ceccatty spent a career on this problem, from a PhD thesis arguing that sponges have a nervous system (Pavans de Ceccatty, 1955), to a position in which he felt that sponges had hormonal (paracrine) signaling, as well as myoid and neuroid aspects but no nervous system (Pavans de Ceccatty, 1979). He said the behavior of sponges has "such a simple pattern," that it "obscures the complexity" of the processes that are involved (Pavans de Ceccatty, 1974b).

In recent years, phylogenomic studies have found that sponges may have branched after ctenophores (Dunn et al., 2008; Ryan et al., 2013; Moroz et al., 2014). Branching order matters because one interpretation of the ctenophore-first hypothesis is that, since ctenophores have neurons, then either nervous systems have evolved repeatedly and independently, or sponges could have lost a nervous system (Leys, 2015; Ryan & Chiodin, 2015) (Figure 2.2). The latter interpretation is intriguing because it provides a different context for exploring behavioural responses by sponges, and has fostered research searching for elements of nervous systems in sponge genomes, and morphological correlations of those (Sakaraya et al., 2007; Kosik, 2009; Conaco et al., 2012; Burkhardt et al., 2014; Burkhardt, 2015; Cavalier-Smith, 2017). The arguments for and against this branching order (sponges or ctenophores as 'first' animals) are many. Phylogenomic analyses with increasing numbers of species provide evidence to support both hypotheses (Dunn et al., 2008; Philippe et al., 2009, 2011; Pick et al., 2010; Moroz et al., 2014; Pisani et al., 2015; Telford et al., 2015; Feuda et al., 2017; Whelan et al., 2015, 2017; Li et al., 2021).

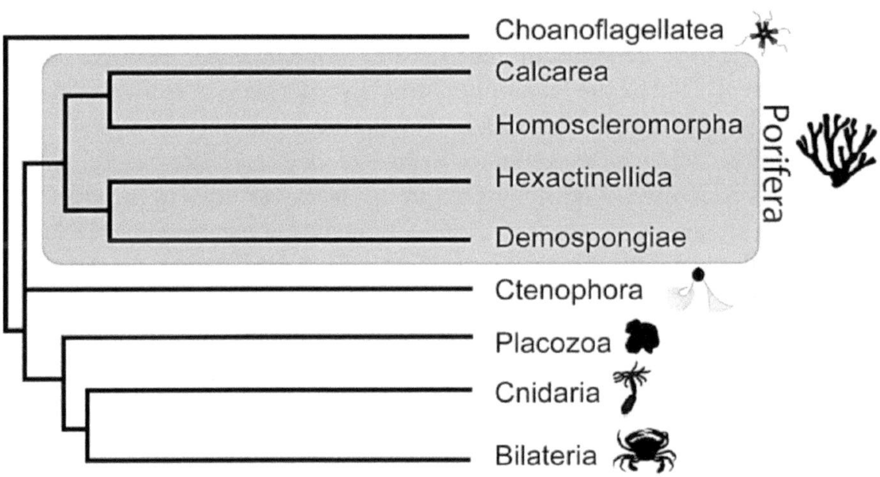

FIGURE 2.2 Phylogenetic relationships of the four sponge classes sponges (Porifera shown in gray) and other non-bilaterian groups withing the Metazoa. The branching of sponges and ctenophores is shown as undetermined.

Analyses of ctenophore genomes have added support to the ctenophore 'first' hypothesis. Ryan et al. (2013) analyzed the genome of *Mnemiopsis leidyi*, a model ctenophore that has been used to study regeneration, axial patterning, and bioluminescence, and found evidence that ctenophores were the sister group to extant animals. They also found neural type gene components (e.g., adhesion and signaling molecules that can be used in forming synapses and in chemical signaling) that were similar to those found in the sponge *Amphimedon queenslandica*, which suggested that sponges have some of the genetic components for a functioning nervous system and may have lost those cell types and some of the associated neuronal genes over time. The genome of another ctenophore *Pleurobrachia bachei* and 10 other ctenophore transcriptomes also suggested that ctenophores are the earliest lineage in Metazoa, based on the absence of *Hox* genes, conserved microRNA machinery, and diminished immune complements in ctenophores (Moroz et al., 2014). Interestingly, the findings of Moroz et al. (2014) also suggested that ctenophores may have independently evolved neural systems, as some of the genes associated with neural tissues in bilaterians and cnidarians (such as *elav* and *musashi*) were not expressed in the neurons of ctenophores. While overall fewer sponge genomes have been available for the analysis (Renard et al., 2018), some new genomes have been sequenced recently (Francis et al., 2017; Kenny et al., 2020). Many studies show that the choice of species, type of model and choice of outgroup drive the outcome of phylogenetic analyses (Pisani et al., 2015; Feuda et al., 2017; Li et al., 2021). For example, in the work of Li et al. (2021), the 'ctenophore first' hypothesis was recovered across all conditions, whereas the 'Porifera first' hypothesis was only recovered under strict conditions, when most outgroups were excluded and if site-heterogeneous CAT models were used. Addition of outgroups however, adds to the chance of long branch attraction, in which the longer branches of ctenophores are drawn to those of the distantly related outgroups (Pick et al., 2010; Nosenko et al., 2013; but see Borowiec et al., 2015). Overall, the discrepancy between the Ctenophora-sister and Porifera-sister hypotheses arises because of a lack of genomic data and restrictions in the different models.

These molecular phylogenomic studies fuel the debate about whether a nervous system could have arisen independently in different lineages and the provocative hypothesis that sponges may have lost neurons over time. If ctenophores came about first then either sponges may have had components of the ctenophore nervous system and lost them, or they could have evolved from an ancestor that lacked neurons entirely but which gave rise to neurons independently in the ctenophore lineage and in cnidarians and other metazoans. Better resolution of the branching of sponges and ctenophores would help us identify whether genes in sponges reflect components of a lost nervous system or novel signaling and coordinating systems in sponges that are analogous to 'conventional' nervous systems, but which did not arise from ancestral (neuronal) signaling systems shared with other metazoans. Much more data are needed about sponge coordination, behavior, and the molecular basis of both in order to understand how they fit into the story of the evolution of the nervous system.

2.3 EVOLUTION OF COORDINATION SYSTEMS AND NEURONS

2.3.1 THEORIES ON THE ORIGINS OF NEURONS

Coordinated behaviors occur in all organisms. Plants can conduct signals between cells (Simons, 1992; Wildon et al., 1992; Léon et al., 2001; Fromm & Lautner, 2007; Fromm et al., 2013), and bacteria have been shown to coordinate proliferation in a biofilm as well as coordinate movements of groups of independent cells (Miller & Bassler, 2001; Claessen et al., 2014). Unicellular fungi such as slime molds can coalesce and coordinate collective behaviors to enhance reproduction (Reid & Latty, 2016). Protists coordinate their ciliary beat for locomotion based on the viscosity of the fluid they move through (Naitoh & Eckert, 1974; Sleigh, 1974), show avoidance behaviors (Dexter et al., 2019), are predators (Roberts et al., 2011; Kirchman, 2018), and respond to light (Foster & Smyth, 1980; Song, 1981; Gualtieri et al., 1992; Barsanti et al., 2000, 2012; Sineshchekov et al., 2002), some using complex contractions in response to changing light (Brunet et al., 2019), and others with lens-like structures that allow detection of polarized light which could help prey detection (Gavelis et al., 2015). Multicellular animals have the benefit of tissues made of cohorts of cell types that act as effectors, such as ciliary arrays or muscle (Brunet & King, 2017). Most of these are coordinated by the nervous system, and its diversity of actions as well as the problems that arise when it malfunctions (Hille, 1992) are what galvanize interest in how nervous systems arose.

In the past, there have been two schools of thought regarding the origin of neurons. One school holds that neurons evolved from a single ancestral cell type, while another holds that neuronal cell types evolved independently (Pantin, 1956; Mackie, 1970, 1990; Moroz, 2009; Nickel, 2010; Arendt, 2021; Leys et al., 2021). All views reflect what is known of cells, tissues, and coordination systems in extant lineages, as it is not known what the ancestor of any of these looked like or what it possessed for coordinating cellular behaviors. Parker (1919) has the most acknowledged and comprehensive analysis on the origin of neurons and the nervous system. He described three stages in the evolution of neurons: an independent effector, a receptor, and 'protoneuron.' The first was a cell similar to those found in the sponges he studied, with a contraction that had "limited… power to spread" (Parker, 1919). In sponges, he found no obvious signs of receptor or conductor/adjustor cell types, only effector cell types, and envisioned this would have been the first stage of neuronal evolution. He called the first cells that evolved with rapid responses 'independent effectors,' cells that have no innervation, and instead sense and respond to stimuli on their own, similar to the independent discharge of nematocysts in many cnidarians. It was proposed that the need to coordinate these independent effectors was the predisposing condition for evolution of the nervous system (reviewed in Mackie, 1990), with 'receptor cells' (sensory cells) evolving from cells in epithelial tissue that were in close contact with the 'effector cells.' It was thought that since movements of cnidarians and ctenophores appeared to be quick compared to "sluggish" sponge contractions, but slow when compared to vertebrate or insect movements, cells in cnidarians could illustrate the next step in the evolution of neurons. His research on cnidarians, which included cutting the bell of a jellyfish and discovering that stimulation of one area could still cause muscle contractions in another, led him to the conclusion that cnidarians have a way to send signals between receptors and effectors, and he coined the term 'protoneuron' to describe it (reviewed in Parker, 1919; Moroz, 2009).

Parker considered that the different cell types described above – effector, independent effector, and protoneuron – evolved independently. However, Passano (1963) pointed out that Parker's three cell types could not have arisen in isolation and needed to be coordinated in order to innervate motor cells and produce any kind of behavior. Pantin (1956) proposed that 'protoneuron' circuits evolved alongside a sheet of coordinated effector cells, such as muscle cells. In all, at this point the idea was well-established that neurons most likely arose from independent cell types, and concurrently with other cells that took on receptor and effector functions.

Other ideas have been raised that are relevant to understanding sponge coordination today. Passano (1963) suggested that neurons may have been stimulated to arise from pacemaker-type cells that were used to generate coordinated contractions. This idea is interesting in light of data showing that all sponges have periodic contractions, typified by the rhythmic pacemaker-like contractions described in *Tethya wilhelma* (Nickel, 2004). Mackie (1990) suggested that neurons derived from electrically coupled myoepithelial-like cells that were capable of receiving, transmitting, and contracting, and which eventually differentiated into different cell types. Data showing that the epithelium of demosponges is contractile (Nickel et al., 2011), and that the epithelium of glass sponges is not only able to propagate electrical signals (Mackie, 1979; Lawn et al., 1981; Leys & Mackie, 1997), but is also contractile (Kahn et al., 2020) lend support to that idea. Many of these ideas, including ones proposed by Parker and Pantin, also left room for non-nervous or neuroid systems, which arose before the nervous system and which continue to exist in animals today (Pavans de Ceccatty, 1979; Mackie, 1990).

The secretory aspect of neurons became a focus in the mid-1900s, and hypotheses were developed to suggest that neurons may have evolved from a sensory cell that developed separate portions of itself for reception and secretion of molecules (Grundfest, 1959, 1965). Because secretion was considered a defining feature of neurons it was necessary to distinguish the communication occurring between neuronal cells from the communication that occurs between all other types of cells. The term "neurotransmitter" was coined in 1936 by Otto Loewi and Henry Dale, who experimented with the effects of drugs on visceral organs (reviewed in Valenstein, 2002). But the challenge arose that none of these characteristics were specific to neuronal cells (Moroz, 2009). Ion channels, and the machinery for adhesion and secretion as found in synapses, are found broadly across eukaryotes (Alie & Manuel, 2010; Burkhardt et al., 2014; Burkhardt, 2015; Liebeskind et al., 2011, 2015, 2016, 2017; Moroz & Kohn, 2016; Sachkova & Burkhardt, 2019; Moroz et al., 2021a). Signaling via neurotransmitters has its origin in chemical signaling in protists (Liebeskind et al., 2015, 2016), some of which have mating pheromones that resemble neurotransmitters, such as the pheromone in *Blepharisma* that is similar to serotonin, or producing and expressing neurotransmitter receptors, such as *Tetrahymena* producing β-endorphin (described in Mackie, 1990).

In recent years, the emphasis has shifted to the discovery of genetic markers for neuronal cells. Homeodomain transcription factors (*Hox*, *Otx*, and *Pax* genes) were first found to be necessary in the specification of the central nervous system in insects and mammals (Hirth et al., 2003; Moroz, 2009). However, the expression patterns of the genes are ineffective in identifying neuronal cells (Moroz, 2009). Another way of identifying potential neurons is to look for genes that code for proteins involved in the structure and function of a synapse, often referred to as 'post' and 'pre' synaptic genes, and largely determined from their presence in vertebrate neurons (Arendt, 2020). However, here the same challenge is encountered, because many 'pre-synaptic genes,' which are related to vesicle trafficking and secretion, are also found in other non-neuronal secretory cell types (Arendt, 2020). 'Post-synaptic genes' are also highly variable, as many neurons can vary the types of receptors or other proteins present in their membranes based on what sort of signaling mechanisms they are involved in. It has also been shown that many gene families that are functional in the synapse have other functions (e.g., vesicle trafficking, adhesion, secretion, etc.) in other organisms (reviewed in Leys, 2015). Therefore, even with advanced molecular approaches it is difficult to find definitive genetic markers for neurons, and to produce a clear definition of a neuronal cell type (Mah & Leys, 2017).

A main problem in developing an effective search image for a 'proto-neuron' is that most studies of nervous systems are conducted on animals with highly specified and organized systems; neurons are already diverse and fully developed. Using the definition of a vertebrate neuron (or the genes expressed in one) to determine what cell gave rise to neurons is inevitably biased (Dunn et al., 2015). It is critical to study the wide range of cell types found in non-bilaterian animals and to have a definition of neurons that encompasses the different cell types that are still being discovered. Apart from the syncytia in Hexactinellida that conduct electrical signals (Mackie et al., 1983; Leys, 1995), we are still far from having a clear understanding of what cell type in a sponge might function analogously to a neuron.

2.4 SPONGE CONTRACTILE BEHAVIOR

A feature that should be mentioned first in any textbook description of Porifera is that all sponges have a very active life, contracting actively every 3 to 20 hours for cellular sponges (Demospongiae, Calcarea, Homoscleromorpha), every two weeks or longer for many Hexactinellida (Figure 2.3). Rhythmic contractions have been found in almost all sponges that have been filmed over time. Small sponges contract more frequently and contraction cycles are shorter (1–3 h long contractions, with 0–2 h between contractions) and larger sponges seem to contract less frequently with longer contraction-expansion cycles (10 h or several day-long contractions with days to weeks between contractions) (Table 2.1 and Figure 2.3). Sponges in the family Tethyidae exhibit very active behavior with *Tectitethya crypta* contracting on a day-night cycle (Reiswig, 1971) and *Tethya wilhelma* contracting multiple times per day and night (Nickel, 2004). *T. wilhelma* shows rhythmic, pacemaker-like body contractions every 1–10 hours that reduce the body volume by up to 73.3% during regular contractions (Figure 2.3(A)) (Ellwanger et al., 2004; Nickel, 2004). Contraction patterns vary between individuals and intervals between contractions are longer in the dark. Nickel (2004) suggested that this could indicate a diurnal rhythm in sponges, that could involve an "intrinsic, endogenous timing or triggering mechanism" or even light sensitivity (Flensburg et al., 2022). Periodic waves of contraction run across the entire body of *Ephydatia fluviatilis* and *E. muelleri* (De Vos & Van de Vyver, 1981; Bond & Harris, 1988; Weissenfels, 1990). Sponges also respond to sediment or caligraphy (sumi) ink by contracting repeatedly, each

TABLE 2.1 Contraction Behavior in Sponges

Species	n	Size range	Depth	Duration of recording	Contraction time	Interval between contractions	Excurrent velocity (cm s-1)	Reference
Verongia gigantea	6	0.6–84.5 L	33–51 m	23–141 h	42 min	19 h	6.8-17.3	1
Tethya crypta	35	2 L	1–6 m	Several months	1-5 days	9-21 days	15-17.7	1
Mycale sp.	10	0.2–2.25 L	17–43 m	698 h	0	0	6.8-7.8	1
Verongia archeri	1	n/a	32 m	46 h	~2 h	n/a	~5	1
Agelas sp.	1	n/a	25 m	48 h	1 h	3-4 h	~12	1
Baikalospongia bacilifera	3	n/a	11–17 m	24–27 h	"Minutes to hours"	n/m	0.2-3.3 (1.9+/-0.9)	2
Tethya wilhelma	17	1 cm	lab	168 –177 h	n/m	Day 215 min; Night 274 min	n/m	3
Dysidea avara	4	n/a	4.7–14.3m	n/a	64 min, 6 min, 0,0	n/a	n/m	4
Chondrosia reniformis	10	n/a	lab	n/a	10 – 60 min	n/a	n/m	4
Cliona orientalis	10 osc		lab	n/a	20.8 min (osc closure)	n/a	6.82	5
Xestospongia muta	29	17.8 mL – 45 L	15–30 m	15–60 h	10-230 min	>2.5 d	0.03-13.66	6
Halichondria panicea	12	0.018 mL	–	n/a	36.5 min	n/a	2.3 +/-1.3	7
Halichondria panicea	3	–	–	270h	37.5min +/- 13.7min	8.2h +/- 2.7h	n/m	7, 11
Lyssacinosida sp. 1	2	–	4000–4100 m	407 d	15.5 ± 1.3 d	33.0 ± 9.2 d	-	8
Hexactinellida sp. 2	6	–	4000–4100 m	568 d	4.4 ± 0.4 d	15.2 ± 1.1 d	-	8
Hyalonema bianchoratum	2	–	4000–4100 m	485 d	3.2 ± 1 d	9.5 ± 2 d	-	8

TABLE 2.1 *(Continued)*

Species	n	Size range	Depth	Duration of recording	Contraction time	Interval between contractions	Excurrent velocity (cm s-1)	Reference
Hyalonema sp. 1	3	–	4000–4100 m	540 d	3.1 ± 1.2 d	10.7 ± 1.3 d	-	8
Ephydatia fluviatilis	3+	1–2 cm	lab	48 h	1h 45 min	2.5 h	n/m	9
Ephydatia muelleri	5	0.5 cm	lab	3 h	20-30 min	n/a	n/m	10

n/m=not measured; n/a=not available or not applicable.

1) Reiswig, 1971. *Mar. Biol. 9*:38–50; 2) Savarese et al., 1997. *Limnol. Oceanogr. 42*(1):171–178; 3) Nickel, 2004. *J. Exp. Biol. 207*:4515–4524; 4) Schläppy et al., 2010. *Limnol. Oceanogr. 55*(3):1289–1300; 5) Strehlow et al., 2016. *PeerJ*, DOI 10.7717/peerj.2761; 6) McMurray et al., 2014. *Aquatic Biol. 23:* 1–13; 7) Kumala et al., 2017. *Mar. Ecol. Progr. Ser. 572*: 117–128; 8) Kahn et al., 2020. *Deep Sea Research II 173:* 104729. 9) Weissenfels, 1984. *Zoomorphology 104:* 292–297; 10) Elliott & Leys, 2007. *J. Exp. Biol. 210*:3736–3748; 11) Kumala et al., 2018. *Front. Mar. Sci. 5*.

wave lasting about 20 minutes and one following the other like convulsions to rid the choanoderm of wastes (Elliott & Leys, 2007).

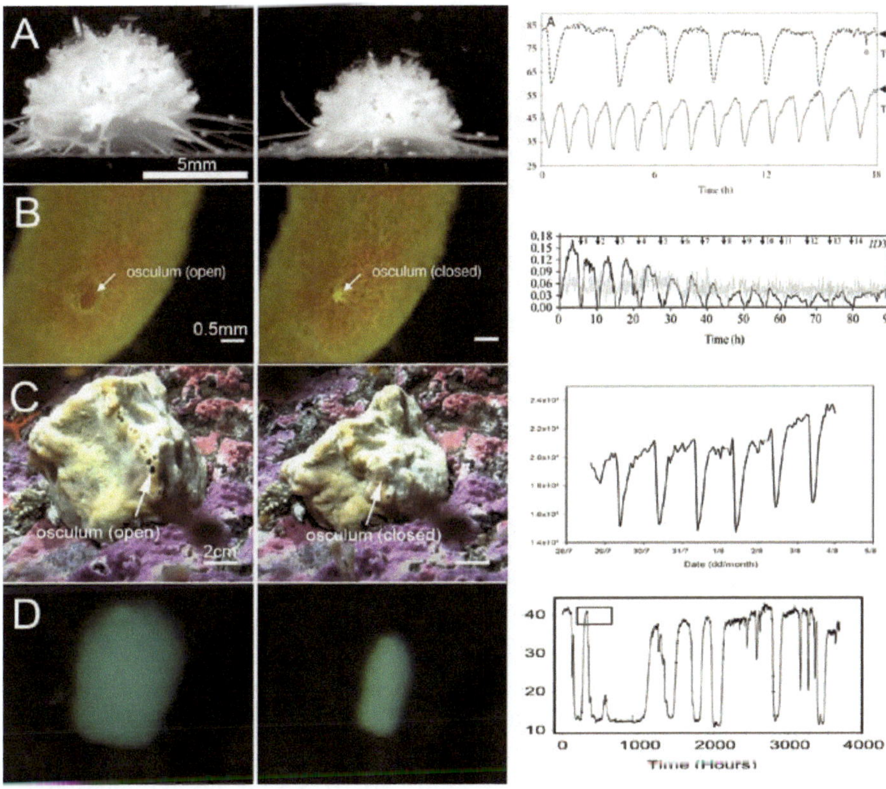

FIGURE 2.3 Contractions carried out by sponges shown by images on left of 'expanded' and 'contracted' states, and examples of the periodicity of contractile behaviors recorded on the right. Recordings show measurements of area of the whole sponge (A), (C), (D) or the osculum (B); values are relative or according to the source. A, *Tethya wilhelma* grown in culture, contracts rhythmically 10–18 times a day. Contractions take 20–24 minutes and expansions 48–54 minutes (n = 17) (adapted from Nickel, 2004; Ellwanger et al., 2007). (B) *Halichondria panicea* explants grown in culture, contract rhythmically 2–5 times a day. Contractions take 37.5 minutes (n = 29) (adapted from Kumala et al., 2018). (C) *Suberites concinnus* imaged *in situ* at the Folger pinnacle in Barkley Sound, BC, using the Ocean Networks Canada observatory. Contractions occur once a day during spring and summer months and take 9–10 hours (n = 14) (Leys unpublished). (D) A glass sponge identified only as '*Hexactinellida* sp. 2' imaged at 4,000 m at Station M, 220 km west of California takes ~24 h to contract, is contracted for ~ 45 h, and takes ~40 h to expand, for a total contraction time of 4.4 days (n = 46) (adapted from Leys et al., 2019; Kahn et al., 2020).

The demosponge *Halichondria panicea* was studied as small explants which are very active, contracting in response to sediment as well as to a range of stimuli (Figure 2.3(B)) (Kumala et al., 2017; Kumala & Canfield, 2018; Goldstein et al., 2020). A fist-sized demosponge, *Suberites concinnus*, filmed *in situ* at the Ocean Networks Canada underwater observatory, contracts nearly daily in spring and summer months with each contraction taking up to 10 hours (~3–4 hours to contract and ~5–6 h to expand) (Figure 2.3(C)). A range of oceanographic parameters collected from the same platform hint at a correlation with changes in pressure (Leys et al., 2019) which accompanied storm activity. Reiswig (1971) also showed that osculum size of *Tectitethya crypta* changed depending on stormy and calm weather.

Calcareous sponges also contract: *Clathrina* shrinks to a fraction of its size with populations that were tracked over months showing daily expansions and contractions but also budding and fusing over days to weeks (Gaino et al., 1991). *Leucosolenia botryoides* undergoes full body contractions (Bond, 2013). It was not known that glass sponges could do this too until analysis of hourly images captured by still cameras placed at 4,000 m depth in the abyssal plain off the coast of California revealed that all the glass sponges as well as a carnivorous sponge, *Cladorhiza kensmithi*, contracted periodically (Figure 2.3(D)) (Kahn et al., 2020). The contraction cycles varied from 9.5 to 33 hours for different glass sponges with 18 hours as the shortest time to fully contract and 167 hours (nearly 7 days!) the longest time; typically contractions took several days for a sponge to reduce its size and a little longer to expand to its normal size. These are very slow changes but nevertheless the sponges they observed contracted by 80% of their former dimensions, a remarkable change in size (Kahn et al., 2020). So far, no explanations for the periodicity seen in the deep-sea sponge contractions have been found, nor has a specific trigger been identified, however one hypothesis is that seasonal fallout of marine snow can be seen slowly darkening the surface of the sponges, and possibly, even at that depth in the deep-sea, a two day contraction may be effective at cleaning up the canals (Kahn et al., 2020).

2.4.1 TYPES OF CONTRACTIONS SHOWN BY SPONGES

Sponges have two overt behaviors: changing the strength of filtration (increasing and decreasing the amount of water filtered), and contracting their body; it is very likely that both behaviors use the same effectors: the contractile epithelial cells. For animals that have no nervous system or muscle tissue, sponges nevertheless show a remarkable range of contraction behaviors. Different sponge species vary in the density of their skeleton (many or few spicules), density of the tissues and the stiffness of the extracellular matrix, and in whether they have a narrow or wide osculum. These characteristics mean that some sponges are flexible enough to contract their whole bodies whereas others can only contract canal tissues; however, as far as we know all sponges contract, and many do this overtly and rhythmically (see for example Dahihande & Thakur, 2019; Kahn et al., 2020).

The excurrent vent or osculum is frequently formed by thinner tissues than the rest of the body, and these semi-transparent membranes are flexible, and vibrate with turbulent water nearby, as well as being fairly quick to contract and close. Similarly, ostia on surface tissues are all contractile. Some have raised edges that can be dropped to cover the incurrent openings (Reiswig, 1979b). Consequently, a range of behaviors have been described in sponges depending on the species and the structure examined. Freshwater sponges have received the most attention because they live in lakes on all continents and are thus easy to find and obtain; they grow well in the lab into tiny sponges that filter, and they are transparent for microscopy. Nonetheless, a variety of experiments on marine and freshwater sponges have also been conducted that demonstrate sponge reactions, and we summarize these below to demonstrate the breadth of sponge behaviors.

To assess the effect of mechanical stimulation on sponges, Parker (1910) carried out experiments on the sponge *Stylotella heliophila*, currently known as *Hymeniacidon heliophila.* He found that exposing the sponge to a current of water and then stopping the current caused the osculum to close (Parker, 1910). This showed that in *S. heliophila* the mechanical stimulation of the water current keeps the osculum open. He lowered a tube over the finger of sponge with the osculum and showed that in still water the osculum closed in 7 minutes (Figure 2.4(A)). He inserted a line to allow water to flow into the tube surrounding the osculum, and the osculum reopened 14 minutes later with a current of water flowing out (Figure 2.4(B)). He then inserted wool to isolate just the upper portion of the osculum and showed that in still water the osculum closed (Figure 2.4(C)) but if a current was directed over the very tip of the osculum, the osculum opened (Figure 2.4(D)). Interestingly, when he touched or 'stroked' the oscular membrane it did not contract, but he found that a mechanical stimulus only within the first 5 mm of the osculum tip caused a reaction. Later, both Parker (1919) and McNair (1923) carried out similar experiments on other marine demosponges as well as the freshwater sponge *E. fluviatilis* and found that the response to an external current depended on the species: in some, like *E. fluviatilis*, the osculum contracted with current, while in others, like *H. heliophila*, it opened with current.

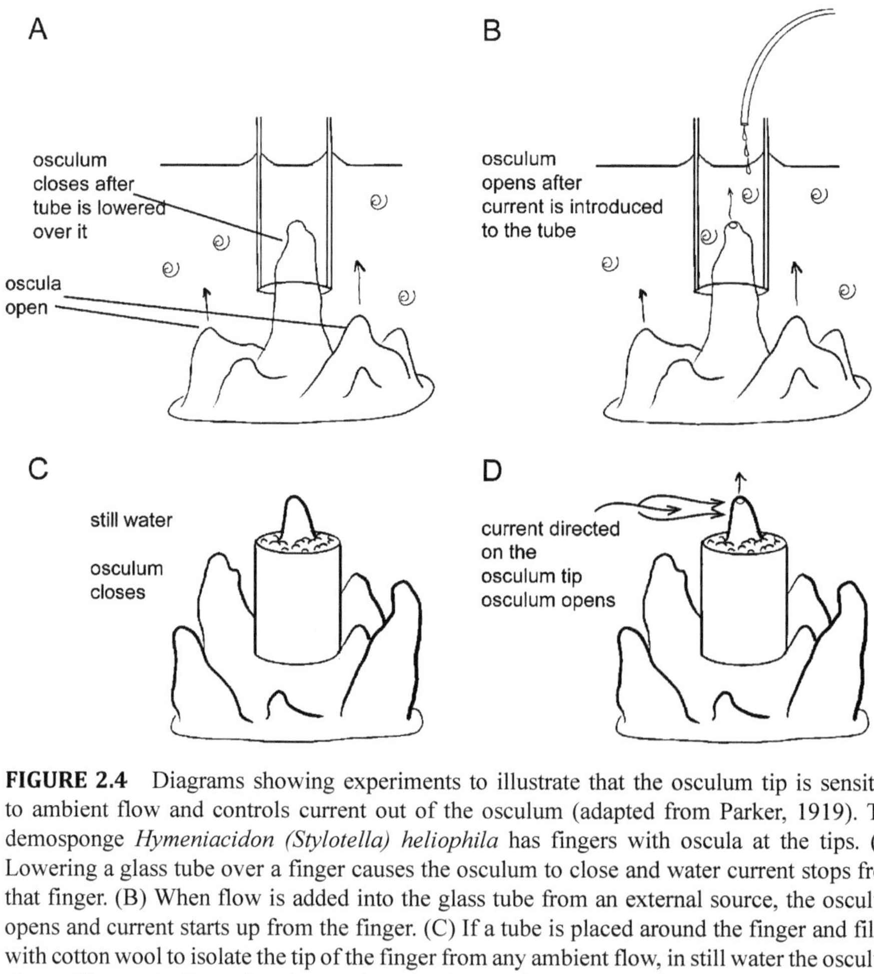

FIGURE 2.4 Diagrams showing experiments to illustrate that the osculum tip is sensitive to ambient flow and controls current out of the osculum (adapted from Parker, 1919). The demosponge *Hymeniacidon (Stylotella) heliophila* has fingers with oscula at the tips. (A) Lowering a glass tube over a finger causes the osculum to close and water current stops from that finger. (B) When flow is added into the glass tube from an external source, the osculum opens and current starts up from the finger. (C) If a tube is placed around the finger and filled with cotton wool to isolate the tip of the finger from any ambient flow, in still water the osculum closes. If water is directed at the osculum tip, the osculum opens and current comes out.

The experiments described above show that the upper part of the osculum is sensitive to water movement, either around it or through it. Changes in current cause the sponge to close or to open its osculum. When sponges close their osculum, it is often thought that they must have to stop their flagella pumps because otherwise pressure would build up. However, as seen in the next experiments by Parker and others, that is not the case.

2.4.2 SPONGES CONTROL EXCURRENT FLOW BY CONSTRICTING CANALS, OSTIA, AND OSCULA

To test whether sponges could control their flagella pumps, Parker (1910, 1919) prevented flow through the osculum and ostia by way of ligatures (Figure 2.5). He determined that the pressure generated by the choanocyte pumps in *H. heliophila* was 3.4 mm H_2O and the pressure that the osculum could withstand was 11–15 mm H_2O; he concluded that contractions of the osculum could readily prevent water flowing through the sponge. The latter experiment was repeated on a number of different species (range 1.3–2.9 mm H_2O, n=7), which confirmed that sponges have very low pressure pumps that filter large volumes of water. Importantly, flow through the sponge can be controlled entirely by contractions of the osculum and sphincters in the canals (Parker, 1914) even though the choanocytes continue to pump. Recent work on explants of the temperate intertidal demosponge *Halichondria panicea* confirms Parker's observations (Kumala et al., 2017; Kumala & Canfield, 2018; Goldstein et al., 2020). Kumala et al. (2017) noted that the contractions of the osculum coincided with change in filtration activities and when the sponges were held in stagnant, filtered water, the osculum remained closed for long periods of time.

Other parts of the sponge also contract. Reiswig (1971) documented this behavior in many individuals of *Mycale*, *Verongula gigantea* (published as *Verongia gigantea*), and *Tectitethya crypta* (published as *Tethya crypta*) over different seasons *in situ*. Reiswig used flowmeters to document filtration, and as with Kumala's work on *H. panicea* (Kumala et al., 2017), he associated contractions of the atrial pores with arrests of filtration. He also studied the body wall of *V. gigantea* using histology and found much thicker walls in specimens that were contracted (Reiswig, 1971); he concluded that the exhalant canals do not passively deflate, instead the canal walls thicken (the mesohyl area becomes larger) by the active contraction of exhalant canals.

Incurrent openings, known as ostia, and groups of these, known as pore fields, also contract. Elliott & Leys (2007) filmed the surface of the freshwater sponge *E. muelleri* and found that fields of ostia contracted just before the excurrent canal system contracted to expel wastes. While an individual ostium took about 40 s to close, areas of up to 35 ostia closed synchronously, as part of a global response to either shaking or adding ink particles; as soon as the excurrent canals relaxed, the ostia reopened almost all at once (Elliott & Leys, 2007).

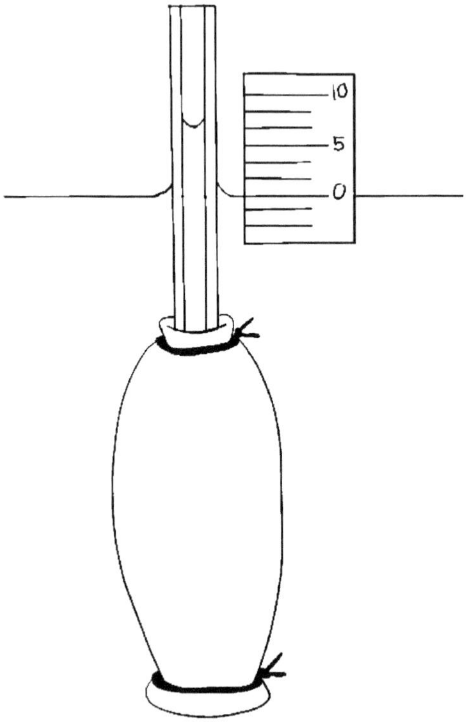

FIGURE 2.5 Parker's method of measuring the pressure developed by the sponge choanocyte chambers (adapted from Parker, 1919). A tube is inserted into the osculum of a finger removed from *Hymeniacidon heliophila.* A ligature is tied around the top of the osculum and the base of the finger to prevent water flowing out the opening at the bottom. The sponge pumps drive water in the tube higher than the outside water to a level which can be measured by a scale. The current pressure for seven species he measured ranged from 1.3 to 2.9 mm H_2O.

In summary, sponges can control the water flow through their bodies by contracting the osculum, canals, and ostia; there is no evidence, so far, that they can control flagella beat rate.

2.5 UNDERSTANDING BEHAVIOR AND COORDINATION IN SPONGES

2.5.1 COORDINATION PATHWAYS IN SPONGES

Since sponges can respond to environmental cues, and change their shape and size over minutes to months (Pavans de Ceccatty et al., 1960; Nickel,

2004, 2010; Dahihande & Thakur, 2019; Kahn et al., 2020), as well as translocate (Morganti et al., 2021b), it is thought there must be some way of coordinating these activities. An alternative view, that cells react individually such that cumulatively they make the whole animal respond, as in 'swarm behavior' (Elgeti et al., 2015) has not been considered except in the case of larval photoreception discussed in Section 2.8.

In the absence of neurons and true muscle, what coordination pathways do sponges use? Pavans de Ceccatty (1979) considered there to be two categories: (i) long-distance signaling via hormones, such as chemicals released by sponges to deter predators; and (ii) short distance signaling which would happen either between cells that were in contact, by tugging or by paracrine signaling, or by cells releasing molecules that affected other cells within a distance of 30–100 cell distances. He also referred to myoid and/or neuroid conduction (presumably short distance electrical signaling through epithelia or cells that were coupled), but he was explicit that he did not consider a nervous system: "it has become clear that the network found in some sponges can no longer be qualified as a nervous system, as was formerly done by myself and others" (Pavans de Ceccatty, 1979, pp. 56–58). One of the greatest challenges to those studying coordination is understanding what sponge cells do and how they interact with one another. In the next sections we give an overview of the sponge body to indicate where different tissues are, we then introduce sponge cell types, and from there we describe the epithelial tissues which are currently considered to be the main effectors of contractions.

2.5.1.1 SPONGE TISSUES

There are over 9,500 described species of sponges (de Voogd et al., 2022) in 4 classes, each of which share characteristic cell and tissue types. Sponges have essentially three layers – an outer epithelium, an inner epithelium lining internal canals, and between those a cellular mesohyl (Leys & Hill, 2012). The mesohyl is not in direct contact with the water, and forms a true "internal milieu" (Pavans de Ceccatty, 1974a). Holes in the surface of the sponges are formed by cells called porocytes that allow the influx of water into epithelial lined canals and often into an epithelial-lined space, the subdermal space, that lies beneath the cortex or surface of the sponge (Leys & Hill, 2012) (Figure 2.1). From the subdermal space, the water is drawn into the body of the sponge via chambers that are lined with flagellated cells called choanocytes, which are active in pumping (Leys & Hill, 2012). These chambers are connected to excurrent channels that all eventually converge into a main excurrent canal or atrium and from there to the osculum (Leys & Hill, 2012).

Sponges have two epithelial types: the pinacoderm and choanoderm. These terms derive from their structure and function: pinacocytes form a 'plate-like' ('tablet') epithelium; choanocytes are 'collar' ('funnel') cells which line the feeding chambers. Only very few of the calcareous sponges have little else but those two layers. Most sponges also have a mesohyl (middle 'matter') formed by extracellular matrix (largely types of collagen) with cells that can be structural or be nutritive, store materials, or carry out immune-like functions (Simpson, 1984). Many sponges also host microbial and or algal symbionts in the mesohyl, and in many species the symbionts are also intracellular (Busch et al., 2022; Maldonado et al., 2022).

Across all Porifera there are approximately 10–13 generic cell types that are fairly easy to identify morphologically (Pavans de Ceccatty, 1979; Simpson, 1984). Following Simpson (1984) these are: pinacocytes, choanocytes, lophocytes and collencytes (both secrete collagen but the former is shaped like a brush and the latter is star-shaped), trophocytes (nutrient storing cells), megasclerocytes (secrete large spicules), microsclerocytes (secrete small spicules), fiber cells (a group of cell types and stages that are involved in secreting skeletal parts including spongin), actinocytes (formerly called myocytes, which may be contractile), cells with extensions (formerly called 'neuroid' cells, but with unknown function), special cells (which comprises a range of cells with inclusions), and gametes. These cell types are named from an understanding of their morphology. Recent work has used genomes to identify groups of cells that cluster together based on their transcriptome (termed 'metacells') and to annotate them based on their expression of functional (marker) genes (Sebé-Pedrós et al., 2018). Using this method, the broad categories of choanocytes, archaeocytes, and pinacocytes were found, as well as cells of 'intermediate' character, as defined by shared sets of genes expressed. Single cell sequencing is a powerful technique where the biology of the animal and cell types are quite well characterized, but it meets its limitations in interpreting sponge cell types because of the paucity of morphological and functional studies on sponge cell types.

Based on morphology/functional work on sponge cell types, within a sponge taxon the number of known cell types increases, as is seen in other phyla; for example, whereas there are approximately 10–13 types of cells that are 'common' across all sponges, in freshwater sponges up to 31 cell types have been distinguished (Simpson, 1984). Although there is little recent work on this problem, a few excellent studies have uncovered interesting findings.

For example, in the freshwater sponge, in addition to cells that make spicules, another three or four mesohyl cell types are involved in moving the spicules around to their respective positions as the sponge develops (Nakayama et al., 2015; Kishimoto et al., 2019).

2.5.1.2 *PINACODERM EPITHELIA: STRUCTURE AND AS CONDUCTION PATHWAYS*

The structure of sponge epithelia has been studied in some detail across a diversity of species (Boury-Esnault, 1973; Maldonado & Riesgo, 2007; Leys et al., 2009; Leys & Riesgo, 2011; Renard et al., 2021). The outer epithelial layer is formed by exopinacocytes. These are typically pentagonal shaped on their surface and overlap one-another slightly at their margins. In the majority of the class Demospongiae these usually have a deeper cell body and a flatter surface, and are called T-shaped, but in freshwater demosponges and the class Homoscleromorpha they are flat, called 'fusiform' (Boury-Esnault, 1973); the reason for this difference is unclear and merits investigation. In Calcarea, exopinacocytes can be both flat and T-shaped in the same epithelium (Boury-Esnault, 1973), while in Hexactinellida the entire sponge is syncytial, and so the outer surface is referred to as the 'dermal membrane' (Reiswig, 1979a). In contrast to exopinacocytes, the endopinacocytes that line the canals and subdermal space of cellular sponges (Demospongiae, Homoscleromorpha, and Calcarea) are always fusiform (flat) and according to Boury-Esnault (1973), exo- and endopinacocytes are two different categories of cells.

A T-shaped pinacocyte has a surface that is 3.5–30 μm in diameter, 0.08–2 μm thick, and it reaches 9–30 μm down into the mesohyl. Fusiform or flat pinacocytes in for example, *Oscarella loblularis*, are 7–8 μm wide and 2–2.5 μm thick where the nucleus lies, and at the thinnest area the cells are no more than 0.3 μm (Boury-Esnault, 1973). In the freshwater sponge *E. muelleri* fusiform exopinacocytes are pentagonal shaped, 42–45 μm across and 1.5 μm thick. Endopinacocytes lining the subdermal space are long and thin cells, 40–44 μm long, 11–12 μm wide (Elliott, 2009). In the calcareous sponges of the genus *Sycon* T-shaped and fusiform exopinacocytes can occur side-by-side and are small, approximately 4.8–8.5 μm by 7.3–10.5 μm in diameter (Boury-Esnault, 1973). One suggestion is that the T-shaped pinacocytes might be contracted fusiform pinacocytes (E. Lanna, personal communication), but this hypothesis must be tested experimentally by making sponge epithelia contract and studying the tissues by electron microscopy.

Exopinacocytes are held together by parallel membrane junctions formed by two cells slightly overlapping one-another, the junction forming a space 10-20 nm, such that the cells form a continuous cover over the sponge surface (Boury-Esnault, 1973; Green & Bergquist, 1979; Adams, 2010; Adams et al., 2010). Molecules involved in both sealing and adhesion and in polarizing epithelia in other animals have been identified in sponge genomes and transcriptomes (Fahey & Degnan, 2010; Riesgo et al., 2014; Mitchell & Nichols, 2019; Renard et al., 2021), but most studies have not distinguished whether these molecules are present in exo- or endopinacocytes, largely because both are so thin and studies to visualize expression in the epithelia have not yet been done. Earlier work often suggested that sponge epithelia were transient structures, with openings into the mesohyl from canals (e.g., Weissenfels, 1983). However, studies using live imaging have shown that connections are not tenuous, rather the pinacoderm is normally static over time – as shown by images of cell contacts retaining position in the exopinacoderm over time (e.g., Adams et al., 2010). In regenerating tissues cells move out of the epithelial sheet (e.g., Bond & Harris, 1988; Bond, 1992). This is not to say that epithelial connections are not able to release and reform in adult non-regenerating tissues, just that this has not yet been demonstrated, and so the view should be that sponge epithelia are normally sealed and function as an integral unit.

Many types of junctions exist between sponge pinacocytes, but all those responsible for sealing epithelial cells have not been identified. Many papers report the absence of desmosome junctions in sponges (Revel, 1966; Green & Bergquist, 1979, 1982; Lethias et al., 1983) and yet septate-type junctions are known in calcareous sponges (Ledger, 1975), demosponge gemmule coats (de Vos, 1977), and larval epithelia of Homoscleromorpha (Boury-Esnault et al., 2003). The presence of occluding junctions in demosponges has been shown to form an effective seal preventing the passage of very small molecules into mesohyl. In *Spongilla lacustris* the epithelium showed resistances up to 5,500 ohms/cm^2, equal to that of amphibian epithelia, and sponge epithelia with high resistance also prevented the passage of molecules as small as 5 kDa (1.5 nm) (Adams et al., 2010). In contrast, there is still no morphological or molecular evidence for gap junctions, which are aqueous connections between cells, and which would allow electrical current to pass between cells (Green & Bergquist, 1982).

2.5.2 *HOW SIGNALS MIGHT PROPAGATE ALONG EPITHELIA*

Based on the structure of sponge epithelia described above it is envisioned that a stimulus received by the sponge would contact the exopinacoderm or endopinacoderm and stimulate a cell to either: (i) contract and tug on a neighboring cell triggering that to contract; or (ii) contract and secrete small molecules which might be received by neighboring cells in a paracrine manner causing others to contract. In either case it is anticipated that the signal might travel to the choanocyte chambers and either cause the chamber to contract, or cause a change in beat of the flagella. While we know that chambers do contract, so far there is no evidence that cellular sponges (non-hexactinellids) are able arrest their flagella. Instead, as described in Section 2.4, it is most likely that most signaling occurs by waves of contractions of endopinacocytes, closing sphincters, sieve cells, ostia, and the osculum.

Waves of contractions move not only along the canals, but often across canals. As contractions move down canals the chambers also contract and flagella can be seen protruding out of the apopyle (exit aperture) (Elliott & Leys, 2007). As the chambers are so constricted, the flagella beat may be constrained. Weissenfels commented that the choanocytes stop beating when the chamber contracts (Weissenfels, 1984, 1990). We might envision that as the contraction passes over the chamber there is an influx of calcium into the choanocyte cell body arresting flagella. It is known that calcium (or divalent cations such as magnesium) is required for contractions to take place (De Vos & Van de Vyver, 1981; Elliott & Leys, 2010), and because amoeboid cells in the mesohyl stop crawling as a wave of contraction moves across the tissue (Elliott & Leys, 2007), it is thought that signals released by cells as they contract trigger influx of calcium into neighboring cells: those cells in epithelia contract, and those that are wandering, stop moving.

All cells contain actin, for integrity and motility, but for contractions to effect substantive shape changes in cells, presumably there should be more actin as well as myosin in the cell and the organization should be such that it effects change across a sheet of cells. Parker (1910) considered cells in the mesohyl to be the effectors, and called them myocytes. But Pavans de Ceccatty (1974a), having studied the ultrastructure of a range of sponge tissues, saw actin filaments across myocytes in thick-walled sponges and across pinacocytes in thin-walled sponges and concluded that in the former, elongate cells referred to as myocytes were involved in the contractions while in the thin-walled sponges he thought the pinacoderm was contractile. Even though most evidence now suggests that the endopinacoderm is the major contractile tissue, as commented by Nickel et al. (2011) it is likely that actin-dense cells (myocytes sensu Bagby, 1966; actinocytes sensu Boury-Esnault & Rützler, 1997) in the mesohyl are involved as antagonists allowing the sponge to expand after waves of contraction pass.

The actin cytoskeleton of pinacocytes across a diversity of sponges is understudied. Work has focused on freshwater sponges because of their transparency. There, we see actin tracts traversing a sheet of basopinacocytes, the layer attached to the substrate (Pavans de Ceccatty, 1981, 1986; Wachtmann et al., 1990) and a similar array of actin tracts stretches across the endopinacocytes lining the subdermal space (Elliott & Leys, 2007; Colgren & Nichols, 2021). Exopinacocytes do not have these actin tracts. Endopinacocytes lining canals have some actin filaments that appear to go around the circumference of the canal (Elliott & Leys, 2007).

With only two possible contractile tissues in sponges (exo and endopinacoderm), it is surprising that so little is known about how they contract, but this reflects the difficulty of creating effective preparations to study this. Nickel et al. (2011) used a microCT imaging approach to quantify the relative area of canal space and mesohyl occupied in expanded and contracted sponges of *T. wilhelma.* They found that in an expanded sponge the canal took up about 72% of the sponge (mesohyl 28%, skeleton 3%), while in a contracted sponge the canal space took up only 32% of the sponge (mesohyl 60%, skeleton 7%) suggesting that contractions eliminated the canal spaces and did not compress the mesohyl. The area occupied by the pinacoderm was 2.75 times smaller in the contracted compared to the expanded sponge. Together that work gave strong support for a contractile endopinacoderm.

Recent work on *Ephydatia muelleri* (Colgren & Nichols, 2021) also suggests that the endopinacoderm of the subdermal cavity has a major role in contractions. These authors first confirmed that the endopinacocytes, not the exopinacocytes, have actin tracts that shorten by nearly 20% during contractions of the subdermal space. Their experiments showed that contractions of the endopinacoderm happened with the release of calcium from internal stores, which prompted them to hypothesize that the endopinacocytes function in a manner similar to muscle cells in other animals. They found that endopinacocytes express striated muscle myosin heavy chain (stMHC) as well as a myocardin-related transcription factor (MRTF) homolog. They carried out RNAseq analysis on sponges treated with an MRTF inhibitor and determined that endopinacocytes are specified by a MRTF, implying they are the sponge equivalent of muscle (Colgren & Nichols, 2021). It is interesting to note that Bidder (1923) commented that the subdermal space functioned like 'muscular pumps.' These experiments illustrate how practical the lab grown species *E. muelleri* is for teasing apart mechanisms, but it would be very interesting to learn how well these findings apply across the diversity of sponges, i.e., is the subdermal space a muscular pump in most sponges, and how does it work?

Taken together the above results strongly suggest that the endopinacoderm plays a major role in effecting contractions across the sponge. The role of mesohyl cells (myocytes/actinocytes) as antagonists still needs investigation as indicated previously (Nickel et al., 2011), and this would especially be the case in sponges where incurrent canals expand while excurrent canals contract, and vice-versa as in the expansion-contraction response of the sneeze shown in *E. muelleri* (Elliott & Leys, 2007).

2.5.3 THE INVOLVEMENT OF CHOANOCYTE CHAMBERS IN CONTRACTIONS

Choanocytes are the functional unit of the pump and form cup-shaped 'chambers' that effectively create a single layered pump (Asadzadeh et al., 2019, 2020). All studies so far indicate that the flagella beat incessantly. Although there are thousands of choanocytes in each cubic millimeter of sponge tissue, all driving water through the sponge, the pressure generated is less than 0.7 mm H_2O (Parker, 1911, 1919; Bidder, 1923; Larsen & Riisgård, 1994; Asadzadeh et al., 2019, 2020). Because of the low pressure, sphincters on the canals and osculum can easily stop the flow, so there is no need to control flagella beat rate to stop the water moving through the sponge (Parker, 1919). But very little is known about the physiology of sponge choanocyte chambers and its structure is only beginning to be properly understood.

While there are typically several to many entrances to chambers, called 'prosopyles,' which can be formed either by pinacocytes or by the choanocytes themselves, there is only one exit, called the 'apopyle,' which is usually formed by L-shaped cells called 'cone cells.' One side of the 'L' contacts and makes a seal with a cellular or glycocalyx gasket that seals around and between the collars, while the other side, together with other cone cells, forms a flap that connects to endopinacocytes lining the excurrent canals (Weissenfels, 1981; Langenbruch & Weissenfels, 1987). Most interestingly, in many sponge, cone cells bear a single cilium which beats in a whip-like manner with forward stroke out of the chamber and recovery into the chamber. The beat of cilia on the apopyle of *E. muelleri* is slow, so much slower than the sinusoidal whip of flagella that it can be watched kick back and forth under the 40x objective of a microscope (Figure 2.6). Another feature of sponge chambers is lateral vanes projecting out either side of each flagellum and in some cases definitely making tight contact with the inside of the collar (Weissenfels, 1992; Leys et al., 2011; Mah et al., 2014; Lavrov et al., 2022). This structure is seen where fixatives have included dyes that attach to mucus such as ruthenium red.

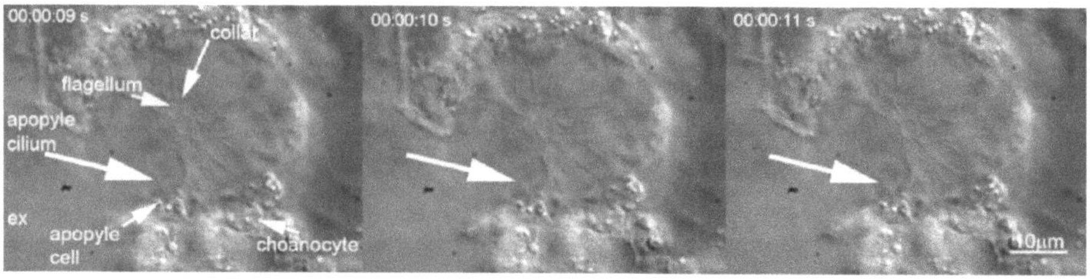

FIGURE 2.6 Beating of the apopyle cilium in *Ephydatia muelleri*. A sequence of images captured at 30 fps showing the beat of one cilium in and out of the focal plane during the same time that flagella in the collar move so fast that they describe a line across the circular opening of the collar. Ex, excurrent canal.

The overall picture of a 'typical' sponge choanocyte chamber therefore is of a two-part chamber, with a space below the gasket, and a space above the gasket through which water exits out of the apopyle. Differences in chambers include choanocyte shape (flat, spherical/cylindrical); variability in choanocyte number (fewer than 20 to more than 200); how chambers are held within the sponge tissues – whether they sit on a thin or thick mesohyl, very little, or no mesohyl; and how they are attached to the excurrent canals – for example, how many apopyle (cone) cells there are and whether there are additional cells such as central cells or sieve cells in the outer part of the chamber (Connes et al., 1971; Reiswig & Brown, 1977; Langenbruch & Jones, 1989; Vacelet et al., 1989; Sciscioli et al., 1997; Hammel & Nickel, 2014). The incurrent side of a chamber (where water enters the choanocyte chamber) can be supported in different ways: choanocytes can lie on a collagenous mesohyl which is backed by endopinacocytes; or choanocytes can lie with their basal side almost open to the incurrent water (Langenbruch & Jones, 1990; Langenbruch & Scalera-Liaci, 1986; Langenbruch & Weissenfels, 1987; Langenbruch et al., 1985; Langenbruch, 1983, 1988; Ludeman et al., 2017).

When the freshwater sponge *E. muelleri* contracts, the chambers are compressed, causing choanocyte flagella to protrude from the apopyle

(Elliott & Leys, 2007). Presumably the choanocytes receive a signal, and as the chamber contracts the flagella beat is stopped (Weissenfels, 1990). Other cells could be involved in stopping flow/contracting epithelia around the chamber and one that has been highlighted is a sieve cell on the exhalent side of the chamber in *T. wilhelma* (Hammel & Nickel, 2014). Central cells have also been proposed to control flow by projecting portions into the collar and wrapping around the flagellum (Connes et al., 1971; Reiswig & Brown, 1977).

2.5.4 PHYSIOLOGICAL BASIS OF CONTRACTIONS

Electrical conduction has only been demonstrated in glass sponges, class Hexactinellida, which have syncytial tissues (Leys & Mackie, 1997; Leys et al., 1999) (*see* Section 2.9). Because the contractile behaviors shown by sponges are slow, and because gap junctions are not known in cellular sponges, it is not expected that electrical signals propagate in sponges other than hexatinellids. Nevertheless, transcriptome and genome analysis of a growing number of cellular sponges (Demospongiae, Calcarea, and Homoscleromorpha) have revealed a range of molecules from families that are used in structures characteristic of neurons (e.g., synapses, ion channels, ionotropic or metabotropic receptor molecules), including recent suggestions that there could be potential neuroid cell types in the choanocyte chambers of the freshwater sponge *Spongilla lacustris* (Musser et al., 2021). To better understand how signals could propagate through sponge tissues, in the next section we review studies that have focused on the physiology of signaling in sponges.

Whereas earlier work largely focused on exposing sponges to a range of chemicals, Prosser et al. (1962) explored the reactions of different sponge species to both mechanical and electrical stimuli and tested the effect of sodium, potassium, and calcium on contractions. They found that responses to mechanical stimuli varied, with some closing oscula, others closing ostia, and some not responding at all. For example, the oscula of *Lissodendoryx hawaiiana* (published as *Damiriana hawaiiana*) contracted by 0.05–0.1 mm, while no response was seen in *Hymeniacidon chloris* (Prosser et al., 1962). No sponges responded to electrical stimuli to the ostia. Varying the concentrations of ions commonly used in nerve or muscle cell action potentials altered the type of response, but they found that addition of potassium to the external environment did not affect the contraction time of the response of the sponges *Tedania*, *Microciona* or *Spongilla*. But when calcium was removed, responsiveness was completely abolished (Prosser et al., 1962). They concluded that sodium/potassium mediated action potentials were unlikely to be the means of propagating contractions in sponges, but that calcium signaling was involved.

So far there is only one report that suggests electrical current passes between the cells of cellular sponges. Lowenstein (1967) recorded current between aggregating cells of *Clathria (Clathria) prolifera* (published as *Microciona prolifera*) and *Haliclona* sp. after the cells had been in contact for 15–30 minutes. Unfortunately, the setup was difficult and so no membrane potential is reported and it is not known whether this exchange of ions would occur *in vivo* (see discussion in Leys & Meech, 2006; Nickel, 2010). Other attempts to record from cellular sponges have been without success, because of the thinness of pinacocytes, and because mucus on their surface prevents the effective attachment of electrodes (Leys, unpublished data).

2.5.5 RESPONSES OF SPONGES TO COMMON SIGNALING MOLECULES

In lieu of electrical signaling, it is widely assumed sponges use chemical signals to coordinate behavior, and it has been shown by many authors that a wide range of substances cause sponges to contract (Parker, 1919; Emson, 1966; Ellwanger et al., 2004; Ellwanger & Nickel, 2006; Ellwanger et al., 2007; reviewed in Nickel, 2010). Whole body contractions are stimulated by substances that are commonly used as signaling molecules in plants and animals, such as nitric oxide (NO) and glycine, extracellular cyclic adenosine monophosphate (cAMP), and even substances known for being neuroactive in vertebrates as well as being signaling molecules used across eukaryotes, such as serotonin, gamma-amino butyric acid (GABA) and glutamate.

1. **Nitric Oxide (NO):** Ellwanger & Nickel (2006) applied NOC-12, which releases NO, to *T. wilhelma* and found the sponge immediately contracted, regardless of where it was in the rhythmic contraction cycle. NOC-12 increased the contraction cycle duration, decreased the amplitude (extent of contraction), and disturbed the endogenous contraction rhythm of *T. wilhelma*. Elliott & Leys (2010) studied NO signaling in the freshwater sponges *E. muelleri* and *S. lacustris* using NAPDH-diaphorase staining and a cGMP accumulation assay which indicates nitric oxide synthase (NOS) activity, the enzyme

that synthesizes NO. NOS activity was found in the choanocyte chambers, osculum, excurrent canals, and apical pinacoderm, but the most heavily stained region was the osculum, and here it was mainly the tip of the osculum that stained most, suggesting the greatest NOS activity there. The cGMP assay showed immunoreactivity in cells also in the osculum, which could indicate that these cells would be active targets for NO signaling in the sponge. When cells in the osculum were exposed to SNAP (a NO donor), they accumulated cGMP, whereas there was no detectable increase in cGMP in the choanosome or pinacoderm (Elliott & Leys, 2010). They also found that the osculum in *E. muelleri* contracted with the application of SNAP, whereas no response was triggered by SNAP in *T. wilhelma* (Ellwanger & Nickel, 2006). L-NAME, a NO synthase inhibitor, prevented contractions in *E. muelleri* (Colgren & Nichols, 2021).

While it was determined that sponge cells could respond to molecules similar to NO, other studies have looked for the presence of the enzyme NOS, which would be an indicator that NO is used as a signaling molecule. Giovine et al. (2001) found NOS activity in cells obtained from demosponges *Axinella polypoides* and *Petrosia ficiformis* using a citrulline assay. This assay involves exposing the dissociated sponge cells to radioactively labeled arginine, which is then converted to citrulline by NOS and the citrulline is then measured to determine the activity of NOS. They also used the NADPH-diaphorase staining assay to visualize the enzymatic activity of NOS (Giovine et al., 2001).

In all cases NO appears to modulate contractions, and the most active tissues appear to be those in the osculum. NO signaling has also been found to be involved in the settlement and metamorphosis of sponge larvae (Ueda et al., 2016) (*see* Section 2.8).

2. **cAMP:** cAMP was identified as a second messenger in a chemosensory transduction pathway in single cell eukaryotes (Yang et al., 1997), and has been shown to induce contractions in *T. wilhelma* (Ellwanger & Nickel, 2006) and to affect the locomotion of dissociated cells of a sponge (Cattaneo-Vietti et al., 1996; Gaino & Magnino, 1999). Little else is known of its effect on sponge contractility or other behaviors.

3. **GABA and Glutamate:** The neurotransmitters GABA and glutamate have different effects on different sponges. Glutamic acid was

found to be high in tissue samples of both *E. muelleri* and *S. lacustris* (Elliott & Leys, 2010). Glutamate, which is an excitatory neurotransmitter in bilaterian nervous systems, stimulates contractions and/or decreases excurrent flow in many sponges (Emson, 1966; Ellwanger et al., 2007; Elliott & Leys, 2010; Goldstein et al., 2020), but GABA, a well-known inhibitory neurotransmitter, has different effects in different sponges. In *E. muelleri* pre-incubation in GABA for 20 minutes prior to addition of glutamate prevented any contractions by the sponges (Elliott & Leys, 2010), while in *T. wilhelma* direct addition of GABA modulated contractions (Ellwanger & Nickel, 2006; Ellwanger et al., 2007) and it induced contractions in both the calcareous sponge *Clathrina* and the homoscleromorph sponge *Oscarella* (Nickel, 2010). It is possible the difference in responses of these species to GABA could be due to different experimental approaches, but given the long divergence time between these species it is also likely these different responses reflect different receptors, or different distribution of receptors along the tissues. The glutamate-induced contractions are typical of sponge behavior in natural conditions, and the cumulative data strongly suggest that glutamate is a major signaling molecule that controls whole-body contractions in sponges.

Only one paper has examined where glutamate receptors might be expressed in sponge tissues. Ramoino et al. (2007) found that the exo- and endopinacocytes as well as choanocytes and archaeocytes showed immunoreactivity to an antibody raised to human GABAB R2, while other cells in the mesohyl did not label. Western blots showed immunorecognition of a band of the expected size (105 kDa) for both GABAB R1 and GABAB R2, but given the difference in sequence between human and sponge GABA receptors, these results need confirmation by using the same antibodies across different species, and by using custom antibodies made to the sponge GABA receptor proteins. These results are nevertheless intriguing and it would be very informative to have more studies focusing on expression patterns of both glutamate and GABA receptors in a range of sponges.

4. **Dopamine and Serotonin:** Aminergic signaling, using dopamine or serotonin, is not found in any of the non-bilaterian phyla (Leys, 2015; Moroz, 2015). However, while serotonergic and dopaminergic receptors are absent from the genomes of sponges, curiously dopamine and 5HT (serotonin) have both been found in chemical extracts

from some sponges (Liu et al., 2004). Furthermore, in *T. wilhelma* application of serotonin caused the sponge to contract (Ellwanger & Nickel, 2006). Other studies have examined the localization of serotonin in tissues. In earlier work, Lentz (1966) showed immunohistochemical localization of serotonin to cells in the mesohyl of the calcareous sponge *Sycon ciliatum.* Later, immunoreactivity to serotonin was shown in cells around the periphery of larvae of the sponge *Tedania ignis* (Caribbean fire sponge), and in cells with stellate-like projections found in the mesohyl of the juvenile (Weyrer et al., 1999). Because of the understanding that sponges lack aminergic receptors (Leys et al., 2019), the immunoreactive studies were considered cautiously. Recently, however, anti-dopamine immunoreactivity was shown at a region around the base of cilia in the larva of the freshwater sponge *Eunapius fragilis* (Sokolova & Voronezhskaya, 2021). In *Eunapius*, the dopamine was suggested to control ciliary beat, as in sea urchin and other invertebrate larvae (Wada et al., 1997). The immunoreactive regions are immediately adjacent to the Golgi apparatus, shown by a Golgi marker 58K, which lies at the side of the nucleus just below the ciliary basal body. Sokolova & Voronezhskaya (2021) infer that as serotonin and dopamine are involved in regulating ciliary beat in a range of invertebrate larvae (Goldberg et al., 2011), dopamine may play a similar role in sponge larvae.

5. **Neuropeptides:** Placozoans, cnidarians, and ctenophores use a range of neuropeptides for different behaviors, and recently neuropeptides have also been found in sponges (Yañez-Guerra et al., 2022). Because in placozoans application of different peptides results in distinct behaviors (Varoqueaux et al., 2018), it would be very interesting to learn more about the role of sponge peptides in sponge behavior using similar approaches.

The focus on most of the above-mentioned substances as stimulators of 'smooth muscle' (e.g., Parker, 1910) or central nervous system (Ellwanger & Nickel, 2006) reflects the general thinking that contractions must be caused by some neuro or muscular mechanism common to other metazoans, which is not necessarily the case. One thing to keep in mind is that adding chemicals to sponges in a bath medium and observing their reactions is not necessarily informative about what the sponges typically respond to *in situ*. However, exactly how chemicals applied in a bath preparation enter through the tissues and trigger down-stream effects is still completely unknown and there remains a vast amount to be discovered regarding intercellular signaling, networks of signaling (combinations of the above signaling molecules perhaps), and in particular about the localization of the receptors for these varied signaling molecules. One hypothesis for how signaling occurs suggests that external stimuli might trigger release of paracrine signals into the mesohyl, which act on receptors on endopinacocytes causing contraction (Ellwanger & Nickel, 2006) (Figure 2.7). Endogenous contraction rhythms may be controlled similarly by paracrine signals released from mobile cells in the mesohyl.

2.5.6 SEEKING ELEMENTS OF A 'PROTOSYNAPSE' IN SPONGES

Synapses are highly specialized regions of juxtaposed membranes between two cells; they are easily identifiable in neuronal tissues. Morphologically they show a very regular nanometer-wide spacing between the membranes of the two cells, as well as regions of darker stained cytoplasm (osmiophilic) indicating a density of scaffolding proteins that support a cytoplasm that contains a number of very small vesicles, called dense core or light core, either dark or light staining, depending on the neurotransmitters they contain. Although no synapses have been identified in sponges so far, families of many of the molecules involved in the structure of synapses in other animals have been identified in sponge genomes and transcriptomes. This includes an almost complete set of postsynaptic protein orthologs found in *Amphimedon queenslandica* (Sakaraya et al., 2007), and similarly in other sponges (e.g., Riesgo et al., 2014; Musser et al., 2021). Most of these proteins belong to families of secretory, adhesion, and vesicle transport proteins that are common to eukaryotes. Phylogenetic analysis of sponge SNARE proteins, said to be 'synaptic' proteins involved in the docking of synaptic vesicles at the cell membrane, showed these belong to typical vesicle transport families involved in trafficking vesicles from the Golgi to the membrane, and are not members of the neuronal SNARE families (Kenny et al., 2020). By studying the expression profiles of the synaptic gene homologues, it was also found that these genes were organized into independent communities and displayed a lack of global coregulation suggesting that synaptic genes may have evolved from pre-existing cellular systems (Conaco et al., 2012). In short, a synapse has not been found in sponges, and genes said to be 'synaptic' are not from specifically neuronal families.

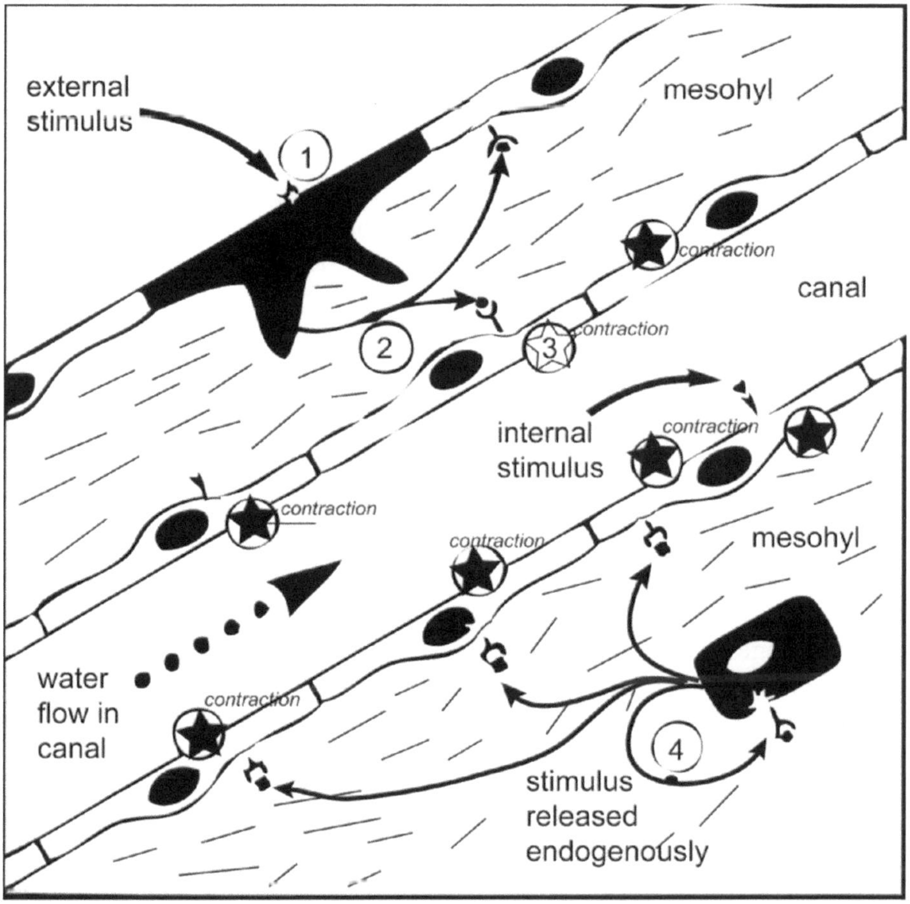

FIGURE 2.7 A diagram illustrating a hypothesis for how signals are received and propagated in a sponge (modified from Ellwanger & Nickel, 2006). An external stimulus: (1) is received by a receptor on the surface of the pinacoderm, which triggers the release of a molecule; (2) that diffuses through the mesohyl and is received by endopinacocytes; (3) which contract (star) by an intracellular signaling pathway. Endopinacocytes could release molecules and propagate this reaction. Stimuli could be received directly by endopinacocytes triggering similar contractions. Endogenous rhythms of contractions could be triggered by the release of substances from cells in the mesohyl; (4) which could be received by endopinacocytes and trigger contractions. Signals could be amplified if the molecules released contacted other cells in the mesohyl triggering further release of signaling molecules.

2.6 SENSORY ORGANS

According to Libbie Hyman (1940, p. 276) organs are groups of tissues that together, carry out a function. There is no doubt that under this criterion sponges possess several organs, and at least two that are sensory: the choanocyte chamber and the osculum.

Choanocyte chambers with the choanoderm, gasket, and apopyle cells, carry out the specialized function of pumping and capture of particulate food and creating the pressure drop over the chamber wall so so that water is drawn in, stalls as it passes through the filter and slides out of the apopyle. That chambers are functional units is supported by evidence that choanocyte chambers are formed as units in one location and are transported as a whole through the sponge mesohyl to be inserted into canals (Weissenfels, 1981; Kahn & Leys, 2016). The cone cells are ciliated and while it is very possible they are sensory, no work so far has examined that hypothesis. In *T. wilhelma*, at the exit of the choanocyte chamber, just beyond the ciliated cone cells, there is a perforated cell (generically called a sieve cell) termed a reticuloapopylocyte for its holes and location near the apopylar (cone) cells (Hammel & Nickel, 2014). In sponges generally, sieve cells are cells that either alone or together form holes around an opening. They are found at ostia and throughout the incurrent side of the sponge canal system where they form sieve plates in many species (Simpson, 1984, p. 266). In some sponges, oscula are also formed by a number of cells that reduce the diameter of the opening, in essence a sieve plate (e.g., *Geodia barretti*), and so the excurrent side of the canal system can also require regulation. The sieve cell in *T. wilhelma* was proposed as a flow regulating cell (Hammel & Nickel, 2014). The rationale is that contractions in *T. wilhelma* involve massive reduction in canal size, increased back pressure (essentially resistance through the hydraulic system), which would affect the delicate tissues of the chambers unless there was a means to protect them. The sieve cell at the chamber exit is proposed to regulate flow into/out of the chamber during these regular dramatic changes in pressure that happen when canals collapse during contractions (Hammel & Nickel, 2014).

A second sponge organ is the osculum. In many sponges, all the cells in the osculum have 4–6 μm long cilia, called primary cilia (Ludeman et al., 2014). Primary cilia are non-motile cellular structures that can act as chemosensors and mechanosensors (Praetorius & Spring, 2005). They lack a central microtubule pair, radial spokes linking the microtubules, and the dynein motors that normally allow movement (Satir et al., 2010). These sensory organelles have been identified in many cell types including chondrocytes (Poole et al., 1997; Muhammad et al., 2012), smooth muscle cells (Poole et al., 1997), olfactory cells (Satir, 2017), photoreceptors and secretory cells (Praetorius & Spring, 2005). The chemo-sensory function of the cilium is based on what receptors are present on the cilium itself. It has been shown that primary cilia can express receptors common to the central nervous system (Praetorius & Spring, 2005) and which have been linked to hedgehog signaling in mice (Satir, 2017). Due to the fact that sponge genomes contain a variety of signaling molecules and proteins that are related to signaling in nervous systems, it is anticipated that primary cilia receive signals and produce reaction cascades.

In *E. muelleri* responses to mechanical stimuli that typically trigger coordinated expansion and contraction events (called 'sneezes') are inhibited by pharmacological substances known to block TRP channels (Ludeman et al., 2014). Removal of the osculum also inhibited the sneeze response in these freshwater sponges, implying that the sensory cilia in the osculum might be involved in coordinating the response. It was also discovered that the transcriptomes of many sponges possess homologs of *pkd1* and *pkd2* (common to TRP channels). It is apparent that much more work is needed to characterize the function and mechanisms of action of primary cilia in sponges.

2.7 WHOLE BODY LOCOMOTION – 'CRAWLING' – BY SPONGES

A fascinating and very much overlooked and understudied behavior of sponges is translocation, the ability of a sponge to move its whole body over some distance. A recent study highlighted this little-known sponge behavior when a group of researchers studying the ecosystem of a seamount, 600–1,000 m deep, discovered spherical sponges about 20–30 cm in diameter at the end of long trails of spicules; it appeared as though the sponges were on the move (Morganti et al., 2021a). The observation was clever because it brought back to light a number of previous studies that have examined this seemingly esoteric behavior in sponges. Nickel (2006) reviewed much of this work when re-examining the active locomotion of *T. wilhelma.* Observations documenting sponges moving across aquaria and in the ocean are scattered across 150 years (Nickel, 2006) and several studies have made important discoveries about the mechanisms and functions of translocation.

For example, Bond & Harris (1988) put explants of a number of marine and one freshwater species attached to different substrates into tanks, and used video microscopy to track and measure the activity of the tissues. They found that a sponge can cover 4mm/day. Most move at 20–160 μm hr^{-1} and movement of the whole sponge occurs by the crawling of its cells over the substrate, much like cells in tissue culture (Bond & Harris, 1988). Sponge cells, like most cells in metazoans, are capable of amoeboid locomotion, and the concerted amoeboid locomotion of all the cells at the base of the sponge makes it move from one spot to another.

Whereas previous work had implied that some sponges in the family Tethyidae move by extending tendrils filled with spicules that touch the substrate and pull the sponge (Fishelson, 1981), when Bond & Harris (1988) watched sponges of the genus *Tethya* they did not see the tendrils pull pebbles towards the body, or contract to roll the sponge. Instead, cells moved out along a "sliding ladder" of spicules and touched the substrate. If the sponge moved in the direction of the tendril the tissue was resorbed, but tendrils were extended equally in all directions, and those in other directions were either retracted or broke off. They concluded that tendrils in *Tethya* attach and release but provide no pull. Nickel (2004), studying *T. wilhelma* also by time-lapse video, agreed and suggested that the tendrils might be used to sense the surrounding environment.

All the crawling sponges that Bond & Harris (1988) studied also had contractile waves that travelled over the whole body, like those described in Section 2.5. These convulsions took 40–60 minutes and took place every 2–19 hours depending on the species. In all sponges they studied, Bond & Harris (1988) found the waves of contraction were not correlated with sponge locomotion. Interestingly, in *E. fluviatilis* they found that during the waves of contraction, cells in the mesohyl stopped crawling as the wave passed (as was also noticed by Elliott & Leys (2007)), but the crawling activity of the leading-edge cells, those moving the sponge forward, did not stop. Both Bond & Harris (1988) and Nickel (2006) conclude that contractile waves are not correlated with locomotion, thus peristalsis is not the force driving the sponge forward.

Why move, especially when the sponge has not developed muscular tissues to do this with? Morganti et al. (2021a) call sponges 'opportunistic feeders,' meaning that they filter food from the water but also can take advantage of dissolved organic matter (DOM) and possibly also they could get food from sediment (fallout of marine snow) or from other matter (Morganti et al., 2021a). Morganti et al. (2021a) also studied images and contours of the seamount where they saw *Geodia* trails, and concluded that sponges moved uphill, therefore against gravity and they suggested two hypotheses: either the sponges were moving find a region with better food – although the substrate was full of spicules it was not dense with sediment; alternatively, as they saw many small individuals on the spicule trails, they suggested that by moving the sponges both generate asexual fragments and leave a nursery of substrate for those to grow up in. In other words, movement allows the sponges to create a nursery for young sponges in a region where food is limited.

The idea that movement is ecologically relevant for sponges had occurred earlier to Maldonado & Uriz (1999). They carried out an insightful set of experiments to test whether sponges could sense and move in response to unfavorable environments (Maldonado & Uriz, 1999). Explants of the demosponge *Scopalina lophyrapoda* were allowed to attach to a range of substrates and and then, once attached to their substrates, they were put in starved *vs.* fed, directional flow, silt *vs.* no silt, and among a low and a high density of clones and of different individuals. This gamut of tests showed that for *Scopalina,* starvation was a “powerful stimulus” to move farther than individuals that were well fed. Silt in the water made sponges move further than sponges that were in no silt, and sponges moved perpendicular to the direction of flow. Interestingly *Scopalina* preferentially moved towards non-conspecifics and if an encounter with another individual occurred, then the sponge moved in a different direction. Chance encounters with a conspecific usually resulted in fusion of the two. Taken together with the studies of Bond & Harris (1988), this work strongly indicates that sponges, like other so-called sessile invertebrates, move to the environment that best allows them to feed.

Do all sponges move? Essentially yes, if the skeleton is not fused to the seafloor. Bond (2013) found that the calcareous sponge *Leucosolenia botryoides* also moves at 25–130 μm hr^{-1}. It uses spicules projecting from the sides to anchor the body wall as it extends the tube (‘ascon’-type sponge) forward. Bell & Barnes (2002) found that individuals of *Hymeniacidon perleve* and *Suberites ficus* survived for months after becoming dislodged from their substrates and carried by currents. Glass sponges also move. Tubular forms roll and plate-like forms drift over the seafloor at rates driven by ocean currents (Kahn et al., 2020). Also, slow changes in size shown over seasons that were previously ascribed to growth could also be movement (Kahn et al., 2012). As noted by Bond & Harris (1988), a lot of the movements described as asexual budding and growth may in fact be local locomotion.

2.8 LARVAL BEHAVIOR

Sponges can reproduce asexually and sexually (Lanna et al., 2024, this volume). Sexual reproduction in some species gives rise to a new juvenile through direct development (e.g., *Tetilla japonica* Watanabe, 1978), but by far the most common known mechanism of reproduction is by indirect development via the formation of motile, swimming, larvae. Larvae are very often brooded, but some sponges release eggs and sperm into the water where eggs are fertilized and develop into a free-swimming larva (*see* Lanna et al., 2024, this volume). While sponge larvae are typically ciliated, motile, and small, from about 30 microns in length to as large as 6 mm (Rützler et al., 2003), a great diversity of larval types have been described (Maldonado & Bergquist, 2002; Maldonado, 2006). These forms include: amphiblastulae—ciliated at one pole and bare at the other (Calcaronea); coeloblastulae—hollow ciliated spheres (Calcinea); cinctoblastulae—ciliated hollow blastulae but with different cilia at the posterior pole, and with special refractile cells around the mid-section (Homoscleromorpha); hopliotomella—non-ciliated spheres with spicules that project out to retard sinking (some Demospongiae); trichimella—lipid-rich elongate larvae with multiciliated cells in a band at the lipid-rich apical pole (Hexactinellida); and parenchymella—elongate to spherical ciliated with uniform cilia or different cilia at the posterior pole (most Demospongiae). With the range of larvae comes a number of documented behaviors from responsiveness to light (towards or away from), contractions, and creeping. Much of this behavior has been reviewed elsewhere (Maldonado & Bergquist, 2002; Maldonado, 2006) so here we focus on taxis and settlement.

2.8.1 PHOTOTAXIS BY SPONGE LARVAE

The release of larvae can be triggered by light, in some species that is by the onset of darkness the night before release, and in others it is by the onset of daylight the day before release (Amano, 1986). It is unknown whether light triggers the parent sponge to move the larvae out (possibly by contracting) or whether light is detected by the nascent larvae and that triggers them to beat their cilia and/or contract themselves to get out of the parent. Larval release can also be triggered by calm water, or a slight increase in water temperature (Maldonado, 2006).

Generally, in laboratory experiments, larvae first swim away from the parent, then change trajectory and, after 12–24 hours, swim down to the substrate. Most larvae rotate due to the metachronal waves of their ciliary beat, but some also ‘cork-screw’ and others ‘skate’ along the substrate bouncing off its surface (Uriz et al., 2001; Holstein et al., 2010). Warburton (1966) saw larvae ‘bounce’ off the air-water interface and found that if he repeatedly inverted a corked tube filled with larvae, the larvae constantly swam upwards; he called this geotaxis. Maldonado (2006) suggests that because larvae have no statocysts or specific gravi-receptors, sinking or movement to the bottom of a tube is more likely to be a combination of phototaxis and photokinesis, due to the different activity of the cilia at different light intensities (Maldonado et al., 1997). Many sponge larvae have spicules at one pole and it is possible that they act as statocysts, but responsiveness to light is the behavior that has been most experimented with.

Some of the earliest observations on how sponge larvae respond to light were on *Sycon* sp. (Minchin, 1896). Minchin said that the whole shape of the larva "with pigment, lens-like body, and central granular cells, gives strongly the impression of a primitive light-perceiving organ" (Minchin, 1896, p. 47). Bidder suggested that for upward swimming light hits the "posterior naked patch of rounded cells," and that as the larva tilts, light reaches the base of the cilia and stimulates them to "more violent action" which rights the larva (Bidder, 1937, p. 139). He also described a way the larva could use the light cue to settle in the dark, saying that with time from release the larva loses its ciliated cells gradually from the middle progressively up to the anterior pole, which reduced and then stopped the tilting so that in the later stage larva, light would only impinge on the 'anterior' ciliated cells causing them to beat more strongly and push the larva back down. Minchin (1896) also suggested that a dark pigment patch found in the center of the *Sycon* larva could function like a retina at the base of the ciliated cells.

Elliott et al. (2004) confirmed that *Sycon coactum* indeed swims away from light and does this increasingly as larvae get older. Bidder did not refer to the earlier work of Duboscq & Tuzet (1933) who described four cross cells at the equator of the amphiblastula larvae; they speculated that these cells were photoreceptors. In *Sycon ciliatum* the cross cells express gene markers for eye development in other animals, such as *PaxB* (Fortunato et al., 2014), and there is some evidence that there may be a photoreceptor-like structure in those cells (Amano & Hori, 1992).

Bidder (1937) called the *Sycon* larva a 'sponge visual organ.' More recent work has called the posterior ring of cilia in the parenchymella larvae of demosponges, a 'sponge eye' (Rivera et al., 2012). In many parenchymella larvae the cilia at the posterior pole are longer than those covering the rest of the larva (Leys & Degnan, 2001; Maldonado et al., 2003; Collin et al., 2010). Those long cilia surround a dark ring that is formed by membrane protrusions of the ciliated cells that contain dark pigment granules. If the larva tilts, the protrusions shade the base of the cilia from light. Maldonado

(2003) studied larvae that were photopositive and others photonegative, and some, such as the late stage larva of *Cacospongia mollior*, which switched from positive to negative after 4–6 hours of swimming. Other larvae, like those of *Haliclona (Soestella) caerulea* (published as *Sigmadocia caerulea*) and *Ircinia oros* swam away from light for their whole larval life of up to 5 days (Maldonado et al., 2003).

In *A. queenslandica* larvae, only negative phototaxis was documented, but there could have been an earlier stage when larvae were photopositive that was not recorded (Leys & Degnan, 2001). In all of these cases for photonegative responses, turning of the larva away from light can be explained by the differential bending or straightening of the long cilia that form a tufted ring at the posterior pole. As the larva corkscrews through the water, the cilia around the pigment ring experience light sequentially. For these larvae, when the light hits the cilia they straighten and or keep beating while the cilia on the side opposite are shaded from light by the pigment and bend over the posterior pole and stop beating. This means that the beating cilia, like oars pulling only on one side of a boat, turn the larva. In the case of negatively phototactic tufted parenchymella larvae, all beating cilia drive the larva in the direction away from the light (Leys & Degnan, 2001; Maldonado et al., 2003). Other larvae turn the opposite way, towards the light. For instance, the parenchymellae of *Neopetrosia proxima* when bathed in light, swim towards it: in these larvae the cilia that experience light (as the larva rotates) bend over the pole, and those shaded from light (as the larva rotates) beat actively; in this way, the beat of cilia as they sequentially rotate out of light, drives the larva towards light. *Neopetrosia proxima* lives at the edge of mangrove roots in the sunlight, and so this mechanism is thought to ensure that larvae, whose adult parents have cyanobacterial symbionts that rely on sunshine, do not end up deep in the shade of the mangroves (Collin et al., 2010).

Most photoreceptors in metazoans typically use some form of opsin. Opsins are found in cnidarians, ctenophores, and even in placozoans. Nonetheless, in placozoans it is not clear which cells might be photoreceptive or what the response to light is. However, opsins have not been found in any sponge genome or transcriptome. One way to determine what the photopigment might be is to test spectral sensitivity. So far only one experiment has been done on sponge larvae to test spectral sensitivity to different wavelengths of light. Leys et al. (2002b) used short band pass filters combined with a graded neutral density filter to test responsiveness of the posterior pole cilia to different intensities and wavelengths of light in *A. queenslandica* (published as *Reniera* sp.). They took advantage of the fact that the larva can be made to stick to a new coverslip by its anterior pole by squirting the larva down onto the coverslip. In this way the larvae were immobilized naturally, and the posterior pole cilia illuminated. They found that the cilia opened/straightened most in response to blue light at 440 nm and also somewhat to orange-red at 600 nm (Leys et al., 2002b) and concluded that most likely the photoreceptor in *Amphimedon* was a flavin-like molecule. A typical spectral sensitivity curve for opsins did not fit these data, however it was later suggested by K. Foster that by plotting light energy rather than wavelength the response at 600 nm could reflect an opsin-like molecule (Figure 7 in Leys & Meech, 2006).

When the genome for *A. queenslandica* became available it was confirmed that no opsins were present, but instead two photolyases were found: Cryptochrome 1 and 2 (CRY1 and CRY2). Photolyases are light-mediated DNA repair molecules, but Rivera et al. (2012) found no evidence for photolyase activity in either CRY. *In situ* hybridization showed that CRY2 was expressed in a ring around the posterior pole of the larva. By expressing the sponge CRY2 in insect cells they tested spectral sensitivity and found that CRY2 had a flavin-cofactor that responded to 450 nm with a minor response at 600 nm. In all, the evidence supports the conclusion that *A. queenslandica* most likely uses a cryptochrome to sense and respond to light typical of the wavelengths experienced in its natural habitat (Leys & Degnan, 2001; Leys et al., 2002a; Rivera et al., 2012). Other sponge larvae might experience other wavelengths however, and more experiments testing spectral sensitivity of different sponge larvae are very much needed.

Recent work has explored how *A. queenslandica* larvae sense light and settle. Jindrich et al. (2017) found that *A. queenslandica* larvae express a number of genes involved in setting circadian rhythms in other animals. In addition to the two CRY genes, there are two *Timeout* genes, two *PAR-bZIP* and seven *bHLH-PAS* transcription factor genes, but no clock genes. Only the CRYs and *PAR-bZIP* transcription factors cycle with day-night and the oscillating expression pattern is lost in either constant dark or light. These genes are also expressed at the posterior pole of the larva suggesting that the pole 'eye' is a light organ in the tufted parenchymella larva (Jindrich et al., 2017).

2.8.2 CHEMICAL SENSITIVITY AND SETTLEMENT

When sponge larvae settle they have an 'inspection phase' in which they stop and start, repeatedly testing the substrate, an activity seen in other invertebrate larvae (Maldonado, 2006). Larvae of *Halichondria* are more torpedo-shaped and 'skate,' frequently touching the surface (Holstein et al., 2010). Tufted parenchymella larvae can spin for minutes to hours during 'inspection stops,' and if frequently disturbed, they may not settle. However, both texture (crevices) and chemicals can enhance settlement and metamorphosis (Maldonado, 2006). Both CsCl and KCl increased chances of settlement in *Aplysilla* sp. together with a biofilmed substrate, but only CsCl caused settlement and metamorphosis on its own, although in the experiments it was used at relatively high concentrations (Woollacott & Hadfield, 1996). In their experiments 30 mM KCl with no biofilm caused 0.9% settlement compared to ~44% settlement with a biofilm. For CsCl, 30 mM without a biofilm caused ~20% settlement, but with a biofilm 30 mM CsCl caused almost 70% of larvae to settle. In other invertebrates these ions have varying effects on settlement and metamorphosis, in many triggering settlement, but in others Cs^+ and K^+ in fact inhibit metamorphosis (Woollacott & Hadfield, 1996). The suggested action of K^+ and Cs^+ is by depolarizing sensitive cells, reducing the threshold for response, or by modulating Na^+/K^+ ATPase and/or kinase C activity, and indirectly affecting depolarization. These cations may also directly affect the tissues via other pathways. The latter seems a good possibility for sponge larvae in that obvious sensory receptors are absent. However, over the last two decades ultrastructure of the anterior poles of the tufted parenchymella larva has highlighted potential receptor cells.

While the parenchymella larva appears uniformly ciliated, the makeup of the ciliated epithelial layer varies by region. In the anterior hemisphere, in addition to the columnar ciliated cells, there are three other well-described cell types: (i) cuboidal shaped cells at the very pole that have recessed cilia; (ii) flask-shaped cells with recessed cilia found around the anterior third of the larva; and (iii) globular cells that lack cilia and extrude portions of themselves up in between the cilia of neighboring cells. The polarized distribution of these cells corresponds to the region the larva uses to 'inspect' when rotating in one spot before settling.

Experiments on *A. queenslandica* larvae suggest the flask cells are involved in signaling to induce settlement (Nakanishi et al., 2015). *A. queenslandica* larvae settle in the presence of the encrusting alga *Amphiroa fragilissima*. At settlement, flask cells show an increase in internal calcium, but when treated with BAPTA-AM which chelates calcium in cells, larvae did not settle as well in the presence of the algal substrate as control larvae did (Nakanishi et al., 2015). In many marine invertebrate larvae NO is an important signaling molecule that enhances settlement and metamorphosis, and as discussed in Section 2.5, NO signaling has been found to be involved in modulating whole body contractions in some sponges (Ellwanger & Nickel, 2006). Ueda et al. (2016) found that addition of the NO synthase inhibitor L-NAME reduced settlement and metamorphosis of larvae, as did PTIO, a NO scavenger. They found that the NO donor SNP induced significantly higher levels of metamorphosis over control treatments; another NO donor, SNAP, did not have the same effect, with no more larvae metamorphosing compared to controls. They tested whether NO worked through MAPK/ERK signaling by using the inhibitor UO126 and found metamorphosis was inhibited. Furthermore DAF-FM, which detects endogenous NOS, clearly labelled globular cells. Their model is that the calcareous alga, common to the boulders on the reef flat where *A. queenslandica* lives, produces a signal that activates calcium signaling by the flask cells. The flask cells signal via Notch Delta or some other diffusible molecule which the globular cells receive, and which induces them to release NO. NO signaling in other animals triggers a huge range of activities, but among them are cell motility, cell adhesion, and cell differentiation.

A very interesting finding is that typically the substrate for NO signaling is not synthesized by *A. queenslandica*; for NOS to be expressed in the larva it needs arginine (Song et al., 2021). Song et al. (2021) tested whether the larva could take up arginine from seawater and found that the amount taken up was insufficient for the amount of NOS expressed by the larva. Arginine is produced from citrulline, and by adding ^{13}C-citrulline and ^{12}C-citrulline to filtered seawater and to symbiont-enriched media they determined that the bacterial symbionts take up the citrulline and convert it to arginine (Song et al., 2021). The resulting hypothesis is that NOS signaling is possible in the larva by obtaining arginine from its symbionts.

These data together show that metamorphosis is triggered by a cascade of small molecule signaling, using pathways known from other invertebrates, but here they act in the absence of neurons. One interesting additional finding is that the larval responses to light and chemical cues may be linked. Say & Degnan (2020) found that in constant light the larvae do not settle in response to the chemical cues. The larvae descend to the bottom at twilight and then begin the inspection phase allowing contact and interaction of the cells at the anterior pole with the substrate, setting into motion the chemical signaling that ends with metamorphosis of the larva (Say & Degnan, 2020).

2.9 GLASS SPONGES AND ELECTRICAL SIGNALING

Glass sponges (Hexactinellida) are unusual among sponges, being the only group that can propagate electrical signals, which cause the immediate arrest of their flagellated pumps (Leys & Mackie, 1997; Leys et al., 1999; reviewed in Leys et al., 2007). Glass sponges can do this because their tissues are syncytial, which allows the electrical signals to traverse great distances unimpeded by membrane boundaries which exist in cellular sponges (Mackie & Singla, 1983; Leys, 1995). Gary Silver, a graduate student at the University of Victoria (Canada), first noticed the arrest behavior when divers approached 'boot' sponges he had outfitted with flow meters underwater (Mackie, 1979). It was fortuitous that he was studying *Rhabdocalyptus dawsoni* because that sponge is highly irritable, and relatively easy to detect the arrest behavior in. Other glass sponges, for example *Aphrocallistes vastus*, also arrest pumping when disturbed, but their tolerance to irritants is much higher, resulting in fewer arrests whether studied *in situ* or in tanks (e.g., Tompkins-MacDonald & Leys, 2008; Grant et al., 2018, 2019).

Arrests in *R. dawsoni* are all-or-none, meaning the action potential propagates across the whole sponge acting on all effectors, which are the flagella (Mackie, 1979). The first work used mechanical and electrical stimuli and recorded arrests in pumping showing a conduction velocity of 0.22 cm s^{-1} (Lawn et al., 1981). Nearly 20 years later, development of an aggregate graft preparation made it finally possible to record electrical propagation through the sponge using extracellular electrodes (Leys & Mackie, 1997). This work found that the action potential (AP) was 5 s long, and had a conduction velocity of 0.26 cm s^{-1} with an absolute refractory period of 29 s, the time within which another action potential could not be initiated, and a relative refractory period of 150 s, within which the conduction velocity was slowed relative to normal. Further experiments showed that the sponge's arrests had no rhythmicity, and were not sensitive to light but were sensitive to both temperature and particulates in the water (Leys et al., 1999). In that work it was found that neither pumping nor arrests occurred at temperatures above 12°C (Leys et al., 1999) but in a later study it was shown that the sponge stopped pumping below 7°C and it was unable to arrest pumping when temperatures were higher than 12°C (Leys & Meech, 2006).

Recording electrical signals from glass sponges is far from trivial. So far, the Bamfield Marine Sciences Center (Canada) is the only location where glass sponges have been maintained in flowing seawater aquaria in good condition for these types of experiments. Graft preparations developed by Leys & Mackie (1997) allowed extracellular recordings, but to date no successful intracellular recordings have been managed, and patch clamp electrophysiology has been attempted with mixed results. The syncytial tissues are particularly difficult to penetrate, being sticky and clogging electrodes; they are also pliable, so that electrodes that do penetrate do not remain in for more than a couple of seconds (Leys, personal observations).

Using the graft preparation with extracellular suction electrodes, a range of ion exchange experiments were carried out by exchanging the entire bath water in a flow-through system with chilled media containing agonists or antagonists. The results were summarized in Leys et al. (1999) and are reviewed elsewhere (Leys & Meech, 2006; Leys et al., 2007). In short, divalent cations (Co^{2+}, Mn^{2+}) blocked the action potential reversibly, and the calcium channel antagonist nimodipine decreased the conduction velocity suggesting a role for calcium influx. In contrast, sodium channel blockers and reduction of sodium by 75% showed little effect, indicating sodium influx is unlikely to be involved. The potassium channel blocker TEA+ also blocked the AP. While the data also support the possibility of a H+ influx, the conclusions drawn were that the glass sponge action potential is most likely driven by an influx of calcium and reset by potassium. How this affects the arrest of flagella beating is not exactly known, but if calcium is involved as the depolarizing stimulus it is expected that calcium entry to the collar bodies causes arrests of flagella beating. The relative refractory period means that successive stimuli have delayed responses. That means that sites near the stimulus will stay arrested for longer, while sites further away will start pumping earlier before another action potential reaches them. Functionally this means that a small stimulus might arrest the whole sponge but regions far away from the stimulus will begin pumping earlier than those closer, and so the arrest of flow through the whole sponge will appear shorter. That seems to be what happens in *A. vastus*.

Studies carried out *in situ* at 200 m deep sponge reefs in British Columbia, Canada, showed that *A. vastus* arrests its pumping frequently, but starts pumping much sooner than *R. dawsoni* (Grant et al., 2018, 2019). *In situ* studies on glass sponges show that they arrest pumping with increasing sediment concentration, which normally happens at a change of tides, for example where currents accelerate over a ridge. They also arrest when sediments around them are disturbed experimentally, and the inference is that they will arrest pumping whenever sediments are resuspended by human activities of trawl fishing or excavation (Grant et al., 2019). Frequent arrests may affect the overall time spent filter feeding by these sponges, so a balance between sensitivity and recovery from arrests must be struck, especially given low food availability in the deep sea. Although recordings have only been made from a few glass sponge species *in situ*, it is anticipated that all glass sponges are able to arrest their pumps because all are syncytial and syncytial tissues enable electrically propagated signals that can trigger the arrests. However, while sponges like *Aphrocallistes vastus* that live in relatively shallow waters of the continental shelf and near shore areas may be adapted to periodic increases in sediments, turning the pump off and restarting quickly, other species that inhabit deep regions that rarely encounter sediment disturbances may not have the same rapid return to pumping. There are many open questions about the pumping, responsiveness, and mechanism of electrical signaling in glass sponges.

2.10 WHERE TO NEXT? PATHWAYS OF FUTURE STUDY

This review shows that there are significant gaps in our understanding of how sponges coordinate behavior, giving plenty of room for new research. Here we highlight some of the more promising paths. As a background for any of the directions, it will be hugely informative to have high-quality genomes from more sponge species. So much has been learned from the first sponge genome sequenced, that of *A. queenslandica* (Srivastava et al., 2010). More genomes will allow experimental approaches by different research groups on a range of species. More sponge genomes will encourage a comparative perspective, and with more genomes it is likely that some practical 'model' species will be developed that can be studied by many groups worldwide. A step in this direction is being made by the Aquatic Symbioses project to sequence 50 new sponge genomes to the same high quality (Sanger, 2021).

How signals are transmitted along and across sponge tissues to coordinate contractile behavior is still unknown. It is expected that paracrine signaling plays a role and various signaling molecules have been implicated, such as ATP, cAMP, glutamate, and NO; of these glutamate is the only molecule that can consistently trigger the 'sneeze' behavior. Where receptors are on sponge tissues, and how these molecules might interact in the signaling pathways to generate coordinated behavior is an area ripe for study. One approach would be to use various pharmacological blockers and analogs. Custom-made antibodies are becoming more affordable, and so together with the increasing availability of transcriptomic data, it would be very useful to be able to localize where receptors are on sponge tissues and to determine which cells are involved in signaling, and which cells are involved in contractions.

Understanding other types of signaling pathways would also be informative. For example, the recent finding that peptides are present in sponges (Yañez-Guerra et al., 2022), and the suggestion that monoamines might be used to control ciliary activity in sponge larvae as happens in other metazoan larvae (Sokolova & Voronezhskaya, 2021) are discoveries that urgently need to be followed up with behavioral studies. Could some of these molecules be used in signaling at a cellular level in sponges to control flagella beat in the chambers?

A fourth very interesting avenue of research would be to better understand how primary cilia in the osculum of sponges provide feedback leading to changes in the contraction state of the sponge's canals. A better understanding of the location of primary cilia in different sponge species would be useful, along with identifying ion channels involved (e.g., are TRP channels generally involved?) in sensing flow through the sponge. An approach that could be effective would be to begin by repeating some of Parker's field and tank experiments on *H. heliophila* (Parker, 1910), and then to take that work to the molecular level with RNAseq studies. The possible involvement of vasodilators, oxygen sensors (e.g., Riesgo et al., 2022) and the role of NO in sending signals from the osculum to the canals should be investigated.

Another line of inquiry should be the role of pH and temperature as signaling mechanisms, again potentially to control flagella beat or beat rate in the sponge (Giovine et al., 2001; Zocchi et al., 2001). Larval photoreceptors need broad study across a diversity of sponge species in different classes and ecological habitats. Open questions include understanding what the photoreceptor molecules are, where they are located in the larval body, and how broadly they are found across sponge groups. Spectral sensitivity experiments coupled with transcriptomic data are needed on a range of sponge larvae to better understand the diversity of photoreceptors that exist in the phylum.

Finally, the glass sponge action potential needs further investigation. Much remains to be learned about the mechanism of excitation of the glass sponge syncytium, and how the action potential reaches collar-flagella units. Is calcium involved in arresting the flagella beat as proposed, and do arrests spread through the whole sponge as proposed or are there partial arrests? Do all glass sponges propagate electrical signals? All glass sponges studied so far are syncytial, and therefore have unimpeded pathways for electrical signaling, but whether they can and do propagate action potentials through the syncytium needs confirmation by bringing other species into the laboratory for electrophysiological studies. A first step would be to carry out an exhaustive in silico study on the diversity of ion channels in a range of glass sponge species, followed by in vitro work using patch clamp electrophysiology.

Sponges, Porifera, are an understudied phylum, and these are just a few of the many aspects that require attention. Now is an excellent time for the student who is looking to make exciting new discoveries in the field of sponge behavior and the mechanisms of coordination of sponge behavior.

ACKNOWLEDGMENTS

We thank the members of the Leys and Kahn labs for discussion of concepts presented here. We acknowledge the scientists, engineers, and staff at the Monterey Bay Aquarium Research Institute for enabling and sharing observations of abyssal sponge behaviors. Funding to support this work came from a Natural Sciences and Engineering Research Council, Canada, Discovery Grant to SPL.

KEYWORDS

- **electrical signaling**
- **pH**
- **Porifera**
- **signaling molecules**
- **sponges**
- **subdermal space**

REFERENCES

Adams, E. D. M., (2010). *Physiology and Morphology of Epithelia in the Freshwater Demosponge, Spongilla Lacustris*. MSc Thesis. University of Alberta.

Adams, E. D. M., Goss, G. G., & Leys, P. S., (2010). Freshwater sponges have functional, sealing epithelia with high transepithelial resistance and negative transepithelial potential. *PLoS One, 5*, e15040.

Alie, A., & Manuel, M., (2010). The backbone of the post-synaptic density originated in a unicellular ancestor of choanoflagellates and metazoans. *BMC Evol. Biol., 10*, 34.
Amano, S., & Hori, I., (1992). Metamorphosis of calcareous sponges I. Ultrastructure of free-swimming larvae. *Invertebr. Reprod. Dev., 21*, 81–90.
Amano, S., (1986). Larval release in response to a light signal by the intertidal sponge *Halichondria panicea. Biol. Bull., 171*, 371–378.
Arendt, D., (2020). The evolutionary assembly of neuronal machinery. *Curr. Biol., 30*, R603–R616.
Arendt, D., (2021). Elementary nervous systems. *Phil. Trans. Roy. Soc. Lond. B., 376*, 20200347.
Asadzadeh, S. S., Kiørboe, T., Larsen, P. S., Leys, S. P., Yahel, G., & Walther, H. J., (2020). Hydrodynamics of sponge pumps and evolution of the sponge body plan. *ELife, 9*, e61012.
Asadzadeh, S. S., Larsen, P. S., Riisgård, H. U., & Walther, H. J., (2019). Hydrodynamics of the leucon sponge pump. *J. R. Soc. Interface, 16*, 20180630.
Bagby, R. M., (1966). The fine structure of myocytes in the sponges *Microciona prolifera* (Ellis and Solander) and *Tedania ignis* (Duchassaing and Michelotti). *J. Morphol., 118*, 167–182.
Barsanti, L., Evangelista, V., Passarelli, V., Frassanito, A. M., & Gualtieri, P., (2012). Fundamental questions and concepts about photoreception and the case of *Euglena gracilis. Integr. Biol., 4*, 22–36.
Barsanti, L., Passarelli, V., Walne, P. L., & Gualtieri, P., (2000). The photoreceptor protein of *Euglena gracilis. FEBS Letters, 482*, 247–251.
Bell, J. J. & D. K. A. Barnes (2002). Density, distribution and decline of two species of unattached demosponge. *Sarsia 87,* 110–118.
Bidder, G. P., (1923). The relation of the form of a sponge to its currents. *Quart. J. Microsc. Sci., 67*, 293–323.
Bidder, G. P., (1937). The perfection of sponges. *Proc. Linn. Soc. Lond., 149*, 119–146.
Bond, C., & Harris, K. A., (1988). Locomotion of sponges and its physical mechanism. *J. Exp. Zool., 246*, 271–284.
Bond, C., (1992). Continuous cell movements rearrange anatomical structures in intact sponges. *J. Exp. Zool., 263*, 284–302.
Bond, C., (2013). Locomotion and contraction in an asconoid calcareous sponge. *Invertebr. Biol., 132*, 283–290.
Borojevic, R., (1970). Différenciation cellulaire dans l'embyogenèse et la morphogenèse chez les spongiaires. *Zool. Soc. Lond., 25*, 467–490.
Borowiec, M. L., Lee, E. K., Chiu, C., & Plachetzki, D. C., (2015). Extracting phylogenetic signal and accounting for bias in whole-genome data sets supports the Ctenophora as sister to remaining Metazoa. *BMC Genom., 16*, 987.
Boury-Esnault, N., & Rützler, K., (1997). Thesaurus of sponge morphology. *Smithson. Contrib. Zool., 596*, 1–55.
Boury-Esnault, N. (1973). L'exopinacoderme des Spongiaires. *Bulletin du muséum nationale d'histoire naturelle 117*, 1193–1206.
Boury-Esnault, N., Ereskovsky, A., Bézac, C., & Tokina, D., (2003). Larval development in the Homoscleromorpha (Porifera, Demospongiae). *Invertebr. Biol., 122*, 187–202.
Brien, P., (1967). Les éponges. Leur nature metazoaire, leur gastrulation, leur état colonial. *Ann. Soc. R. Zool. Belg., 97*, 197–235.

Brunet, T., & Arendt, D., (2016). From damage response to action potentials: Early evolution of neural and contractile modules in stem eukaryotes. *Philos. Trans. R. Soc. Lond. B, 371*, 20150043.

Brunet, T., & King, N., (2017). The origin of animal multicellularity and cell differentiation. *Dev. Cell, 43*, 124–140.

Brunet, T., Larson, B. T., Linden, T. A., Vermeij, M. J. A., McDonald, K., & King, N., (2019). Light-regulated collective contractility in a multicellular choanoflagellate. *Science, 366*, 326–334.

Brusca, R., Moore, W., & Shuster, S., (2016). *Invertebrates*. Sinauer Associates, OUP.

Burkhardt, P., (2015). The origin and evolution of synaptic proteins – choanoflagellates lead the way. *J. Exp. Biol., 218*, 506–514.

Burkhardt, P., Grønborg, M., McDonald, K., Sulur, T., Wang, Q., & King, N., (2014). Evolutionary insights into premetazoan functions of the neuronal protein homer. *Mol. Biol. Evol., 31*, 2342–2355.

Busch, K., Marulanda-Gómez, A., Morganti, T., Bayer, K., & Pita, L., (2024). Sponge symbiosis: Microbes make an essential part of what it means to be a sponge. In: *Frontiers in Invertebrate Physiology*. Saleuddin, S., Leys, S.P., Roer, R.D., and Wilkie, I. C. Apple Academic Press. Vol. 1 Non-Bilaterian Phyla.

Byrum, C. A., & Martindale, M. Q. (2004). Gastrulation in the Cnidaria and Ctenophora. In: Stern, C., (ed.), *Gastrulation: From Cells to Embryos* (pp. 33–50). Cold Spring Harbor Press.

Cattaneo-Vietti, R., Bavestrello, G., Cerrano, C., et al., (1996). Optical fibers in an Antarctic sponge. *Nature, 383*, 397–398.

Cavalier-Smith, T., (2017). Origin of animal multicellularity: Precursors, causes, consequence – the choanoflagellate/sponge transition, neurogenesis and the Cambrian explosion. *Philos. Trans. R. Soc. Lond. B, 372*, 20150476.

Claessen, D., Rozen, D. E., Kuipers, O. P., Sogaard-Andersen, L., & van Wezel, G. P. (2014). Bacterial solutions to multicellularity: A tale of biofilms, filaments and fruiting bodies. *Nat. Rev. Micro., 12*, 115–124.

Colgren, J. and S. A. Nichols (2022). MRTF specifies a muscle-like contractile module in Porifera. *Nature Communications 13,* 4134. BioRxiv: 2021.2004.2011.439235.

Collin, R., Mobley, A. S., Busutil, L. L., Leys, S. P., Cristina, D. M., & Thacker, W. R., (2010). Phototactic responses of larvae from the marine sponges *Neopetrosia proxima* and *Xestospongia bocatorensis* (Haplosclerida: Petrosiidae). *Invertebr. Biol., 129*, 121–128.

Conaco, C., Bassett, D. S., Zhou, H., et al., (2012). Functionalization of a protosynaptic gene expression network. *Proc. Natl. Acad. Sci. U S A, 109*, 10612–10618.

Connes, R., Diaz, J. P., & Paris, J., (1971). Choanocytes et cellule centrale chez la démosponge *Suberites massa* Nardo. *C.R. Acad. Sc. Paris, 273*, 1590–1593.

Dahihande, A. S., & Thakur, L. N., (2019). Temperature- and size-associated differences in the skeletal structures and osculum cross-sectional area influence the pumping rate of contractile sponge *Cinachyrella* cf. *cavernosa*. *Mar. Ecol., 40*, e12565.

De Voogd, N. J., Alvarez, B., Boury-Esnault, N., et al., (2022). *World Porifera Database*. Accessed at https://www.marinespecies.org/porifera doi: 10.14284/359. (accessed on 03 May 2023).

De Vos, L., & G. Van de Vyver, (1981). Etude de la contraction spontanée chez l'éponge d'eau douce *Ephydatia fluviatilis* cultivée en vitro. *Ann. Soc. R. Zool. Belg., 111*, 21–31.

De Vos, L., (1977). Morphogenesis of the collagenous shell of the gemmules of a freshwater sponge *Ephydatia fluviatilis*. *Arch. Biol., 88*, 479–494.

Dexter, J. P., Prabakaran, S., & Gunawardena, J., (2019). A complex hierarchy of avoidance behaviors in a single-cell eukaryote. *Curr Biol., 29*, 4323–4329.e4322.

Duboscq, O., & Tuzet, O., (1933). Quelques structure des amphiblastules d'éponges calcaires. *C.R. Acad. Sc. Paris, 197*, 561–563.

Dunn, C. W., Hejnol, A., Matus, D. Q., et al., (2008). Broad phylogenomic sampling improves resolution of the animal tree of life. *Nature, 452*, 745–749.

Dunn, C. W., Leys, S. P., & Haddock, S. H. D., (2015). The hidden biology of sponges and ctenophores. *Trends Ecol. Evol., 30*, 282–291.

Elgeti, J., Winkler, R. G., & Gompper, G., (2015). Physics of microswimmers—Single particle motion and collective behavior: A review. *Rep. Prog. Phys., 78*, 056601.

Elliott, G. R. D., & Leys, S. P., (2007). Coordinated contractions effectively expel water from the aquiferous system of a fresh water sponge. *J. Exp. Biol., 210*, 3736–3748.

Elliott, G. R. D., & Leys, S. P., (2010). Evidence for glutamate, GABA and NO in coordinating behavior in the sponge, *Ephydatia muelleri* (Demospongiae, Spongillidae). *J. Exp. Biol., 213*, 2310–2321.

Elliott, G. R. D., (2009). *The Cytology and Physiology of Coordinated Behavior in a Freshwater Sponge, Ephydatia muelleri*. PhD Thesis. University of Alberta.

Elliott, G. R. D., MacDonald, T. A., & Leys, S. P., (2004). Sponge larval phototaxis: A comparative study. *Mus. Ist. Biol. Univ. Genova, 68*, 291–300.

Ellwanger, K., & Nickel, M., (2006). Neuroactive substances specifically modulate rhythmic body contractions in the nerveless metazoon *Tethya wilhelma* (Demospongiae, Porifera). *Front. Zool., 3*. doi: 10.1186/1742-9994-1183-1187.

Ellwanger, K., et al. (2004). Glutamate, GABA and serotonin induce contractions in the sponge *Tethya wilhelma* (Porifera: Demospongiae). *Jahrestagung der Deutschen Zoologischen Gesellschaft*. Rostock, Zoologisches Institut der Universität Rostock. Abstractband, 157.

Ellwanger, K., Eich, A., & Nickel, M., (2007). GABA and glutamate specifically induce contractions in the sponge *Tethya wilhelma. J. Comp. Physiol. A, 193*, 1–11.

Emson, R. H., (1966). The reactions of the sponge *Cliona celata* to applied stimuli. *Comp. Biochem. Physiol., 18*, 805–827.

Ereskovsky, A. V., & Dondua, K. A., (2006). The problem of germ layers in sponges (*Porifera*) and some issues concerning early metazoan evolution. *Zool. Anz., 245*, 65–76.

Fahey, B., & Degnan, B. M., (2010). Origin of animal epithelia: Insights from the sponge genome. *Evo & Devo, 12*, 601–617.

Feuda, R., Dohrmann, M., Pett, W., et al., (2017). Improved modeling of compositional heterogeneity supports sponges as sister to all other animals. *Curr. Biol., 27*, 3864–3870. e3864.

Fishelson, L., (1981). Observations on the moving colonies of the genus *Tethya* (Demospongiae, Porifera). *Zoomorphology, 98*, 89–99.

Fortunato, S. A. V., Leininger, S., & Adamska, M., (2014). Evolution of the pax-six-eya-dach network: The calcisponge case study. *EvoDevo, 5*, 23.

Foster, K. W., & Smyth, D. R., (1980). Light antennas in phototactic algae. *Microbiol. Rev., 44*, 572–630.

Francis, W. R., Eitel, M., Vargas, S., et al., (2017). The Genome of the Contractile Demosponge *Tethya wilhelma* and the Evolution of Metazoan Neural Signaling Pathways. BioRxiv: 120998.

Fromm, J., & Lautner, S., (2007). Electrical signals and their physiological significance in plants. *Plant Cell Environ., 30*, 249–257.

Flensburg S. B., Garm A., & Funch P., (2022). The contraction-expansion behavior in the demosponge *Tethya wilhelma* is light controlled and follows a diurnal rhythm. *J Exp Biol. 225*(24):jeb244751.

Fromm, J., Hajirezaei, M. R., Becker, V. K., & Lautner, S., (2013). Electrical signaling along the phloem and its physiological responses in the maize leaf. *Front Plant Sci., 4*, 10.3389/fpls.2013.00239.

Gaino, E., & Magnino, G., (1999). Dissociated cells of the calcareous sponge *Clathrina*: A model for investigating cell adhesion and cell motility *in vitro*. *Microsc. Res. Tech., 44*, 279–292.

Gaino, E., Pansini, M., Pronzato, R., & Cicogna, F., (1991). Morphological and structural variations in *Clathrina clathrus* (Porifera, Calcispongiae). In: Reitner, J., & Keupp, (eds.), *Fossil and Recent Sponges* (pp. 360–371). Springer-Verlag.

Gavelis, G. S., Hayakawa, S., White, III. R. A., et al., (2015). Eye-like ocelloids are built from different endosymbiotically acquired components. *Nature, 523*, 204.

Giovine, M., Pozzolini, M., Favre, A., et al., (2001). Heat stress-activated, calcium-dependent nitric oxide synthase in sponges. *Nitric Oxide – Biol. Chem., 5*, 427–431.

Goldberg, J. I., Rich, D. R., Muruganathan, S. P., et al., (2011). Identification and evolutionary implications of neurotransmitter–ciliary interactions underlying the behavioral response to hypoxia in *Lymnaea stagnalis* embryos. *J. Exp. Biol., 214*, 2660–2670.

Goldstein, J., Bisbo, N., Funch, P., & Riisgård, U. H., (2020). Contraction-expansion and the effects on the aquiferous system in the demosponge *Halichondria panicea*. *Front Mar. Sci., 7*, 10.3389/fmars.2020.00113.

Grant, N., Matveev, E., Kahn, A. S., & Leys, P. S., (2018). Suspended sediment causes feeding current arrests *in situ* in the glass sponge *Aphrocallistes vastus*. *Mar. Environ. Res., 137*, 111–120.

Grant, N., Matveev, E., Kahn, A. S., et al., (2019). Effect of suspended sediments on the pumping rates of three species of glass sponge *in situ*. *Mar. Ecol. Prog. Ser., 615*, 79–100.

Green, C. R., & Bergquist, R. P., (1979). Cell membrane specialisations in the *Porifera*. *Colloq. Int. Cent. Natl. Rech. Sci., Biologie des Spongiaires* (Vol. 29, pp. 153–158).

Green, C. R., & Bergquist, R. P., (1982). Phylogenetic relationships within the invertebrata in relation to the structure of septate junctions and the development of occluding junctional types. *J. Cell Sci., 53*, 279–305.

Gruhl, A., & Okamura, B., (2015). Tissue characteristics and development in Myxozoa. In: Okamura, B., Gruhl, A., & Bartholomew, J. L., (eds.), *Myxozoan Evolution, Ecology and Development* (pp. 155–174). Springer International Publishing.

Grundfest, H., (1959). Evolution of conduction in the nervous system. In: Bass, A., (ed.), *Evolution of Nervous Control from Primitive Organisms to Man* (pp. 43–86). AAAS.

Grundfest, H., (1965). Evolution of electrophysiological properties among sensory receptor systems. In: Pringle, J., (ed.), *Essays on Physiological Evolution* (pp. 107–138). Pergamon Press.

Gualtieri, P., Pelosi, P., Passarelli, V., & Barsanti, L., (1992). Identification of a rhodopsin photoreceptor in *Euglena gracilis*. *Biochim. Biophys. Acta, 1117*, 55–59.

Hammel, J. U., & Nickel, M., (2014). A new flow-regulating cell type in the demosponge *Tethya wilhelma* – functional cellular anatomy of a leuconoid canal system. *PLoS One, 9*, e113153.

Hille, B., (1992). *Ionic Channels of Excitable Membranes*. Sinauer.
Hirth, F., Kammermeier, L., Frei, E., Walldorf, U., Noll, M., & Reichert, H., (2003). An urbilaterian origin of the tripartite brain: Developmental genetic insights from *Drosophila. Development, 130*, 2365–2373.
Holstein, K., Rivera, A., Windsor, P., et al., (2010). Expansion, diversification and expression of T-box family genes in Porifera. *Dev. Genes Evol., 220*, 251–262.
Hyman, L. H., (1940). *The Invertebrates: Protozoa Through Ctenophora* (Vol. 1). New York and London.
Jékely, G., (2021). The chemical brain hypothesis for the origin of nervous systems. *Philos. Trans. R. Soc. Lond. B, 376*, 20190761.
Jindrich, K., Roper, K. E., Lemon, S., Degnan, B. M., Reitzel, A. M., & Degnan, S. M., (2017). Origin of the animal circadian clock: Diurnal and light-entrained gene expression in the sponge *Amphimedon queenslandica. Front. Mar. Sci., 4*. 10.3389/fmars.2017.00327.
Jones, W. C., (1962). Is there a nervous system in sponges? *Biol. Rev. Camb. Philos. Soc., 37*, 1–50.
Kahn, A. S., & Leys, Leys, S. P. (2016). The role of cell replacement in benthic–pelagic coupling by suspension feeders. *Roy. Soc. Open Sci., 3*, 160484.
Kahn, A. S., Pennelly, C. W., McGill, P. R., & Leys, S. P., (2020). Behaviors of sessile benthic animals in the abyssal northeast Pacific Ocean. *Deep Sea Res. II, 173*, 104729.
Kahn, A., Ruhl, H., & Smith, K., (2012). Temporal changes in deep-sea sponge populations are correlated to changes in surface climate and food supply. *Deep-Sea Res., 70*, 36–41.
Kenny, N. J., Francis, W. R., Rivera-Vicéns, R. E., et al., (2020). Tracing animal genomic evolution with the chromosomal-level assembly of the freshwater sponge *Ephydatia muelleri. Nat. Commun., 11*, 3676.
Kirchman, D. L., (2018). *Predation and Protists. Processes in Microbial Ecology*. Oxford University Press.
Kishimoto, K., Sugano-Yasunaga, W., Taniguchi, A., Agata, K., Nonaka, S., & Funayama, N., (2019). Skeleton construction upon local regression of the sponge body. *Dev. Growth Diff., 61*, 485–500.
Kosik, K. S., (2009). Exploring the early origins of the synapse by comparative genomics. *Biol. Lett., 5*, 108–111.
Kumala, L., & Canfield, D. E., (2018). Contraction dynamics and respiration of small single-osculum explants of the demosponge *Halichondria panicea. Front Mar. Sci., 5*.
Kumala, L., Riisgård, H. U., & Canfield, D. E., (2017). Osculum dynamics and filtration activity in small single-osculum explants of the demosponge *Halichondria panicea. Mar. Ecol. Progr. Ser., 572*, 117–128.
Langenbruch, P. F., & Jones, C. W., (1990). Body structure of marine sponges. VI. Choanocyte chamber structure in the *Haplosclerida* (Porifera, Demospongiae) and its relevance to the phylogenesis of the group. *J. Morph., 204*, 1–8.
Langenbruch, P. F., & Scalera-Liaci, L., (1986). Body structure of marine sponges IV. Aquiferous system and choanocyte chambers in *Haliclona elegans* (Porifera, Demospongiae). *Zoomorphology, 106*, 205–211.
Langenbruch, P. F., & Weissenfels, N., (1987). Canal systems and choanocyte chambers in freshwater sponges (Porifera, Spongillidae). *Zoomorphology, 107*, 11–16.
Langenbruch, P. F., (1983). Body structure of marine sponges I. Arrangement of the flagellated chambers in the canals system of *Reniera* sp. *Mar. Biol., 75*, 319–325.

Langenbruch, P. F., (1988). Body structure of marine sponges V. Structure of choanocyte chambers in some Mediterranean and Caribbean haplosclerid sponges (Porifera). *Zoomorphology, 108*, 13–21.
Langenbruch, P. F., Simpson, T. L., & Scalera-Liaci, L., (1985). Body structure of marine sponges III. The structure of choanocyte chambers in *Petrosia ficiformis* (Porifera, Demospongiae). *Zoomorphology, 105*, 383–387.
Langenbruch, P., & Jones, C. W., (1989). A new type of central cell in the choanocyte chambers of *Pellina fistulosa* (*Porifera, Demospongiae*). *Zoomorphology, 109*, 11–14.
Lanna, E., Riesgo, A., Koutsouveli, V., & Leys, S. P., (2024). Physiology of reproduction in Porifera. In: *Frontiers in Invertebrate Physiology*. Saleuddin, S., Leys, S.P., Roer, R.D., and Wilkie, I. C. Apple Academic Press. Vol. 1 Non-Bilaterian Phyla.
Larsen, P. S., & Riisgård, H. U., (1994). The sponge pump. *J. Theor. Biol., 168*, 53–63.
Lavrov, A. I., Bolshakov, F. V., Tokina, D. B., & Ereskovsky, A. V., (2022). Fine details of the choanocyte filter apparatus in asconoid calcareous sponges (Porifera: Calcarea) revealed by ruthenium red fixation. *Rep. Sci. Res. Challenger Zool., 150*, 125984.
Lawn, I. D., Mackie, G. O., & Silver, G., (1981). Conduction system in a sponge. *Science, 211*, 1169–1171.
Ledger, P. W., (1975). Septate junctions in the calcareous sponge *Sycon ciliatum*. *Tissue Cell, 7*, 13–18.
Lentz, T. L., (1966). Histochemical localization of neurohumors in a sponge. *J. Exp. Zool., 162*, 171–179.
Léon, J., Rojo, E., & Sánchez-Serrano, J. J., (2001). Wound signaling in plants. *J Exp. Biol., 52*, 1–9.
Lethias, C., Garrone, R., & Mazzorana, M., (1983). Fine structure of sponge cell membranes: Comparative study with freeze-fracture and conventional thin section methods. *Tiss. Cell, 15*, 523–535.
Leys, S. P., & Hill, A., (2012). The physiology and molecular biology of sponge tissues. *Adv. Mar. Biol., 62,* 1–56.
Leys, S. P., & Mackie, O. G., (1997). Electrical recording from a glass sponge. *Nature, 387*, 29–30.
Leys, S. P., (1995). Cytoskeletal architecture and organelle transport in giant syncytia formed by fusion of hexactinellid sponge tissues. *Biol. Bull., 188*, 241–254.
Leys, S. P., (2004). Gastrulation in sponges. In: Stern, C., (ed.), *Gastrulation: From Cells to Embryo* (pp. 23–31). Cold Spring Harbor Press.
Leys, S. P., (2015). Elements of a 'nervous system' in sponges. *J. Exp. Biol., 218*, 581–591.
Leys, S. P., Mackie, G. O., & Meech, W. R., (1999). Impulse conduction in a sponge. *J. Exp. Biol., 202*, 1139–1150.
Leys, S. P., Mackie, G. O., & Reiswig, H., (2007). The biology of glass sponges. *Adv. Mar. Biol., 52*, 1–145.
Leys, S. P., Mah, J. L., McGill, P. R., Hamonic, L., De Leo, F. C., & Kahn, S. A., (2019). Sponge behavior and the chemical basis of responses: A post-genomic view. *Integr. Comp. Biol., 59*, 751–764.
Leys, S. P., Yahel, G., Reidenbach, M. A., Tunnicliffe, V., Shavit, U., & Reiswig, M. H., (2011). The sponge pump: The role of current induced flow in the design of the sponge body plan. *PLoS One, 6*(12), e27787.
Leys, S., & Degnan, B., (2001). Cytological basis of photoresponsive behaviour in a sponge larva. *Biol. Bull., 201*, 323–338.

Leys, S., & Meech, R., (2006). Physiology of coordination in sponges. *Can J. Zool., 84*, 288–306.

Leys, S., & Riesgo, A., (2011). Epithelia, an evolutionary novelty of metazoans. *J. Exp. Zool. B. Mol. Dev. Evol., 314B*. doi: 10.1002/jez.b.21442.

Leys, S., (2007). Sponge coordination, tissues, and the evolution of gastrulation. In: Custodio, M., Lobo-Hadju, G., Lobo-Hadju, E., & Muricy, G., (eds.), *Porifera Research: Biodiversity, Innovation and Sustainability; 7th International Porifera* Congress (pp. 53–59). Museu Nacional.

Leys, S. P., Cronin, T., Degnan, B. M., & Marshall, J., (2002). Spectral sensitivity in a sponge larva. *J. Comp. Phys. A, 188*, 199–202.

Leys, S., Mah, J., & Esposito, E., (2021). Evolution of the sensory/neural cell types. In: Leys, S., & Hejnol, A., (eds.), *Origin and Evolution of Metazoan Cell Types* (pp. 101–128). CRC Press.

Leys, S., Nichols, S., & Adams, E., (2009). Epithelia and integration in sponges. *Integr. Comp. Biol., 49*, 167–177.

Li, Y., Shen, X. X., Evans, B., Dunn, C. W., & Rokas, A., (2021). Rooting the animal tree of life. *Mol. Biol. Evol., 38*, 4322–4333.

Liebeskind, B. J., Hillis, D. M., & Zakon, H. H., (2011). Evolution of sodium channels predates the origin of nervous systems in animals. *Proc. Natl. Acad. Sci. U S A., 108*, 9154–9159.

Liebeskind, B. J., Hillis, D. M., & Zakon, H. H., (2015). Convergence of ion channel genome content in early animal evolution. *Proc. Natl. Acad. Sci. U S A., 112*, E846.

Liebeskind, B. J., Hillis, D. M., Zakon, H. H., & Hofmann, A. H., (2016). Complex homology and the evolution of nervous systems. *Trends Ecol. Evol., 31*, 127–135.

Liebeskind, B. J., Hofmann, H. A., Hillis, D. M., & Zakon, H. H., (2017). Evolution of animal neural systems. *Ann. Rev. Ecol. Evol. Sys., 48*, 377–398.

Liu, H., Mishima, Y., Fujiwara, T., et al., (2004). Isolation of Araguspongine M, a new stereoisomer of an Araguspongine/Xestospongin alkaloid, and dopamine from the marine sponge *Neopetrosia exigua* collected in Palau. *Mar. Drugs, 2*, 154–163.

Loewenstein, W. R., (1967). On the genesis of cellular communication. *Dev. Biol., 15*, 503–520.

Lu, T. M., Kanda, M., Furuya, H., & Satoh, N., (2019). Dicyemid mesozoans: A unique parasitic lifestyle and a reduced genome. *Genom. Biol. Evol., 11*, 2232–2243.

Ludeman, D. A., Farrar, N., Riesgo, A., Paps, J., & Leys, P. S., (2014). Evolutionary origins of sensation in metazoans: Functional evidence for a new sensory organ in sponges. *BMC Evol. Biol., 14*, 3.

Ludeman, D. A., Reidenbach, M. A., & Leys, P. S., (2017). The energetic cost of filtration by demosponges and their behavioral response to ambient currents. *J. Exp. Biol., 220*, 995–1007.

Mackie, G. O., & Singla, L. C., (1983). Studies on hexactinellid sponges. I Histology of *Rhabdocalyptus dawsoni* (Lambe, 1873). *Philos. Trans. Roy. Soc. Lond. B, 301*, 365–400.

Mackie, G. O., (1970). Neuroid conduction and the evolution of conducting tissues. *Quart. Rev. Biol., 45*, 319–332.

Mackie, G. O., (1979). In: Lévi, C., & Boury-Esnault, N., (eds), Is There a Conduction System in Sponges? (Vol. 291, pp. 145–151). Colloques int. du C.N.R.S.

Mackie, G. O., (1990). The elementary nervous system revisited. *Amer. Zool., 30*, 907–920.

Mackie, G. O., Lawn, I. D., & Pavans de Ceccatty, M. (1983). Studies on hexactinellid sponges. II. Excitability, conduction and coordination of responses in *Rhabdocalyptus dawsoni* (Lambe, 1873). *Philos. Trans. R. Soc. Lond. Biol. Sci. No., 301*, 401–418.
Mah, J. L., & Leys, P. S., (2017). Think like a sponge: The genetic signal of sensory cells in sponges. *Dev. Biol., 431*, 93–100.
Mah, J. L., Christensen-Dalsgaard, K. K., & Leys, P. S., (2014). Choanoflagellate and choanocyte collar-flagellar systems and the assumption of homology. *Evol. Dev., 16*, 25–37.
Maldonado, M., & Bergquist, R. P., (2002). Phylum Porifera. In: Young, C. M., Sewell, M. A., & Rice, M. E., (eds.), *Atlas of Marine Invertebrate Larvae* (pp. 21–50). Academic Press.
Maldonado, M., & Riesgo, A., (2007). Maldonado, M. and A. Riesgo (2007). Intra-epithelial spicules in a homosclerophorid sponge. *Cell and Tissue Research 328*, 639–650.
Maldonado, M., & Uriz, J. M., (1999). An experimental approach to the ecological significance of microhabitat-scale movement in an encrusting sponge. *Mar. Ecol. Prog. Ser., 185*, 239–255.
Maldonado, M., (2006). The ecology of the sponge larva. *Can. J. Zool., 84*, 175–194.
Maldonado, M., et al. (2024). Nitrogen And Phosphorus Cycling Through Marine Sponges: Physiology, cytology, genomics, and ecological implications. *Frontiers in Invertebrate Physiology.* A. Saleuddin, S. P. Leys, R. D. Roer and I. Willkie, Apple Academic Press. Vol. 1 Non-bilaterian Phyla.
Maldonado, M., Durfort, M., McCarthy, D. A., & Young, M. C., (2003). The cellular basis of photobehavior in the tufted parenchymella larva of demosponges. *Mar. Biol., 143*, 427–441.
Maldonado, M., George, S. B., Young, C. M., & Vaquerizo, I., (1997). Depth regulation in parenchymella larvae of a demosponge: Relative roles of skeletogenesis, biochemical changes and behavior. *Mar. Ecol. Prog. Ser., 148*, 115–124.
Maldonado, M., Ribes, M., & Van, D. F. C., (2012). Chapter 3 – nutrient fluxes through sponges: Biology, budgets, and ecological implications. In: Becerro, M., Uriz, M., Maldonado, M., & Turon, X., (eds.), *Adv. Mar. Biol.* (pp. 113–182). Academic Press.
McMurray, S. E., Pawlik, J. R., & Finelli, M. C., (2014). Trait-mediated ecosystem impacts: How morphology and size affect pumping rates of the Caribbean giant barrel sponge. *Aquat. Biol., 23*, 1–13.
McNair, G. T., (1923). Motor reactions of the fresh-water sponge *Ephydatia fluviatilis*. *Biol. Bull., 44*, 153–166.
Miller, D. J., & Ball, E. E., (2005). Animal evolution: The enigmatic phylum Placozoa revisited. *Curr. Biol., 15*, R26–R28.
Miller, M., & Bassler, B., (2001). Quorum sensing in bacteria. *Annu. Rev. Microbiol., 55*, 165–199.
Minchin, E. A., (1896). Note on the larva and post larval development of *Leucosolenia variabilis*, H. sp., with remarks on the development of other asconidae. *Proc. R. Soc. Lond. B, 60,* 42–52.
Mitchell, J. M., & Nichols, A. S., (2019). Diverse cell junctions with unique molecular composition in tissues of a sponge (Porifera). *EvoDevo, 10*, 26.
Morganti, T. M., Purser, A., Rapp, H. T., et al., (2021a). In situ observation of sponge trails suggests common sponge locomotion in the deep central Arctic. *Curr. Biol., 31*, R368–R370.
Morganti, T. M., Ribes, M., Moskovich, R., Weisz, J. B., Yahel, G., & Coma, R., (2021b). *In situ* pumping rate of 20 marine demosponges is a function of osculum area. *Front. Mar. Sci., 8*.

Moroz, L. L., & Kohn, B. A., (2016). Independent origins of neurons and synapses: Insights from ctenophores. *Phil. Trans. Roy. Soc. B: Biol. Sci., 371*, 20150041.
Moroz, L. L., (2009). On the independent origins of complex brains and neurons. *Brain Behav. Evol., 74*, 177–190.
Moroz, L. L., (2015). Convergent evolution of neural systems in ctenophores. *J. Exp. Biol., 218*, 598–611.
Moroz, L. L., Kocot, K. M., & Citarella, M. R., et al., (2014). The ctenophore genome and the evolutionary origins of neural systems. *Nature, 510*, 109–114.
Moroz, L. L., Nikitin, M. A., Poličar, P. G., Kohn, A. B., & Romanova, D. Y., (2021a). Evolution of glutamatergic signaling and synapses. *Neuropharmacology, 199*, 108740.
Moroz, L. L., Romanova, D. Y., & Kohn, K. B., (2021b). Neural versus alternative integrative systems: Molecular insights into origins of neurotransmitters. *Philos. Trans. R. Soc. Lond., B, Biol. Sci., 376*, 20190762.
Muhammad, H., Rais, Y., Miosge, N., & Ornan, M. E., (2012). The primary cilium as a dual sensor of mechanochemical signals in chondrocytes. *Cell. Mol. Life Sci., 69*, 2101–2107.
Musser, J. M., Schippers, K. J., Nickel, M., et al., (2021). Profiling cellular diversity in sponges informs animal cell type and nervous system evolution. *Science, 374*, 717–723.
Naitoh, Y., & Eckert, R., (1974). The control of ciliary activity in protozoa. In: Sleigh, M. A., (ed.), *Cilia and Flagella* (p. 305–352). Academic Press.
Nakanishi, N., Sogabe, S., & Degnan, M. B., (2014). Evolutionary origin of gastrulation: Insights from sponge development. *BMC Biol., 12*, 1–9.
Nakanishi, N., Stoupin, D., Degnan, S. M., & Degnan, M. B., (2015). Sensory flask cells in sponge larvae regulate metamorphosis via calcium signaling. *Integr. Comp. Biol., 55*, 1018–1027.
Nakayama, S., Arima, K., Kawai, K., et al., (2015). Dynamic transport and cementation of skeletal elements build up the pole-and-beam structured skeleton of sponges. *Curr. Biol., 25*, 2549–2554.
Nichols, S. A., Dirks, W., Pearse, J. S., & King, N., (2006). Early evolution of animal cell signaling and adhesion genes. *Proc. Natl. Acad. Sci. USA, 103*, 12451–12456.
Nichols, S. A., Roberts, B. W., Richter, D. J., Fairclough, S. R., & King, N., (2012). Origin of metazoan cadherin diversity and the antiquity of the classical cadherin/β-catenin complex. *Proc. Natl. Acad. Sci. U S A, 109*, 13046–13051.
Nickel, M., (2004). Kinetics and rhythm of body contractions in the sponge *Tethya wilhelma* (Porifera: Demospongiae). *J. Exp. Biol., 207*, 4515–4524.
Nickel, M., (2006). Like a `rolling stone': Quantitative analysis of the body movement and skeletal dynamics of the sponge *Tethya wilhelma. J. Exp. Biol., 209*, 2839–2846.
Nickel, M., (2010). Evolutionary emergence of synaptic nervous systems: What can we learn from the non-synaptic, nerveless Porifera? *Invertebr. Biol., 129*, 1–16.
Nickel, M., Donath, T., Schweikert, M., & Beckmann, F., (2006). Functional morphology of *Tethya* species (Porifera): 1. Quantitative 3D-analysis of *Tethya wilhelma* by synchrotron radiation based X-ray microtomography. *Zoomorphology, 125*, 209–223.
Nickel, M., Scheer, C., Hammel, J. U., Herzen, J., & Beckmann, F., (2011). The contractile sponge epithelium *sensu lato* – body contraction of the demosponge *Tethya wilhelma* is mediated by the pinacoderm. *J. Exp. Biol., 214*, 1692–1698.
Nosenko, T., Schreiber, F., Adamska, M., et al., (2013). Deep metazoan phylogeny: When different genes tell different stories. *Mol. Phyogenet. Evol., 67*, 223–233.
Pantin, C., (1956). The origin of the nervous system. *Pubbl. Staz. Zool. Napoli., 28*, 171–181.

Parker, G. H., (1910). The reactions of sponges with a consideration of the origin of the nervous system. *J. Exp. Zool., 8,* 765–805.
Parker, G. H., (1911). The origin and significance of the primitive nervous system. *Proc. Amer. Philos. Soc., 50,* 217–225.
Parker, G. H., (1914). On the strength and the volume of the water currents produced by sponges. *J. Exp. Zool., 16,* 443–446.
Parker, G. H., (1919). The elementary nervous system. *Monographs on Exp. Biol.,* 25–49.
Passano, L. M., (1963). Primitive nervous systems. *Proc. Natl. Acad. Sci. U S A, 50,* 306–313.
Pavans de Ceccatty, M., (1955). Le systeme nerveux des éponges calcaires et siliceuses. *Ann. Sci. Natur. Zool., 17,* 203–288.
Pavans de Ceccatty, M., (1974a). Coordination in sponges. The foundations of integration. *Am. Zool., 14,* 895–903.
Pavans de Ceccatty, M., (1974b). The origin of the integrative systems: A change in view derived from research on coelentrates and sponges. *Perspect. Biol. Med., 17,* 379–390.
Pavans de Ceccatty, M., (1979). Cell correlations and integration in sponges. *Colloq. Int. C.N.R.S.,* In: *Biologie Des Spongiaires* (Vol. 291, pp. 123–135).
Pavans de Ceccatty, M., (1981). Demonstration of actin filaments in sponge cells. *Cell Biol. Int. Rep., 5,* 945–952.
Pavans de Ceccatty, M., (1986). Cytoskeletal organization and tissue patterns of epithelia in the sponge *Ephydatia muelleri. J. Morph., 189,* 45–65.
Pavans de Ceccatty, M., Cargouil, M., & Coraboeuf, E., (1960). Les reactions motrices de l'eponge *Tethya lyncurium* (Lmk.) a quelques stimulations experimentales (I). *Vie et Milieu, 11,* 594–600.
Philippe, H., Brinkmann, H., Lavrov, D. V., et al., (2011). Resolving difficult phylogenetic questions: Why more sequences are not enough. *PLoS Biol., 9,* e1000602.
Philippe, H., Derelle, R., Lopez, P., et al., (2009). Phylogeomics revives traditional views on deep animal relationships. *Curr. Biol., 19,* 706–712.
Pick, K. S., Philippe, H., Schreiber, F., et al., (2010). Improved phylogenomic taxon sampling noticeably affects nonbilaterian relationships. *Mol. Biol. Evol., 27,* 1983–1987.
Pisani, D., Pett, W., Dohrmann, M., et al., (2015). Genomic data do not support comb jellies as the sister group to all other animals. *Proc. Natl. Acad. Sci. U S A, 112,* 15402–15407.
Poole, C. A., Jensen, C. G., Snyder, J. A., Gray, C. G., Hermanutz, V. L., & Wheatley, N. D., (1997). Confocal analysis of primary cilia structure and colocalization with the Golgi apparatus in chondrocytes and aortic smooth muscle cells. *Cell Biol. Int., 21,* 483–494.
Praetorius, H. A., & Spring, R. K., (2005). A physiological view of the primary cilium. *Annu. Rev. Physiol., 67,* 515–529.
Prosser, C. L., Nagai, T., & Nystrom, A. R., (1962). Oscular contractions in sponges. *Comp. Biochem. Physiol., 6,* 69–74.
Ramoino, P., Gallus, L., Paluzzi, S., et al., (2007). The GABAergic-like system in the marine demosponge *Chondrilla nucula. Microsc. Res. Tech., 70,* 944–951.
Reid, C. R., & Latty, T., (2016). Collective behavior and swarm intelligence in slime moulds. *FEMS Microbiol. Rev., 40,* 798–806.
Reiswig, H. M., & Brown, J. M., (1977). The central cells of sponges. *Zoomorphology, 88,* 81–94.
Reiswig, H. M., (1971). *In situ* pumping activities of tropical Demospongiae. *Mar. Biol., 9,* 38–50.

Reiswig, H. M., (1979a). In: Lévi, C., & Boury-Esnault, N., (eds.), Histology of Hexactinellida (Porifera) (Vol. 291, pp. 173–180). *Coll. Intern. C.N.R.S.*

Reiswig, H. M., (1979b). A new sponge with rapid contraction systems. *Can. J. Zool. Program Abstracts, 83A.*

Renard, E., Le Bivic, A., & Borchiellini, C., (2021). Origin and evolution of epithelial cell types. In: Leys, S., &. Hejnol, A., (eds.), *Origin and Evolution of Metazoan Cell Types* (p. 26). CRC Press.

Renard, E., Leys, S. P., Wörheide, G., & Borchiellini, C., (2018). Understanding animal evolution: The added value of sponge transcriptomics and genomics. *BioEssays, 40*, 1700237.

Revel, J. P., (1966). Fine structure of intercellular contacts in the sponge, *Microciona prolifera. Biol. Bull., 131*, 402.

Riesgo, A., Farrar, N., Windsor, P. J., Giribet, G., & Leys, P. S., (2014). The analysis of eight transcriptomes from all poriferan classes reveals surprising genetic complexity in sponges. *Mol. Biol. Evol., 31,* 1102–1120.

Riesgo, A., Santodomingo, N., Koutsouveli, V. et al. (2022). Molecular machineries of ciliogenesis, cell survival, and vasculogenesis are differentially expressed during regeneration in explants of the demosponge *Halichondria panicea. BMC Genom. 23*, 858.

Rivera, A. S., Ozturk, N., Fahey, B., et al., (2012). Blue-light-receptive cryptochrome is expressed in a sponge eye lacking neurons and opsin. *J. Exp. Biol., 215*, 1278–1286.

Roberts, E. C., Legrand, C., Steinke, M., & Wootton, C. E., (2011). Mechanisms underlying chemical interactions between predatory planktonic protists and their prey. *J. Plankton Res., 33*, 833–841.

Rützler, K., Van, S. R. W. M., & Alvarez, B., (2003). *Svenzea zeai*, a Caribbean reef sponge with a giant larva, and *Scopalina ruetzleri*: A comparative fine-structural approach to classification (Demospongiae, Halichondrida, Dictyonellidae). *Invertebr. Biol., 122*, 203–222.

Ryan, J. F., & Chiodin, M., (2015). Where is my mind? How sponges and placozoans may have lost neural cell types. *Phil. Trans. Roy. Soc. Lond. B, 370*, 20150059.

Ryan, J. F., Pang, K., Schnitzler, C. E., et al., (2013). The genome of the ctenophore *Mnemiopsis leidyi* and its implications for cell type evolution. *Science, 342*, 1242592.

Sachkova, M., & Burkhardt, P., (2019). Exciting times to study the identity and evolution of cell types. *Development, 146*, dev178996.

Sakaraya, O., Armstrong, K. A., Adamska, M., et al., (2007). A post-synaptic scaffold at the origin of the animal kingdom. *PLoS One, 2*, e506.

Sanger, W., (2021). *Aquatic Symbiosis Genomics.* https://www.sanger.ac.uk/collaboration/aquatic-symbiosis-genomics-project/ (accessed on 12 July 2023).

Satir, P., (2017). CILIA: Before and after. *Cilia, 6*, 1.

Satir, P., Pedersen, L. B., & Christensen, T. S., (2010). The primary cilium at a glance. *J. Cell Sci., 123*, 499–503.

Savarese, M., Patterson, M. R., Chernykh, V. I., & Fialkov, A. V., (1997). Trophic effects of sponge feeding within Lake Baikal's littoral zone. 1. *In situ* pumping rates. *Limnol. Oceanogr., 42*, 171–178.

Say, T. E., & Degnan, M. S., (2020). Molecular and behavioral evidence that interdependent photo – and chemosensory systems regulate larval settlement in a marine sponge. *Mol. Ecol., 29*, 247–261.

Schläppy, M. L., Weber, M., Mendola, D., Hoffmann, F., & De Beer, D., (2010). Heterogeneous oxygenation resulting from active and passive flow in two Mediterranean sponges, *Dysidea avara* and *Chondrosia reniformis*. *Limnol. Oceanogr., 55*, 1289–1300.

Schuchert, P., (1993). *Trichoplax adhaerens* (Phylum Placozoa) has cells that react with antibodies against the neuropeptide RFamide. *Acta Zool., 74*, 115–117.

Sciscioli, M., Lepore, E., Corriero, G., Scalera-Liaci, L., & Gaino, E., (1997). Ultrastructural organization of choanocyte chambers in the haplosclerid *Pellina semitubulosa* (Porifera, Demospongiae): A cue for water flow into the sponge body. *Ital. J. Zool., 64*, 291–296.

Sebé-Pedrós, A., Chomsky, E., Pang, K., et al., (2018). Early metazoan cell type diversity and the evolution of multicellular gene regulation. *Nat. Ecol. Evol., 2*, 1176–1188.

Simons, P., (1992). *The Secret Feelings of Plants* (pp. 29–33). New Scientist.

Simpson, T. L., (1984). *The Cell Biology of Sponges*. Springer Verlag.

Sineshchekov, O. A., Jung, K. H., & Spudich, L. J., (2002). Two rhodopsins mediate phototaxis to low- and high-intensity light in *Chlamydomonas reinhardtii*. *Proc. Natl. Acad Sci U S A, 99*, 8689–8694.

Sleigh, M. A., (1974). *Cilia and Flagella*. Academic Press.

Sokolova, A. M., & Voronezhskaya, E. E., (2021). Dopamine-like immunoreactivity in sponge larvae. *Invert. Zool., 18*, 345–354.

Song, H., Hewitt, O. H., & Degnan, M. S., (2021). Arginine biosynthesis by a bacterial symbiont enables nitric oxide production and facilitates larval settlement in the marine-sponge host. *Curr. Biol., 31*, 433–437.e433.

Song, P. S., (1981). Photosensory transduction in *Stentor coeruleus* and related organisms. *Biochim. Biophys. Acta, 639*, 1–29.

Srivastava, M., Simakov, O., Chapman, J., et al., (2010). The *Amphimedon queenslandica* genome and the evolution of animal complexity. *Nature, 466*, 720–727.

Strehlow, B. W., Jorgensen, D., Webster, N. S., Pineda, M. C., & Duckworth, A., (2016). Using a thermistor flowmeter with attached video camera for monitoring sponge excurrent speed and oscular behavior. *PeerJ, 4*, e2761.

Syed, T., & Schierwater, B., (2002). *Trichoplax adherens*: Discovered as a missing link, forgotten as a hydrozoan, re-discovered as a key to metazoan evolution. *Vie Milieu, 52*, 177–187.

Telford, M. J., Budd, G. E., & Philippe, H., (2015). Phylogenomic insights into animal evolution. *Curr. Biol., 25*, R876–R887.

Tompkins-MacDonald, G., & Leys, S., (2008). Glass sponges arrest pumping in response to sediment: Implications for the physiology of the hexactinellid conduction system. *Mar. Biol., 154*, 973–954.

Ueda, N., Richards, G., Degnan, B., et al., (2016). An ancient role for nitric oxide in regulating the animal pelagobenthic life cycle: Evidence from a marine sponge. *Sci. Rep., 6*, 37546.

Uriz, M. J., Turon, X., & Becerro, A. M., (2001). Morphology and ultrastructure of the swimming larvae of *Crambe crambe* (Demospongiae, Poecilosclerida). *Invertebr. Biol., 120*, 295–307.

Vacelet, J., Boury-Esnault, N., De Vos, L., & Donadey, C., (1989). Comparative study of the choanosome of Porifera: II. The keratose sponges. *J. Morph., 201*, 119–129.

Valenstein, E. S., (2002). The discovery of chemical neurotransmitters. *Brain Cogn., 49*, 73–95.

Varoqueaux, F., Williams, E. A., Grandemange, S., et al., (2018). High cell diversity and complex peptidergic signaling underlie placozoan behavior. *Curr. Biol., 28*, 3495–3501. e3492.

Wachtmann, D., Stockem, W., & Weissenfels, N., (1990). Cytoskeletal organization and cell organelle transport in basal epithelial cells of the freshwater sponge *Spongilla lacustris*. *Cell Tissue Res., 261*, 145–154.

Wada, Y., et al. (1997). Modification of ciliary beating in sea urchin larvae induced by neurotransmitters: beat-plane rotation and control of frequency fluctuation. *Journal of Experimental Biology 200*, 9–18.

Warburton, F. E., (1966). The behavior of sponge larvae. *Ecology, 47*, 672–674.

Watanabe, Y., (1978). The development of two species of *Tetilla* (Demosponge). (Vol. 29, pp. 71–106). Natural Science Report, Ochanomizu University.

Weissenfels, N., (1981). Bau und funktion des Süβwasserschwamms *Ephydatia fluviatilits* L. (Porifera) VIII. Die entstehung und entwicklung der kragengeisselkammern und ihre verbindung mit dem ausfuhrenden kanalsystem. *Zoomorphology, 98*, 35–45.

Weissenfels, N., (1983). Bau und Funktion des Süβwasserschwamms *Ephydatia fluviatilis* (Porifera) X. Der nachweis des offenen mesenchyms durch verfutterung von backerhefe (*Saccharomyces cerevisiae*). *Zoomorphology, 103*, 15–23.

Weissenfels, N., (1984). Bau und funktion des Süβwasserschwamms *Ephydatia fluviatilis* (Porifera) XI. Nachweis einer endogenen kontraktionsrhythmik durch infrarot-reflexion. *Zoomorphology, 104*, 292–297.

Weissenfels, N., (1989). *Biologie und Mikroskopische Anatomie Der Süβwasserschwämme* (Spongillidae). Gustav Fischer Verlag.

Weissenfels, N., (1990). Condensation rhythm of fresh-water sponges (Spongillidae, Porifera). *Eur. J. Cell Biol., 53*, 373–383.

Weissenfels, N., (1992). The filtration apparatus for food collection in freshwater sponges (Porifera, Spongillidae). *Zoomorphology, 112*, 51–55.

Weyrer, S., Rützler, K., & Rieger, R., (1999). Serotonin in Porifera? Evidence from developing *Tedania ignis*, the Caribbean fire sponge (Demospongiae). *Memoirs of the Queensland Museum, 44*, 659–665.

Whelan, N. V., Kocot, K. M., Moroz, L. L., & Halanych, M. K., (2015). Error, signal, and the placement of Ctenophora sister to all other animals. *Proc. Natl. Acad. Sci. U S A, 112*, 5773–5778.

Whelan, N. V., Kocot, K. M., Moroz, T. P., et al., (2017). Ctenophore relationships and their placement as the sister group to all other animals. *Nat. Ecol. Evol., 1*, 1737–1746.

Wildon, D. C., Thain, J. F., Minchin, P. E. H., et al., (1992). Electrical signaling and systemic proteinase inhibitor induction in the wounded plant. *Nature, 360*, 62–65.

Windsor, P. J., (2014). *Evolution of the Sponge Body Plan: Wnt and the Development of Polarity in Freshwater Sponges*. Doctoral Dissertation. University of Alberta.

Windsor, P. J., & Leys, S. P., (2010). Wnt signaling and induction in the sponge aquiferous system: Evidence for an ancient origin of the organizer. *Evol. Dev., 12*, 481–490.

Woollacott, R. M., & Hadfield, G. M., (1996). Induction of metamorphosis in larvae of a sponge. *Invertebr. Biol., 115*, 257–262.

Yañez-Guerra, L. A., et al. (2022). Premetazoan Origin of Neuropeptide Signaling. *Molecular Biology and Evolution. 39, 4, msac051*.

Yang, W. Q., Braun, C., Plattner, H., Purvee, J., & Van, H. J. L., (1997). Cyclic nucleotides in glutamate chemosensory signal transduction of *Paramecium*. *J. Cell Sci., 110*, 2567–2572.

Zocchi, E., Carpaneto, A., Cerrano, C., et al., (2001). The temperature-signaling cascade in sponges involves a heat-gated cation channel, abscisic acid, and cyclic ADP-ribose. *Proc. Natl. Acad. Sci. USA, 98*, 14859–14864.

Zverkov, O. A., Mikhailov, K. V., Isaev, S. V., et al., (2019). Dicyemida and Orthonectida: Two stories of body plan simplification. *Front. Genet., 10*.

CHAPTER 3

Sponge Symbiosis: Microbes Make an Essential Part of What It Means to Be a Sponge

KATHRIN BUSCH,[1,$] ANGELA MARULANDA-GÓMEZ,[1,$]
TERESA MARIA MORGANTI,[2] KRISTINA BAYER,[1] and LUCÍA PITA[1,3]

[1]*Marine Symbioses Unit, Department of Marine Ecology, GEOMAR Helmholtz Center for Ocean Research Kiel, Kiel, Germany*

[2]*Max Planck Institute for Marine Microbiology; HGF MPG Joint Research Group for Deep-Sea Ecology and Technology, Bremen, Germany*

[3]*Department of Marine Biology and Oceanography, Institut de Ciències del Mar, ICM-CSIC, Barcelona, Spain*

[$]*Equal contribution.*

ABSTRACT

Sponges live in tight associations with associated microbes thriving within their bodies. In this chapter, we discuss these intertwined host-symbiont relationships and how microbes influence multiple features of sponge physiology. We start by describing the vast and specific microbial diversity harbored by sponges, which include archaea, bacteria, eukaryote, and virus taxa; how this diversity is influenced by the mode of symbiont transmission; and the environmental factors that shape these microbial communities. Furthermore, we showcase the links between symbiont density and sponge physiology by comparing distinctive traits (e.g., pumping rates, feeding strategies, nutrient cycling) across the continuous spectrum of high (HMA) and low microbial abundance (LMA) sponge holobionts. Second, we present the functions performed by the

Frontiers in Invertebrate Physiology: A Collection of Reviews, Volume 1: Non-Bilaterian Phyla.
Saber Saleuddin, Sally P. Leys, Robert D. Roer, and Iain C. Wilkie, (Eds.)

sponge-associated microbes, focusing on models for metabolic exchanges and nutrition specialization that illustrate the intricate and specialized interactions for the efficient recycling of resources within the host. We also addressed the repercussions of the microbiome-mediated functions at the ecosystem level and the concept of sponges as "nested ecosystems" by giving examples of enhanced productivity and nutrient cycling, which determine the ecological role of the holobiont. From the host perspective, we introduce the immune system as the essential force for establishing and maintaining sponge-microbe interactions, and as the guardian of homeostasis within the holobiont. We summarize the current knowledge on the immune cellular and molecular mechanisms that regulate the crosstalk between the sponge host and its symbiotic partners and identify future perspectives on the emerging research field of sponge immunity. Finally, we examine how sponge holobionts face a changing environment, highlight the relevance of understanding the mechanisms controlling homeostasis and dysbiosis in the holobiont, and how changes in the sponge-associated microbial community can, either lead to disease outbreaks or aid sponges to adapt quickly under stress-related scenarios. Understanding the complexity of sponge holobionts is a major challenge, but establishing sponge model systems and adopting holistic approaches will allow us to continue deciphering the interplay between sponges and microbes.

3.1 INTRODUCTION

3.1.1 SYMBIOSIS: AN OPPORTUNITY FOR INNOVATION

Microbes make an essential part of animals. When the first animals appeared, they had to survive and evolve in a microbial world (McFall-Ngai et al., 2013). Microbial life predates that of animals by at least 3 billion years (Figure 3.1). If we think in these evolutionary terms, it is then not so surprising that animals learned the language of microbes and that certain microbes have evolved to communicate with animals (e.g., Hadfield, 2011; Schroeder & Bäckhed, 2016; Schwartzman & Ruby, 2016), in a way that they are able to establish stable associations that we call symbiosis. Here we use symbiosis (and symbiont) in the loosest definition, *sensu* De Bary (1831–1888), to refer to different organisms that live together in close and stable association, without assuming particular benefits to either of the partners. In the literature, particularly in bipartite (one host-one microbe) systems, symbiosis is also used as a synonym for "mutualism," referring only to those associations that benefit all partners (e.g., the Hawaiian bobtail squid-*Vibrio fischeri*, vascular plant-mycorrhizae, and aphid-*Buchnera* associations). However, we prefer the former definition because most animals harbor diverse microbial communities and the net benefits or costs of the interactions are often context-dependent (i.e., may vary according to the animal life stage or environmental conditions) and, thus, difficult to prove (McFall-Ngai, 2008).

FIGURE 3.1 Animals emerged and evolved in a microbial world. Let us compress the history of our planet Earth in 12 hours, as in a clock face. The first form of life appeared around 2 o'clock. And it was a microbe. Life remained exclusively microbial until 10 o'clock. The timeline of the evolution of life on Earth sets perspective on the role of microbes as an important feature of the environment where multicellular organisms appeared and evolved.

Animal-microbe symbioses were first considered an exception rather than the rule. Lynn Margulis (1938–2011) conveyed how symbiosis determined the origin of the eukaryotic cells, based on her previous work (Sagan, 1967). The organelles (i.e., mitochondria and chloroplasts) descend from free-living eubacterial ancestors that were engulfed by another cell, becoming the first endosymbionts. These symbioses allowed the emergence of complex life on Earth. Margulis was an advocate for symbiosis as a source of innovation in evolution and anticipated that we live on a "symbiotic planet" (Margulis, 1998). However, it took a technological revolution to show how right she was (McFall-Ngai, 2008). New microscopy and molecular techniques (in particular sequencing and the use of 16S rRNA gene as a phylogenetic marker for microbes) revealed the "unseen": animals (and plants) live in association with microbes (McFall-Ngai, 2008; Bosch & McFall-Ngai, 2011; McFall-Ngai et al., 2013). Microbial symbionts offer new opportunities for their hosts as their activity expands the metabolic capability of the animal or results in secondary metabolites that contribute to defense (Venn et al., 2008;

Dubilier et al., 2008; Hudspith et al., 2021). Interestingly, microbes offer the potential for plastic responses to a changing environment because of their rapid evolutionary rates (Webster & Reusch, 2017). In summary, symbiosis with microbes is a source of innovation and contributes to the ecological success of animals and, thus, their evolution.

3.1.2 THE SPONGE HOLOBIONT: THE WHOLE IS MORE THAN THE SUM OF ITS PARTS

Animals outsource key functions, like nutrition or defense, to their microbial partners (i.e., the microbiome) (McFall-Ngai et al., 2013). Thus, to understand animal physiology, we need to study microbes. The term "holobiont" (here as syn. "metaorganism," Bosch & McFall-Ngai, 2011) refers to the unit formed by the host, the associated microbiome, and the interactions among them. The holobiont concept not only acknowledges the complexity of animal physiology in the light of symbiosis, but it also highlights that interactions within the microbiome and between the host and the microbes result in new properties that determine, for example, the ecological role of the animal (Rohwer et al., 2002; Reshef et al., 2006). In this sense, the "holobiont" is a new level of biological organization, and, probably, a new unit of selection (Theis et al., 2016). In this regard, there is still some controversy about whether and how selection can occur at the level of the holobiont, considering its complexity and conflictive selective forces acting at the level of the individual members (Douglas & Werren, 2016).

Sponges beautifully illustrate the holobiont concept and its implications to understand animal physiology (Taylor et al., 2007; Webster & Thomas, 2016; Pita et al., 2018b). Sponges appeared on our planet 600–800 million years ago and, with the help of their microbes, this group still persists today and has conquered (almost) all aquatic habitats, including the most remote ones, like the deep sea (Maldonado et al., 2015). Sponges are efficient filter-feeders, for which bacteria in the water are an important source of food (Pile et al., 1996; Coma et al., 2001; Yahel et al., 2007). Yet, they harbor specific microbiomes that are different from the microbes in the surrounding seawater (Hentschel et al., 2002; Thomas et al., 2016; Moitinho-Silva et al., 2017b). Sponge microbiomes span a wide range of diversity, from sponges dominated by few taxa to species that host some of the most diverse microbiomes (Thomas et al., 2016). In most sponges, symbionts occur extracellularly in close contact with animal cells (Figure 3.2); although there are exceptions to this rule: some sponges, like *Petrosia* species, harbor their microbes packed in modified cells known as bacteriocytes (*see also* Figure 3.4(A)).

Importantly, sponge-associated microbes determine the success and the ecological role of sponges in the ecosystem. For example, the community of associated microbes is capable of a whole nitrogen cycle inside the sponge (Bayer et al., 2008; Radax et al., 2012b; Fiore et al., 2015; Maldonado et al., 2021). Thus, the nitrogen budget, that is the net N consumption or release, is a combination of both, the host and microbial metabolism (Ribes et al., 2012; Morganti et al., 2017). Such nutrient pathways, combined with the enormous pumping capacity of the sponges (e.g., sponges can filter hundreds of liters of water per day; Kahn et al., 2015; Morganti et al., 2021; Reiswig, 1971), determines the role that sponge holobionts play as sink and/or source of nutrients in the ecosystem (reviewed in Maldonado et al., 2012; de Goeij et al., 2013). Whatever happens inside the sponge on a microscale, coupled with the animal activity, scales up to the ecosystem level (reviewed in Pita et al., 2018b).

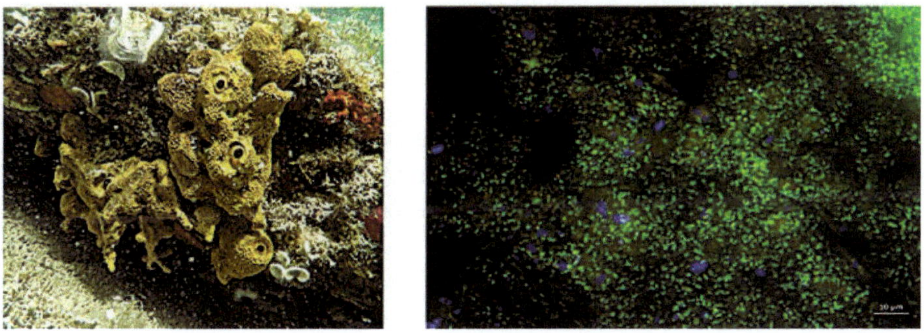

FIGURE 3.2 The sponge holobiont inside out. (A) Underwater photograph of the sponge *Aplysina aerophoba*. (B) Representative image of the mesohyl of *A. aerophoba*. Eubacteria are stained in green; sponge cell nuclei are stained with DAPI (blue).

3.1.3 A BRIEF HISTORY OF SPONGE SYMBIOSIS

The first observation of bacteria associated with sponges comes from Dosse in 1939. However, it was in the 60s and 70s with the seminal works of Jean Vacelet, Clive Wilkinson, and others, that the ubiquity and relevance of the sponge-microbe interactions started to be unveiled. These scientists examined sponges under the electron microscope and, to their surprise, they found a microbial world inside (e.g., Vacelet & Donadey, 1977). They noticed different morphotypes of sponge-associated bacteria, already pointing to diverse microbiota (Vacelet & Donadey, 1977). Vacelet & Gallisian even described "virus-like" particles in the sponge mesohyl (Vacelet & Gallissian 1978)! Their experiment with radio-labeled bacteria suggested that this diversity is not random, but sponges could discriminate host-associated from free-living bacteria (Wilkinson, 1984). They also noticed something striking: not all sponge species harbor the same number of bacteria. Some sponges were found in association with highly dense microbiomes (and they called them "bacteriosponges"), while others only harbored a handful of bacterial cells in their bodies (Vacelet & Donadey, 1977). This dichotomy was re-defined as high-microbial (HMA) and low-microbial abundance (LMA) sponges, which are estimated to contain 10^8–10^9 and 10^5–10^6 microbial cells per gram of sponge, respectively (Hentschel et al., 2003; Gloeckner et al., 2014; Moitinho-Silva et al., 2017c). Today, we know these two groups represent the two opposite extremes of a continuum characterized by symbiont density but also morphological and physiological traits (e.g., Moitinho-Silva et al., 2017c; Rix et al., 2020; Morganti et al., 2021). The pioneers of the sponge microbiology field not only focused on describing the microbial diversity they could see under the microscope, but they also wondered about the evolutionary and ecological roots of these interactions: When were these symbiotic relationships established (Wilkinson, 1987)? How do symbionts resist being digested (Wilkinson et al., 1979)? Are these relationships specific (Wilkinson, 1984)? What resources are exchanged in these associations (Wilkinson, 1979)? They started to tackle the key questions of sponge symbiosis sparking ideas that we are revisiting today with the help of state-of-the-art techniques that were not available at that time.

Molecular techniques confirmed that sponges harbor a tremendous microbial diversity source (Hentschel et al., 2002). Hentschel et al. (2002) were the first to show that the sponge microbial diversity is specific, different from the microbes in the surrounding seawater that serve sponges as an important food source (Hentschel et al., 2002). The field of sponge microbiology consolidated with the 1st symposium on Sponge microbiology in Würzburg (Germany, 2011; Taylor et al., 2011). In the second symposium, in Baltimore (USA, 2014), the germ for one of the most important endeavors for this research community was forged: The Sponge Microbiome Project (Thomas et al., 2016; Moitinho-Silva et al., 2017b). Different labs, from different parts of the world, joined forces to collect 81 sponge species and 804 specimens (Thomas et al., 2016). Samples were then processed and sequenced (*16S* rRNA gene amplicon sequencing, Table 3.1) in a standardized way, under the umbrella of the Earth Microbiome Project (Gilbert et al., 2014; Thompson et al., 2017), enabling a comparative approach to define the diversity, specificity, and commonalities of the microbiota across the phylum Porifera (Thomas et al., 2016; Moitinho-Silva et al., 2017b).

Developing standardized protocols and community efforts greatly advanced our knowledge of sponge-associated microbial diversity, but shedding light on the lifestyle of sponge symbionts and on the molecular mechanisms that control sponge-microbe interactions remains a big challenge. Current efforts are directed towards unveiling still unexplored

TABLE 3.1 The Methodological Toolbox in Sponge Symbiosis*

	Technique	Purpose	Example
Microbial Diversity	*16S* rRNA gene amplicon sequencing.	Biodiversity assessment of complex microbial communities. Since hundreds of samples can be processed simultaneously, this technique allows relatively fast and neat insights into the microbial community composition, including relative abundance data for uncultured microbial taxa.	Sponges harbor species-specific microbial communities composed of generalists and specialists (Thomas et al., 2016).
	16S rRNA gene qPCR.	Quantification of *16S* rRNA gene copy numbers. Provides absolute values of microbial abundance, allowing a more accurate view of microbial dynamics over time than amplicon sequencing.	Absolute numbers of Archaea, *Chloroflexi*, and the candidate phylum *Poribacteria* can be 4–6 orders of magnitude higher in HMA than in LMA sponges. *Actinobacteria* numbers can be 1–2 orders higher in HMA over LMA sponges (Bayer et al., 2014).
Microbial Function	Metagenomics, metagenomic binning, metatranscriptomics.	• **Metagenomics:** Sequencing of total microbial DNA from a sample. It provides information about the metabolic potential of the microbiome. • **Metagenomic Binning:** Extraction of genomic information for a particular microbe from the metagenomic data. • **Metatranscriptomics:** Sequencing of microbial RNA from a sample, providing information about the microbial genes expressed at a particular time point.	Several core functions can be found consistently in sponge microbiomes, even in phylogenetically divergent sponges (Fan et al., 2012; Slaby et al., 2017).

TABLE 3.1 *(Continued)*

	Technique	Purpose	Example
Imaging	SEM, TEM	• **Scanning Electron Microscopy (SEM):** It provides a detailed, almost 3D view of microbial cells within the sponge. • **Transmission Electron Microscopy (TEM):** It provides a 2D image of a tissue section at high resolution (nm). Both techniques reveal the ultrastructure of the sponge and show the microbial cells in their spatial context.	Sponges can be distinguished into HMA and LMA sponges (Vacelet & Donadey, 1977; Gloeckner et al., 2014).
	FISH, FISH-CLEM, FISH-IHC-CLEM	• **Fluorescence** ***In situ*** **Hybridization (FISH):** It allows to mark off of specific nucleic acid sequences with a fluorescent signal for the detection of a specific microbial taxon/microbial activity. An inconvenience of FISH is that the structural context is lost, a problem that can be resolved by correlative light and electron microscopy (CLEM). In combination with immunohistochemistry (IHC), FISH-IHC-CLEM can reveal the localization of metabolic activity (protein) at subcellular resolution.	Sponge symbionts (such as *Chloroflexi* and *Poribacteria*) occur frequently in the sponge extracellular matrix (Bayer et al., 2018; Jahn et al., 2016).
	NanoSIMS	Nanoscale secondary ion mass spectrometry (nano-SIMS) is a high-resolution technique that uses light isotopes and/or rare elements to visualize metabolic processes within a biological sample.	Microbial symbionts are actively involved in the assimilation of DOM, but host filtering cells seem to be the primary site of organic matter uptake (Hudspith et al., 2021; Rix et al., 2020).

TABLE 3.1 *(Continued)*

	Technique	Purpose	Example
Physiology of the Sponge Holobiont	Incubation chambers, in-ex method (VacuSIP), dye front speed method (DFS)	These methods allow the estimation of sponge physiological traits (pumping, respiration, and feeding) *in situ* and the laboratory. • **Incubation Chambers:** Indirect method to estimate the rate of change in nutrient and oxygen concentrations in the incubated water. Rates of production or removal by the holobiont can be estimated when compared to adequate controls. • **In-Ex Method:** Direct method to estimate ingestion/excretion rates of the compound of interest by simultaneously sampling the water inhaled and exhaled by the sponge holobiont. An example of the application of the In-Ex method is the VacuSIP device. • **DFS:** It is a dye technique that allows direct estimation of the volume of water processed by the sponge.	Sponges play a key role in the benthopelagic coupling of energy, nutrients, and organic matter. This role is the result of microbial and host metabolism (Yahel et al., 2005; Morganti et al., 2016; Maier et al., 2020).
Sponge Immune System	Host transcriptomics.	Transcriptomics of the sponge host can be used to detect changes in gene expression in response to certain experimental conditions.	Recognition of and response to microbial-associated molecular patterns via immune receptors differs between analyzed HMA and LMA sponges (Pita et al., 2018; Schmittmann et al., 2021).

*Main techniques used in sponge symbiosis research and applied in the works references in this chapter. The array of techniques is equally diverse as the research questions.

diversity (e.g., that in the deep sea) and moving from diversity to functional aspects of sponge-associated microbes. Disclosing the interactions within the sponge holobiont and their physiological impact is the new frontier. The sponge holobiont can be viewed as a miniature ecosystem (McFall-Ngai et al., 2013; Pita et al., 2018b). Understanding this system requires a holistic approach that integrates ecological perspectives with microbiology, zoology, chemistry, and genetics (Dittami et al., 2021) (Table 3.1).

In this chapter, we discuss how microbes shape sponge physiology. First, we will introduce the enormous microbial diversity harbored in sponges, how it is transmitted, and the connection between microbial diversity and sponge physiology (illustrated by the HMA-LMA spectrum). Second, we will present the functions provided by the microbiome to its hosts. In particular, we will focus on the metabolism, going from the microbial activities within the sponge to the consequences at the ecosystem level. Then, we will look at the immune system as guardian of sponge-microbe interactions and, finally, we will discuss how sponge holobionts face a changing environment.

3.2 SPONGE MICROBIAL DIVERSITY

3.2.1 WHO IS THERE?

Marine sponges host a broad range of microbes including archaea, bacteria, eukaryotes, and viruses. In terms of taxonomic diversity, > 63 microbial phyla have previously been detected within sponges (Thomas et al., 2016; Moitinho-Silva et al., 2017b). Currently, the majority of environmental microbes remain uncultivated and are therefore described by the term “microbial dark matter” (Rinke et al., 2013). *In vitro* cultivation is presently also impossible for the predominant sponge-associated microbial phylotypes, despite considerable efforts and various tested approaches (Sipkema et al., 2011; Steinert et al., 2014; Lavy et al., 2014; Versluis et al., 2017; Gutleben et al., 2020). It was hypothesized that the inability to adequately recreate the sponge microenvironment (i.e., the mesohyl conditions), hamper cultivation success as natural sponge-microbe associations are complex interaction networks involving extensive cross-feeding activities (Pande & Kost, 2017; Gutleben et al., 2020). Thus, the characterization of sponge microbial diversity relies on molecular tools like *16S* rRNA gene amplicon sequencing (Table 3.1). The following paragraphs will provide an overview of some key taxa thriving within sponges.

Members of the domain *Archaea* occur widely in the world’s oceans, despite they were once thought to be found only in extreme environments (Massana et al., 2000; Karner et al., 2001). The first archaeon described from a marine sponge (the sponge *Axinella mexicana*, Laubenfels, 1935) was the psychrophilic thaumarchaeon *Cenarchaeum symbiosum* (Preston et al., 1996; which belongs to the *Nitrosopumilaceae*). Similar to seawater-living *Thaumarchaeota*, *Cenarchaeum symbiosum* could be either a strict autotroph (i.e., using CO_2 as carbon source) or a mixotroph (i.e., using organic material as carbon sources) (Hallam et al., 2006). A broad association of archaea with sponges has since been demonstrated (Holmes & Blanch, 2006; Lee et al., 2011; Polónia et al., 2014a, b), with the four most common archaeal phyla namely *Thaumarchaeota*, *Euryarchaeota*, *Nanoarchaeota*, and *Crenarchaeota*

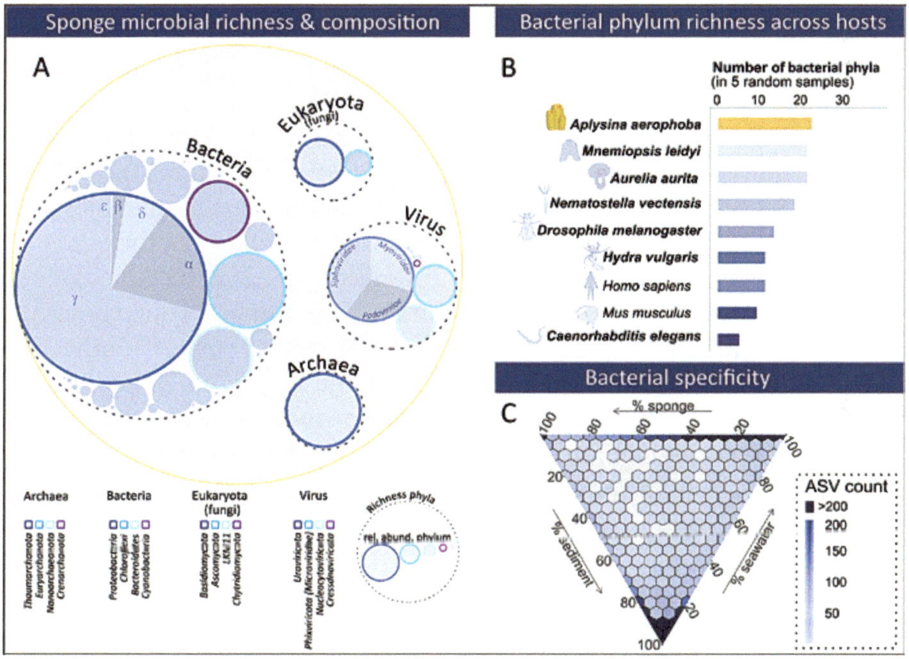

FIGURE 3.3 Sponges are reservoirs of unique microbial diversity. (A) Richness and composition of archaea (after Steinert et al., 2020), bacteria (after Thomas et al., 2016), fungi (after Naim et al., 2017), and viruses (after Laffy et al., 2018). The size of each group circle (archaea, bacteria, fungi, virus) is according to group richness (number of phyla). Blue bubbles inside each group are sized according to the relative abundances of each phylum within the respective group. The four most abundant phyla are highlighted for each group by blue/purple outlines around the blue bubbles. Other, less abundant, phyla have no outlines around the blue bubbles. Note that for some eukaryotic groups their relative abundance is too low for visualization. Pie charts show classes for bacteria; and families of viruses. (B) The number of bacterial phyla indicated for 9 different organisms (sorted in descending order; data and icons based on Rausch et al., 2019). Invertebrates are written in bold, and data from the sponge *Aplysina aerophoba* are presented in yellow. (C) Specificity of sponge-associated bacterial taxa (ASVs) in comparison to environmental reference microbiomes (seawater and sediment).

(Figure 3.3(A)). Importantly, ammonia-oxidizing archaea (AOA) act as the main contributors to nitrification in sponges (Hoffmann et al., 2009; Radax et al., 2012b; Li et al., 2014; Zhang et al., 2014; Tian et al., 2016).

Sponges represent a lush reservoir of many interesting microbes within the domain *Bacteria*. Proteobacteria (mostly *Alpha*- and *Gammaproteobacteria*) represent one of the most successful bacterial phyla within sponges (Figure 3.3(A)), especially lineages belonging to the families *Rhodobacteraceae* and *Rhodospirillaceae* dominate sponge-associated microbial communities (Naim et al., 2014; Karimi et al., 2017, 2019). Another prominent bacterial phylum thriving within sponges is *Chloroflexi*, especially the SAR202 clade (Moitinho-Silva et al., 2017b; Bayer et al., 2018). SAR202 is considered to fulfill important metabolic functions in the sponge holobiont, such as cycling of organic matter (Bayer et al., 2018), and has been shown to dominate the sponge bacterial community consistently across different life stages (embryos, juveniles, adults) of some deep-sea sponges (Busch et al., 2020b). Several bacterial groups typically found in sponges are known for their involvement in sponge holobiont nitrogen cycling (Hoffmann et al., 2009; Fan et al., 2012; Liu et al., 2012; Radax et al., 2012b). Among these, a common group is *Nitrospira*, presumably responsible for the conversion of nitrite to nitrate (i.e., nitrite-oxidation, the second step of nitrification) within sponges (Hentschel et al., 2002; Schmitt et al., 2007; Taylor et al., 2007; Off et al., 2010). Another prominent sponge-associated bacteria group is the candidate phylum *Poribacteria. Poribacteria* were initially considered to live exclusively in sponges – hence the name (Fieseler et al., 2004; Lafi et al., 2009; Taylor et al., 2013; Steinert et al., 2017), but were later found also in other environments (Podell et al., 2018). Many tropical and temperate sponges are inhabited by large abundances of cyanobacterial symbionts, particularly of the genus *Synechococcus* (Unson et al., 1994; Hentschel et al., 2006; Bayer et al., 2014b), and specifically the sponge symbiont *Candidatus* S. spongiarum (Steindler et al., 2005; Erwin & Thacker, 2008). These lineages play a relevant role in carbon and nitrogen fixation (Burgsdorf et al., 2015). And although less ubiquitous, Entotheonella bacteria have received special attention because of their role in the production of sponge-derived natural products and the resulting potential in biotechnological and medical applications (e.g., Bhushan et al., 2017; Freeman et al., 2016; Mori et al., 2018; Piel et al., 2004a) (*see* Section 3.4.3.2). In summary, sponges represent a lush reservoir of many interesting bacterial phyla and they harbor a comparably higher diversity of bacteria in comparison to other invertebrates (Figure 3.3(B)). This bacterial diversity is strikingly high in comparison to the hitherto described archaeal diversity in sponges, and also in comparison to the eukaryotic as well as viral diversity in sponges which will be described in the following paragraphs (Figure 3.3(A)).

Eukaryotic microbes found in sponges include both fungi and microalgae. Fungi are a prominent microbial group in terrestrial ecosystems, while they are far less studied in the marine context (Naim et al., 2017). It is therefore notable that fungal introns have been found in sponge genomes (Rot et al., 2006) and several representatives have been isolated from sponges (Höller et al., 2000; Pivkin et al., 2006; Wang et al., 2008; Baker et al., 2009; Liu et al., 2010; Paz et al., 2010; Wiese et al., 2011; Passarini et al., 2013). The fungal taxa detected in sponges belong to the phyla *Ascomycota* and *Basidiomycota* (Naim et al., 2017; Figure 3.3(A)). The vast majority of fungal reads were yeast and the most abundant representatives belonged to the order *Malasseziales*. Overall, fungi seem to occur in very limited abundance inside sponges compared to other microbes. So far, the microalgae found in sponges are the green-alga *Acrochaete leptochaete*, the diatom *Nanofrustulum* cf. *shiloi*, the rhodophyte *Acrochaetium spongicola*, and the chlorachniophyte *Lotharella oceanica* (Konstantinou et al., 2020). Only a few studies have assessed fungal and other microbial eukaryotic communities in marine sponges by means of high-throughput sequencing technologies (Hardoim et al., 2021); thus, there remains potential for discoveries. For example, the symbiosis of the freshwater sponge *Ephydatia muelleri* (Lieberkühn, 1856) and its *Chlorella*-like algal partner is a promising model for studying animal-algal endosymbiosis (Kenny et al., 2020; Geraghty et al., 2021).

Research on the viral diversity in sponges is only in its infancies (Laffy et al., 2018; Pascelli et al., 2018; Jahn et al., 2019, 2021), but phages are appearing as key players in governing interactions within the holobiont (Jahn et al., 2019, 2021) (*see* Section 3.4.2). Common viral taxa reported in sponges are *Uroviricota* (especially *Siphoviridae*, *Myoviridae*, and *Podoviridae*), *Phixviricota* (*Microviridae*), *Nucleocytoviricota*, and *Cressdnaviricota* (Figure 3.3(A)). Importantly, the viral beta diversity found in sponges is highly unique – even on the sponge individual level, including taxa that have not been reported before in any other habitat (Jahn et al., 2019).

3.2.2 HOW SPECIFIC ARE SPONGE-ASSOCIATED MICROBIAL COMMUNITIES?

The specificity and stability of host-microbe associations are major themes in symbiosis research. This question is particularly relevant in sponges because of their filter-feeding lifestyle. Hentschel et al. (2002) were the first to conclude that sponges of different phylogenetic orders and sampling regions harbor a uniform and phylogenetically complex bacterial community, different from that in the surrounding seawater. The authors defined sponge-specific clusters (SSC) as groups that contain at least three sequences that fulfill the following three criteria: "(i) have been recovered from different sponge species and/or from different geographic locations; (ii) are more closely related to each other than to any other sequence from non-sponge sources; and (iii) cluster together independent of the treeing method used" (Hentschel et al., 2002). In recent years, with novel methodological advances at hand, a higher resolution of bacterial community data could be achieved and SSC-sequences were retrieved (in low abundances) from other environments than sponges (e.g., Webster et al., 2010; Lee et al., 2011; Taylor et al., 2013). Nevertheless, sponge-associated bacterial communities are largely distinct from environmental reference samples in terms of their richness and composition (Thomas et al., 2016; Figure 3.3(C); but note also Cleary et al., 2019). To account for the occurrence of 'sponge-specific' microbes in the environment, the term 'sponge-enriched' is now preferred (Moitinho-Silva et al., 2014; Thomas et al., 2016).

Host identity is one determinant of sponge-associated bacterial diversity, leading to largely species-specific bacterial communities (Pita et al., 2013a; Easson & Thacker, 2014; Thomas et al., 2016; Moitinho-Silva et al., 2017c). In particular, relative abundances of individual bacterial taxa are frequently characteristic of different sponge species. Yet, the core bacterial community, which represents those microbial taxa consistently recovered in individuals of a sponge species, typically consists of a rather small fraction of the whole microbiota per sponge (Erwin et al., 2015; Thomas et al., 2016; Astudillo-García et al., 2017). This highlights that each bacterial community is a mix of generalist (i.e., cosmopolitan symbionts present in multiple sponge species and even in other hosts like corals) and specialists (i.e., symbionts in one or a handful of sponge species but not in the others) (Thomas et al., 2016).

Knowledge on the specificity of sponge-associated archaea is very limited today, although a sponge specificity with intra-individual variations between sponges of the same species has been observed (Steinert et al., 2020). With respect to eukaryotes and viruses, knowledge about their specificity is also still rather limited. No global large-scale studies are available so far, but existing case studies suggest that fungi show no specificity (Naim et al., 2017), while viruses exhibit sponge species-specificity (Jahn et al., 2019).

Intriguingly, within the species-specific pattern, each sponge individual of the same species hosted a unique virome, even if sponge individuals were collected in close proximity (Jahn et al., 2019).

Patterns of microbial diversity are likely influenced by the mode of transmission of microbes. In sponges, both symbiont vertical transmission (from parents to early life stages) and horizontal transmission (from the environment) were observed (Schmitt et al., 2008; Ribes et al., 2015; Björk et al., 2019; Rodrigues de Oliveira et al., 2020) and, at least in shallow waters, it is accepted that the most common transmission mode of microbial taxa is the combination of both, the so-called leaky vertical transmission, or mixed mode of transmission (Schmitt et al., 2008; Ribes et al., 2015; Björk et al., 2019; Rodrigues de Oliveira et al., 2020). Vertically-transmitted microbes may work as a seed for the microbial community in the next sponge generation. Further, the process of vertical transmission links microbes and host evolutionary trajectories (Thacker & Starnes, 2003; Rodrigues de Oliveira et al., 2020).

3.2.3 DO ENVIRONMENTAL FACTORS SHAPE SPONGE MICROBIAL COMMUNITIES?

Interactions of the sponge holobiont with the environment are multidirectional: while the sponge holobiont affects biogeochemical cycling of key nutrients like carbon (C), nitrogen (N), phosphorous (P), and silicate (Si) (reviewed in Maldonado et al., 2012; Pita et al., 2018), also the ecosystem may influence the microbial community composition in sponges. With respect to the latter, several studies have revealed significant differences in microbial community composition over spatial and temporal scales, geographic location, nutrient concentration, or habitat (Anderson et al., 2010; Burgsdorf et al., 2014; Cleary et al., 2020; Fiore et al., 2013b; Luter et al., 2015; Santos et al., 2020; Swierts et al., 2018; Turque et al., 2010; Weigel & Erwin, 2016). Other studies on the contrary, observed stable microbial community compositions between locations, seasons, and depths (Hentschel et al., 2002; Taylor et al., 2005; Erwin et al., 2012, 2015; Björk et al., 2013; Pita et al., 2013a; Simister et al., 2013; Reveillaud, 2014). It was suggested that different host-microbe interaction strengths of different sponge species may explain contrasting results (Lee et al., 2009; Cleary et al., 2013; Thomas et al., 2016).

In an elaborate conceptual study, Björk et al. (2017) observed a negative relationship between the temporal variability of microbial taxa and their mean abundance in a long-term (three years) experiment on shallow water sponges. They concluded that the stability of the microbiome is not related to diversity, but rather to the density of the core microbiome. A similar observation that stability is not directly related to diversity has been described by Glasl et al. (2018). Large-scale surveys within the frame of the EMP have identified salinity as a prominent environmental driver for both, host-associated as well as free-living microbes globally (Thompson et al., 2017). A network analysis of the global shallow water sponge EMP has further revealed that the overall structuring of the microbiome is influenced by abiotic factors, while the core community is driven by biotic interactions (Lurgi et al., 2019).

In both of the above-mentioned large-scale approaches (Thompson et al., 2017; Lurgi et al., 2019), the number of abiotic/environmental factors included was limited to depth, temperature, and salinity. Biogeochemical drivers of the microbial community composition are however not well understood. For shallow-water sponges, microbiome composition was found to be not related to dissolved organic carbon (DOC), particulate organic carbon (POC), phosphate, and NO_x-fluxes, while significant correlations between microbial community structure and NH_4^+ flux were observed (Gantt et al., 2019). In a study on deep-sea sponges, variations in microbial community compositions of three sponge species were most strongly related to variations in depth, nutrient (nitrate, silicate), and oxygen concentrations (based on a set of 9 analyzed environmental parameters in total) (Busch et al., 2020a). Harmonized large-scale studies on environmental drivers of microbiome variations are still lacking, but are key in times of rapid environmental change (Busch, 2020). Further evaluations are not only needed for direct environmental drivers (such as temperature, salinity, and nutrients) but also for "indirect" environmental drivers. One example of putative indirect drivers of microbial community composition is ocean currents which play a relevant role in governing the dispersion of sponge larvae, that in turn can be related to speciation processes and concomitant variations in microbial community composition (Busch et al., 2021).

3.3 HMA-LMA SPONGES: A LINK BETWEEN MICROBIAL DENSITY AND SPONGE PHYSIOLOGY?

Based on the microbial load inside the tissue, sponges are classified as low-microbial abundance (LMA) sponges, or high-microbial abundance (HMA) sponges (Vacelet & Donadey, 1977; Reiswig, 1981; Hentschel et al., 2003;

Figure 3.4). The causes for this HMA-LMA spectrum remain a mystery, as they cannot clearly be attributed to phylogenetic and evolutionary drivers (e.g., Moitinho-Silva et al., 2017b). Several bacterial phyla and taxa are differentially enriched in either one of the two microbial abundance states (Moitinho-Silva et al., 2017c): *Chloroflexi* (e.g., SAR202, *Caldilineaceae*), *Acidobacteria* (e.g., *Solibacteres*, *PAUC37f*, *Sva725*), *Poribacteria*, and *Actinobacteria* (e.g., *Acidimicrobiia*) are typical members of the bacterial signature of HMA sponges, whereas *Proteobacteria* (e.g., G*ammaproteobacteria*), *Bacteroidetes* (e.g., *Flavobacteriia*), and *Planctomycetes* (e.g., *Planctomycetia*) are typical LMA-indicator taxa. No indicator phyla have been identified from the three other microbial groups (i.e., *Archaea*, viruses, eukaryotic microbes) thriving in sponges, yet.

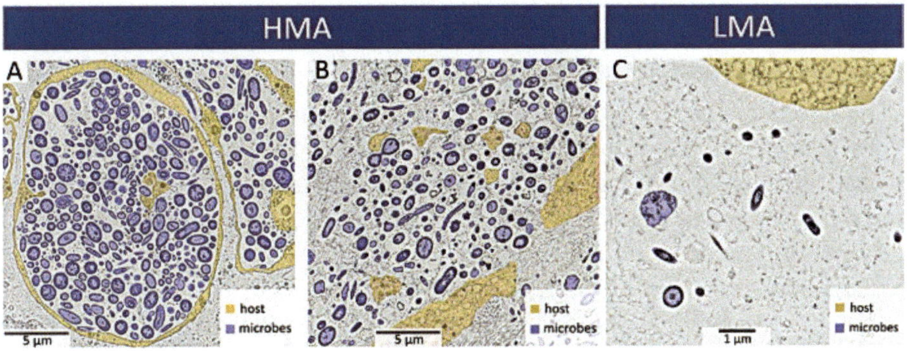

FIGURE 3.4 HMA-LMA dichotomy illustrated by microscopy. (A) Intracellular symbionts within bacteriocytes of an HMA sponge (*Petrosia crassa*). (B) Extracellular symbionts in an HMA sponge. (C) Extracellular symbionts in an LMA sponge.

Source: Reprinted from Busch et al. (2020); https://macau.uni-kiel.de/receive/macau_mods_00000940. https://creativecommons.org/licenses/by/4.0/deed.en

Besides the different microbial community composition and density, HMA and LMA sponges have contrasting aquiferous systems and physiological features, with pumping rate (i.e., the volume of water processed by the sponges over time) generally considered one of the main distinctive traits of these two life strategies. HMA species were observed to pump 52–92% less volume of water than LMAs per unit of sponge volume (Weisz et al., 2008; Morganti et al., 2019). Such difference was attributed to the distinct morphology of the canals (Vacelet & Donadey, 1977; Weisz et al., 2008; Schläppy et al., 2010) and choanocyte chamber density (Poppell et al., 2013). However, the validity of this distinction was recently questioned, as sponge size arose as the main driver of sponge pumping (Morganti et al., 2019). The pumping rate per unit of sponge volume decreases with sponge size, which suggests that the use of single-size normalized values might be misleading when comparing nutrient fluxes and metabolic rates (Morganti et al., 2019). Thus, at least in some instances, the different pumping rates observed between HMA and LMA species might be directly a consequence of the different sizes of analyzed sponges. Other factors such as choanocyte size, structure, and activity should be investigated in light of HMA-LMA status and their pumping divergences.

A second distinctive physiological feature that commonly characterizes the HMA-LMA status is their different feeding strategy. Since the pioneering study by Reiswig in 1974, the HMA-LMA status was widely recognized as different usage of dissolved and particulate organic food by these two sponge groups (reviewed by Maldonado et al., 2012). Due to the wide recognition that microbes play a fundamental role in dissolved organic matter (DOM) cycling (Azam et al., 1983), HMA sponges were thought to be better adapted to tapping DOM, whereas LMA sponges would solely use particulate food to meet their energetic needs. Our knowledge of sponge feeding has significantly improved since Reiswig's study (1974), revealing convergences and divergences in food preferences and acquisition across the HMA-LMA spectrum. The hypothesis that DOM uptake was restricted to HMA species was confused as many studies documented DOM uptake by LMAs and reported, surprisingly, that DOC could represent their main carbon source (~90%), as observed for HMA species (Bart et al., 2020; reviewed by de Goeij et al., 2017; Morganti et al., 2017; Rix et al., 2020). Contrary to what was previously thought, DOM uptake was not restricted to only symbiont cells, but host cells (i.e., choanocytes) can take up DOM in both sponge groups (Achlatis et al., 2019; Rix et al., 2020; Hudspith et al., 2021). Overall, two strategies for exploiting carbon sources can be described for LMA species: some LMA species mostly rely on DOC as the main C input – as HMA species do (de Goeij et al., 2017; Morganti et al., 2017; Rix et al., 2020; Bart et al., 2021), while others heavily rely on filter feeding of particulate organic matter, with negligible DOC ingestion (Yahel et al., 2007; Hoer et al., 2018a; McMurray et al., 2018). Although the HMA-LMA status seems to not apply here, the removal rates and the mechanisms involved in the DOM uptake indeed reflect HMA-LMA traits. DOM is assimilated more efficiently in HMAs (Bart et al., 2020; Rix et al., 2020; but see Campana et al., 2021), but sometimes at lower removal rates due to the discrepancy in pumping rates (Morganti et al., 2017; Rix et al., 2020; Bart et al., 2020). Notably, DOM processing is microbe-mediated in HMA species (microbes making up 65–87% of DOM uptake), while it is sponge host-mediated in LMAs (microbes making up less than 5% of DOM uptake) (Rix et al., 2020). This process corresponds to the significant difference between the host and symbiont biomass across HMA-LMA status (Rix et al., 2020), but does not depend on microbial community structure (Gantt et al., 2019). Although isotope tracer studies demonstrated that both sponge guilds can tap into dissolved organic nitrogen (DON), our knowledge of net DON ingestion and/or excretion *in situ* is limited to a few studies with contrasting results (Ribes et al., 2012; Morganti et al., 2017; Hoer et al., 2018a). Among them, only one study documented a distinct pattern of DON ingestion (HMA) and excretion (LMA) (Morganti et al., 2017).

Both sponge groups are commonly recognized sources of dissolved inorganic nitrogen (DIN) in shallow and deep-sea ecosystems (Hoer et al., 2018b; Maldonado et al., 2021). However, the chemical speciation of DIN fluxes, which has been attributed to different microbial structures and compositions (Ribes et al., 2012; Gantt et al., 2019), generally reflects the HMA-LMA status. Briefly, HMAs typically ingest NH_4^+ and excrete NO_x^- (reviewed by Maldonado et al., 2012) as a result of nitrifying microbial activity (Corredor et al., 1988); while LMAs are net sources of NH_4^+, although NO_x^- excretion has been also observed (Jimenez & Ribes, 2007; Southwell et al., 2008; Campana et al., 2021). The different nitrogen fluxes that characterize the HMA-LMA status are further indicated by their distinct nitrogen trophic level, with HMAs usually having lower N isotopic signal compared to LMAs, indicative of different N cycling pathways within the sponge tissue (Weisz et al., 2007; Freeman et al., 2014, 2020). Interestingly, N fluxes (ingestion and/or excretion) and rates can vary within the same species depending on habitat locations (Fiore et al., 2013a; Archer et al., 2017), seasonality (Bayer et al., 2008), and different initial ambient concentrations (Ribes et al., 2012; Archer et al., 2017; Morganti et al., 2017). They can further be coupled with the inter-individual variability in pumping activity (Reiswig, 1971; Bell et al., 1999; McMurray et al., 2014; Strehlow et al., 2016) and with fluctuations of the microbial community structure (Maldonado et al., 2021). With regards to phosphate, no distinct pattern was observed between HMA and LMA species, with studies documenting phosphate efflux in both sponge types (Jimenez & Ribes, 2007; Yahel et al., 2007; Ribes et al., 2012; López-Acosta et al., 2019; Maldonado et al., 2021; Campana et al., 2021).

Their different strategy in DOM exploitation (Rix et al., 2020) and preferences (Rix et al., 2016b; Campana et al., 2021), together with the observed different nitrogen cycling between sympatric HMA and LMA sponges, might facilitate their co-existence in dense benthic communities thriving in oligotrophic environments such as coral reefs and Mediterranean coralligenous concretions (Morganti et al., 2017; McMurray et al., 2018). It remains unknown whether the DON and NH_4^+ excreted by LMA species may be utilized by HMAs (Morganti et al., 2017), as well as whether the DOC excreted by LMA species may be utilized by HMAs (McMurray et al., 2018). Further studies should elucidate the putative trophic link between HMA and LMA species, which ultimately will valuably aid to understand DOM-DOM (re)-cycling in the ecosystem driven by the "sponge loop" (de Goeij et al., 2013). The sponge loop consists on the transfer of sponge assimilated dissolved organic matter (DOM) to higher trophic levels via sponge detritus production and direct predation on sponge tissue (de Goeij et al., 2013; Rix et al., 2016a). Sponges have been observed to excrete DON and carbon at high rates, sometimes as a function of their ambient concentration (Mueller et al., 2014; McMurray et al., 2016; Morganti et al., 2017). So far, two studies on Caribbean species observed that holobiont-derived DOM had a different molecular profile than ambient DOM, suggesting that sponges may impact local DOM composition. However, it is not clear whether HMA and LMA sponges have different holobiont-derived DOM profiles (Fiore, Cara, Freeman, & Kujawinski, 2017; Olinger et al., 2021). Given the important role of sponges in processing DOM and transferring this matter from/to benthic-pelagic realms, little is known about the impact that sponges have on the composition of DOM and its ecological function in the food web.

3.4 FUNCTIONS OF SPONGE-ASSOCIATED MICROBES

3.4.1 COMMON FUNCTIONS IN THE SPONGE MICROBIOME

One limitation in the study of the sponge microbiome and its contribution to sponge physiology is that most sponge symbionts remain uncultivable. For this reason, not only their diversity but also their lifestyle is interpreted by sequencing-based techniques (mainly metagenomics and metatranscriptomics) and bioinformatic tools, in combination with high-resolution microscopy techniques (Table 3.1). The first studies on the genomic potential of sponge microbiomes aimed at identifying the set of genes and gene functions that are enriched in sponge-associated microbes compared to their free-living counterparts (reviewed by Pita et al., 2018b; Webster & Thomas, 2016). Through these comparative analyses, common patterns of enriched functions arose from different sponge species, despite their species-specific microbial diversity (Thomas et al., 2010; Fan et al.,

2012). This "sponge symbiont functions" likely correspond to specific requirements to survive inside the sponge mesohyl or respond to the symbiosis currency (either between the microbes or with the host). For example, one of the most prevalent enriched functions includes defense mechanisms, as clustered regularly interspaced short palindromic repeats (CRISPR) and restriction-modification systems, which are interpreted as a toolbox against mobile genetic elements (e.g., viruses, transposable elements, plasmids) brought into the sponge body by the sponge's filter activity (Fan et al., 2012; Horn et al., 2016). Sponge microbiomes are also commonly enriched in genes involved in sponge matrix utilization (Kamke et al., 2013; Slaby et al., 2017) Finally, the observed enrichment of eukaryote-like gene repeat proteins may be linked to the need to avoid host phagocytosis (*see* Section 3.4).

These sponge symbiotic functions are not equally distributed among microbial phyla within the sponge microbiome (Robbins et al., 2021). For example, glycosyl hydrolases and carbohydrate esterases, enzymes that could target components of sponge organic components, were largely restricted to the phyla *Poribacteria*, *Latescibacteria*, *Spirochaetota*, *Chloroflexota* classes *UBA2235* and *Anaerolineae* (Robbins et al., 2021). Enrichment of eukaryote-like gene repeat protein was detected consistently across different microbial taxa but was more pronounced in specific lineages such as *Poribacteria*, *Latescibacteriota*, and *Acidobacteriota* (Robbins et al., 2021). Robbins et al. (2021) suggested that these symbiotic functions were either acquired from a common ancestor or resulted from more recent lateral gene transfer events, potentially mediated by mobile genetic elements.

For the rest of this section, we want to focus on studies that point to the currency of symbiosis, that is, the resources exchanged among microbes or between microbes and the host (Wein et al., 2019). We will discuss nutrition specialization, which may explain to a certain extent, the enormous microbial diversity in sponges; we will present detailed models for potential metabolic exchanges within the sponge holobiont; and we will link microbial activity to the functioning of the sponge holobiont within the ecosystem.

3.4.2 HOST-MICROBE AND MICROBE-MICROBE METABOLIC EXCHANGES WITHIN THE SPONGE HOLOBIONT

LMA sponges are typically dominated by a few microbial phylogroups, whereas the microbiome in HMA sponges is more complex. For this reason, LMA sponges serve as a naturally-reduced system to decipher metabolic interactions from metagenomic data. Here we present two detailed models of metabolic reconstructions in LMA sponges, one in the temperate demosponge *Cymbastella concentrica* (Lendenfeld, 1887) (based on Moitinho-Silva et al., 2017a) and the other in the deep sea hexactinellid sponge *Vazella pourtalesii* (Schmidt, 1879) (based on Bayer et al., 2020). We use these two studies to illustrate how specialized metabolisms and niche specializations result in intricated metabolic interactions that efficiently recycle the resources within the holobiont.

3.4.2.1 THE METABOLIC INTERTWINING WITHIN C. CONCENTRICA

Moitinho-Silva et al. (2017a) reconstructed three microbial genome bins from the metagenome of the sponge *C. concentrica*. These genome bins corresponded to an alphaproteobacterial symbiont belonging to the family *Phyllobacteriaceae* (CcPhy), a member of the genus *Nitrospira* (CcNi), and one of the thaumarchaeal order *Nitrosopumilus* (CcThau), as revealed by *16S* rRNA gene phylogeny. These microbes dominated *C. concentrica*'s microbiome and their metabolic activity was demonstrated by metatranscriptomic analysis (Figure 3.5(A)). In addition, the metabolic potentials of the sponge host and an associated diatom were integrated into the model of potential metabolic exchanges within this holobiont (Figure 3.5(A)).

CcPhy exhibited a heterotrophic metabolism fueled by the acquisition of organic compounds derived from the sponge's filtration activity and symbiotic carbon fixation (e.g., by the diatom and CcThau) (Figure 3.5(A)), as suggested by the active expression of many membrane transporters. These organic compounds are likely found widely distributed across the sponge tissue, which may explain the scattered distribution of CcPhy cells throughout the sponge mesohyl, as observed by fluorescence *in situ* hybridization (FISH). CcPhy central metabolism included glycolysis, the TCA pathway, and the pentose phosphate pathway. The expression of genes involved in fatty acid uptake, the glyoxylate cycle, β-oxidation, and the propanoyl-CoA degradation pathway make aerobic growth on fatty acids and acetate possible. CcPhy cells most likely release urea into the sponge mesohyl, as a product of the degradation of creatine to glycine, because they lack urease (the enzyme for urea breakdown). In the absence of oxygen, CcPhy can perform nitrate respiration to nitrite, which is most likely excreted into the mesohyl.

The genomes of CcThau (*Nitrosopumilus*) and CcNi (*Nitrospira*) revealed a chemolithoautotrophic lifestyle with a coupled capacity for nitrification (ammonia-oxidation, nitrite-oxidation). Genes for energy acquisition through ammonia oxidation (i.e., *amoCBA*) were among the most abundant transcribed genes in CcThau, whereas the key enzyme for energy acquisition through nitrite oxidation to nitrate (i.e., nitrite oxidoreductase) was among the most abundant transcripts in CcNi. Remarkably, both symbiont groups responsible for nitrification (CcThau and CcNi) were observed in close proximity to each other using FISH, providing further evidence for the direct interaction of these microbes.

Further, CcThau utilizes urea as a carbon source, which might be transported from the mesohyl into the cells or could also result from the internal degradation of L-arginine. Another piece of evidence for the suggested use of external organic carbon sources comes from genes for creatinine degradation. In addition, CcThau likely uses the HP/HB (hydroxypropionate/hydroxybutyrate) pathway to fix CO_2 autotrophically. CcNi possesses a putative operon with genes related to the transport and decomposition of urea, including those encoding for subunits of urease, and urea ABC transporter among others. CcNi highly expressed a gene encoding for cyanate lyase, which decomposes cyanate to CO_2 and ammonia. This reaction may mediate nitrogen assimilation, cyanate detoxification, and energy acquisition. The authors (Moitinho-Silva et al., 2017a) hypothesize that CcNi cells fix carbon via the reductive Tricarboxylic Acid (rTCA) cycle, similar to other *Nitrospira* members (Daims et al., 2016), despite the lack of key genes of this pathway in the genome bin.

The functional assignment of diatom and sponge genes was quite limited in comparison to the just-discussed microbial pathways. The annotated genes showed that diatom symbionts invested a large transcriptional effort in functions for energy harvesting via photosynthesis and energy consumption via glycolysis. An active urea cycle potentially redistributes and metabolizes inorganic carbon and nitrogen coming from catabolism and/or photorespiration. Diatom expressed functions related to nitrate assimilation and production of ammonia via nitrite generated from the oxygenation of alkyl nitronates, which indicates that these symbionts may simultaneously use multiple sources of nitrogen. Sponge cells abundantly expressed functions of oxidative phosphorylation and glycolysis, but the most abundant expressed functions were related to arginine and proline metabolism that produces L-ornithine (e.g., from L-arginine and glycine). This reaction of glycine amidinotransferase is the first of the two-step creatine biosynthesis. Along with creatine, creatinine is expected to be found in sponge cells, as it is a product of the spontaneous, nonenzymatic conversion of the former. Creatine and creatinine are likely incorporated by CcPhy and CcThau (Figure 3.5).

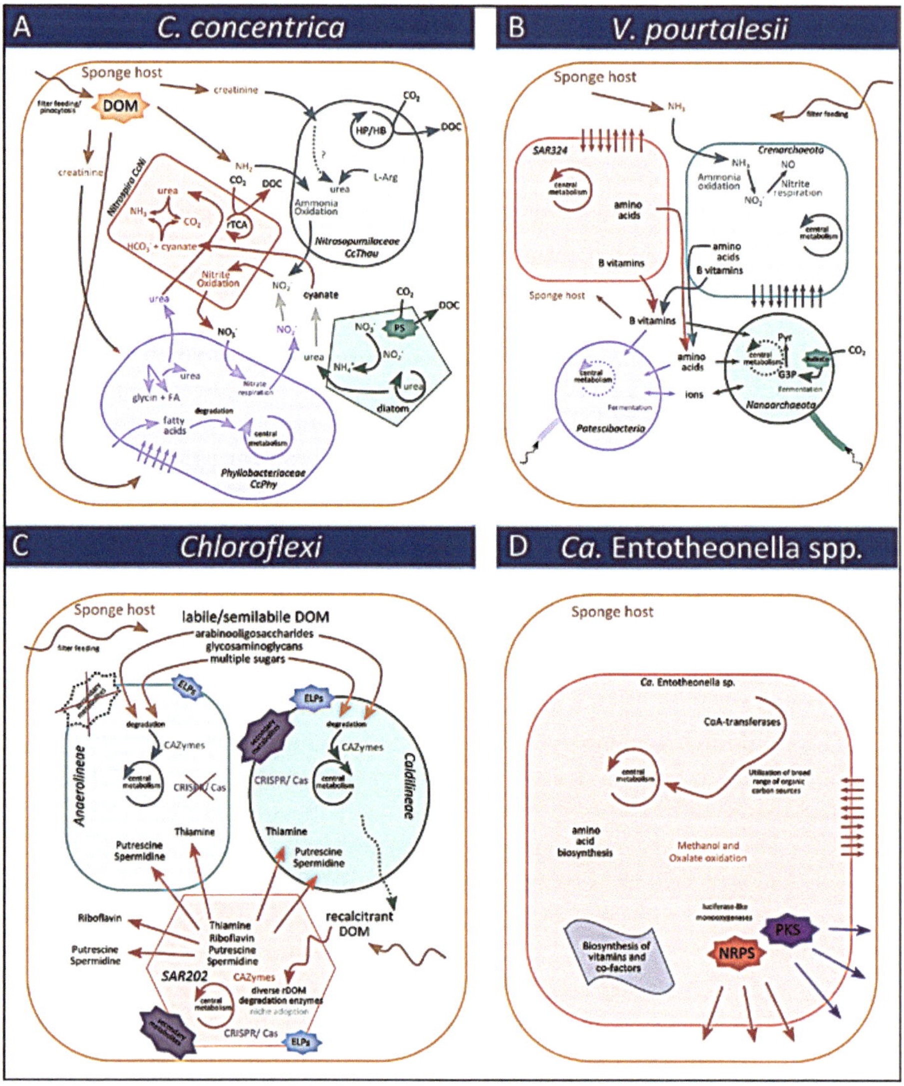

FIGURE 3.5 Host-microbe and microbe-microbe interactions within the sponge holobiont based on the genomic information. (A) A model of metabolic exchanges within the sponge *C. concentrica*, according to Moitinho-Silva et al. (2017). (B) A model of metabolic exchanges within the sponge *V. pourtalesii*, according to Bayer et al. (2020). (C) Nutrition specialization of prevalent sponge-associated *Chloroflexi* taxa, according to Bayer et al. (2018). (D) The lifestyle of *Ca.* Entotheonella spp., according to Lackner et al. (2017).

Taking these findings together, the authors proposed that CcPhy may be a major contributor to the uptake and utilization of DOM (Figure 3.5), which may come from the external environment through the sponge's filtering activity or pinocytosis (de Goeij et al., 2008, 2013), but may also be derived from diatom photosynthetic activity. Creatine and creatinine derived from the sponge metabolism are actively used by CcPhy and CcThau (Figure 3.5). Additionally, the metabolism of CcPhy, diatoms, CcNi, and CcThau are linked to each other by the production and consumption of urea (Figure 3.5). While CcPhy, CcThau, and diatoms are sources of urea (by internal creatinine degradation, via L-arginine degradation and active urea cycle, respectively), CcNi may benefit directly from urea in the extracellular matrix, as it abundantly expresses the genes related to the transport and degradation of urea. Additionally, CcPhy and the diatom might be linked indirectly to CcNi via urea and cyanate (Figure 3.5). The source of ammonia in the sponge mesohyl is most likely related to host pumping activity and/or host metabolism. CcPhy takes up ammonium (NH_4^+) for nitrogen assimilation, while CcThau oxidizes ammonia to generate energy, releasing nitrite, which in turn serves as the energy source for CcNi. The interdependency between CcThau and CcNi is reflected in their spatial distribution, as they occur close together in the sponge mesohyl. The end product of the combined aerobic nitrification by CcThau and CcNi is a nitrate, which would support anaerobic nitrate respiration by CcPhy and the nitrate assimilation by diatoms.

3.4.2.2 THE METABOLIC INTERTWINING WITHIN V. POURTALESII

Bayer et al. (2020) provided insights into the microbial lifestyle in the hexactinellid sponge *V. pourtalesii* by metagenome-guided microbial genome reconstruction, combined with microscopic imaging. The microbial cells associated with *V. pourtalesii* appeared in dense patches within the host, which might be syncytial tissue aggregates, as judged by SEM and TEM imaging. *16S* rRNA gene amplicon sequencing and abundance analyses of metagenomic data revealed that representatives of SAR324 bacteria, *Crenarchaeota*, *Patescibacteria*, and *Nanoarchaeota* are abundant members of the sponge microbiome, thus their genome features and metabolic capacities were examined in detail.

Based on the functional analyses of the metagenome-assembled genomes, the authors divided the four *V. pourtalesii* dominant microbial groups (i.e., SAR324 bacteria, *Crenarchaeota*, *Patescibacteria*, and *Nanoarchaeota*) into two different life strategies: "the givers" and "the takers" (Figure 3.5(B)) (Bayer et al., 2020). SAR324 and crenarchaeal genomes encode for the synthesis of amino acids and several B vitamins. These co-factors are essential for the biosynthesis of proteins and amino and nucleic acids, in both eu- and prokaryotic cells. Since SAR324 and *Crenarchaeota* probably provide such substances to other microbiome members and/or to the host itself, they were named "givers." In contrast, *Nanoarchaeota* and *Patescibacteria* are "takers" because they lack almost all known genes required for the *de novo* biosynthesis of amino acids, vitamins, nucleotides, cofactors, and lipids. Even key pathways of the central metabolism, such as glycolysis, were incomplete. A limited number of transporters may allow the import of substances such as amino acids, sugars, co-factors, and lipids possibly provided by the host's filter-feeding activity or other members of the microbiome such as the "givers" described above. A striking finding in this study was the detection of genes possibly involved in the building of pili-like cell structures, which could enable the "takers" to take up required substances from the extracellular matrix to compensate for the reduced metabolic capacity encoded in their genomes. Respective genes may also be involved in cell-cell contact and could stabilize the symbiotic lifestyle. The reduced genomes of the four main *V. pourtolesii* symbionts, but particularly of *Nanoarchaeota* and *Patescibacteria* (genome size <1 Mb) suggest a strong dependency on a symbiotic lifestyle, which seems to be even obligatory for the takers.

These are probably the two most detailed metabolic reconstructions of interactions within sponge holobionts existing to date. Remarkably, both studies found examples of the spatial distribution of microbes, as revealed by high-resolution microscopy, agreeing with the predicted metabolic interactions. And even if only a few members of the microbiomes were analyzed, these studies offer a new perspective on the nutritional currencies in sponge holobionts. One subsequent question is: To what extent can these examples be extrapolated to other sponge species? Likely, each sponge species has evolved its metabolic trade within the microbiome and between the microbes and host. However, we can also identify common patterns. For example, it seems that sponges found a source of vitamins in their microbiome, whereas sponge-derived organic compounds serve as a food source for symbionts. Nitrogen compounds, both organic and inorganic, arise as a common currency in sponges. The tight interdependency between members of the thaumarchaeal order *Nitrosopumilus* and the bacterial genus *Nitrospira* in nitrification is likely to present in many other sponge species, as these are common members of sponge microbiomes (e.g., Hoffmann et al., 2009; Radax et al., 2012a; Ribes et al., 2012). In general, each sponge holobiont can be seen as a miniature ecosystem, in which common metabolic exchanges have probably resulted from convergent solutions to enhance productivity and nutrient cycling.

3.4.3 FROM MICROBES TO ECOSYSTEMS

Holobionts can be regarded as host-microbe ecosystems integrated, like Russian dolls, in a large ecosystem (McFall-Ngai et al., 2013; Pita et al., 2018b). Sponges are prominent examples of this "nested ecosystems" concept and how microbiome-mediated functions within the holobiont translate into ecosystem functions (Pita et al., 2018b). The sponge holobiont is the result of complex microbe-microbe and host-microbe interactions, with a prominent role of the host as a "steering" agent (Foster et al., 2017) (*see* Section 3.5). At the same time, sponge holobionts do not occur as isolated units. They are embedded in a community of holobionts, forming an ecosystem. Thus, whether the complex interactions within the sponge holobiont result in a "net" benefit, will be also largely influenced by the ecosystem the holobiont belongs to. Here we focus on microbial functions that scale up from the sponge holobiont to the large ecosystem. We present an example that points to microbial-mediated recycling of carbon and then we discuss symbionts that determine the chemical diversity of their sponge host.

3.4.3.1 SPONGE-ASSOCIATED CHLOROFLEXI: DIVERSITY, METABOLIC SPECIALIZATION, AND DOM CYCLING

Many sponges harbor bacteria belonging to the phylum *Chloroflexi*, suggesting that these microbes play a critical ecological role in the sponge holobiont (Schmitt et al., 2011; Bayer et al., 2018). Members of the phylum *Chloroflexi* are indicator species in HMA sponges (Bayer et al., 2014; Moitinho-Silva et al., 2017c), meaning that they are particularly enriched in this sponge type (*see* Section 3.3). The most abundant *Chloroflexi* taxa in marine sponges belong to the classes *Anaerolineae* and *Caldilineae*, as well as the candidate clade SAR202 (Bayer et al., 2018). Metagenomic analysis of the sponge *Aplysina aerophoba* (Nardo, 1833) revealed that SAR202 belongs to a different functional guild than *Caldilineae* and *Anaerolineae*, based on different enrichment of cluster orthologous groups [COGs defined as proteins of similar potential functions, as assigned by an orthology-based approach (Galperin et al., 2015)] (Slaby et al., 2017). Later, detailed genomic data of sponge-associated *Chloroflexi* was generated by a single-cell sequencing approach and metagenome binning from the *A. aerophoba* microbiome (Bayer et al., 2018). The most complete reconstructed microbial genomes, two for each mentioned class/clade, were further analyzed and here we highlight some of their metabolic features (based on Bayer et al., 2018).

Metabolic reconstruction suggests that all three clades of sponge-associated *Chloroflexi* are aerobic and heterotrophic bacteria. Since genes encoding enzymes for rTCA and the Wood-Ljungdahl pathway were largely present, CO_2 fixation might be possible in these bacteria. All three classes/clades are able to import and assimilate ammonia. They also synthesize and degrade fatty acids. Additionally, all genomes encoded several ABC transporters to supplement nutrition and cell growth-related compounds (including oligopeptides, phosphate, amino acids, minerals, and metal ions). Further, all genomes largely encode enzymes needed for the biosynthesis of most amino acids.

Besides these commonalities, class-specific genomic features suggest specific contributions to nutrition and defense (Bayer et al., 2018). The genomes of both, *Caldilineae* and the SAR202 clade, contained gene clusters encoding type 1 polyketide synthetase (PKS) and other secondary metabolites probably defending the microbiomes and/or the host against predators (reviewed in Pawlik, 2011). Bayer et al. (2018) also found a potential for nutritional dependency on vitamin synthesis (Figure 3.5(C)). The SAR202 genomes contained genes encoding for co-factor biosynthesis, including riboflavin (vitamin B2) and thiamine (vitamin B1). The genes for the biosynthesis of these two vitamins were absent in the other two *Chloroflexi* classes and, for the latter, also in free-living SAR202. Interestingly, *Anaerolineae* and *Caldilineeae* genomes encode for an ABC transporter for thiamine so these cells may profit from vitamins produced by SAR202 symbionts. Thus, vitamin exchange, particularly of thiamine, is possibly a key symbiotic currency.

Interestingly, SAR202 may play a dominant role in DOM recycling within the sponge holobiont. Whereas *Anaerolineae* and *Caldilineae* genomes showed an extensive gene repertoire for carbohydrate (mono- and oligosaccharides) uptake and degradation, SAR202 genomes suggest these symbionts use of recalcitrant or refractory DOM (Figure 3.5(C)). SAR202 genomes were enriched in some of the enzymes that are proposed to mediate the degradation of recalcitrant or refractory DOM in the deep sea (Landry et al., 2017), including acyl-CoA-transferase, alkanesulfonate monooxygenase, and alcohol dehydrogenase. Thus, sponge-associated SAR202 microbes could use recalcitrant DOM derived from the metabolism of other bacteria or the seawater, contributing to its recycling into POM within the sponge holobiont. Via the "sponge loop" (de Goeij et al., 2013; Rix et al., 2016a), the activity of SAR202 symbionts may indeed promote productivity at the ecosystem level.

3.4.3.2 DEFENSE SYMBIOSIS

Sponges produce a specific and diverse chemical arsenal to avoid predators, competitors, microbial infection, or fouling. Thus, the chemistry of sponges contributes to their success in the ecosystem (Pawlik, 1993; Rohde et al., 2015). However, it is the interest in sponges as a rich source of marine natural products, with potential medical and biotechnological applications (Erwin et al., 2010), which has driven the search for symbionts as producers of sponge secondary metabolites. Here we present a few cases in which the microbial origin of the main chemicals in the sponge was confirmed, but we focus on the sponge-microbe interaction.

The sponge *Theonella swinhoei* (Gray, 1868) lives in association with filamentous microbes classified as *Candidatus (Ca.)* Enthotheonella sp., which are only distantly related to any cultivated bacterium and have been assigned to the candidate phylum 'Tectomicrobia' (Wilson et al., 2014). *T. swinhoei* sponges are chemically diverse, in a way that different sponges present different arrays of secondary metabolites, and the *Ca.* Entotheonella symbionts seem to be the reason. In *T. swinhoei* WA chemotype, "*Ca.* Entotheonella serta" TSWA1 was proposed as a producer of an actin-binding polyketide misakinolide (Ueoka et al., 2015). In the *T. swinhoei Y* chemotype, at least two *Ca.* Entotheonella symbionts are most likely to produce key secondary metabolites: (i) "*Ca.* Entotheonella factor" TSY1 was proposed to produce almost all known bioactive polyketides and modified peptides (e.g., onnamides, polytheonamides, keramamides, pseudotheonamides, cyclotheonamides, and nazumamides) (Piel et al., 2004b; Brück et al., 2008; Freeman et al., 2012, 2017); (ii) "*Candidatus* Entotheonella gemina TSY2 is most likely involved in producing the antialgae Kasumigamide, a compound which might be distributed by lateral gene transfer to other sponge microbiome members (Nakashima et al., 2016)." Related *Ca.* Entotheonella were also detected by PCR in a wide range of other sponge species and for example linked to natural product biosynthesis in the sponge *Discodermia calyx* (Freeman et al., 2016; Nakashima et al., 2016).

The genomes of *Ca.* Entotheonella spp. are relatively large (~9 Mb) and encode a complex metabolism (Lackner et al., 2017) (Figure 3.5(D)). Most likely, *Ca.* Entotheonella spp. cells perform aerobic metabolism but are facultative anaerobes (Liu et al., 2016; Lackner et al., 2017). Moreover, the genomes encode for the biosynthesis of several vitamins (e.g., nicotine amide, flavin, folate, pyridoxal phosphate, and cobalamin) and rare co-factors such as F_{420}, pyrroloquinoline quinone, and mycothiol (Lackner et al., 2017). Thus, *Ca.* Entotheonella sp. is able to use a broad range of organic carbon sources like organic acids, alcohols, polyols, polysaccharides, or even complex aromatic compounds. Interestingly, the typical symbiont functions of sponge-associated microbes (*see* Section 3.4.1) were lacking in *Ca.* Entotheonella, suggests that other mechanisms may be in place to enable these symbionts to colonize *T. swinhoei*.

The genomes of *Ca.* Enthotheonella serta and *Ca.* Entotheonella factors were rich in nonribosomal peptide synthetase (NRPS) and PKS gene clusters compared with the *Ca.* Entotheonella genima (Mori et al., 2018). This highlights the chemical richness and variability of *Ca.* Entotheonella microbes. For *T. swinhoei Y*, matrix-assisted laser desorption-ionization imaging mass spectrometry (MALDI-IMS) allowed the localization within the sponge tissue of two of the bioactive compounds, the PKS-NRPS hybrid products, onnamide A and cyclotheonamide A. These metabolites were found in areas of pores, chambers, and the exterior part of the yellow chemotype of the sponge (Lackner et al., 2017). Importantly, the distribution of these two bioactive compounds co-localized with *Ca.* Entotheonella symbionts, as judged by CARD-FISH (Lackner et al., 2017). The authors conclude that this spatial pattern agrees with the role of *Ca.* Entotheonella-derived compounds in protecting the most exposed areas of the sponge against predators or opportunistic microbes (Lackner et al., 2017).

Sponges belonging to one *Haliclona* clade revealed a very special example of a chemical-producing symbiont. The producer of renieramycin in these sponges is an intracellular bacterium, *Candidatus* Endohaliclona renieramycinifaciens, hosted in specialized bacteriocytes (Tianero et al., 2019). The gene producing the renieramycin is in a complex plasmid, but otherwise, this bacterium has an extremely reduced genome that suggests an obligate symbiotic lifestyle (Tianero et al., 2019). These genomic features and the specificity of the interaction suggest an intimate symbiosis, driven by defense, in which host and bacterial evolutionary trajectories are now intertwined.

Although they have different lifestyles, *T. swinhoei* and *Haliclona* specialized symbionts work as factories of sponge bioactive compounds. However, in other sponge species, building the chemical diversity takes a village. In *Mycale* (*Carmia*) *hentscheli* (Bergquist & Fromont, 1988), the biosynthetic gene clusters of PKS of the main chemical compounds identified in the sponge were detected in the microbial metagenomes (Rust et al., 2020). Subsequently, genome binning revealed 22 microbial taxa as bioactive producers, assigned to 11 different bacterial phyla, including *Proteobacteria*, *Actinobacteria*, *Verrucomicrobia*, *Acidobacteria*, and *Planctomycetes*, among others (Rust et al., 2020). In this case, the defensive function responds to a community effort and the selective force on the specific microbe may be more relaxed, allowing for more flexible interactions (Wein et al., 2019).

To sum up, nutrition and defense emerge as key currencies in sponge symbiosis, shaping sponge microbiomes and determining holobiont ecological roles. The examples presented here cover a diverse array of interactions: specialist symbiotic partners (e.g., *Ca.* Entotheonella spp.), a generalist sponge symbiont that stands out because of a certain metabolic capacity (e.g., SAR202), or microbe-microbe exchanges that secure nutrient cycling (e.g., nitrification by members of the *Thaumarchaeota* and *Nitrospira*). Thus, nutrition and defense steer the interactions within the sponge holobiont; but the underlying mechanisms and evolutionary forces differ among different members of the community or depending on the sponge species considered.

3.5 THE SPONGE IMMUNE SYSTEM

3.5.1 DIFFERENTIATING FRIEND FROM FOE

Immunity is the capability of all multicellular organisms to defend themselves against foreign bodies (or "non-self"), including pathogens while tolerating what belongs to the "self." The ability of sponges to eliminate non-self particles was described for the first time more than 100 years ago by Elias Metchnikoff while investigating inflammation processes. Metchnikoff observed that sponge motile phagocytotic cells encapsulate foreign material within cell aggregates and then eliminate it (Metchnikoff, 1893). Later on, transplantation experiments using sponge cell suspension and tissue fragments showed that sponges fuse with tissue fragments transplanted from the same individual (autographs) and reject fragments coming from another individual of the same species (allografts) (*see* Glossary; e.g., Hildemann et al., 1979; Moscona, 1968; Smith & Hildemann, 1986). Different immune responses against tissue or cells from the same species can bring allograft rejection if recognized as "non-self." For example, when sponge cells from two different species were mixed, cells from the same species were reaggregated only among them and formed primmorphs (defined a multi-cellular aggregates formed from dissociated single cells with proliferative activity reorganized in a spherical body with a continuous pinacoderm, which separates the internal mass of cells from the external environment; e.g., Custodio et al., 1998). In tissue explants, some species responded to allografts by building physical barriers, while others mounted inflammatory responses to destroy the non-self-tissue (e.g., Hildemann et al., 1979; Smith & Hildemann, 1986). These studies highlight the high precision of sponges for recognizing "self" from "non-self."

In recent years, the holobiont concept and the ubiquity of animal-microbe interactions have challenged the traditional perspective of the immune system as a defense mechanism. Instead, immunity is now regarded as the guardian of homeostasis within the holobiont (*see* Section 3.6). But how do animals conceal fighting pathogens while nurturing their microbiota? Immune receptors, signaling pathways, and antimicrobial peptides (AMPs) mediate the immune response to pathogen-derived as well as symbiont-derived molecules; the former context produces inflammation, and the latter promotes tolerance (Chu & Mazmanian, 2013). To illustrate, in *Hydra* spp. polyps overexpressing the AMP periculin-1a were significantly less colonized by bacteria compared to controls (Fraune et al., 2010). This antimicrobial peptide-based protection strategy in *Hydra* was suggested to control and shape the microbial community during polyps development (Fraune et al., 2010). Another example comes from the symbiosis between the Hawaiian squid *Euprymna scolopes* and the bioluminescent bacterium *Vibrio fischeri*, in which symbiont-derived molecules induce a sequence of events to promote colonization and direct the development of the host light organ (reviewed in Nyholm & McFall-Ngai, 2021). Although the exact underlying control mechanisms for specificity remain largely unknown, increasing evidence shows that the immune system is essential for establishing and maintaining host-microbe interactions within the holobiont (Kitano & Oda, 2006; McFall-Ngai, 2007; Nyholm & Graf, 2012; Bosch, 2014; Gerardo et al., 2020).

In sponges, the filter-feeding lifestyle might pose the need for them to distinguish between food, pathogens, and symbionts. The early evidence of specific microbial recognition by sponges came from the pioneering immunological compatibility experiments of Wilkinson in the early 1980s

(Wilkinson, 1984). He demonstrated different agglutination reactions of bacteria in sponge antisera, depending on the isolation source and phylogeny of the bacteria (Wilkinson, 1984). Wilkinson et al. were also the first to provide evidence that sponges discriminate between seawater bacteria and sponge-associated bacteria (1984). By incubating sponges *in situ* with tritium-labeled bacteria and tracking the bacteria in the sponge tissue using high-resolution radioautography, the authors showed that seawater bacteria were retained by the sponge and frequently found in vacuoles of choanocyte and pinacocyte cells, whereas sponge-associated bacteria were expelled in the exhalant current and rarely incorporated into any sponge phagocytic cell (Wilkinson et al., 1984). Decades later, Wehrl et al. (2007) built on this work and performed further feeding experiments using culturable marine bacteria (i.e., defined as food bacteria), microbial seawater consortia, and sponge symbiont consortia. The estimated uptake rates of seawater microbes were nearly two orders of magnitude higher compared to symbionts, whereas there was no obvious discrimination among the different types of food bacteria or the different strains of symbionts (Wehrl et al., 2007). These studies demonstrated the capacity of sponges to differentiate between their bacterial symbionts and food bacteria.

Despite this promising start, the field of sponge symbiosis has focused on the microbial side of the interaction, and for a long time, the role of the sponge host in the symbiosis remained overlooked. How immunity regulates sponge-microbe crosstalk is an emerging frontier of research. In the following section, we will present the current knowledge of the cellular and molecular mechanisms involved in sponge-microbe interactions.

3.5.2 IMMUNITY IN SPONGE-MICROBE INTERACTIONS: CELLULAR MECHANISMS

Phagocytosis is a universal and conserved means of both nutrition and internal defense in the animal kingdom (reviewed by Hartenstein & Martinez, 2019); but it also plays a role in symbiosis (McFall-Ngai, 2007; Nyholm & Graf, 2012). The observations of sponge microbial discrimination (*see* Section 3.1) point to phagocytosis as a key process to understanding the relationship of sponges with microbes. Most of our current knowledge on sponge phagocytosis comes from feeding experiments, which measured clearance rates, examined the fate (i.e., ingestion and digestion process) of different bacteria and particles, and used microscopy observations to identify the cells involved in phagocytosis (e.g., Imsiecke, 1993; Leys &

Eerkes-Medrano, 2006; Maldonado et al., 2010; Weissenfels, 1976). In the deep-sea sponge *Geodia barretti* (Bowerbank, 1858), electron micrographs revealed that around 4% of the sponge's symbionts were phagocytized by amoeboid cells, likely supplementing the sponge's nutrition (Leys et al., 2018). Yet, it remains unclear if there is a real differentiation between immune and feeding phagocytosis in sponges, and to what extent these processes are involved in the discrimination of bacteria by the sponge. With ameboid cells patrolling the sponge mesohyl, the ability to evade phagocytosis seems an evident prerequisite for symbionts to persist within the sponge holobiont.

In recent years, the first mechanisms for evasion of phagocytosis by sponge symbionts were revealed (Nguyen et al., 2014; Jahn et al., 2019). The genomes of sponge-associated bacteria are enriched in ARPs when compared to free-living bacteria (e.g., Thomas et al., 2010). In human intracellular symbionts and intracellular pathogens, ARPs modulate host phagocytosis, facilitating infection and bacterial intracellular survival (e.g., Al-Khodor et al., 2008; Habyarimana et al., 2008; Pan et al., 2008). Based on the role of ARPs in other systems, Thomas et al. (2010) hypothesized that these proteins could represent a mechanism that mediates the discrimination of associated and non-associated bacteria in sponges. To test this hypothesis, Nguyen et al. (2014) used the free-living amoeba *Acanthamoeba castellanii* as a model organism to approximate sponge cell phagocytosis as it has physiological and structural similarities to amoebocytes. Genetically-modified *Escherichia coli* expressing sponge symbiont-derived ankyrin-repeat proteins were phagocytized by *A. castellanii* but not digested. The bacteria accumulated in large vacuoles of the amoeba cells (Nguyen et al., 2014). Further experimental evidence supporting the key role of ankyrin repeats in phagocytosis comes from a surprising finding. Sponge viromes comprised bacteriophages encoding proteins with auxiliary ankyrin repeats (termed "Ankyphages") (Jahn et al., 2019). Bacteria decorated with the phage-encoded ankyrin repeat protein (i.e., ANKp) down-regulated their proinflammatory signaling in human murine macrophages and escaped phagocytosis (Jahn et al., 2019). This study marks a milestone in the research of inter-kingdom interactions within the sponge holobiont (Jahn et al., 2019). Although phagocytosis seems to be an important cellular mechanism for the establishment and maintenance of symbioses, we are still missing the characterization of how bacteria manipulate this process in sponge cells.

3.5.3 IMMUNITY IN SPONGE-MICROBE INTERACTIONS: MOLECULAR MECHANISMS

The molecular mechanisms of innate immunity rely on the recognition of microbe-associated molecular patterns (MAMPs) by pattern recognition receptors (PRRs), which activate signaling pathways that produce AMPs and other effectors. PRRs sense a variety of MAMPs, such as lipopolysaccharide (LPS), peptidoglycan (PNG), or flagellin, which are absent in eukaryotic organisms; and they also detect damage signals coming from the animal cell (Janeway & Medzhitov, 2002). The next step after the receptors recognize and signal the occurrence of microbes or damaged cells is to eliminate the potential threats, repair tissue damage, or promote the tolerance of symbionts. Common effector pathways for eliminating infected or stressed target cells include cytokines and chemokines (Magor & Magor, 2001).

3.5.3.1 PATTERN RECOGNITION RECEPTORS (PRRS)

In sponges, genomic and transcriptomic data has revealed high diversity and complexity of pattern recognition receptors (PRRs) remarkably similar in structure to vertebrate genes (Figure 3.6). Among these, nucleotide-binding and oligomerization domain (NOD)-like receptors (NLRs), scavenger receptor cysteine-rich (SRCRs) domains, and a receptor type similar to toll-like receptors (TLRs) (Srivastava et al., 2010; Hentschel et al., 2012; Yuen et al., 2014; Riesgo et al., 2014). NLRs and SRCRs are highly diversified in sponges compared to other animal groups (Buckley & Rast, 2015). Another additional expanded receptor family in sponges are G-protein coupled receptors (GPCRs), which are not traditionally considered PRRs but appear to mediate sponge response to different stimuli (e.g., Guzman & Conaco, 2016; Pita et al., 2018a). Yet, the function of all these receptors in sponges is still poorly understood due to a lack of experimental evidence. Despite this limitation, the conserved domain structure and diversification of some of the PRRs suggest that these receptors are good candidates for mediating the sponge-microbe crosstalk (Degnan, 2015).

The NLR genes are a family of PRRs characterized by the presence of a nucleotide-binding domain (NACHT) and a C-terminal leucine-rich repeat (LRR) domain. In sponges, NLRs are strikingly diverse. At least 135 NLR genes were identified in the Great Barrier Reef sponge *Amphimedon queenslandica* (Hooper & van Soest, 2006) (based on Yuen et al., 2014), in contrast to the 20 NLR genes in the human genome (which is almost 20 times larger than *A. queenslandica* genome) (Buckley & Rast, 2015). The reference transcriptome of the sponge *Dysidea avara* (Schmidt, 1862) also included a high abundance of NLRs (i.e., 80 *bona fide* NLRs), of which two were activated in response to LPS and PNG (Pita et al., 2018a). However, in the transcriptomes of other sponges, the repertoire of NLRs is less diverse (e.g., in *A. aerophoba* (Pita et al., 2018a) and *Vaceletia sp.* (Germer et al., 2017)). Overall, the complex NLR repertoire in sponges and its similarity in structure with the vertebrate counterparts suggests their potential to recognize a broad spectrum of microbial ligands (Degnan, 2015).

The SRCR genes are a family of extracellular PPRs characterized by the presence of scavenger receptor cysteine-rich domains, but their function is not necessarily conserved. In vertebrates, these cell surfaces or secreted proteins enhance the phagocytic clearance of microbes, environmental particles, and DNA (reviewed by Sarrias et al., 2004). The first report of an SRCR molecule in sponges was a macrophage class A scavenger receptor (SR-A) in the species *Geodia cydonium* (Linneaus, 1767), (based on Pancer et al., 1997) and its function was related to adhesion and cell recognition (Pahler et al., 1998). However, sponges host hundreds of genes within the SRCR family, that appear in combination with many different domains (Ryu et al., 2016). In *A. queenslandica* SRCR domains were enriched in the choanocytes (Sebé-Pedrós et al., 2018). Experimental work revealed that SRCRs take part in sponge-microbe interactions. For example, *A. queenslandica* juveniles exposed to enriched bacterial suspensions showed differential gene expression of 60 SRC domains (Yuen, 2016), and an SRCR domain-containing gene was also differentially expressed in *A. aerophoba* in response to MAMPs (Pita et al., 2018a). Gene screening on individuals of the Mediterranean sponge *Petrosia ficiformis* (Poiret, 1789) showed upregulation of one SRCR-domain containing gene in sponge harboring cyanobacterial symbionts, compared to aposymbiotic (i.e., cyanobacteria-free) specimens (Steindler et al., 2007). Like in vertebrates, most likely the SRCR genes in sponges play complex roles in immunity.

The TLR genes are transmembrane PRRs comprising an extracellular domain (i.e., LRRs), responsible for the recognition of MAMPs, and an intracellular domain (i.e., TIR: Toll/interleukin-1 receptor), in charge of inducing a signaling cascade. No conventional TLRs are found in sponges (reviewed in Hentschel et al., 2012). Sponges harbor genes containing a TIR domain, homologous to that in vertebrate TLRs (Riesgo et al., 2014), but it is associated with immunoglobulin (Ig) domains instead of the typical LRR motifs (reviewed in Hentschel et al., 2012). It remains unclear whether these types

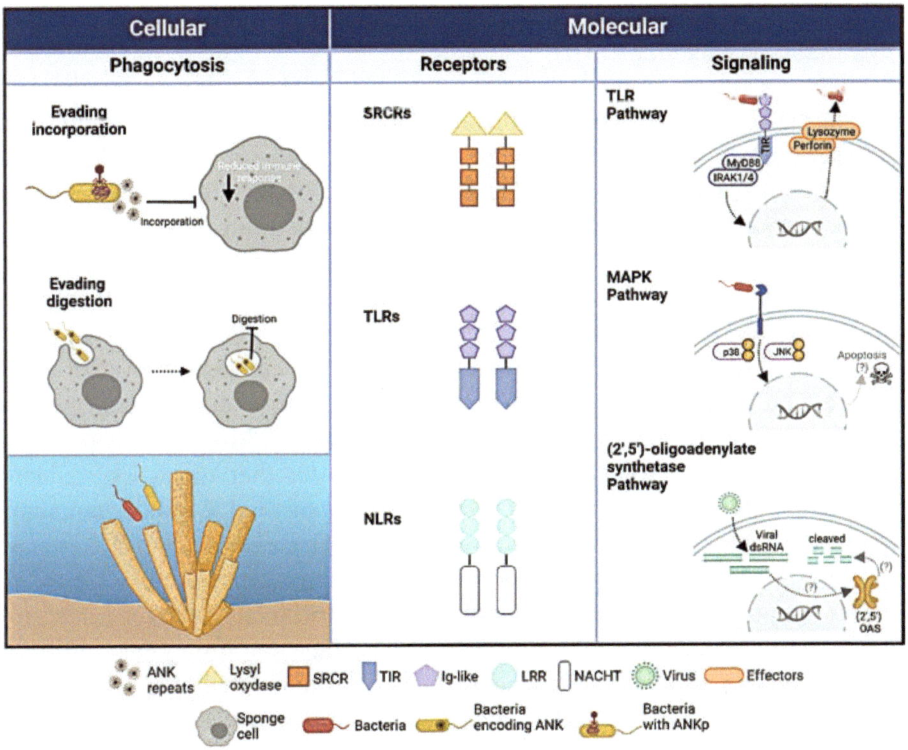

FIGURE 3.6 Cellular and molecular components of sponge immunity. Left panel: symbiont control on sponge phagocytosis, based on findings in Jahn et al. (2019) and Nguyen et al. (2014). Middle panel: main pattern recognition receptors in sponges, according to Hentschel et al. (2012). Right panel: exemplary signaling pathways involved in microbial recognition in sponges, reconstructed based on different studies (see text for references). Question marks represent the potential output of the pathway, inferred from gene annotation.

of receptors are involved in the recognition of MAMPs, but all components of the TLR signaling cascade are present in the sponge (Riesgo et al., 2014).

3.5.3.2 SIGNALING IN RESPONSE TO MICROBIAL CUES

The TLR pathway is one of the most important and well-characterized pathways contributing to innate immunity (Rosenstiel et al., 2009). TLRs recruit cytoplasmic TIR domain-containing adaptor proteins (e.g., MyD88: myeloid differentiation primary response 88 and TRIF: TIR-domain-containing adapter-inducing interferon-β), which trigger a downstream signal that leads to the activation of nuclear factor kappa B (NF-κB), interferon regulatory factors (IRFs), or mitogen-activated protein (MAP) kinases (Takeda, 2005; Kawasaki & Kawai, 2014).

Despite the different structures of poriferan TLRs, the major elements of the TLR pathway have been repeatedly reported in the genomes and transcriptomes of different sponge species (e.g., Germer et al., 2017; Pita et al., 2018a; Riesgo et al., 2014; Schmittmann et al., 2021). Experimental work supports the involvement of the TLR cascade in the sponges' response to MAMPs (Figure 3.6). For instance, incubation of *Suberites domuncula* (Olivi, 1792) with LPS showed a strong up-regulation of the TLR adapter molecule MyD88 (Wiens et al., 2005). This adapter binds to a receptor on the sponge cell surface (i.e., LPS-interacting protein), which results in an elevated expression of a perforin-like executing molecule with antibacterial properties (i.e., a macro-phage protein) (Wiens et al., 2005). Moreover, exposing *S. domuncula* to a microbial lipoprotein (LP) revealed that a TLR-like receptor and an IRAK-4l protein are potential mediators of the LP-stimulated signal pathway in this sponge (Wiens et al., 2007). The expression of these two proteins leads to the expression of a caspase-like molecule and the initiation of the apoptotic machinery (Wiens et al., 2007). Interestingly, TLR transcripts were more abundant in cells on the sponge surface and lining the aquiferous canals (Wiens et al., 2007). These studies demonstrated that sponges are equipped with a TLR-mediated defense mechanism against MAMPs, which ultimately might trigger apoptosis.

A second signaling pathway known to be involved in the response to LPS is the mitogen-activated protein kinase (MAPK) pathway (Figure 3.6). *S. domuncula* possesses two MAPKs with high homology to the mammalian p38 and JNK protein kinases (82% and 75% similarity with humans, respectively) which showed to react upon ligation to LPS (Böhm et al., 2000, 2001). A lectin was deduced to be the potential receptor responsible for activating this signaling cascade in sponge primmorphs (Schröder et al., 2003). Furthermore, differential gene expression analysis also demonstrated that sponges respond to microbial elicitors by the transduction of signals mediated by serine-threonine protein kinases, which probably yield apoptosis and regulate metabolic processes (Pita et al., 2018a).

A third pathway also implicated in the immune system of sponges is the (2′-5′) oligoadenylate synthetase [(2-5)A synthetase] pathway (Figure 3.6). This is an efficient defense mechanism against invading microorganisms that are activated in the presence of certain classes of RNA, mainly double-stranded RNA (Kuusksalu et al., 1995). (2–5)A synthetases cloned from *S. domuncula*, *G. cydonium*, and *Halichondria panicea* (Pallas, 1766) exhibit high similarity with vertebrate enzymes, and the level and activity of these enzymes increase after bacterial infection of the sponge (Grebenjuk et al., 2002). Based on the knowledge of the vertebrate system, the activation of the (2-5)A synthetase system in response to foreign bacteria was proposed to be initiated by the JAK-STAT pathway (Müller & Müller, 2003).

3.5.3.3 EFFECTORS OF THE RESPONSE TO MICROBIAL ENCOUNTERS

Elimination of foreign bacteria in sponges is likely to be driven by the activation of endocytosis and the subsequent production of lysosomal enzymes (e.g., cathepsin) (Krasko et al., 1997; Thakur et al., 2005). In *S. domuncula* the *lysozyme* gene is strongly expressed in gray cells that are embedded in the sponge's mesohyl. The enzymes produced by these cells were shown to only digest extracellular bacteria but not the ones occurring in bacteriocytes (Thakur et al., 2005).

An additional effector molecule involved in sponge immunity is the perforin-like protein. In mammals, perforin is a specific cytotoxic mechanism found in T-cells, which operates based on the secretion of lytic proteins, causing cell membrane damage and thus inducing apoptosis (Lowin et al., 1994). In the marine sponge *Tethya lyncurium* (Pallas, 1766) a pore-forming protein (i.e., *Tethya* hemolysin) was detected and cloned. Tests conducted in erythrocytes of different animals revealed that this perforin rapidly lysed red blood cell membranes (Mangel et al., 1995). Primmorphs of *S. domuncula* also has the capacity to produce a perforin-like antibacterial protein, whose sequence is highly similar to the human perforin I protein (Thakur et al., 2003). The expression of this perforin-like molecule is stimulated after exposure to LPS (Wiens et al., 2005). Interestingly, the expression of the perforin-like molecule in *S. domuncula* was observed to occur when the sponge was exposed to *E. coli* LPS but not when presented with Gram-negative sponge-associated bacteria. This finding indicates that the stimulation of the perforin-like molecule is not initiated by LPS itself, and provides further evidence of the potential of sponges to differentiate between associated and non-associated bacteria (Gardères et al., 2015).

3.5.4 LINK BETWEEN CELLULAR AND MOLECULAR MECHANISMS: OPEN QUESTIONS

Over the last few years, molecular data have revealed that sponges possess the essential genes commonly associated with innate immunity. However, the validation of these gene functions and their connection to cellular and physiological responses is still lacking. The signaling cascades and effectors mediating the response to microbes in symbiotic and non-symbiotic contexts are likely wider than the ones investigated so far and presented here. Sponge cellular mechanisms have been mainly studied in the context of developing a sponge cell culture and examining sponge regeneration (e.g., Alexander et al., 2014; Ereskovsky et al., 2019; Schippers et al., 2011); yet the cellular mechanisms involved in the immune response that occurs in intact and functional sponge specimens still needs to be decoded.

Experimental evidence integrating both the molecular and the cellular level will give insights into the genetic repertoire, the potential molecular pathways, and the type of cells that are driving the immune response in sponges. This is relevant since it is still difficult to assign specific functions to sponge cells. For example, it is not clear if the phagocytic activity of archeocytes is different from the one of choanocytes (i.e., "immune phagocytosis" from "feeding phagocytosis," respectively; Hartenstein & Martinez, 2019). It is hypothesized that archeocytes are responsible for providing immunity in sponges (Funayama et al., 2005), but this remains to be tested. Therefore, it is essential to characterize relevant immune cellular responses (e.g., phagocytosis, apoptosis, or cell proliferation) using different microbial challenges (e.g., sponge-associated *vs.* non-associated, potential pathogens, bacteria with different structures) and under controlled conditions. This experimental evidence would help to unravel the gene functions and the molecular mechanisms responsible for triggering specific cellular responses. Linking the cellular and molecular levels would broaden our understanding of immunity and its role to guard stable and specific interactions between sponges and their microbes.

3.6 THE SPONGE HOLOBIONT IN HEALTH AND DISEASE

Human activities are modifying the marine environment at a pace never seen before (Waters et al., 2016). We are witnessing an unprecedented increase in disease and disease-like symptoms, many of which affect the inhabiting benthic organisms, including corals, algae, and sponges, and even result in mass mortality events (Webster, 2007; Garrabou et al., 2009; Bourne et al., 2009; Burge et al., 2014; Hughes et al., 2018; Gómez-Gras et al., 2021). Disease outbreaks seem to respond to multiple factors and complex underlying processes, including physiological stress (Coma et al., 2009), infection by a specific pathogen (Webster et al., 2002), or proliferation of opportunistic microbes (Maldonado et al., 2010a). Interestingly, emerging evidence points to a misbalance of host-microbe interactions as the main driver of disease susceptibility in humans (Bäckhed et al., 2012) as well as in marine organisms (Egan and Gardiner, 2016).

Homeostasis is a fundamental property of life: the capacity to keep stable internal conditions, via feedback loops, despite disturbances. Homeostasis is usually orchestrated by the immune system (Palmer, 2018) and, therefore, it is urgent to gain a better understanding of the mechanisms controlling homeostasis in sponges (*see* Section 3.5). In addition, microbes are also suspected to play a key role in homeostasis control. A particularity in sponges is that both, host and microbial cells, are connected through the extracellular matrix (Figure 3.2.); thus, not only the animal but also the symbionts have opportunities to directly shape the internal conditions of the mesohyl.

Understanding sponge health and disease thus requires, once again, adopting the perspective of the holobiont. The sponge holobiont ecosystem is in a dynamic equilibrium, resulting from host-microbe and microbe-microbe interactions (Pita et al., 2018b). These interactions respond to changes in the environment in a way that they may undergo readjustments to preserve functional integrity (Pita et al., 2018b) (Figure 3.7). Beyond a threshold, i.e., if the environmental pressure is too high or is maintained over time, the equilibrium is at risk (Figure 3.7). For example, the giant barrel sponge *Xestospongia muta* only recovered from bleaching as long as the elevated temperature remained below a certain threshold (López-Legentil et al., 2008).

At present, research is focusing on describing the natural variability of sponge microbial diversity and function, in order to distinguish stress related changes. In sponges, each species harbors a specific microbiota that is stable overall (*see* Section 3.2). However, some sponge species harbor more variable microbiotas than others (e.g., Thomas et al., 2016; Weigel & Erwin, 2016), or variability may be influenced by season (e.g., Björk et al., 2013; Erwin et al., 2012). In addition to natural variability, different stressors are investigated under controlled aquarium conditions. The most studied stressor is warming, which even at sub-lethal levels, can alter sponge-associated bacterial (Ramsby et al., n.d.; Webster et al., 2008; Simister et al., 2012b; Posadas et al., 2021) and viral diversity (Laffy et al., 2019), as well as host gene expression and physiology (Pantile & Webster, 2011; Massaro et al., 2012; Webster et al., 2013; Koutsouveli et al., 2020; Posadas et al., 2021). The combination of warming with acidification may exacerbate the negative effects of warming on the sponge holobiont (Lesser et al., 2016), although in some cases may mitigate thermal stress, depending on the species considered

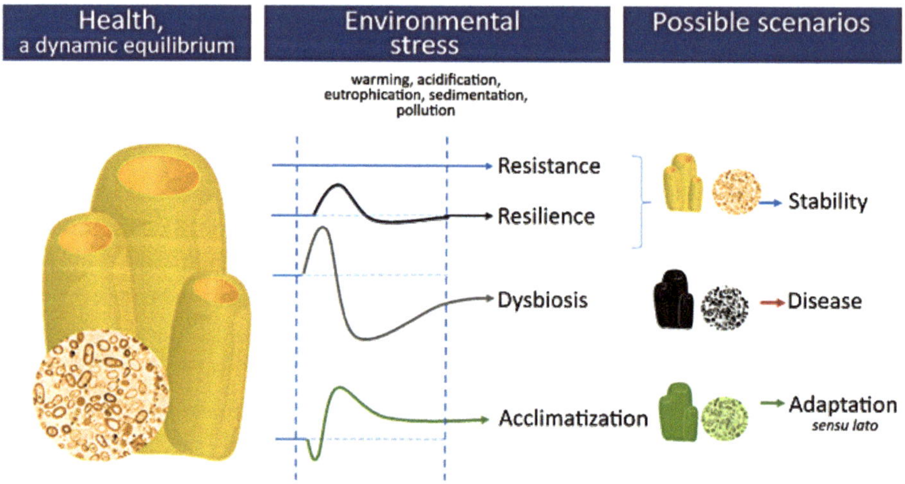

FIGURE 3.7 The sponge holobiont under a changing environment. Host-microbe interactions occur in a dynamic equilibrium that determines sponge health and also its fate under environmental pressure. Different possible scenarios may occur depending on the stressor but also on the characteristics of the holobiont, in terms of the resilience of the interactions, their flexibility, and/or the relative role of host *vs.* microbial-related factors in determining health.

Source: Figure adapted from Pita et al. (2018b); https://creativecommons.org/licenses/by/4.0/

(Bennett et al., 2017). Furthermore, several experiments tested the effect of additional local environmental pressures provoking poor water quality (e.g., Webster et al., 2001; Simister et al., 2012c; Tian et al., 2014; Pineda et al., 2016, 2017a, b; Luter et al., 2019), although only oil pollution had a significant detrimental effect on sponge health (Luter et al., 2019).

Stress can induce dysbiosis (defined as compositional (and functional) alterations of the microbiota in relation to stress or disease, which may increase disease susceptibility (Egan & Gardiner, 2016). We expect that dysbiosis will look different depending on the species or even specimen considered (e.g., in terms of which microbial groups are enriched/depleted in sick *vs* healthy individuals), but we can also find some common patterns (reviewed in Pita et al., 2018b). Dysbiosis in sponges is often defined by significant changes in the structure of the sponge-associated microbial community, shifting from a specific to a generalist-dominated community in both stressed (Ramsby et al., n.d.; Simister et al., 2012c) as well as sick individuals (Webster et al., 2008; Olson et al., 2014; Blanquer et al., 2016; Luter et al., 2017; Deignan et al., 2018). Particularly in response to stress, there is an increase in beta diversity (Erwin et al., 2012; Lesser et al., 2016; Pineda et al., 2017b), that is, increased dissimilarity in the microbial community of the different samples. This pattern is attributed to stress-related non-deterministic changes, following the Anna Karenina principle applied to host-associated microbial communities: "All healthy microbiomes are similar; each dysbiotic microbiome is dysbiotic in its way" (Zaneveld et al., 2017). In terms of function, dysbiosis has been correlated with enrichment in cell motility, chemotaxis, or virulence genes (Fan et al., 2013; Tian et al., 2014). Thus, opportunistic microbes or polymicrobial infections follow the onset of disease.

Fortunately, not all changes in the microbiota are necessarily bad. It has been proposed that microbes could give the animal an additional opportunity to adapt to the rapid changes in our oceans, in a process called microbiome-mediated transgenerational acclimatization (Webster & Reusch, 2017). The microbial genetic information within the holobiont can change through different mechanisms: the introduction of new microorganisms from the environment, the shifts in dominance of existing microbes of the microbiota, or through mutation and/or horizontal gene transfer (reviewed in Webster & Reusch, 2017). Microbes duplicate faster and mutate at a higher rate than animals and they can more easily exchange DNA with others. Thus, microbes can quickly change the phenotype of the holobiont (Reshef et al., 2006). If these changes are vertically transmitted, they will enable transgenerational acclimatization (Webster & Reusch, 2017) (Figure 3.7).

One approach to evaluating the contribution of the microbiome to cope with environmental change is controlled experiments in aquaria. Three Mediterranean sponges (*D. avara*, *Agelas oroides* (Schmidt, 1864), and *Chondrosia reniformis* (Nardo, 1847)) showed species-specific differences in the acquisition of new microbes during an experiment on ocean acidification (OA): high levels of acquisition in *D. avara*, moderate in *A. oroides*, and null in *C. reniformis* (Ribes et al., 2016). This variability was inversely correlated with growth rate, severely reduced in the sponge with no capacity to acquire new symbionts (Ribes et al., 2016). Thus, this observation suggests that flexibility in the acquisition of new microbial partners may indeed be a strategy to tolerate acidification (Ribes et al., 2016). The experiment by Luter et al. (2020) aimed at explicitly testing the potential of microbiome-mediated transgenerational acclimatization in the Great Barrier Reef sponge *Phyllospongia* (formerly named *Carteriospongia*) *foliascens* (Pallas, 1766). The authors showed that adults of *P. foliascens* transfer 90% of the microbiome to their offspring (Luter et al., 2020). When rearing adults under different IPCC future scenarios of temperature (ambient, 2050 prediction (+1°C), 2100 prediction (+2°C)), sponges looked healthy and the microbial diversity of adults and their larvae did not change significantly according to treatment (Luter et al., 2020). However, they could detect certain microbial taxa enriched in warming conditions, and some of these were vertically transmitted, suggesting the potential of symbiont shuffling as an acclimatization strategy (Luter et al., 2020). In addition, recruits transferred to different conditions than their parents showed the potential for symbiont switching (Luter et al., 2020). Nevertheless, host genotype was a strong determinant of the sponge microbial diversity (Luter et al., 2020), a reminder of the intertwining between ecological and evolutionary aspects of host-microbe interactions in determining the fate of the holobiont in a rapidly changing environment.

Another opportunity to test for the role of microbes in acclimatization is given by "natural experiments," that is, evaluating sponge holobionts at locations with strong natural gradients of environmental conditions. For example, Schuster et al. (2021a) investigated the microbiome of sponges thriving in a semi-enclosed marine lake, which experienced seasonal anoxia The abundance of microbes like *Thaumarchaeota* correlated with anoxic conditions and may confer deoxygenation tolerance to those individuals (Schuster et al., 2021). At shallow volcanic CO_2 seeps, the increase in partial CO_2 pressure locally reduces the pH at values close to those expected under future climate scenarios (pH $\leq$ 7.7), providing the opportunity to test acclimatization to OA (Hall-Spencer et al., 2008; Fabricius et al., 2011). At these sites, the diversity of sponge species is lower than in adjacent sites with "normal" conditions (pH ~ 8) (Fabricius et al., 2011; Goodwin et al., 2014), although those sponges able to survive at the CO_2 seeps could reach high abundances (Morrow et al., 2015). The microbial communities differed in individuals of the same species thriving at different sites (Morrow et al., 2015). A common feature in diverse sponge species was an increase in the relative abundance of cyanobacteria at low pH conditions (Morrow et al., 2015). However, the microbial community did not change upon transplantation experiments between CO_2 seeps and adjacent control sites; thus, in this case, microbial flexibility cannot be linked to increased tolerance (Kandler et al., 2018). Metagenomic analysis revealed shifts in the metabolic potential of sponge-associated microbes in CO_2 seeps (Botté et al., 2019). The sponge species most tolerant to seep conditions seemed to increase the efficiency for carbon fixation and nitrogen metabolism, particularly mediated by members of the *Thaumarchaeota* (Botté et al., 2019). Thus, the energetic balance may be a factor of stress (Botté et al., 2019). The potential role of host mechanisms in acclimatization at these natural seeps has not yet been evaluated.

An important question today is to what extent microbial or host-related factors are responsible for holobiont health and for the disbalance that happens upon perturbation (Posadas et al., 2021). In terms of stress and disease, three scenarios are possible: (i) the microbial activity, much more plastic than the animal, respond to change and breaks the currency of the symbiosis; (ii) the host, in particular the host immune system, is compromised due to stress (e.g., by energetic constraints) and loses control of host-microbe interactions; (iii) both host and microbial activities are affected by the environment in a way that breaks the agreement of the symbioses. Deciphering the underlying processes (and which scenario dominates) is crucial to understanding sponge health and to developing effective monitoring and restoration strategies to secure the future of our marine ecosystems.

3.7 CONCLUSION

Sponge holobionts are prominent examples of how microbes shape animal physiology. Sponges host one of the richest microbiotas, including sponge-specific microbial taxa. The microbial diversity in sponges results from the combination of host evolutionary trajectories, stochasticity, and the environment. However, some sponge-associated microbial groups, like viruses and microbial eukaryotes, remain underexplored. The HMA-LMA spectrum provides an opportunity to investigate how the density of the microbiome may affect animal immunity, metabolism, and ecological role. In mammals, for example, different microbial densities have been correlated with gut structure, which responds as well to diet (Contijoch et al., 2019). In sponges, both HMA and LMA strategies seem successful, in a way that, at least in shallow waters, they often colonize similar habitats and live close to each other. However, we still do not understand what factors lead to one or the other.

In recent years, the combination of deep sequencing with high-resolution imaging is unveiling the lifestyle of the elusive sponge-associated microbes. The focus is moving towards understanding resource exchanges – within the microbiome and between the host and the microbes – and how these interactions are linked to the ecological role of sponges and affected by the environment. This knowledge is fundamental for understanding the benefits and costs of sponge-microbe symbioses and for being able to predict future health outcomes of sponges in a changing environment.

The most recent insights into potential interactions within sponge microbiomes revealed some commonalities, like the exchange of vitamins or coupled nitrogen metabolism, but also revealed sponges as unique microbial ecosystems. Each sponge hides its relationship secrets. In this regard, how each sponge host shapes its microbiome like an "ecosystem engineer" (Foster et al., 2017) needs to be addressed. As in other animal systems (Nyholm & Graf, 2012; Dierking & Pita, 2020), we proposed that sponge immunity is the guardian of holobiont homeostasis, although our knowledge of sponge immunology is still in its infancy.

A major challenge, both on the host and microbial side, is the functional validation of genomic-based predictions. Developing sponge model systems for symbiosis may help to fill this gap (Pita et al., 2016). And we mean "sponge model system*s*," in the plural because we can learn much from the natural variability within the whole phylum, as shown already with the HMA-LMA spectrum. In some cases, reductionist models like *E. muelleri*, or *A. queenslandica* may serve to decode host gene functions; in other cases, complex sponge holobionts may bring to light new advances and perspectives on animal-microbe interactions (e.g., in the inter-kingdom interaction ankyphages-bacteria-host (Jahn et al., 2019)). Embracing the complexity of the sponge holobiont may seem a daunting task, but only a truly holistic approach will enable us to understand sponge physiology.

ACKNOWLEDGMENTS

We thank the Marine Symbioses Unit at GEOMAR Helmholtz Center for Ocean Research Kiel for fruitful discussions on the lifestyle of sponge-associated microbes and the implications for the sponge holobiont. AMG was funded by the International Max Planck Research School for Evolutionary Biology and the CRC1182 "Origin and Function of Metaorganisms" (TPB1, PI: Ute Hentschel Humeida).

KEYWORDS

- **allografts**
- **alloimmune responses**
- **autografts**
- **cluster orthologous groups (COGs)**
- **currency**
- **dysbiosis**
- **holobiont**
- **homeostasis**
- **microbiome**
- **primmorphs**
- **sponge pumping rate**
- **sponge-loop**
- **symbiosis**

REFERENCES

Achlatis, M., Pernice, M., Green, K., De Goeij, J. M., Guagliardo, P., Kilburn, M. R., Hoegh-Guldberg, O., & Dove, S., (2019). Single-cell visualization indicates direct role of sponge host in uptake of dissolved organic matter. *Proc. R. Soc. B Biol. Sci., 286*, 20192153. doi: 10.1098/rspb.2019.2153.

Alexander, B. E., Liebrand, K., Osinga, R., et al., (2014). Cell turnover and detritus production in marine sponges from tropical and temperate benthic ecosystems. *PLoS One, 9*, e109486. doi: 10.1371/journal.pone.0109486.

Al-Khodor, S., Price, C. T., Habyarimana, F., Kalia, A., & Abu, K. Y., (2008). A Dot/Icm-translocated ankyrin protein of *Legionella pneumophila* is required for intracellular proliferation within human macrophages and protozoa. *Mol. Microbiol., 70*, 908–923. doi: 10.1111/j.1365-2958.2008.06453.x.

Anderson, A. S., Northcote, P. T., & Page, J. M., (2010). Spatial and temporal variability of the bacterial community in different chemotypes of the New Zealand marine sponge *Mycale hentscheli. FEMS Microbiol. Ecol., 72*, 328–342. doi: 10.1111/j.1574-6941.2010.00869.x.

Archer, S. K., Stevens, J. L., Rossi, R. E., Matterson, K. O., & Layman, A. C., (2017). Abiotic conditions drive significant variability in nutrient processing by a common Caribbean sponge, *Ircinia felix. Limnol. Oceanogr., 62*, 1783–1793. doi: 10.1002/lno.10533.

Astudillo-García, C., Bell, J. J., Webster, N. S., Glasl, B., Jompa, J., Montoya, J. M., & Taylor, W. M., (2017). Evaluating the core microbiota in complex communities: A systematic investigation. *Environ. Microbiol., 19*, 1450–1462. doi: 10.1002/art.34459.

Azam, F., Fenchel, T., Field, J. G., Gray, J. S., Meyer-Reil, L. A., & Thingstad, F., (1983). The ecological role of water - column microbes in the sea. *Mar. Ecol. Prog. Ser., 10*, 257–263. doi: 10.3354/meps010257.

Bäckhed, F., Fraser, C. M., Ringel, Y., et al., (2012). Defining a healthy human gut microbiome: Current concepts, future directions, and clinical applications. *Cell Host Microbe, 12*, 611–622. doi: 10.1016/j.chom.2012.10.012.

Baker, P. W., Kennedy, J., Dobson, A. D. W., & Marchesi, R. J., (2009). Phylogenetic diversity and antimicrobial activities of fungi associated with *Haliclona simulans* isolated from Irish coastal waters. *Mar. Biotechnol., 11*, 540–547. doi: 10.1007/s10126-008-9169-7.

Bart, M. C., De Kluijver, A., Hoetjes, S., Absalah, S., Mueller, B., Kenchington, E., Rapp, H. T., & De Goeij, J. M., (2020). Differential processing of dissolved and particulate organic matter by deep-sea sponges and their microbial symbionts. *Sci. Rep., 10*, 17515. doi: 10.1038/s41598-020-74670-0.

Bart, M. C., Mueller, B., Rombouts, T., et al., (2021). Dissolved organic carbon (DOC) is essential to balance the metabolic demands of four dominant North-Atlantic deep-sea sponges. *Limnol. Oceanogr., 66*, 925–938. doi: 10.1002/lno.11652.

Bayer, K., Busch, K., Kenchington, E., Beazley, L., Franzenburg, S., Michels, J., Hentschel, U., & Slaby, M. B., (2020). Microbial strategies for survival in the glass sponge *Vazella pourtalesii* S. *mSystems, 5*, e00473–20. doi: 10.1128/mSystems.00473-20.

Bayer, K., Jahn, M. T., Slaby, B. M., Moitinho-Silva, L., & Hentschel, U., (2018). Marine sponges as *Chloroflexi* hot spots: Genomic insights and high-resolution visualization of an abundant and diverse. *mSystems, 3*, 1–19.

Bayer, K., Kamke, J., & Hentschel, U., (2014). Quantification of bacterial and archaeal symbionts in high and low microbial abundance sponges using real-time PCR. *FEMS Microbiol. Ecol., 89*, 679–690. doi: 10.1111/1574-6941.12369.

Bayer, K., Schmitt, S., & Hentschel, U., (2008). Physiology, phylogeny and *in situ* evidence for bacterial and archaeal nitrifiers in the marine sponge *Aplysina aerophoba*. *Environ. Microbiol., 10*, 2942–2955. doi: 10.1111/j.1462-2920.2008.01582.x.

Bayne, C. J., (1990). Phagocytosis and non-self recognition in invertebrates. *Bioscience, 40*, 723–731.

Bell, A., Bergquist, P., & Battershill, C., (1999). Feeding biology of *Polymastia croceus*. *Mem. Queensl. Museum, 44*, 51–56.

Bennett, H. M., Altenrath, C., Woods, L., Davy, S. K., Webster, N. S., & Bell, J. J., (2017). Interactive effects of temperature and pCO_2 on sponges: From the cradle to the grave. *Glob. Chang. Biol., 23*, 2031–2046. doi: 10.1111/gcb.13474.

Bhushan, A., Peters, E. E., & Piel, J., (2017). Entotheonella bacteria as source of sponge-derived natural products: Opportunities for biotechnological production. In: Müller, W., Schröder, H., & Wang, X., (eds.), *Blue Biotechnology. Progress in Molecular and Subcellular Biology*. Springer, Cham.

Björk, J. R., Díez-Vives, C., Astudillo-García, C., Archie, E. A., & Montoya, M. J., (2019). Vertical transmission of sponge microbiota is inconsistent and unfaithful. *Nat. Ecol. Evol., 3*. doi: 10.1038/s41559-019-0935-x.

Björk, J. R., Diéz-Vives, C., Coma, R., Ribes, M., & Montoya, J., (2013). Specificity and Temporal Dynamics of Complex Bacteria-Sponge Symbiotic Interactions. *Ecology, 94*, 2781–2791.

Björk, J., O'Hara, R., Ribes, M., Coma, R., & Montoya, J., (2017). The Dynamic Core Microbiome: Structure, Dynamics and Stability. *bioRxiv*, 137885. doi: 10.1101/137885.

Blanquer, A., Uriz, M. J., Cebrian, E., & Galand, E. P., (2016). Snapshot of a bacterial microbiome shift during the early symptoms of a massive sponge die-off in the western Mediterranean. *Front. Microbiol.* doi: 10.3389/fmicb.2016.00752.

Böhm, M., Hentschel, U., Friedrich, A., Fieseler, L., Steffen, R., Gamulin, V., Müller, I., & Müller, W., (2001). Molecular response of the sponge *Suberites domuncula* to bacterial infection. *Mar. Biol., 139*, 1037–1045. doi: 10.1007/s002270100656.

Böhm, M., Schröder, H. C., Müller, I. M., Müller, W. E. G., & Gamulin, V., (2000). The mitogen-activated protein kinase p38 pathway is conserved in metazoans: Cloning and activation of p38 of the SAPK2 subfamily from the sponge *Suberites domuncula*. *Biol. Cell, 92*, 95–104. doi: 10.1016/S0248-4900(00)89017-6.

Bosch, T. C. G., & McFall-Ngai, J. M., (2011). Metaorganisms as the new frontier. *Zoology, 114*, 185–190. doi: 10.1016/j.zool.2011.04.001.

Bosch, T. C. G., (2014). Rethinking the role of immunity: Lessons from hydra. *Trends Immunol., 35*, 495–502. doi: 10.1016/j.it.2014.07.008.

Botté, E. S., Nielsen, S., Abdul, W. M. A., Webster, J., Robbins, S., Thomas, T., & Webster, S. N., (2019). Changes in the metabolic potential of the sponge microbiome under ocean acidification. *Nat. Commun., 10*, 1–10. doi: 10.1038/s41467-019-12156-y.

Bourne, D. G., Garren, M., Work, T. M., Rosenberg, E., Smith, G. W., & Harvell, D. C., (2009). Microbial disease and the coral holobiont. *Trends Microbiol., 17*, 554–562. doi: 10.1016/j.tim.2009.09.004.

Brück, W. M., Sennett, S. H., Pomponi, S. A., Willenz, P., & McCarthy, J. P., (2008). Identification of the bacterial symbiont Entotheonella sp. in the mesohyl of the marine sponge *Discodermia* sp. *ISME J., 2*, 335–339. doi: 10.1038/ismej.2007.91.

Buckley, K. M., & Rast, P. J., (2015). Diversity of animal immune receptors and the origins of recognition complexity in the deuterostomes. *Dev. Comp. Immunol., 49*, 179–189. doi: 10.1016/j.dci.2014.10.013.

Burge, C. A., Mark, E. C., Friedman, C. S., et al., (2014). Climate change influences on marine infectious diseases: Implications for management and society. *Ann. Rev. Mar. Sci., 6*, 249–277. doi: 10.1146/annurev-marine-010213-135029.

Burgsdorf, I., Erwin, P. M., López-Legentil, S., Cerrano, C., Haber, M., Frenk, S., & Steindler, L., (2014). Biogeography rather than association with cyanobacteria structures symbiotic microbial communities in the marine sponge *Petrosia ficiformis*. *Front. Microbiol., 5*, 1–11. doi: 10.3389/fmicb.2014.00529.

Burgsdorf, I., Slaby, B. M., Handley, K. M., et al., (2015). Lifestyle evolution in cyanobacterial symbionts of sponges. *MBio, 6*. doi: 10.1128/mBio.00391-15.

Busch, K., (2020). *Biodiversity of Deep-Sea Sponge Microbiomes in the Ecosystem Context*. PhD thesis. Christian-Albrechts University of Kiel. https://nbn-resolving.org/urn:nbn:de:gbv:8-mods-2020-00511-6 (accessed on 03 May 2023).

Busch, K., Hanz, U., Mienis, F., Mueller, B., Franke, A., Martyn, R. E., Tore, R. H., & Hentschel, U., (2020a). On giant shoulders: How a seamount affects the microbial community composition of seawater and sponges. *Biogeosciences, 17*, 3471–3486. doi: 10.5194/bg-17-3471-2020.

Busch, K., Taboada, S., Riesgo, A., Koutsouveli, V., Ríos, P., Cristobo, J., Franke, A., et al., (2021). Population connectivity of fan-shaped sponge holobionts in the deep Cantabrian Sea. *Deep Sea Research Part I: Oceanographic Research Papers, 167*, 103427.

Busch, K., Wurz, E., Rapp, H. T., Bayer, K., Franke, A., & Hentschel, U., (2020b). *Chloroflexi* dominate the deep-sea golf ball sponges *Craniella zetlandica* and *Craniella infrequens* throughout different life stages. *Front. Mar. Sci., 7*, 1–12. doi: 10.3389/fmars.2020.00674.

Campana, S., Hudspith, M., Lankes, D., et al., (2021). Processing of naturally sourced macroalgal- and coral-dissolved organic matter (DOM) by high and low microbial abundance encrusting sponges. *Front. Mar. Sci., 8*, 1–13. doi: 10.3389/fmars.2021.640583.

Chu, H., & Mazmanian, K. S., (2013). Innate immune recognition of the microbiota promotes host-microbial symbiosis. *Nat. Immunol., 14*, 668–675. doi: 10.1038/ni.2635.

Cleary, D. F. R., Becking, L. E., De Voogd, N. J., Pires, A. C. C., Polónia, A. R. M., Egas, C., & Gomes, N. C. M., (2013). Habitat- and host-related variation in sponge bacterial symbiont communities in Indonesian waters. *FEMS Microbiol. Ecol., 85*, 465–482. doi: 10.1111/1574-6941.12135.

Cleary, D. F. R., Ferreira, M. R. S., Bat, N. K., Polónia, A. R. M., Gomes, N. C. M., & De Voogd, N. J., (2020). Bacterial composition of sponges, sediment and seawater in enclosed and open marine lakes in ha long bay Vietnam. *Mar. Biol. Res., 16*, 18–31. doi: 10.1080/17451000.2019.1702214.

Cleary, D. F. R., Swierts, T., Coelho, F. J. R. C., et al., (2019). The sponge microbiome within the greater coral reef microbial metacommunity. *Nat. Commun., 10*, 1–12. doi: 10.1038/s41467-019-09537-8.

Coma, R., Ribes, M., Gili, J. M., & Hughes, N. R., (2001). The ultimate opportunists: Consumers of seston. *Mar. Ecol. Prog. Ser., 219*, 305–308. doi: 10.3354/meps219305.

Coma, R., Ribes, M., Serrano, E., Jiménez, E., Salat, J., & Pascual, J., (2009). Global warming-enhanced stratification and mass mortality events in the Mediterranean. *Proc. Natl. Acad. Sci. U. S. A., 106*, 6176–6181.

Contijoch, E. J., Britton, G. J., Yang, C., et al., (2019). Gut microbiota density influences host physiology and is shaped by host and microbial factors. *Elife, 8*, 1–26. doi: 10.7554/eLife.40553.

Corredor, J. E., Wilkinson, C. R., Vicente, V. P., More, J. M., & Otero, E., (1988). Nitrate Release by Caribbean Reef Sponges, *Limnology and Oceanography, 33*, 114–120.

Daims, H., Lücker, S., & Wagner, M., (2016). A new perspective on microbes formerly known as nitrite-oxidizing bacteria. *Trends Microbiol., 24*, 699–712. doi: 10.1016/j.tim.2016.05.004.

De Goeij, J. M., Lesser, M. P., & Pawlik, R. J., (2017). Nutrient fluxes and ecological functions of coral reef sponges in a changing ocean. In: Carballo, J., & Bell, J., (eds.), *Climate Change, Ocean Acidification and Sponges: Impacts Across Multiple Levels of Organization* (pp. 373–410). Springer International Publishing.

De Goeij, J. M., Moodley, L., Houtekamer, M., Carballeira, N. M., & Van, D. F. C., (2008). Tracing C-13-enriched dissolved and particulate organic carbon in the bacteria-containing coral reef sponge *Halisarca caerulea*: Evidence for DOM feeding. *Limnol. Oceanogr., 53*, 1376–1386. doi: 10.4319/lo.2008.53.4.1376.

De Goeij, J. M., Van, O. D., Vermeij, M. J. A., Osinga, R., Middelburg, J. J., De Goeij, A. F. P. M., & Admiraal, W., (2013). Surviving in a marine desert: The sponge loop retains resources within coral reefs. *Science, 342*, 108–110. doi: 10.1126/science.1241981.

Degnan, S. M., (2015). The surprisingly complex immune gene repertoire of a simple sponge, exemplified by the NLR genes: A capacity for specificity? *Dev. Comp. Immunol., 48*, 269–274. doi: 10.1016/j.dci.2014.07.012.

Deignan, L. K., Pawlik, J. R., & Erwin, M. P., (2018). Agelas wasting syndrome alters prokaryotic symbiont communities of the Caribbean brown tube sponge, *Agelas tubulata. Microb. Ecol.*, 1–8. doi: 10.1007/s00248-017-1135-3.

Dierking, K., & Pita, L., (2020). Receptors mediating host-microbiota communication in the metaorganism: The invertebrate perspective. *Front. Immunol., 11*, 1251. doi: 10.3389/fimmu.2020.01251.

Dittami, S. M., Arboleda, E., Auguet, J. C., et al., (2021). A community perspective on the concept of marine holobionts: Current status, challenges, and future directions. *PeerJ, 9*, 1–34. doi: 10.7717/peerj.10911.

Dosse, G., (1939). Bakterien- und Pilzbefunde sowie pathologische und fäulnisvorgänge in Meeres- und Süsswasserschwämmen. Untersuchungen im Zusammenhang mit dem gegenwärtigen Sterben der Badeschwämme in Westindien. *Z Parasitkde Bd II*, 331–356.

Douglas, A. E., & Werren, H. J., (2016). Holes in the hologenome: Why host-microbe symbioses are not holobionts. *MBio, 7*, 1–7. doi: 10.1128/mBio.02099-15.

Dubilier, N., Bergin, C., & Lott, C., (2008). Symbiotic diversity in marine animals: The art of harnessing chemosynthesis. *Nat. Rev. Microbiol., 6*, 725–740. doi: 10.1038/nrmicro1992.

Easson, C. G., & Thacker, W. R., (2014). Phylogenetic signal in the community structure of host-specific microbiomes of tropical marine sponges. *Front. Microbiol., 5*, 532. doi: 10.3389/fmicb.2014.00532.

Egan, S., & Gardiner, M., (2016). Microbial dysbiosis: Rethinking disease in marine ecosystems. *Front. Microbiol., 7*, 991. doi: 10.3389/fmicb.2016.00991.

Ereskovsky, A. V., Tokina, D. B., Saidov, D. M., Baghdiguian, S., Le Goff, E., & Lavrov, I. A., (2020). Transdifferentiation and mesenchymal-to-epithelial transition during regeneration in Demospongiae (Porifera). *J. Exp. Zool. Part B Mol. Dev. Evol., 334*, 37–58. doi: 10.1002/jez.b.22919.

Erwin, P. M., & Thacker, W. R., (2008). Cryptic diversity of the symbiotic cyanobacterium *Synechococcus spongiarum* among sponge hosts. *Mol. Ecol., 17*, 2937–2947. doi: 10.1111/j.1365-294X.2008.03808.x.

Erwin, P. M., Coma, R., López-Sendino, P., Serrano, E., & Ribes, M., (2015). Stable symbionts across the HMA-LMA dichotomy: Low seasonal and inter-annual variation in sponge-associated bacteria from taxonomically diverse hosts. *FEMS Microbiol., 91*, 1–11. doi: 10.1093/femsec/fiv115.

Erwin, P. M., López-legentil, S., & Schuhmann, W. P., (2010). The pharmaceutical value of marine biodiversity for anti-cancer drug discovery. *Ecol. Econ., 70*, 445–451. doi: 10.1016/j.ecolecon.2010.09.030.

Erwin, P. M., Pita, L., López-Legentil, S., & Turon, X., (2012). Stability of sponge-associated bacteria over large seasonal shifts in temperature and irradiance. *Appl. Environ. Microbiol., 78*, 7358–7368. doi: 10.1128/AEM.02035-12.

Fabricius, K. E., Langdon, C., Uthicke, S., et al., (2011). Losers and winners in coral reefs acclimatized to elevated carbon dioxide concentrations. *Nat. Clim. Chang., 1*, 165–169. doi: 10.1038/nclimate1122.

Fan, L., Liu, M., Simister, R., Webster, N. S., & Thomas, T., (2013). Marine microbial symbiosis heats up: The phylogenetic and functional response of a sponge holobiont to thermal stress. *ISME J., 7*, 991–1002. doi: 10.1038/ismej.2012.165.

Fan, L., Reynolds, D., Liu, M., Stark, M., Kjelleberg, S., Webster, N. S., & Thomas, T., (2012). Functional equivalence and evolutionary convergence in complex communities of microbial sponge symbionts. *Proc. Natl. Acad. Sci., 109*, E1878–E1887. doi: 10.1073/pnas.1203287109.

Fieseler, L., Horn, M., Wagner, M., & Hentschel, U., (2004). Discovery of the novel candidate phylum "*Poribacteria*" in marine sponges. *Appl. Environ. Microbiol., 70*, 3724–3732. doi: 10.1128/AEM.70.6.3724.

Fiore, C. L., Baker, D. M., & Lesser, M. P., (2013a). Nitrogen biogeochemistry in the Caribbean sponge, *Xestospongia muta*: A source or sink of dissolved inorganic nitrogen? *PLoS One, 8*, e72961. doi: 10.1371/journal.pone.0072961.

Fiore, C. L., Freeman, C. J., & Kujawinski, E., (2017). Sponge exhalent seawater contains a unique chemical profile of dissolved organic matter. *PeerJ*, 1–22. doi: 10.7717/peerj.2870.

Fiore, C. L., Jarett, J. K., & Lesser, M. P., (2013b). Symbiotic prokaryotic communities from different populations of the giant barrel sponge, *Xestospongia muta*. *Microbiology Open, 2*, 938–952. doi: 10.1002/mbo3.135.

Fiore, C. L., Labrie, M., Jarett, J. K., Lesser, M. P., Jarettt, J. K., & Lesser, P. M., (2015). Transcriptional activity of the giant barrel sponge, *Xestospongia muta* holobiont: Molecular evidence for metabolic interchange. *Front. Microbiol., 6*, 364. doi: 10.3389/fmicb.2015.00364.

Foster, K. R., Schluter, J., Coyte, K. Z., & Rakoff-Nahoum, S., (2017). The evolution of the host microbiome as an ecosystem on a leash. *Nature, 548*, 43–51. doi: 10.1038/nature23292.

Fraune, S., Augustin, R., Anton-Erxleben, F., Wittlieb, J., Gelhaus, C., Klimovich, V. B., Samoilovich, M. P., & Bosch, T. C. G., (2010). In an early branching metazoan, bacterial colonization of the embryo is controlled by maternal antimicrobial peptides. *Proc. Natl. Acad. Sci. U. S. A., 107*, 18067–18072. doi: 10.1073/pnas.1008573107.

Freeman, C. J., Easson, C. G., & Baker, M. D., (2014). Metabolic diversity and niche structure in sponges from the Miskito Cays, Honduras. *PeerJ, 2014*, 1–20. doi: 10.7717/peerj.695.

Freeman, C. J., Easson, C. G., Matterson, K. O., Thacker, R. W., Baker, D. M., & Paul, J. V., (2020). Microbial symbionts and ecological divergence of Caribbean sponges: A new perspective on an ancient association. *ISME J., 14*, 1571–1583. doi: 10.1038/s41396-020-0625-3.

Freeman, M. F., Gurgui, C., Helf, M. J., et al., (2012). Metagenome mining reveals polytheonamides as posttranslationally modified ribosomal peptides. *Science, 338*, 387–390. doi: 10.1126/science.1226121.

Freeman, M. F., Helf, M. J., Bhushan, A., Morinaka, B. I., & Piel, J., (2017). Seven enzymes create extraordinary molecular complexity in an uncultivated bacterium. *Nat. Chem., 9*, 387–395. doi: 10.1038/nchem.2666.

Freeman, M. F., Vagstad, A. L., & Piel, J., (2016). Polytheonamide biosynthesis showcasing the metabolic potential of sponge-associated uncultivated "Entotheonella" bacteria. *Curr. Opin. Chem. Biol., 31*, 8–14. doi: 10.1016/j.cbpa.2015.11.002.

Funayama, N., Nakatsukasa, M., Kuraku, S., Takechi, K., Dohi, M., Iwabe, N., Miyata, T., & Agata, K., (2005). Isolation of Ef silicatein and Ef lectin as molecular markers sclerocytes and cells involved in innate immunity in the freshwater sponge *Ephydatia fluviatilis*. *Zoolog. Sci., 22*, 1113–1122. doi: 10.2108/zsj.22.1113.

Galperin, M., Makarova, K. S., Wolf, Y. I., & Koonin, E. V., (2015). Expanded microbial genome coverage and improved protein family annotation in the COG database. *Nucleic Acids Res, 43*(Database issue), D261–269. doi: 10.1093/nar/gku1223.

Gantt, S. E., McMurray, S. E., Stubler, A. D., Finelli, C. M., Pawlik, J. R., & Erwin, M. P., (2019). Testing the relationship between microbiome composition and flux of carbon and nutrients in Caribbean coral reef sponges. *Microbiome, 7*, 1–13. doi: 10.1186/s40168-019-0739-x.

Gardères, J., Bedoux, G., Koutsouveli, V., Crequer, S., Desriac, F., & Le Pennec, G., (2015). Lipopolysaccharides from commensal and opportunistic bacteria: Characterization and response of the immune system of the host sponge *Suberites domuncula*. *Mar. Drugs, 13*, 4985–5006. doi: 10.3390/md13084985.

Garrabou, J., Coma, R., Bensoussan, N., et al., (2009). Mass mortality in Northwestern Mediterranean rocky benthic communities: Effects of the 2003 heat wave. *Glob. Chang. Biol., 15*, 1090–1103. doi: 10.1111/j.1365-2486.2008.01823.x.

Gerardo, N. M., Hoang, K. L., & Stoy, S. K., (2020). Evolution of animal immunity in the light of beneficial symbioses. *Philos. Trans. R. Soc. Lond. B. Biol. Sci., 375*, 20190601. doi: 10.1098/rstb.2019.0601.

Germer, J., Cerveau, N., & Jackson, J. D., (2017). The holo-transcriptome of a calcified early branching metazoan. *Front. Mar. Sci., 4*, 1–19. doi: 10.3389/fmars.2017.00081.

Gilbert, J. A., Jansson, J. K., & Knight, R., (2014). The earth microbiome project: Successes and aspirations. *BMC Biol., 12*, 1–4. doi: 10.1186/s12915-014-0069-1.

Glasl, B., Smith, C. E., Bourne, D. G., & Webster, S. N., (2018). Exploring the diversity-stability paradigm using sponge microbial communities. *Sci. Rep., 8*, 1–9. doi: 10.1038/s41598-018-26641-9.

Gloeckner, V., Wehrl, M., Moitinho-Silva, L., et al., (2014). The HMA-LMA dichotomy revisited: An electron microscopical survey of 56 sponge species. *Biol. Bull., 227*, 78–88.

Gómez-Gras, D., Linares, C., Dornelas, M., et al., (2021). Climate change transforms the functional identity of Mediterranean coralligenous assemblages. *Ecol. Lett., 24*, 1038–1051. doi: 10.1111/ele.13718.

Goodwin, C., Rodolfo-Metalpa, R., Picton, B., & Hall-Spencer, M. J., (2014). Effects of ocean acidification on sponge communities. *Mar. Ecol., 35*, 41–49. doi: 10.1111/maec.12093.

Grebenjuk, V. A., Kuusksalu, A., Kelve, M., Schütze, J., Schröder, H. C., & Müller, W. E. G., (2002). Induction of (2'-5') oligoadenylate synthetase in the marine sponges *Suberites domuncula* and *Geodia cydonium* by the bacterial endotoxin lipopolysaccharide. *Eur. J. Biochem., 269*, 1382–1392. doi: 10.1046/j.1432-1033.2002.02781.x.

Gutleben, J., Loureiro, C., Ramírez, R. L. A., Shetty, S., Wijffels, R. H., Smidt, H., & Sipkema, D., (2020). Cultivation of bacteria from *Aplysina aerophoba*: Effects of oxygen and nutrient gradients. *Front. Microbiol., 11*, 1–19. doi: 10.3389/fmicb.2020.00175.

Guzman, C., & Conaco, C., (2016). Gene expression dynamics accompanying the sponge thermal stress response. *PLoS One, 11*, e0165368. doi: 10.1371/journal.pone.0165368.

Habyarimana, F., Al-Khodor, S., Kalia, A., Graham, J. E., Price, C. T., Garcia, M. T., & Kwaik, A. Y., (2008). Role for the ankyrin eukaryotic-like genes of *Legionella pneumophila* in parasitism of protozoan hosts and human macrophages. *Environ. Microbiol., 10*, 1460–1474. doi: 10.1111/j.1462-2920.2007.01560.x.

Hadfield, M. G., (2011). Biofilms and marine invertebrate larvae: What bacteria produce that larvae use to choose settlement sites. *Ann. Rev. Mar. Sci., 3*, 453–470. doi: 10.1146/annurev-marine-120709-142753.

Hallam, S. J., Konstantinidis, K. T., Putnam, N., et al., (2006). Genomic analysis of the uncultivated marine crenarchaeote *Cenarchaeum symbiosum*. *Proc. Natl. Acad. Sci. U. S. A., 103*, 18296–18301. doi: 10.1073/pnas.0608549103.

Hall-Spencer, J. M., Rodolfo-Metalpa, R., Martin, S., et al., (2008). Volcanic carbon dioxide vents show ecosystem effects of ocean acidification. *Nature, 454*, 96–99. doi: 10.1038/nature07051.

Hardoim, C. C. P., Lôbo-Hajdu, G., Custódio, M. R., & Hardoim, R. P., (2021). Prokaryotic, fungal, and unicellular eukaryotic core communities across three sympatric marine sponges from the Southwestern Atlantic coast are dominated largely by deterministic assemblage processes. *Front. Microbiol., 12*, 674004. doi: 10.3389/fmicb.2021.674004.

Hartenstein, V., & Martinez, P., (2019). Phagocytosis in cellular defense and nutrition: A food-centered approach to the evolution of macrophages. *Cell Tissue Res., 377*, 527–547. doi: 10.1007/s00441-019-03096-6.

Hentschel, U., Fieseler, L., Wehrl, M., Gernert, C., Steinert, M., Hacker, J., & Horn, M., (2003). Microbial diversity of marine sponges. In: Müller, W. E. G., (ed.), *Molecular Marine Biology of Sponges* (pp. 60–88). Springer.

Hentschel, U., Horn, M., Friedrich, A. B., et al., (2002). Molecular evidence for a uniform microbial community in sponges from different oceans. *Appl. Environ. Microbiol., 68*, 4431–4440. doi: 10.1128/AEM.68.9.4431.

Hentschel, U., Piel, J., Degnan, S. M., & Taylor, W. M., (2012). Genomic insights into the marine sponge microbiome. *Nat. Rev. Microbiol., 10*, 641–654. doi: 10.1038/nrmicro2839.

Hentschel, U., Usher, K. M., & Taylor, W. M., (2006). Marine sponges as microbial fermenters. *FEMS Microbiol. Ecol., 55*, 167–177. doi: 10.1111/j.1574-6941.2005.00046.x.

Hildemann, W. H., Johnson, I. S., & Jokiel, L. P., (1979). Immunocompetence in the lowest metazoan phylum: Transplantation immunity in sponges. *Science, 204*, 420–422. doi: 10.1126/science.441730.

Hoer, D. R., Gibson, P. J., Tommerdahl, J. P., Lindquist, N. L., & Martens, C. S., (2018a). Consumption of dissolved organic carbon by Caribbean reef sponges. *Limnol. Oceanogr., 63*, 337–351. doi: 10.1002/lno.10634.

Hoer, D. R., Tommerdahl, J. P., Lindquist, N. L., & Martens, C. S., (2018b). Dissolved inorganic nitrogen fluxes from common Florida Bay (USA) sponges. *Limnol. Oceanogr.*, *63*, 2563–2578. doi:doi: 10.1002/lno.10960.

Hoffmann, F., Radax, R., Woebken, D., et al., (2009). Complex nitrogen cycling in the sponge *Geodia barretti*. *Environ. Microbiol.*, *11*, 2228–2243. doi: 10.1111/j.1462-2920.2009.01944.x.

Höller, G., Wright, A. D., Matthée, G. F., Konig, G. M., Draeger, S., Aust, H. J., & Schulz, B., (2000). Fungi from marine sponges: Diversity, biological activity and secondary metabolites. *Mycol. Res.*, *104*, 1354–1365. doi: 10.1017/S0953756200003117.

Holmes, B., & Blanch, H., (2006). Genus-specific associations of marine sponges with group I crenarchaeotes. *Mar. Biol.*, *150*, 759–772. doi: 10.1007/s00227-006-0361-x.

Horn, H., Slaby, B. M., Jahn, M. T., Bayer, K., Moitinho-Silva, L., Förster, F., Abdelmohsen, U. R., & Hentschel, U., (2016). An enrichment of CRISPR and other defense-related features in marine sponge-associated microbial metagenomes. *Front. Microbiol.*, *7*. doi: 10.3389/fmicb.2016.01751.

Hudspith, M., Rix, L., Achlatis, M., et al., (2021). Subcellular view of host–microbiome nutrient exchange in sponges: Insights into the ecological success of an early metazoan–microbe symbiosis. *Microbiome, 9*, 44. doi: 10.1186/s40168-020-00984-w.

Hughes, T. P., Anderson, K. D., Connolly, S. R., et al., (2018). Spatial and temporal patterns of mass bleaching of corals in the Anthropocene. *Science, 359*, 80–83. doi: 10.1126/science.aan8048.

Imsiecke, G., (1993). Ingestion, digestion, and egestion in *Spongilla lacustris* (Porifera, Spongillidae) after pulse feeding with *Chlamydomonas reinhardtii* (Volvocales). *Zoomorphology, 113*, 233–244. doi: 10.1007/BF00403314.

Jahn, M. T., Arkhipova, K., Markert, S. M., et al., (2019). A phage protein aids bacterial symbionts in eukaryote immune evasion. *Cell Host Microbe, 26*, 542–550.e5. doi: 10.1016/j.chom.2019.08.019.

Jahn, M. T., Lachnit, T., Markert, S. M., Stigloher, C., Pita, L., Ribes, M., Dutilh, B. E., & Hentschel, U., (2021). Lifestyle of sponge symbiont phages by host prediction and correlative microscopy. *ISME J.* doi: 10.1038/s41396-021-00900-6.

Janeway, C. A., & Medzhitov, R., (2002). Innate immune recognition. *Annu. Rev. Immunol.*, *20*, 197–216. doi: 10.1146/annurev.immunol.20.083001.084359.

Jimenez, E., & Ribes, M., (2007). Sponges as a source of dissolved inorganic nitrogen: Nitrification mediated by temperate sponges. *Limnol. Oceanogr.*, *52*, 948–958.

Kahn, A. S., Yahel, G., Chu, J. W. F., Tunnicliffe, V., & Leys, P. S., (2015). Benthic grazing and carbon sequestration by deep-water glass sponge reefs. *Limnol. Oceanogr.*, *60*, 78–88. doi: 10.1002/lno.10002.

Kamke, J., Sczyrba, A., Ivanova, N., Schwientek, P., Rinke, C., Mavromatis, K., Woyke, T., & Hentschel, U., (2013). Single-cell genomics reveals complex carbohydrate degradation patterns in poribacterial symbionts of marine sponges. *ISME J.*, *7*, 2287–2300. doi: 10.1038/ismej.2013.111.

Kandler, N. M., Wahab, M. A. A., Noonan, S. H. C., Bell, J. J., Davy, S. K., Webster, N. S., & Luter, M. H., (2018). *In situ* responses of the sponge microbiome to ocean acidification. *FEMS Microbiol. Ecol.*, *94*. doi: 10.1093/femsec/fiy205.

Karimi, E., Keller-Costa, T., Slaby, B. M., Cox, C. J., Da Rocha, U. N., Hentschel, U., & Costa, R., (2019). Genomic blueprints of sponge-prokaryote symbiosis are shared by low abundant and cultivatable *Alphaproteobacteria*. *Sci. Rep.*, *9*, 1–15. doi: 10.1038/s41598-019-38737-x.

Karimi, E., Ramos, M., Gonçalves, J. M. S., Xavier, J. R., Reis, M. P., & Costa, R., (2017). Comparative metagenomics reveals the distinctive adaptive features of the *Spongia officinalis* endosymbiotic consortium. *Front. Microbiol., 8*, 1–16. doi: 10.3389/fmicb.2017.02499.

Karner, M. B., Delong, E. F., & Karl, M. D., (2001). Archaeal dominance in the mesopelagic zone of the Pacific Ocean. *Nature, 409*, 507–510.

Kawasaki, T., & Kawai, T., (2014). Toll-like receptor signaling pathways. *Front. Immunol., 5*, 1–8. doi: 10.3389/fimmu.2014.00461.

Kitano, H., & Oda, K., (2006). Robustness trade-offs and host-microbial symbiosis in the immune system. *Mol. Syst. Biol., 2*, 1–10. doi: 10.1038/msb4100039.

Konstantinou, D., Kakakiou, R. V., Panteris, E., Voultsiadou, E., & Gkelis, S., (2020). Photosynthetic sponge-associated eukaryotes in the Aegean Sea: A culture-dependent approach. *J. Eukaryot. Microbiol., 67*, 660–670. doi: 10.1111/jeu.12818.

Koutsouveli, V., Manousaki, T., Riesgo, A., et al., (2020). Gearing up for warmer times: Transcriptomic response of *Spongia officinalis* to elevated temperatures reveals recruited mechanisms and potential for resilience. *Front. Mar. Sci., 6*, 1–14. doi: 10.3389/fmars.2019.00786.

Kuusksalu, A., Pihlak, A., Müller, W. E. G., & Kelve, M., (1995). The (2'-5') oligoadenylate synthetase is present in the lowest multicellular organisms, the marine sponges: Demonstration of the existence and identification of its reaction products. *Eur. J. Biochem., 232*, 351–357. doi: 10.1111/j.1432-1033.1995.351zz.x.

Lackner, G., Peters, E. E., Helfrich, E. J. N., Piel, J., Edzard, E., Helfrich, E. J. N., & Piel, J., (2017). Insights into the lifestyle of uncultured bacterial natural product factories associated with marine sponges. *Proc. Natl. Acad. Sci., 114*, E347–E356. doi: 10.1073/pnas.1616234114.

Laffy, P. W., Botté, E. S., Wood-Charlson, E. M., Weynberg, K. D., Rattei, T., & Webster, S. N., (2019). Thermal stress modifies the marine sponge virome. *Environ. Microbiol. Rep., 11*, 690–698. doi: 10.1111/1758-2229.12782.

Laffy, P. W., Wood-Charlson, E. M., Turaev, D., et al., (2018). Reef invertebrate viromics: Diversity, host specificity and functional capacity. *Environ. Microbiol., 20*, 2125–2141. doi: 10.1111/1462-2920.14110.

Lafi, F. F., Fuerst, J. A., Fieseler, L., Engels, C., Goh, W. W. L., & Hentschel, U., (2009). Widespread distribution of *Poribacteria* in Demospongiae. *Appl. Environ. Microbiol., 75*, 5695–5699. doi: 10.1128/AEM.00035-09.

Landry, Z., Swa, B. K., Herndl, G. J., Stepanauskas, R., & Giovannoni, J. S., (2017). SAR202 genomes from the dark ocean predict pathways for the oxidation of recalcitrant dissolved organic matter. *MBio, 8*, 1–19. doi: 10.1128/mBio.00413-17.

Lavy, A., Keren, R., Haber, M., Schwartz, I., & Ilan, M., (2014). Implementing sponge physiological and genomic information to enhance the diversity of its culturable associated bacteria. *FEMS Microbiol. Ecol., 87*, 486–502. doi: 10.1111/1574-6941.12240.

Lee, O. O., Wang, Y., Yang, J., Lafi, F. F., Al-Suwailem, A., & Qian, & P, Y., (2011). Pyrosequencing reveals highly diverse and species-specific microbial communities in sponges from the Red Sea. *ISME J., 5*, 650–664. doi: 10.1038/ismej.2010.165.

Lee, O. O., Wong, Y. H., & Qian, & P, Y., (2009). Inter- and intraspecific variations of bacterial communities associated with marine sponges from San Juan Island, Washington. *Appl. Environ. Microbiol., 75*, 3513–3521. doi: 10.1128/AEM.00002–09.

Lesser, M. P., Fiore, C., Slattery, M., & Zaneveld, J., (2016). Climate change stressors destabilize the microbiome of the Caribbean barrel sponge, *Xestospongia muta*. *J. Exp. Mar. Bio. Ecol., 475*, 11–18. doi: 10.1016/j.jembe.2015.11.004.

Leys, S. P., & Eerkes-Medrano, I. D., (2006). Feeding in a calcareous sponge: Particle uptake by pseudopodia. *Biol. Bull., 211*, 157–171. doi: 10.2307/4134590.

Leys, S. P., Kahn, A. S., Fang, J. K. H., Kutti, T., & Bannister, J. R., (2018). Phagocytosis of microbial symbionts balances the carbon and nitrogen budget for the deep-water boreal sponge *Geodia barretti*. *Limnol. Oceanogr., 63*, 187–202. doi: 10.1002/lno.10623.

Li, Z. Y., Wang, Y. Z., He, L. M., & Zheng, J. H., (2014). Metabolic profiles of prokaryotic and eukaryotic communities in deep-sea sponge *Lamellomorpha* sp. indicated by metagenomics. *Sci. Rep., 4*, 1–12. doi: 10.1038/srep03895.

Liu, F., Li, J., Feng, G., & Li, Z., (2016). New genomic insights into "Entotheonella" symbionts in *Theonella swinhoei*: Mixotrophy, anaerobic adaptation, resilience, and interaction. *Front. Microbiol., 7*. doi: 10.3389/fmicb.2016.01333.

Liu, M., Fan, L., Zhong, L., Kjelleberg, S., & Thomas, T., (2012). Metaproteogenomic analysis of a community of sponge symbionts. *ISME J., 6*, 1515–1525. doi: 10.1038/ismej.2012.1.

Liu, W. C., Li, C. Q., Zhu, P., Yang, J. L., & Cheng, D. K., (2010). Phylogenetic diversity of culturable fungi associated with two marine sponges: *Haliclona simulans* and *Gelliodes carnosa*, collected from the Hainan Island coastal waters of the South China Sea. *Fungal Divers., 42*, 1–15. doi: 10.1007/s13225-010-0022-8.

López-Acosta, M., Leynaert, A., Chavaud, L., Amice, E., Bihannic, I., Le Bec, T., & Maldonado, M., (2019). *In situ* determination of Si, N, and P utilization by the demosponge *Tethya citrina*: A benthic-chamber approach. In: Montoya, J. M., (ed.), *PLoS One* (Vol. 14, p. e0218787). doi: 10.1371/journal.pone.0218787.

López-Legentil, S., Song, B., McMurray, S. E., & Pawlik, R. J., (2008). Bleaching and stress in coral reef ecosystems: Hsp70 expression by the giant barrel sponge *Xestospongia muta*. *Mol. Ecol., 17*, 1840–1849. doi: 10.1111/j.1365-294X.2008.03667.x.

Lowin, B., Hahne, M., Mattmann, C., & Tschopp, J., (1994). Cytolytic T-cell cytotoxicity is mediated through perforin and Fas lytic pathways. *Nature, 370*, 650–652. doi: 10.1038/370650a0.

Lurgi, M., Wemheuer, B., Webster, N. S., & Montoya, M. J., (2019). Modularity and predicted functions of the global sponge-microbiome network. *Nat. Commun.* doi: 10.1038/s41467-019-08925-4.

Luter, H. M., Andersen, M., Versteegen, E., Laffy, P., Uthicke, S., Bell, J. J., & Webster, S. N., (2020). Cross-generational effects of climate change on the microbiome of a photosynthetic sponge. *Environ. Microbiol., 22*, 4732–4744. doi: 10.1111/1462-2920.15222.

Luter, H. M., Bannister, R. J., Whalan, S., Kutti, T., Pineda, M., & Webster, S. N., (2017). Microbiome analysis of a disease affecting the deep-sea sponge *Geodia barretti*. *FEMS Microbiol. Ecol., 93*, fix074. doi: 10.1093/femsec/fix074.

Luter, H. M., Whalan, S., Andreakis, N., Abdul, W. M., Botté, E. S., Negri, A. P., & Webster, S. N., (2019). The effects of crude oil and dispersant on the larval sponge holobiont. *mSystems, 4*, 1–17. doi: 10.1128/msystems.00743-19.

Luter, H. M., Widder, S., Botté, E. S., Abdul, W. M., Whalan, S., Moitinho-Silva, L., Thomas, T., & Webster, S. N., (2015). Biogeographic variation in the microbiome of the ecologically important sponge, *Carteriospongia foliascens*. *PeerJ, 3*, e1435. doi: 10.7717/peerj.1435.

Magor, B. G., & Magor, E. K., (2001). Evolution of effectors and receptors of innate immunity. *Dev. Comp. Immunol., 25*, 651–682. doi: 10.1016/S0145-305X(01)00029-5.

Maldonado, M., Aguilar, R., Bannister, R., et al., (2015). Sponge grounds as key marine habitats: A synthetic review of types, structure, functional roles and conservation concerns. In: *Marine Animal Forests* (pp. 1–26).

Maldonado, M., López-Acosta, M., Busch, K., et al., (2021). A microbial nitrogen engine modulated by bacteriosyncytia in hexactinellid sponges: Ecological implications for deep-sea communities. *Front. Mar. Sci., 8*. doi: 10.3389/fmars.2021.638505.

Maldonado, M., Ribes, M., & Van, D. F. C., (2012). Nutrient fluxes through sponges: Biology, budgets, and ecological implications. In: *Advances in Marine Biology* (pp. 113–182).

Maldonado, M., Sánchez-Tocino, L., & Navarro, C., (2010a). Recurrent disease outbreaks in corneous demosponges of the genus *Ircinia*: Epidemic incidence and defense mechanisms. *Mar. Biol., 157*, 1577–1590. doi: 10.1007/s00227-010-1431-7.

Maldonado, M., Zhang, X., Cao, X., Xue, L., Cao, H., & Zhang, W., (2010b). Selective feeding by sponges on pathogenic microbes: A reassessment of potential for abatement of microbial pollution. *Mar. Ecol. Prog. Ser., 403*, 75–89. doi: 10.3354/meps08411.

Mangel, A., Leitao, J., Batel, R., Zimmermann, H., Müller, W. E. G., & Schröder, C. H., (1995). Purification and characterization of two exopolyphosphatases from the marine sponge *Tethya lyncurium*. *Eur. J. Biochem., 1245*, 17–28. doi: 10.1016/0304-4165(95)00067-L.

Margulis, L., (1998). *Symbiotic Planet: A New Look at Evolution*, B. Books [ed.].

Massana, R., Delong, E. F., & Pedrós-Alió, C., (2000). A few cosmopolitan phylotypes dominate planktonic archaeal assemblages in widely different oceanic provinces. *Appl. Environ. Microbiol., 66*, 1777–1787. doi: 10.1128/AEM.66.5.1777-1787.2000.

Massaro, A. J., Weisz, J. B., Hill, M. S., & Webster, S. N., (2012). Behavioral and morphological changes caused by thermal stress in the great barrier reef sponge *Rhopaloeides odorabile*. *J. Exp. Mar. Bio. Ecol., 416, 417*, 55–60. doi: 10.1016/j.jembe.2012.02.008.

McFall-Ngai, M., (2007). Adaptive immunity: Care for the community. *Nature, 445*, 153. doi: 10.1038/445153a.

McFall-Ngai, M., (2008). Are biologists in "future shock"? Symbiosis integrates biology across domains. *Nat. Rev. Microbiol., 6*, 789–792.

McFall-Ngai, M., Hadfield, M. G., Bosch, T. C. G., et al., (2013). Animals in a bacterial world, a new imperative for the life sciences. *Proc. Natl. Acad. Sci. U. S. A., 110*, 3229–3236. doi: 10.1073/pnas.1218525110.

McMurray, S. E., Johnson, Z. I., Hunt, D. E., Pawlik, J. R., & Finelli, M. C., (2016). Selective feeding by the giant barrel sponge enhances foraging efficiency. *Limnol. Oceanogr.* 1271–1286. doi: 10.1002/lno.10287.

McMurray, S. E., Stubler, A. D., Erwin, P. M., Finelli, C. M., & Pawlik, R. J., (2018). A test of the sponge-loop hypothesis for emergent Caribbean reef sponges. *Mar. Ecol. Prog. Ser., 588*, 1–14. doi: 10.3354/meps12466.

McMurray, S., Pawlik, J., & Finelli, C., (2014). Trait-mediated ecosystem impacts: How morphology and size affect pumping rates of the Caribbean giant barrel sponge. *Aquat. Biol., 23*, 1–13. doi: 10.3354/ab00612.

Moitinho-Silva, L., Bayer, K., Cannistraci, C. V., Giles, E. C., Ryu, T., Seridi, L., Ravasi, T., & Hentschel, U., (2014). Specificity and transcriptional activity of microbiota associated with low and high microbial abundance sponges from the Red Sea. *Mol. Ecol., 23*, 1348–1363. doi: 10.1111/mec.12365.

Moitinho-Silva, L., Díez-Vives, C., Batani, G., Esteves, A. I., Jahn, M. T., & Thomas, T., (2017a). Integrated metabolism in sponge–microbe symbiosis revealed by genome-centered metatranscriptomics. *ISME J., 11*, 1651–1666. doi: 10.1038/ismej.2017.25.

Moitinho-Silva, L., Nielsen, S., Amir, A., et al., (2017b). The sponge microbiome project. *Giga Sci., 6*. doi: 10.1093/gigascience/gix077.

Moitinho-Silva, L., Steinert, G., Nielsen, S., et al., (2017c). Predicting the HMA-LMA status in marine sponges by machine learning. *Front. Microbiol., 8*. doi: 10.3389/fmicb.2017.00752.

Morganti, T. M., Ribes, M., Moskovich, R., Weisz, J. B., Yahel, G., & Coma, R., (2021). *In situ* pumping rate of 20 marine demosponges is a function of osculum area. *Front. Mar. Sci., 8*. doi: 10.3389/fmars.2021.583188.

Morganti, T. M., Ribes, M., Yahel, G., & Coma, R., (2019). Size is the major determinant of pumping rates in marine sponges. *Front. Physiol., 10*. doi: 10.3389/fphys.2019.01474.

Morganti, T., Coma, R., Yahel, G., & Ribes, M., (2017). Trophic niche separation that facilitates co-existence of high and low microbial abundance sponges is revealed by *in situ* study of carbon and nitrogen fluxes. *Limnol. Oceanogr., 62*, 1963–1983. doi: 10.1002/lno.10546.

Mori, T., Cahn, J. K. B., Wilson, M. C., et al., (2018). Single-bacterial genomics validates rich and varied specialized metabolism of uncultivated Entotheonella sponge symbionts. *Proc. Natl. Acad. Sci. U. S. A., 115*, 1718–1723. doi: 10.1073/pnas.1715496115.

Morrow, K. M., Bourne, D. G., Humphrey, C., et al., (2015). Natural volcanic CO_2 seeps reveal future trajectories for host-microbial associations in corals and sponges. *ISME J., 9*, 894–908. doi: 10.1038/ismej.2014.188.

Moscona, A. A., (1968). Cell aggregation: Properties of specific cell-ligands and their role in the formation of multicellular systems. *Dev. Biol., 18*, 250–277. doi: 10.1016/0012-1606(68)90035-3.

Mueller, B., De Goeij, J. M., Vermeij, M. J. A., Mulders, Y., Van, D. E. E., Ribes, M., & Van, D. F. C., (2014). Natural diet of coral-excavating sponges consists mainly of dissolved organic carbon (DOC). *PLoS One, 9*. doi: 10.1371/journal.pone.0090152.

Müller, W. E. G., & Müller, M. I., (2003). Origin of the metazoan immune system: Identification of the molecules and their functions in sponges. *Integr. Comp. Biol., 292*, 281–292.

Naim, M. A., Morillo, J. A., Sørensen, S. J., Waleed, A. A. S., Smidt, H., & Sipkema, D., (2014). Host-specific microbial communities in three sympatric North Sea sponges. *FEMS Microbiol. Ecol., 90*, 390–403. doi: 10.1111/1574-6941.12400.

Naim, M. A., Smidt, H., & Sipkema, D., (2017). Fungi found in Mediterranean and North Sea sponges: How specific are they? *PeerJ, 2017*, 1–19. doi: 10.7717/peerj.3722.

Nakashima, Y., Egami, Y., Kimura, M., Wakimoto, T., & Abe, I., (2016). Metagenomic analysis of the sponge *Discodermia* reveals the production of the cyanobacterial natural product kasumigamide by 'Entotheonella.' *PLoS One, 11*, e0164468. doi: 10.1371/journal.pone.0164468.

Nguyen, M. T. H. D., Liu, M., & Thomas, T., (2014). Ankyrin-repeat proteins from sponge symbionts modulate amoebal phagocytosis. *Mol. Ecol., 23*, 1635–1645. doi: 10.1111/mec.12384.

Nyholm, S. V., & McFall-Ngai, J. M., (2021). A lasting symbiosis: How the Hawaiian bobtail squid finds and keeps its bioluminescent bacterial partner. *Nat. Rev. Microbiol., 10*. doi: 10.1038/s41579-021-00567-y.

Nyholm, V. S., & Graf, J., (2012). Knowing your friends: Invertebrate innate immunity fosters beneficial bacterial symbioses. *Nat. Rev. Microbiol., 10*, 815–827. doi: 10.1038/nrmicro2894.

Off, S., Alawi, M., & Spieck, E., (2010). Enrichment and physiological characterization of a novel *Nitrospira*-like bacterium obtained from a marine sponge. *Appl. Environ. Microbiol., 76*, 4640–4646. doi: 10.1128/AEM.00320-10.

Olinger, L. K., Strangman, W. K., McMurray, S. E., & Pawlik, R. J., (2021). Sponges with microbial symbionts transform dissolved organic matter and take up organohalides. *Front. Mar. Sci., 8*, 1–13. doi: 10.3389/fmars.2021.665789.

Olson, J. B., Thacker, R. W., & Gochfeld, J. D., (2014). Molecular community profiling reveals impacts of time, space, and disease status on the bacterial community associated with the Caribbean sponge *Aplysina cauliformis*. *FEMS Microbiol. Ecol., 87*, 268–279. doi: 10.1111/1574-6941.12222.

Pahler, S., Blumbach, B., Müller, I., & Müller, W. E. G., (1998). Putative multiadhesive protein from the marine sponge *Geodia cydonium*: Cloning of the cDNA encoding a fibronectin-, an SRCR-, and a complement control protein module. *J. Exp. Zool., 282*, 332–343. doi: 10.1002/(SICI)1097-010X(19981015)282:3<332::AID-JEZ6>3.0.CO;2-N.

Palmer, C. V., (2018). Immunity and the coral crisis. *Commun. Biol., 1*, 91. doi: 10.1038/s42003-018-0097-4.

Pan, X., Lührmann, A., Satoh, A., Laskowski-Arce, M. A., & Roy, R. C., (2008). Ankyrin repeat proteins comprise a diverse family of bacterial type IV effectors. *Science, 320*, 1651–1654. doi: 10.1126/science.1158160.

Pancer, Z., Münkner, J., Müller, I., & Müller, W. E. G., (1997). A novel member of an ancient superfamily: Sponge (*Geodia cydonium*, Porifera) putative protein that features scavenger receptor cysteine-rich repeats. *Gene, 193*, 211–218. doi: 10.1016/S0378-1119(97)00135-2.

Pande, S., & Kost, C., (2017). Bacterial unculturability and the formation of intercellular metabolic networks. *Trends Microbiol., 25*, 349–361. doi: 10.1016/j.tim.2017.02.015.

Pantile, R., & Webster, N., (2011). Strict thermal threshold identified by quantitative PCR in the sponge *Rhopaloeides odorabile*. *Mar. Ecol. Prog. Ser., 431*, 97–105. doi: 10.3354/meps09128.

Pascelli, C., Laffy, P. W., Kupresanin, M., Ravasi, T., & Webster, S. N., (2018). Morphological characterization of virus-like particles in coral reef sponges. *PeerJ, 2018*, 1–21. doi: 10.7717/peerj.5625.

Passarini, M. R. Z., Santos, C., Lima, N., Berlinck, R. G. S., & Sette, D. L., (2013). Filamentous fungi from the Atlantic marine sponge *Dragmacidon reticulatum*. *Arch. Microbiol., 195*, 99–111. doi: 10.1007/s00203-012-0854-6.

Pawlik, J. R., (1993). Marine invertebrate chemical defenses. *Chem. Rev., 93*, 1911–1922. doi: 10.1021/cr00021a012.

Pawlik, J. R., (2011). The chemical ecology of sponges on Caribbean reefs: Natural products shape natural systems. *Bioscience, 61*, 888–898. doi: 10.1525/bio.2011.61.11.8.

Paz, Z., Komon-Zelazowska, M., Druzhinina, I. S., et al., (2010). Diversity and potential antifungal properties of fungi associated with a Mediterranean sponge. *Fungal Divers., 42*, 17–26. doi: 10.1007/s13225-010-0020-x.

Piel, J., Hui, D., Wen, G., Butzke, D., Platzer, M., Fusetani, N., & Matsunaga, S., (2004). Antitumor polyketide biosynthesis by an uncultivated bacterial symbiont of the marine sponge *Theonella swinhoei*. *Proc. Natl. Acad. Sci. U. S. A., 101*, 16222–16227. doi: 10.1073/pnas.0405976101.

Pile, A. J., Patterson, M. R., & Witman, D. J., (1996). *In situ* grazing on plankton <10 pm by the boreal sponge *Mycale lingua*. *Mar. Ecol. Prog. Ser., 141*, 95–102.

Pineda, M. C., Strehlow, B., Duckworth, A., Doyle, J., Jones, R., & Webster, S. N., (2016). Effects of light attenuation on the sponge holobiont- implications for dredging management. *Sci. Rep., 6*, 39038. doi: 10.1038/srep39038.

Pineda, M. C., Strehlow, B., Kamp, J., Duckworth, A., Jones, R., & Webster, N. S., (2017a). Effects of combined dredging-related stressors on sponges: A laboratory approach using realistic scenarios. *Sci. Rep., 7*, 5155. doi: 10.1038/s41598-017-05251-x.

Pineda, M. C., Strehlow, B., Sternel, M., Duckworth, A., Den, H. J., Jones, R., & Webster, N. S., (2017b). Effects of sediment smothering on the sponge holobiont with implications for dredging management. *Sci. Rep., 7*, 1–15. doi: 10.1038/s41598-017-05243-x.

Pita, L., Fraune, S., & Hentschel, U., (2016). Emerging sponge models of animal-microbe symbioses. *Front. Microbiol., 7*, 2102. doi: 10.3389/fmicb.2016.02102.

Pita, L., Hoeppner, M. P., Ribes, M., & Hentschel, U., (2018a). Differential expression of immune receptors in two marine sponges upon exposure to microbial-associated molecular patterns. *Sci. Rep., 8*, 16081. doi: 10.1038/s41598-018-34330-w.

Pita, L., Rix, L., Slaby, B. M., Franke, A., & Hentschel, U., (2018b). The sponge holobiont in a changing ocean: From microbes to ecosystems. *Microbiome, 6*, 46. doi: 10.1186/s40168-018-0428-1.

Pita, L., Turon, X., López-Legentil, S., & Erwin, P. M. P. M., (2013). Host rules: Spatial stability of bacterial communities associated with marine sponges (*Ircinia* spp.) in the Western Mediterranean Sea. *FEMS Microbiol. Ecol., 86*, 268–276. doi: 10.1111/1574-6941.12159.

Pivkin, M. V., Aleshko, S. A., Krasokhin, V. B., & Khudyakova, V. Y., (2006). Fungal assemblages associated with sponges of the southern coast of Sakhalin Island. *Russ. J. Mar. Biol., 32*, 207–213. doi: 10.1134/S1063074006040018.

Podell, S., Blanton, J. M., Neu, A., Agarwal, V., Biggs, J. S., Moore, B. S., & Allen, E. E., (2018). Pangenomic comparison of globally distributed *Poribacteria* associated with sponge hosts and marine particles. *ISME J.* doi: 10.1038/s41396-018-0292-9.

Polónia, A. R. M., Cleary, D. F. R., Duarte, L. N., De Voogd, N. J., & Gomes, N. C. M., (2014a). Composition of Archaea in seawater, sediment, and sponges in the Kepulauan Seribu reef system, Indonesia. *Microb. Ecol., 67*, 553–567. doi: 10.1007/s00248-013-0365-2.

Polónia, A. R. M., Cleary, D. F. R., Freitas, R., De Voogd, N. J., & Gomes, N. C. M., (2014b). The putative functional ecology and distribution of archaeal communities in sponges, sediment and seawater in a coral reef environment. *Mol. Ecol.,* 409–423. doi: 10.1111/mec.13024.

Poppell, E., Weisz, J., Spicer, L., Massaro, A., Hill, A., & Hill, M., (2013). Sponge heterotrophic capacity and bacterial community structure in high- and low-microbial abundance sponges. *Mar. Ecol.* 414–424. doi: 10.1111/maec.12098.

Posadas, N., Baquiran, J. I. P., Nada, M. A. L., Kelly, M., & Conaco, C., (2021). Microbiome diversity and host immune functions influence survivorship of sponge holobionts under future ocean conditions. *ISME J.* doi: 10.1038/s41396-021-01050-5.

Preston, C. M., Wu, K. E. Y., Molinskit, T. F., & Delong, F. E., (1996). A psychrophilic crenarchaeon inhabits a marine sponge: *Cenarchaeum symbiosum* gen. Nov, sp. Nov. *Proc. Natl. Acad. Sci. U. S. A., 93*, 6241–6246.

Radax, R., Hoffmann, F., Rapp, H. T., Leininger, S., & Schleper, C., (2012a). Ammonia-oxidizing archaea as main drivers of nitrification in cold-water sponges. *Environ. Microbiol., 14*, 909–923. doi: 10.1111/j.1462-2920.2011.02661.x.

Radax, R., Rattei, T., Lanzen, A., Bayer, C., Rapp, H. T., Urich, T., & Schleper, C., (2012b). Metatranscriptomics of the marine sponge *Geodia barretti*: Tackling phylogeny and function of its microbial community. *Environ. Microbiol., 14*, 1308–1324. doi: 10.1111/j.1462-2920.2012.02714.x.

Ramsby, B., Hoogenboom, M. O., Whalan, S., & Webster, N. S. (2018). Elevated seawater temperature disrupts the microbiome of an ecologically important bioeroding sponge. *Mol. Ecol., 1*. doi: 10.1111/mec.14544.

Reiswig, H. M., (1971). Particle feeding in natural populations of three marine demosponges. *Biol. Bull., 141*, 568–591.

Reiswig, H. M., (1974). Water transport, respiration and energetics of 3 tropical marine sponges. *J. Exp. Mar. Bio. Ecol., 14*, 231–249. doi: 10.1016/0022-0981(74)90005-7.

Reiswig, H., (1981). Partial carbon and energy budgets of the bacteriosponge *Verongia* in Barbados. *P.S.Z.N.I Mar. Ecol., 2*, 273–293.

Reshef, L., Koren, O., Loya, Y., Zilber-Rosenberg, I., & Rosenberg, E., (2006). The coral probiotic hypothesis. *Environ. Microbiol., 8*, 2068–2073. doi: 10.1111/j.1462-2920.2006.01148.x.

Reveillaud, J., Maignien, L., Murat, E. A., Huber, J. A., Apprill, A., Sogin, M. L., & Vanreusel, A., (2014). Host-specificity among abundant and rare taxa in the sponge microbiome. *ISME J.,* 1–12. doi: 10.1038/ismej.2013.227.

Ribes, M., Calvo, E., Movilla, J., Logares, R., Coma, R., & Pelejero, C., (2016). Restructuring of the sponge microbiome favors tolerance to ocean acidification. *Environ. Microbiol. Rep., 8*, 536–544. doi: 10.1111/1758-2229.12430.

Ribes, M., Dziallas, C., Coma, R., & Riemann, L., (2015). Microbial diversity and putative diazotrophy in high- and low- microbial-abundance Mediterranean sponges. *Appl. Environ. Microbiol., 81*, 5683–5693. doi: 10.1128/AEM.01320-15.

Ribes, M., Jimenez, E., Yahel, G., et al., (2012). Functional convergence of microbes associated with temperate marine sponges. *Environ. Microbiol., 14,* 1224–1239. doi: 10.1111/j.1462-2920.2012.02701.x.

Riesgo, A., Farrar, N., Windsor, P. J., Giribet, G., & Leys, P. S., (2014). The analysis of eight transcriptomes from all *poriferan* classes reveals surprising genetic complexity in sponges. *Mol. Biol. Evol., 31*, 1102–1120. doi: 10.1093/molbev/msu057.

Rinke, C., Schwientek, P., Sczyrba, A., et al., (2013). Insights into the phylogeny and coding potential of microbial dark matter. *Nature, 499*, 431–437. doi: 10.1038/nature12352.

Rix, L., De Goeij, J. M., Mueller, C. E., et al., (2016a). Coral mucus fuels the sponge loop in warm- and cold-water coral reef ecosystems. *Sci. Rep., 6*, 18715. doi: 10.1038/srep18715.

Rix, L., De Goeij, J. M., Van, O. D., et al., (2016b). Differential recycling of coral and algal dissolved organic matter via the sponge loop. *Funct. Ecol., 31*, 778–789. doi: 10.1111/1365-2435.12758.

Rix, L., Ribes, M., Coma, R., et al., (2020). Heterotrophy in the earliest gut: A single-cell view of heterotrophic carbon and nitrogen assimilation in sponge-microbe symbioses. *ISME J.* doi: 10.1038/s41396-020-0706-3.

Robbins, S. J., Song, W., Engelberts, J. P., et al., (2021). A genomic view of the microbiome of coral reef demosponges. *ISME J.* doi: 10.1038/s41396-020-00876-9.

Rodrigues De, O. B. F., Freitas-Silva, J., Sanchez-Robinet, C., & Silva, L. M., (2020). Transmission of the sponge microbiome: Moving towards a unified model. *Environ. Microbiol. Rep.* doi: 10.1111/1758-2229.12896.

Rohde, S., Nietzer, S., & Schupp, J. P., (2015). Prevalence and mechanisms of dynamic chemical defenses in tropical sponges. *PLoS One, 10*, 1–19. doi: 10.1371/journal.pone.0132236.

Rohwer, F., Seguritan, V., Azam, F., & Knowlton, N., (2002). Diversity and distribution of coral-associated bacteria. *Mar. Ecol. Prog. Ser., 243*, 1–10. doi: 10.3354/meps243001.

Rosenstiel, P., Philipp, E. E. R., Schreiber, S., & Bosch, T. C. G., (2009). Evolution and function of innate immune receptors - Insights from marine invertebrates. *J. Innate Immun., 1*, 291–300. doi: 10.1159/000211193.

Rot, C., Goldfarb, I., Ilan, M., & Huchon, D., (2006). Putative cross-kingdom horizontal gene transfer in sponge (*Porifera*) mitochondria. *BMC Evol. Biol., 6*, 1–11. doi: 10.1186/1471-2148-6-71.

Rust, M., Helfrich, E. J. N., Freeman, M. F., et al., (2020). A multiproducer microbiome generates chemical diversity in the marine sponge *Mycale hentscheli. Proc. Natl. Acad. Sci. U. S. A., 117*, 9508–9518. doi: 10.1073/pnas.1919245117.

Ryu, T., Seridi, L., Moitinho-Silva, L., et al., (2016). Hologenome analysis of two marine sponges with different microbiomes. *BMC Genomics, 17*, 158. doi: 10.1186/s12864-016-2501-0.

Sagan, L., (1967). On the origin of mitosing cells. *J. Theor. Biol., 14*, 225–274.

Santos F. M. R., Cleary, D. F. R., Coelho, F. J. R. C., Gomes, N. C. M., Huang, Y. M., Polónia, A. R. M., & De Voogd, N. J., (2020). Geographical location and habitat predict variation in prokaryotic community composition of *Suberites diversicolor. Ann. Microbiol., 70*. doi: 10.1186/s13213-020-01546-z.

Schippers, K. J., Martens, D. E., Pomponi, S. A., & Wijffels, H. R., (2011). Cell cycle analysis of primary sponge cell cultures. *Vitr. Cell. Dev. Biol. - Anim., 47*, 302–311. doi: 10.1007/s11626-011-9391-x.

Schläppy, M. L., Schöttner, S. I., Lavik, G., Kuypers, M. M. M., De Beer, D., & Hoffmann, F., (2010). Evidence of nitrification and denitrification in high and low microbial abundance sponges. *Mar. Biol., 157,* 593–602. doi: 10.1007/s00227-009-1344-5.

Schmitt, S., Angermeier, H., Schiller, R., Lindquist, N., & Hentschel, U., (2008). Molecular microbial diversity survey of sponge reproductive stages and mechanistic insights into vertical transmission of microbial symbionts. *Appl. Environ. Microbiol., 74*, 7694–7708. doi: 10.1128/AEM.00878–08.

Schmitt, S., Deines, P., Behnam, F., Wagner, M., & Taylor, W. M., (2011). *Chloroflexi* bacteria are more diverse, abundant, and similar in high than in low microbial abundance sponges. *FEMS Microbiol. Ecol., 78*, 497–510. doi: 10.1111/j.1574-6941.2011.01179.x.

Schmitt, S., Weisz, J. B., Lindquist, N., & Hentschel, U., (2007). Vertical transmission of a phylogenetically complex microbial consortium in the viviparous sponge *Ircinia felix. Appl. Environ. Microbiol., 73*, 2067–2078. doi: 10.1128/AEM.01944-06.

Schmittmann, L., Franzenburg, S., & Pita, L., (2021). Individuality in the immune repertoire and induced response of the sponge *Halichondria panicea. Front. Immunol., 12,* 1–13. doi: 10.3389/fimmu.2021.689051.

Schröder, H. C., Ushijima, H., Krasko, A., Gamulin, V., Thakur, N. L., Diehl-Seifert, B., Müller, I. M., & Müller, W. E. G., (2003). Emergence and disappearance of an immune molecule, an antimicrobial lectin, in basal metazoa. A tachylectin-related protein in the sponge *Suberites domuncula. J. Biol. Chem., 278*, 32810–32817. doi: 10.1074/jbc.M304116200.

Schroeder, B. O., & Bäckhed, F., (2016). Signals from the gut microbiota to distant organs in physiology and disease. *Nat. Med., 22*, 1079–1089. doi: 10.1038/nm.4185.

Schuster, A., Strehlow, B. W., Eckford-Soper, L., McAllen, R., & Canfield, E. D., (2021). Effects of seasonal anoxia on the microbial community structure in demosponges in a marine lake in Lough Hyne, Ireland. *mSphere, 6*, 1–24. doi: 10.1128/mSphere.00991-20.

Schwartzman, J. A., & Ruby, G. E., (2016). A conserved chemical dialog of mutualism: Lessons from squid and vibrio. *Microbes Infect., 18*, 1–10. doi: 10.1016/j.micinf.2015.08.016.

Sebé-Pedrós, A., Chomsky, E., Pang, K., et al., (2018). Early metazoan cell type diversity and the evolution of multicellular gene regulation. *Nat. Ecol. Evol., 2*, 1176–1188. doi: 10.1038/s41559-018-0575-6.

Simister, R. L., Deines, P., Botté, E. S., Webster, N. S., & Taylor, M. W., (2012a). Sponge-specific clusters revisited: A comprehensive phylogeny of sponge-associated microorganisms. *Environ. Microbiol., 14*, 517–524. doi: 10.1111/j.1462-2920.2011.02664.x.

Simister, R., Taylor, M. W., Rogers, K. M., Schupp, P. J., & Deines, P., (2013). Temporal molecular and isotopic analysis of active bacterial communities in two New Zealand sponges. *FEMS Microbiol. Ecol., 85*, 195–205. doi: 10.1111/1574-6941.12109.

Simister, R., Taylor, M. W., Tsai, P., & Webster, N., (2012c). Sponge-microbe associations survive high nutrients and temperatures. *PLoS One, 7*, e52220. doi: 10.1371/journal.pone.0052220.

Simister, R., Taylor, M. W., Tsai, P., Fan, L., Bruxner, T. J., Crowe, M. L., & Webster, N., (2012b). Thermal stress responses in the bacterial biosphere of the great barrier reef sponge, *Rhopaloeides odorabile*. *Environ. Microbiol., 14*, 3232–3246. doi: 10.1111/1462-2920.12010.

Sipkema, D., Schippers, K., Maalcke, W. J., Yang, Y., Salim, S., & Blanch, W. H., (2011). Multiple approaches to enhance the cultivability of bacteria associated with the marine sponge *Haliclona (gellius)* sp. *Appl. Environ. Microbiol., 77*, 2130–2140. doi: 10.1128/AEM.01203-10.

Slaby, B. M., Hackl, T., Horn, H., Bayer, K., & Hentschel, U., (2017). Metagenomic binning of a marine sponge microbiome reveals unity in defense but metabolic specialization. *ISME J., 11*, 2465–2478. doi: 10.1038/ismej.2017.101.

Smith, L. C., & Hildemann, H. W., (1986). Allogeneic cell interactions during graft rejection in *Callyspongia diffusa* (Porifera, Demospongia); a study with monoclonal antibodies. *Proc. R. Soc. Lond. B. Biol. Sci., 226*, 465–477. doi: 10.1098/rspb.1986.0004.

Southwell, M. W., Weisz, J. B., Martens, C. S., et al., (2008). *In situ* fluxes of dissolved inorganic nitrogen from the sponge community on Conch Reef, Key Largo, Florida. *Limnol. Oceanogr., 53*, 986–996. doi: 10.4319/lo.2008.53.3.0986.

Srivastava, M., Simakov, O., Chapman, J., et al., (2010). The *Amphimedon queenslandica* genome and the evolution of animal complexity. *Nature, 466*, 720–726. doi: 10.1038/nature09201.

Steindler, L., Huchon, D., Avni, A., & Ilan, M., (2005)., 16S rRNA phylogeny of sponge-associated cyanobacteria. *Appl. Environ. Microbiol., 71*, 4127–4131. doi: 10.1128/AEM.71.7.4127-4131.2005.

Steindler, L., Schuster, S., Ilan, M., Avni, A., Cerrano, C., & Beer, S., (2007). Differential gene expression in a marine sponge in relation to its symbiotic state. *Mar. Biotechnol., 9*, 543–549. doi: 10.1007/s10126-007-9024-2.

Steinert, G., Busch, K., Bayer, K., et al., (2020). Compositional and quantitative insights into bacterial and archaeal communities of South Pacific deep-sea sponges (Demospongiae and Hexactinellida). *Front. Microbiol., 11*, 1–16. doi: 10.3389/fmicb.2020.00716.

Steinert, G., Gutleben, J., Atikana, A., Wijffels, R. H., Smidt, H., & Sipkema, D., (2017). Coexistence of poribacterial phylotypes among geographically widespread and phylogenetically divergent sponge hosts. *Environ. Microbiol. Rep., 10*, 80–91. doi: 10.1111/1758-2229.12609.

Steinert, G., Whitfield, S., Taylor, M. W., Thoms, C., & Schupp, J. P., (2014). Application of diffusion growth chambers for the cultivation of marine sponge-associated bacteria. *Mar. Biotechnol., 16*, 594–603. doi: 10.1007/s10126-014-9575-y.

Strehlow, B. W., Jorgensen, D., Webster, N. S., Pineda, M., & Duckworth, A., (2016). Using a thermistor flowmeter with attached video camera for monitoring sponge excurrent speed and oscular behaviour. *PeerJ, 4*, e2761. doi: 10.7717/peerj.2761.

Swierts, T., Cleary, D. F. R., & De Voogd, N. J., (2018). Prokaryotic communities of Indo-Pacific giant barrel sponges are more strongly influenced by geography than host phylogeny. *FEMS Microbiol. Ecol., 94*, 1–12. doi: 10.1093/femsec/fiy194.

Takeda, K., (2005). Evolution and integration of innate immune recognition systems: The toll-like receptors. *J. Endotoxin Res., 11*, 51–55. doi: 10.1179/096805105225006687.

Taylor, M. W., Hill, R. T., & Hentschel, U., (2011). Meeting report: 1st international symposium on sponge microbiology. *Mar. Biotechnol., 13*, 1057–1061. doi: 10.1007/s10126-011-9397-0.

Taylor, M. W., Radax, R., Steger, D., & Wagner, M., (2007). Sponge-associated microorganisms: Evolution, ecology, and biotechnological potential. *Microbiol. Mol. Biol. Rev., 71*, 295–347. doi: 10.1128/MMBR.00040–06.

Taylor, M. W., Schupp, P. J., De Nys, R., Kjelleberg, S., & Steinberg, D. P., (2005). Biogeography of bacteria associated with the marine sponge *Cymbastela concentrica. Environ. Microbiol., 7*, 419–433. doi: 10.1111/j.1462-2920.2004.00711.x.

Taylor, M. W., Tsai, P., Simister, R. L., Deines, P., Botte, E., Ericson, G., Schmitt, S., & Webster, S. N., (2013). "Sponge-specific" bacteria are widespread (but rare) in diverse marine environments. *ISME J., 7*, 438–443. doi: 10.1038/ismej.2012.111.

Thacker, W., & Starnes, S., (2003). Host Specificity of the Symbiotic Cyanobacterium Oscillatoria Spongeliae in Marine Sponges, *Dysidea* spp., *Marine Biology*, 643–648. doi: 10.1007/s00227-002-0971-x.

Thakur, N. L., Hentschel, U., Krasko, A., Pabel, C. T., Anil, A. C., & Müller, W. E. G., (2003). Antibacterial activity of the sponge *Suberites domuncula* and its primmorphs: Potential basis for epibacterial chemical defense. *Aquat. Microb. Ecol., 31*, 77–83. doi: 10.3354/ame031077.

Thakur, N. L., Perović-Ottstadt, S., Batel, R., Korzhev, M., Diehl-Seifert, B., Müller, I. M., & Müller, W. E. G., (2005). Innate immune defense of the sponge *Suberites domuncula* against gram-positive bacteria: Induction of lysozyme and AdaPTin. *Mar. Biol., 146*, 271–282. doi: 10.1007/s00227-004-1438-z.

Theis, K. R., Dheilly, N. M., Klassen, J. L., et al., (2016). Getting the hologenome concept right: An eco- evolutionary framework for hosts and their microbiomes. *mSystems, 1*, e00028–16. doi: 10.13140/RG.2.1.2745.7048.

Thomas, T., Moitinho-Silva, L., Lurgi, M., et al., (2016). Diversity, structure and convergent evolution of the global sponge microbiome. *Nat. Commun., 7*, 11870. doi: 10.1038/ncomms11870.

Thomas, T., Rusch, D., DeMaere, M. Z., et al., (2010). Functional genomic signatures of sponge bacteria reveal unique and shared features of symbiosis. *ISME J., 4*, 1557–1567. doi: 10.1038/ismej.2010.74.

Thompson, L. R., Sanders, J. G., McDonald, D., et al., (2017). A communal catalogue reveals earth's multiscale microbial diversity. *Nature*. doi: 10.1038/nature24621.

Tian, R. M., Sun, J., Cai, L., Zhang, W. P., Zhou, G. W., Qiu, J. W., & Qian, Y. P., (2016). The deep-sea glass sponge *Lophophysema eversa* harbors potential symbionts responsible for the nutrient conversions of carbon, nitrogen and sulfur. *Environ. Microbiol., 18*, 2481–2494. doi: 10.1111/1462–2920.13161.

Tian, R., Wang, Y., Bougouffa, S., Gao, Z., Cai, L., Zhang, W., Bajic, V., & Qian, P., (2014). Effect of copper treatment on the composition and function of the bacterial community in the sponge *Haliclona cymaeformis*. *MBio, 5*, e01980–14. doi: 10.1128/mBio.01980-14. Editor.

Tianero, M. D., Balaich, J. N., & Donia, S. M., (2019). Localized production of defense chemicals by intracellular symbionts of *Haliclona* sponges. *Nat. Microbiol., 4*, 1149–1159. doi: 10.1038/s41564-019-0415-8.

Turque, A. S., Batista, D., Silveira, C. B., et al., (2010). Environmental shaping of sponge associated archaeal communities. *PLoS One*, 5. doi: 10.1371/journal.pone.0015774.

Ueoka, R., Uria, A. R., Reiter, S., et al., (2015). Metabolic and evolutionary origin of actin-binding polyketides from diverse organisms. *Nat. Chem. Biol., 11*, 705. doi: 10.1038/nchembio.1870.

Unson, M. D., Holland, N. D., & Faulkner, J. D., (1994). A brominated secondary metabolite synthesized by the cyanobacterial symbiont of a marine sponge and accumulation of the crystalline metabolite in the sponge tissue. *Mar. Biol., 119*, 1–11. doi: 10.1007/BF00350100.

Vacelet, J., & Donadey, C., (1977). Electron microscope study of the association between some sponges and bacteria. *J. Exp. Mar. Biol. Ecol., 30*, 301–314.

Vacelet, J., & Gallissian, & M, F., (1978). Virus-like particles in cells of the sponge *Verongia cavernicola* (Demospongiae, Dictyoceratida) and accompanying tissues changes. *J. Invertebr. Pathol., 31*, 246–254. doi: 10.1016/0022-2011(78)90014-9.

Venn, A. A., Loram, J. E., & Douglas, E. A., (2008). Photosynthetic symbioses in animals. *J. Exp. Bot., 59,* 1069–1080. doi: 10.1093/jxb/erm328.

Versluis, D., McPherson, K., Van, P. M. W. J., Smidt, H., & Sipkema, D., (2017). Recovery of previously uncultured bacterial genera from three Mediterranean sponges. *Mar. Biotechnol., 19*, 454–468. doi: 10.1007/s10126-017-9766-4.

Wang, G., Li, Q., & Zhu, P., (2008). Phylogenetic diversity of culturable fungi associated with the Hawaiian sponges *Suberites zeteki* and *Gelliodes fibrosa*. Antonie van Leeuwenhoek. *Int. J. Gen. Mol. Microbiol., 93*, 163–174. doi: 10.1007/s10482-007-9190-2.

Waters, C. N., Zalasiewicz, J., Summerhayes, C., et al., (2016). The Anthropocene is functionally and stratigraphically distinct from the Holocene. *Science, 351*, 137–148. doi: 10.1126/science.aad2622.

Webster, N. S., & Reusch, T. B. H., (2017). Microbial contributions to the persistence of coral reefs. *ISME J., 11*, 2167–2174. doi: 10.1038/ismej.2017.66.

Webster, N. S., & Thomas, T., (2016). The sponge hologenome. *MBio, 7*, e00135–16. doi: 10.1128/mBio.00135-16.

Webster, N. S., (2007). Sponge disease: A global threat? *Environ. Microbiol., 9*, 1363–1375. doi: 10.1111/j.1462-2920.2007.01303.x.

Webster, N. S., Negri, A. P., Webb, R. I., & Hill, T. R., (2002). A spongin-boring α-proteobacterium is the etiological agent of disease in the great barrier reef sponge *Rhopaloeides odorabile*. Mar. *Ecol. Prog. Ser., 232*, 305–309. doi: 10.3354/meps232305.

Webster, N. S., Taylor, M. W., Behnam, F., Lücker, S., Rattei, T., Whalan, S., Horn, M., & Wagner, M., (2010). Deep sequencing reveals exceptional diversity and modes of transmission for bacterial sponge symbionts. *Environ. Microbiol., 12,* 2070–2082. doi: 10.1111/j.1462-2920.2009.02065.x.

Webster, N. S., Webb, R. I., Ridd, M. J., Hill, R. T., & Negri, P. A., (2001). The effects of copper on the microbial community of a coral reef sponge. *Environ. Microbiol., 3,* 19–31. doi: 10.1046/j.1462-2920.2001.00155.x.

Webster, N. S., Xavier, J. R., Freckelton, M., Motti, C. A., & Cobb, R., (2008). Shifts in microbial and chemical patterns within the marine sponge *Aplysina aerophoba* during a disease outbreak. *Environ. Microbiol., 10,* 3366–3376. doi: 10.1111/j.1462-2920.2008.01734.x.

Webster, N., Pantile, R., Botté, E., Abdo, D., Andreakis, N., & Whalan, S., (2013). A complex life cycle in a warming planet: Gene expression in thermally stressed sponges. *Mol. Ecol.* doi: 10.1111/mec.12213.

Wehrl, M., Steinert, M., & Hentschel, U., (2007). Bacterial uptake by the marine sponge *Aplysina aerophoba. Microb. Ecol., 53,* 355–365. doi: 10.1007/s00248-006-9090-4.

Weigel, B. L., & Erwin, M. P., (2016). Intraspecific variation in microbial symbiont communities of the sun sponge, *Hymeniacidon heliophila,* from intertidal and subtidal habitats. *Appl. Environ. Microbiol., 82,* 650–658. doi: 10.1128/AEM.02980-15.

Wein, T., Romero, P. D., Blow, F., Woehle, C., Jami, E., Reusch, T. B. H., Martin, W. F., & Dagan, T., (2019). Currency, exchange, and inheritance in the evolution of symbiosis. *Trends Microbiol., 27,* 836–849. doi: 10.1016/j.tim.2019.05.010.

Weissenfels, N., (1976). Bau und funktion des Süßwasserschwamms *Ephydatia fluviatilis* L. (Porifera) – III. Nahrungsaufnahme, verdauung und defäkation. *Zoomorphologie, 85,* 73–88. doi: 10.1007/BF00995405.

Weisz, J. B., Hentschel, U., Lindquist, N., & Martens, S. C., (2007). Linking abundance and diversity of sponge-associated microbial communities to metabolic differences in host sponges. *Mar. Biol., 152,* 475–483. doi: 10.1007/s00227-007-0708-y.

Weisz, J. B., Lindquist, N., & Martens, S. C., (2008). Do associated microbial abundances impact marine demosponge pumping rates and tissue densities? *Oecologia, 155,* 367–376. doi: 10.1007/s00442-007-0910-0.

Wiens, M., Korzhev, M., Krasko, A., et al., (2005). Innate immune defense of the sponge *Suberites domuncula* against bacteria involves a MyD88-dependent signaling pathway: Induction of a perforin-like molecule. *J. Biol. Chem., 280,* 27949–27959. doi: 10.1074/jbc.M504049200.

Wiens, M., Korzhev, M., Perovic-Ottstadt, S., Luthringer, B., Brandt, D., Klein, S., & Müller, W. E. G., (2007). Toll-like receptors are part of the innate immune defense system of sponges (Demospongiae: Porifera). *Mol. Biol. Evol., 24,* 792–804. doi: 10.1093/molbev/msl208.

Wiese, J., Ohlendorf, B., Blümel, M., Schmaljohann, R., & Imhoff, F. J., (2011). Phylogenetic identification of fungi isolated from the marine sponge *Tethya aurantium* and identification of their secondary metabolites. *Mar. Drugs, 9,* 561–585. doi: 10.3390/md9040561.

Wilkinson, C. R., (1979). Nutrient translocation from symbiotic cyanobacteria to coral reef sponges. In: Levi, C., & Boury-Esnault, N., (eds.), *Biologie des Spongiaires* (pp. 373–380).

Wilkinson, C. R., (1984). Immunological evidence for the Precambrian origin of bacterial symbioses in marine sponges. *Proc. R. Soc. London – Biol. Sci., 220,* 509–517. doi: 10.1098/rspb.1984.0017.

Wilkinson, C. R., (1987). Significance of microbial symbionts in sponge evolution and ecology. *Symbiosis (Philadelphia, PA), 4*, 135–145.

Wilkinson, C. R., Garrone, R., & Vacelet, J., (1979). Marine sponges discriminate between food bacteria and bacterial symbionts: Electron microscope radioautography and *in situ* evidence. *Proc R Soc L. B, 205*, 519–528. doi: 10.1098/rspb.1979.0086.

Wilkinson, C. R., Garrone, R., & Vacelet, J., (1984). Marine sponges discriminate between food bacteria and bacterial symbionts: Electron microscope radioautography and *in situ* evidence. *Proc. R. Soc. London - Biol. Sci., 220*, 519–528. doi: 10.1098/rspb.1984.0018.

Wilson, M. C., Mori, T., Ruckert, C., et al., (2014). An environmental bacterial taxon with a large and distinct metabolic repertoire. *Nature, 506*, 58–62. doi: 10.1038/nature12959.

Yahel, G., Whitney, F., Reiswig, H. M., Eerkes-medrano, D. I., & Leys, P. S., (2007). *In situ* feeding and metabolism of glass sponges (Hexactinellida, Porifera) studied in a deep temperate fjord with a remotely operated submersible. *Limnol. Oceanogr., 52*, 428–440.

Yuen, B., Bayes, J. M., & Degnan, M. S., (2014). The characterization of sponge NLRs provides insight into the origin and evolution of this innate immune gene family in animals. *Mol. Biol. Evol., 31*, 106–120. doi: 10.1093/molbev/mst174.

Zaneveld, J. R., McMinds, R., & Vega, T. R., (2017). Stress and stability: Applying the anna Karenina principle to animal microbiomes. *Nat. Microbiol., 2*, 17121. doi: 10.1038/nmicrobiol.2017.121.

Zhang, F., Pita, L., Erwin, P. M., Abaid, S., López-Legentil, S., & Hill, T. R., (2014). Symbiotic archaea in marine sponges show stability and host specificity in community structure and ammonia oxidation functionality. *FEMS Microbiol. Ecol.*, 1–9. doi: 10.1111/1574-6941.12427.

CHAPTER 4

Nitrogen and Phosphorus Cycling Through Marine Sponges: Physiology, Cytology, Genomics, and Ecological Implications

MANUEL MALDONADO,[1] KRISTINA BAYER,[2] and MARIA LOPEZ-ACOSTA[3]

[1]*Department of Marine Ecology, Center for Advanced Studies of Blanes (CEAB-CSIC), Blanes, Girona, Spain*

[2]*Marine Symbioses Unit, GEOMAR Helmholtz Center for Ocean Research Kiel, Kiel, Germany*

[3]*Department of Oceanology, Instituto de Investigaciones Marinas (IIM-CSIC), Vigo, Pontevedra, Spain*

ABSTRACT

Several inorganic compounds of nitrogen (N) and phosphorus (P) are key to ocean ecology because, among other effects, they sustain primary production. After discovering in the 1980s that sponges can be both source and sink of such nutrients, much has been learned, including that fluxes derive from the metabolic integration of the sponge tissues and the assemblage of prokaryotic microbes living in them (i.e., the microbiota). The advent of molecular techniques revealed exceptional phylogenetic biodiversity in the microbiota and allowed the identification of genes coding for enzymes transforming N and P compounds. However, the accumulated information remains relatively inarticulate and its ecological dimension is uncharted. Herein we summarize the basics of N and P cycling in the marine environment to further address nutrient flux rates compiled from 92 sponge species. Ammonium release or

Frontiers in Invertebrate Physiology: A Collection of Reviews, Volume 1: Non-Bilaterian Phyla.
Saber Saleuddin, Sally P. Leys, Robert D. Roer, and Iain C. Wilkie, (Eds.)

consumption, followed by nitrite release, emerged as the most common fluxes in sponges. Phosphate release was also prevalent. A difficulty with the available information is a bias towards tropical shallow-water demosponges and the use of non-comparable units. A total of 63 prokaryotic phyla are known from sponge microbiomes. Collectively, they have the genetic potential for all aerobic and anaerobic N transformations, facilitating the formation of closed circuits for N to recycle within the holobionts (i.e., sponge + microbiota). Often, such circuits are fueled by important production/consumption of ammonium. Phosphorus cycling remains understudied, with evidence of phosphate and (organic) phosphonate utilization. Phosphate does not appear to limit sponge microbiomes, with polyphosphate (Poly-P) probably serving more as energy storage than as a P reservoir. Dissimilatory phosphite oxidation (DPO), which would explain the phosphate efflux from the sponges, has not been detected and the causes of the efflux (perhaps anoxic Poly-P degradation) remain uncertain. A relevant benefit provided by the microbiome, in addition to recycling sponge N wastes and provisioning vitamins and some organic C and N compounds through fixation, is to serve as energetically inexpensive particulate food, liberating sponges from strict dependence on inputs of external food. To facilitate co-existence and cooperation between aerobic and anaerobic microbial lineages, sponges modulate pumping activity and have evolved special cells (bacteriocytes) to enclose microbes. Species-specific metabolic integration between sponges and their microbiome yields singular holobionts with remarkable roles in the benthic-pelagic coupling of N and P cycles. Some sponge aggregations can achieve higher denitrification rates per unit area than sediments; others have higher ammonium consumption rates than eutrophic phytoplanktonic communities. Through their microbiomes, some sponge species may also cope with low oxygen conditions and modify local N and P nutrient concentrations, unchaining a cascade of ecological changes that may lead to the exclusion of competitors. Identified gaps in knowledge related to: (i) how the nutrients going in and out of the holobiont are quantitatively connected to the microbial processes occurring inside; (ii) how microbes interact with each other; and (iii) how sponges co-evolved to facilitate co-existence and functional networking in the microbiome.

4.1 INTRODUCTION

4.1.1 SPONGES AS ECOSYSTEM ENGINEERS

Sponges (Phylum Porifera) are a group of sessile invertebrates containing over 9,450 described species (de Voogd et al., 2021). They are essentially marine, except for a family specialized in freshwater. According to the fossil record, sponges are the oldest extant animals, a lineage often interpreted —not without controversy— as the first animals diverging from an ancestor shared with protists. Their distinct body architecture, based on an intricate system of aquiferous canals through which ambient seawater is pumped for feeding and respiration, is unique among animals, as it is their simple histology based on totipotent cells. Despite anatomical simplicity (or maybe because of it), sponges are a successful group ecologically. They are ubiquitous in modern ocean environments and they also thrived during the past 500 million years throughout many different paleo-ocean scenarios (Laubenfels, 1955; Wiedenmayer, 1994; Finks et al., 2004). In the modern ocean, the importance of sponges as relevant players for ecosystem functioning was initially addressed by pioneering studies on coral reefs (Reiswig, 1971a, 1981; Wilkinson, 1987; Diaz & Rützler, 2001; Wulff, 2001) and Antarctic sublittoral bottoms (Dayton et al., 1974; Dayton, 1989). Today, the idea that sponges are crucial to the ecological functioning of a large variety of marine benthic communities is well established (reviewed in Bell, 2008; Maldonado et al., 2012, 2017; de Goeij et al., 2017; Folkers & Rombouts, 2020). During the last decade, we have learned that the functional contributions of sponges often operate through species-specific complex networks of biogeochemical processes and organismal interactions, which, in most cases, remain only incipiently deciphered. During the last decades, the multiplication of scuba diving observations in shallow waters followed by the generalized use of remotely operated vehicles (ROVs) for deep-sea exploration have made it evident that an abundance of sponges in either shallow and deep-water benthic communities is often associated with a comparatively high diversity of other macro-organisms. It was then deduced that the three-dimensional body form of many sponges increases the 3D structure of the habitats, providing a multiplicity of physical refugees that, in turn, attract a large variety of small adult organisms and juvenile stages of others, and their respective predators. Some illustrative figures in this sense are: (i) that up to 100 species have been described in association with the shallow-water, Mediterranean tube-like sponge *Aplysina aerophoba* (Voultsiadou-Koukoura et al., 1987); (ii) that about 242 species were associated to the sponge aggregations on the outer shelf and upper slope of Faroe Islands (Klitgaard, 1995); (iii) that, in Caribbean coral reefs, up to 30 species of fish alone were found to dwell in sponges (Tyler & Böhlke, 1972); and (iv) that, for the total of tropical coral reefs, about 284 guest species have been described to dwell in sponges (Pérez-Botello & Simões,

2021). However, while it may be true that sponges foster general benthic biodiversity by passively increasing the 3D structure of habitats, the newly emerging view is that the mechanisms behind the enrichment of diversity are more complex than merely providing physical refuge. A growing number of studies report that sponges interact with a variety of dissolved organic and inorganic nutrients, affecting the cycling of chemical elements as important as carbon, silicon, nitrogen, phosphorus, and sulfur. The key point is that, wherever sponges become abundant, the small physiological activity of each individual is added to that of the collective, resulting in net fluxes that may reach relevant magnitudes at both local and regional scales, creating distinct patterns in the benthic-pelagic coupling of those important chemical elements. For instance, an illustrative example is the role of sponges in silicon cycling. The incorporation of dissolved silicic acid (i.e., biologically available silicon dissolved in the seawater) by the sponges to build their siliceous skeletons causes the transfer of large amounts of silicon from the seawater to the benthos. Once sponges die, their silicon, in the form of skeletal pieces, is dropped into the sediments and forms large deposits (Maldonado et al., 2019). While those skeletal deposits represent relevant regional sinks of silicon, the partial dissolution of the skeletal pieces before becoming permanently buried in the sediment causes an efflux of silicic acid from the bottom (Maldonado et al., 2021a). This efflux enriches the bottom water around the sponges with biologically available dissolved silicon, a pivotal nutrient facilitating not only the growth of sponges but also that of diatoms, which are important primary producers. Thus, sponges are habitat engineers not only because they add tridimensionality to benthic habitats, but also because they directly affect the availability of nutrients that are crucial to the development of bacterioplankton and phytoplankton communities and that, in the end, regulate food and energy transfer from the seawater to the food webs (Bell, 2008; Maldonado et al., 2012; de Goeij et al., 2013, 2017). The exact mechanisms that determine the capacity of sponges to drive such ecological effects at the community level are various, depending on species-specific strategies, and, in most cases, are only poorly understood. In this chapter, we review what is known about the causes and effects of the fluxes of inorganic nitrogen and phosphorus detected in marine sponges. We have integrated multidimensional knowledge from various disciplines (i.e., sponge physiology, microbiology, metagenomics, biochemistry) in order to provide a comprehensive context to discuss the state-of-the-art knowledge on this topic and to address future guidelines.

4.1.2 SPONGES AS USERS OF ORGANIC AND INORGANIC NUTRIENTS

Sponges incorporate organic carbon (C), organic nitrogen (N), and organic phosphate (P) during the process of feeding, like any other heterotrophic organism. They are essentially filter feeders, except for a carnivorous family (cladorhizid demosponges) that preys on tiny crustaceans. During filter feeding, they retain mainly microplankton (<50–60 μm), including viruses, bacterioplankton, picoflagellates, and microphytoplankton. To obtain enough particulate food by filter feeding, sponges pump through phenomenal volumes of seawater daily (i.e., 1,000 times their body volume), retaining high-efficiency food "particles" <10 μm in size, which they digest. This feeding activity is not ecologically inconsequential, since, for instance, it typically depletes the water layer over coral reefs in picoplankton, which in turn, enhances light penetration and several other cascading effects (Reiswig, 1971a; Richter & Wunsch, 1999; Scheffers et al., 2004; Lesser, 2006; Perea-Blázquez et al., 2012; Valentine & Butler, 2019).

In addition to particulate food, many sponges are also able to incorporate dissolved organic matter (DOM) from seawater, a complementary nutritional source of varying importance, depending on the species (e.g., de Goeij et al., 2017; Bart et al., 2020). DOM, defined as organic macromolecules that pass through a 0.22 μm pore-size filter, is abundant in seawater and contributes a huge reservoir of DOC. In most benthic communities, macroalgae, invertebrates, and vertebrates regularly release to the seawater a variety of mucous secretions and aged cells replaced from their bodies during tissue maintenance tasks (Wild et al., 2004, 2008; de Goeij et al., 2009; Maldonado, 2015; Dash et al., 2018). Only a part of this organic matter is labile DOC (i.e., biologically available), while the bulk of it consists of refractory molecules, particularly in deep-water environments. The portion of organic detritus dissolved directly as labile C enriches the seawater with small sugars, lipids, peptides, and amino acids, all of which can readily be consumed by the planktonic populations of bacteria, archaea, and primary producers. Until the end of the 20th century, the paradigm was that DOC was utilized only by microbial communities, which would introduce the energy of this resource into the food web. However, it is now clear that the nourishment of many sponges —as well as other invertebrates and fish— also relies on DOC incorporation. Part of the incorporated DOC appears to be used directly by the microbial populations living in the sponge mesohyl, but direct DOC incorporation by the sponge cells themselves (i.e., not transfer from microbes) has been demonstrated as well (e.g., Achlatis et al., 2019). Again, as a result of such physiological activity, sponges cause DOC depletion in the seawater of coral reefs and, particularly, in sponge-dominated reef cavities (de Goeij & Van Duyl, 2007; de Goeij et al., 2008; Nelson et al., 2011). It has also been demonstrated that some sponge species allocate an important part of the energy obtained from the DOC to growth and body maintenance, which may include regular replacement of cells in their epithelia and the mesohyl (de Goeij et al., 2009; Alexander et al., 2014). Labeling of the replacement cells shed by the sponges revealed that those cells and/or their remnants are used as a food source by detritivorous invertebrates of the local food web. By converting DOC (which most invertebrates are unable to take up) into particulate organic matter available to detritivorous, sponges enhance the transfer of matter and energy from the seawater to the food webs, a process known as "the sponge loop" (de Goeij et al., 2013) and that facilitates organismal abundance and diversity in the benthic communities. While the effects of the sponge loop have only been demonstrated in Caribbean coral reefs to date, it is predicted that the mechanism may operate in many other sponge-rich communities, including deep-sea sponge aggregations (de Goeij et al., 2017; Maldonado et al., 2017; Bart et al., 2021).

Regarding inorganic nutrients, it is long known that most sponge species (but not all) incorporate silicon (Si) in the form of dissolved silicic acid to elaborate biogenic silica, the main material making their siliceous skeletons (Simpson & Volcani, 1981; Uriz et al., 2003; Maldonado et al., 2020). It has also been demonstrated that some sponges can fix dissolved, inorganic C and N, but those capabilities, unlike in the case of silicon, do not derive from the sponge cells themselves but from microbes living symbiotically within the sponges and that further transfer the fixed C and N to their host (Wilkinson et al., 1999; Taylor et al., 2007; Weisz et al., 2010; Fiore et al., 2013; Zhang et al., 2019a; van Duyl et al., 2020). These microbial assemblages typically include photoautotrophs such as cyanobacteria (Wilkinson & Fay, 1979; Mohamed et al., 2008), methanotrophic and sulfur bacteria (Vacelet et al., 1995; Arellano et al., 2013), and all forms of nitrifying microbes (Hoffmann et al., 2009; van Duyl et al., 2020), among others chemoautotrophs. Sponges also show incorporation/release of inorganic forms of phosphorous and again, those fluxes appear to be mediated by the symbiotic microbial populations integrated into the sponge metabolism (Zhang et al., 2015). The origin and ecological consequences of the sponge-mediated N and P fluxes will constitute the main subject of this chapter. To provide a didactic context in which to place this topic, the basics of N and P cycling in the marine environment are first summarized in the two following sections.

4.2 THE BASICS OF THE MARINE NITROGEN CYCLE

There is an abundance of inorganic nitrogen gas (N_2) dissolved in the seawater of the oceans (63% on average), which arrives by diffusion from the atmosphere. However, this N remains unavailable to most marine organisms until autotrophic microorganisms convert it into a biologically available form (typically ammonium, nitrate, or nitrite; Table 4.1) that can then be incorporated by the consumers of the food webs. Conversely, when the organisms that have incorporated the bioavailable N in their bodies as part of organic molecules (e.g., amino acids, DNA, etc.) die, those organic molecules are degraded by a variety of prokaryotic microbes and fungi, releasing back to the seawater bioavailable inorganic N forms. Therefore, the enzyme-mediated microbial cycling between biologically available and unavailable N forms is crucial to the ecological functioning of the ocean. The core of N transformations in the ocean occurs via several major pathways, namely nitrification, denitrification, dissimilatory nitrate reduction to ammonium (DNRA), assimilatory nitrate reduction (ANR), nitrogen fixation, anaerobic ammonia oxidation (anammox), nitrogen fixation and ammonification. In some marine habitats, the coexistence of microbes with different biochemical abilities enables these pathways to couple to with each other, closing a circuit in one or another way for N to recycle (Figure 4.1).

4.2.1 *NITRIFICATION*

Autotrophic nitrification is a two-step process (Figure 4.1) carried out by autotrophic nitrifiers that use C from CO_2 or carbonates, rather than from organic matter, and are obligate aerobes. The first step of nitrification oxidizes ammonium to nitrate through two consecutive reactions carried out by ammonia-oxidizing bacteria (AOB) and ammonia-oxidizing archaea (AOA), that have the enzymes ammonia mono-oxygenase (*amoCAB* genes) and hydroxylamine dehydrogenase (*hao* genes).

$$NH_3^+ + 2H^+ + O_2 + 2e^- \rightarrow {}^{\text{ammonia monooxygenase}} \rightarrow NH_2OH + H_2O \rightarrow {}^{\text{hydroxylamine dehydrogenase}} \rightarrow NO_2^- + 4e^- + 5H^+$$

Two of the four electrons released in the second reaction are used for a subsequent process of NH_3^+ oxidation, and the two others are used in electron transport, generating ATP energy for cell growth or metabolic maintenance. The second step of nitrification oxidizes nitrite to nitrate and it is carried

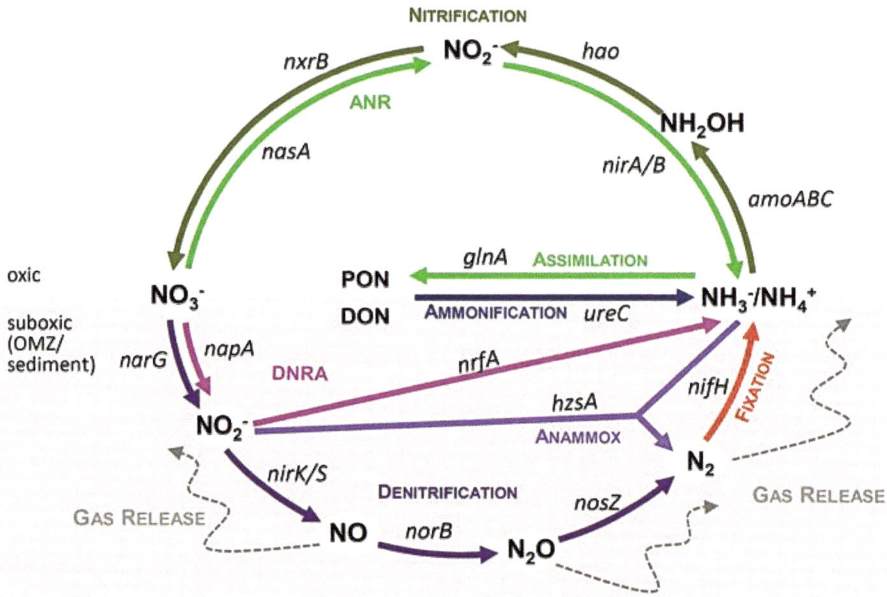

FIGURE 4.1 Summary of the microbial N cycle in the marine environment. Inputs and outputs of N to the marine biogeochemical cycle are beyond the main subject of the current approach and the cycle shown is simplified to contain what is addressed in the review. The colored arrows represent the main pathways in the N cycle, labeled in upper case and including the genes encoding the key enzymes for the respective conversions that are commonly used as markers. ANR: assimilatory nitrate reduction; DNRA: dissimilatory nitrate reduction to ammonia. Ammonification, nitrate reduction, and assimilation can occur both anaerobically and aerobically. Some gas compounds may be released during some pathways. Genes coding for the enzymes mediating in the biochemical transformations are in italics and their functions and phylogenetic distribution are explained in the main text sections.

out by strains of nitrite-oxidizing bacteria (NOB) that are different from those oxidizing ammonium to nitrite, except for the rare exception of some microbes that can conduct complete nitrification by themselves (Daims et al., 2015; Van Kessel et al., 2015). NOB are distinctively provided with the enzyme nitrite oxidoreductase (*nxrAB* genes).

$$NO_2^- + H_2O \rightarrow^{\text{nitrite oxidoreductase}} \rightarrow NO_3^- + 2H^+ + 2e^-$$

There is also heterotrophic nitrification, which in the marine environment is of marginal importance compared to its autotrophic counterpart and that can be carried out in two modes. In the most common mode, a variety of marine heterotrophic bacteria have ammonia- and hydroxylamine-oxidizing enzymes similar to those of the autotrophic nitrifiers and oxidize the ammonium to nitrate (Li et al., 2017). However, these are less efficient than the autotrophs. The second heterotrophic pathway is organic and appears limited to fungi. It involves the oxidation of amines or amides to nitrate, through a chain of reactions that are not coupled to ATP synthesis and produce no metabolic energy (Robertson & Groffman, 2015).

4.2.2 DENITRIFICATION

Denitrification is the reduction of nitrate to the N gases, NO, N_2O, and N_2 (Figure 4.1). The process involves four consecutive reactions, respectively mediated by the enzymes nitrate reductase (*narGHI, napAB* genes), nitrite reductase (*nirK/S* genes), NO reductase (*norBC* genes), and nitrous oxide reductase (*nosZ* genes).

$$NO_3^- + 2H^+ + 2e^- \rightarrow \text{nitrate reductase} \rightarrow NO_2^- + H_2O$$

$$NO_2^- + 2H^+ + 2e^- \rightarrow \text{nitrite reductase} \rightarrow NO + H_2O$$

$$2NO + 2H^+ + 2e^- \rightarrow \text{nitric oxide reductase} \rightarrow N_2O + H_2O$$

$$N_2O + 2H^+ + 2e \rightarrow \text{nitrous oxide reductase} \rightarrow N_2 + H_2O$$

At any step in the denitrification process, intermediate products can be exchanged with the "environment," being this either the intercellular medium in the case of prokaryotes in symbiosis with macro-organisms or the ambient seawater, in the case of free-living prokaryotes. A wide variety of mostly facultative anaerobic, heterotrophic bacteria (although denitrifiers can also be chemo- and photolithotrophs) can denitrify, using NO_3^- rather than oxygen as a terminal electron acceptor during respiration. Since nitrate is a less-efficient electron acceptor than oxygen gas (O_2), most denitrifiers undertake denitrification only if oxygen is at a very low concentration or unavailable. The process phosphorylates ATP, rendering metabolic energy. Importantly, denitrification allows, along with anammox, the fixed N to reenter the seawater as N_2 gas, which eventually diffuses in the atmosphere. In the absence of denitrification and anammox, N_2 fixers would eventually exhaust the atmospheric and marine N_2 pools and the biosphere would be oversupplied with nitrate.

4.2.3 DISSIMILATORY NITRATE REDUCTION TO AMMONIUM (DNRA)

The DNRA pathway involves the transformation of nitrate to nitrite mediated by a nitrate reductase (*narGHI*, *napAB* genes) and then to ammonium (Figure 4.1), through the catalysis of a cytochrome C reductase (*nrfAH* genes) or an NADH-dependent nitrite reductase (*nirBD* genes). The process is, indeed, the anaerobic respiration of chemoorganoheterotrophic microbes that use nitrate rather than O_2 as an electron acceptor. Some fungi, foraminifers, and diatoms appear capable of performing DNRA as well (Kamp et al., 2015).

$$NO_3^- \rightarrow \text{nitrate reductase} \rightarrow NO_2^- \rightarrow \text{cytochrome C nitrite reductase} \rightarrow NH_4^+$$

The pathway consumes more electrons than denitrification. Thus, DNRA is only favored *in situ*ations when the ratio of carbon to nitrate is high, such as during the oxidation of high inputs of organic matter in the system.

4.2.4 ASSIMILATORY NITRATE REDUCTION (ANR)

As in the DNRA pathway, assimilatory nitrate reduction (ANR) uses nitrate as an N source to generate ammonium (Figure 4.1). However, unlike in the DNRA pathway, the ammonium produced will ultimately be incorporated (i.e., assimilated) into organic compounds, such as amino acids, nucleotides, and so on. Both DNRA and ANR pathways involve the active transport of nitrate into the cell, its transitory storage, and its subsequent intracellular reduction to ammonium by two sequential reductive reactions. The controls behind the partitioning of the intracellular nitrate between dissimilation and assimilation remain to be elucidated. Nitrate is assimilated by prokaryotes, but also by fungi, algae, and plants (Kamp et al., 2015). Heterotrophic bacteria are known to possess three different nitrate-assimilatory enzyme systems, which are genetically different from those of the dissimilatory system. They include an assimilatory nitrate reductase (*nasAB* genes), a ferredoxin-nitrate reductase (*narB* genes), and a (NAD(P)H)-dependent nitrate reductase (*nr* genes). In the general pathway, nitrate is first reduced to nitrite in the cytosol by an NADH-dependent assimilatory nitrate reductase. Nitrite is subsequently reduced to ammonium by a ferredoxin-dependent assimilatory nitrite reductase (*NirA*, *NIT-6* genes).

$$NO_3^- \rightarrow \text{assimilatory nitrate reductase} \rightarrow NO_2^- \rightarrow \text{assimilatory nitrite reductase} \rightarrow NH_4^+$$

The produced ammonium serves as a source of nitrogen for biosynthesis. For instance, the glutamine synthetase enzyme incorporates this ammonium as the amide group of glutamine using glutamate as substrate. The glutamate synthase transfers the amide group onto a 2-oxoglutarate molecule producing two glutamates. Further transaminations are carried out to make other amino acids (most commonly asparagine) from glutamine.

4.2.5 ANAEROBIC AMMONIUM OXIDATION (ANAMMOX)

Anammox is the conversion of ammonium and nitrite to N_2 (Figure 4.1). The bioenergetics and biochemistry of the process are not completely clarified yet and two different reaction mechanisms, going through hydrazine as an intermediate step, have been proposed (e.g., Ni & Zhang, 2013). In general terms, the process can be described as:

$$NH_4^+ + NO_2^- \rightarrow\rightarrow N_2H_2 \rightarrow\rightarrow N_2^+ 2H_2O$$

Importantly, the reactions are carried out by enzymes not located on the prokaryotic membrane but within distinct organelles of Planctomycetes bacteria, called anammoxomes. To date, two enzymes unique to anammox bacteria have been identified: hydrazine synthase (*hzs* genes) and hydrazine dehydrogenase (*hdh* genes).

4.2.6 NITROGEN FIXATION

Nitrogen fixation is the process of converting inert nitrogen gas N_2 into biologically available nitrogen (Figure 4.1). N_2 gas is a very stable compound because of the triple bond between the two nitrogen atoms. To break this bond is energetically costly, requiring eight electrons and at least 16 ATP molecules. Thus, only a few prokaryotes (though of diverse phylogenetic adscription) are able to undertake this process, which can be carried out in either oxic or anoxic conditions and through both phototrophic and chemotrophic metabolisms. A shared feature of nitrogen fixers is the occurrence of the nitrogenase enzyme complex (*nifGKH*, *anfG*, *vnfDKGH* genes), which catalyzes the reduction of N_2 to ammonia.

$$N_2 + 16ATP + 8H^+ + 8e^- \rightarrow 2NH_3 + H_2 + 16ADP + 16Pi$$

4.2.7 AMMONIFICATION

When organisms die, fungi and prokaryotes decompose their cells and/or tissues. During this process, the large organic macromolecules containing nitrogen are depolymerized into simpler organic molecules (amino acid, nucleotides, urea, etc.) still containing amine or amide groups (of general formula R-NH_2). These low-molecular-weight compounds are further degraded, producing ammonium, through a mineralization process known as ammonification (Figure 4.1). As posed by Romillac (2018), ammonification is the "last" step of the global nitrogen cycle involving an organic compound and is the intermediary process between the depolymerization of large organic molecules and the nitrification pathway.

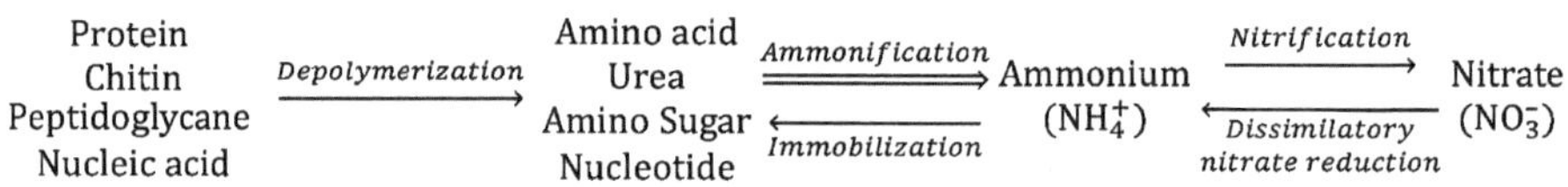

4.3 THE BASICS OF MARINE PHOSPHORUS CYCLE

Among other things, phosphorus (P) in the form of phosphate is a relevant marine nutrient. It is also the fundamental building block of phosphate-phosphate bonds in ATP and other energy-carrying molecules essential in life metabolism. Moreover, P is an effective electron donor in anaerobic carbon fixation. However, despite its biological relevance, the cycling of P in the marine environment (Figure 4.2) is still poorly understood, compared to that of C, N, and Si elements, particularly in relation to the biological component of benthic communities.

4.3.1 MARINE P POOLS

Phosphorous is supplied to the oceans mainly via rivers after weathering of apatite minerals, but entries through submarine groundwater discharges and atmospheric deposition are also important. But P is not only present in seawater as dissolved inorganic forms (DIP) resulting from mineral weathering. Numerous forms of dissolved organic P (DOP) also occur. The pool of DIP is dominated by phosphates (P_i), with oxidation stage P^{+5} (Table 4.1 and Figure 4.2). Phosphates include orthophosphate (PO_4^{3-}), hydrogen phosphate

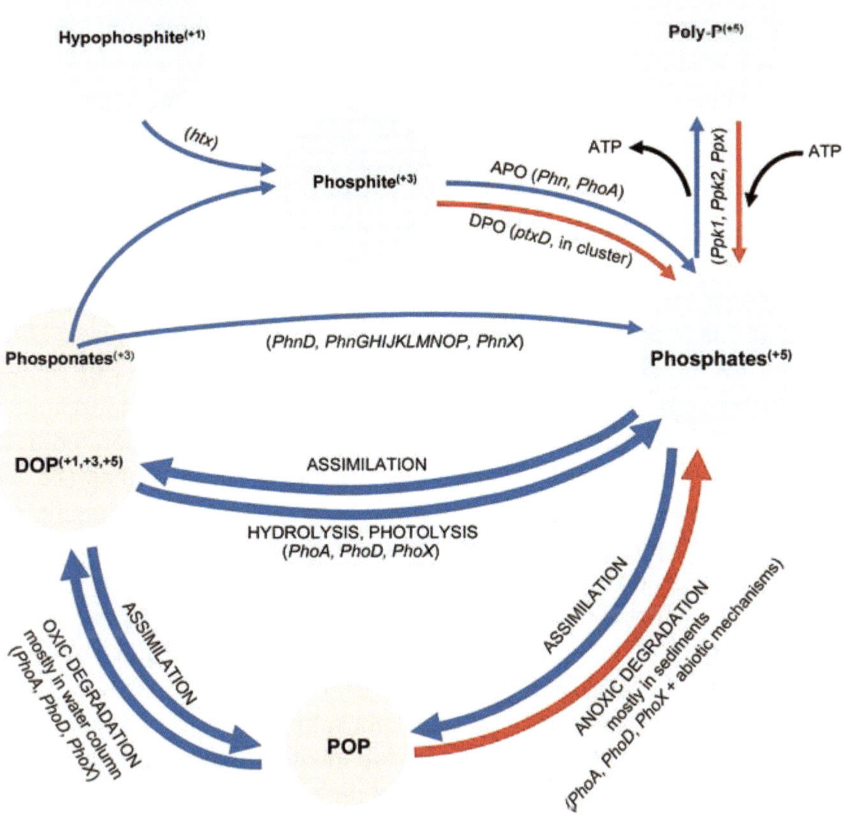

FIGURE 4.2 Summary of the microbial phosphorous (P) cycle in the marine environment. Inputs and outputs of P to the marine biogeochemical cycle are beyond the main subject of the current approach and the cycle shown is simplified to contain what is addressed in the review. Blue and red arrows indicate aerobic and anaerobic pathways, respectively. Thicker arrows stand for the classical view of the P cycle. P compounds on a bluish background are organic forms, while those on a brownish background are inorganic. DOP and POP stand for the respective pools of dissolved and particulate organic phosphorus compounds. P oxidation degree in the compounds is indicated by a super index ($^{+\#}$). Names of major pathways are given in the upper case, with APO and DPO referring to assimilatory and dissimilatory phosphite oxidation, respectively. The name of the identified genes encoding the main enzymes that mediate in the different pathways is indicated in italics.

(HPO_4^{2-}), and dihydrogen phosphate ($H_2PO_4^-$). In addition, DIP also includes inorganic polyphosphate (Poly-P), phosphite, and hypophosphite (Table 4.1 and Figure 4.2), along with the corresponding radicals of phosphorus oxide (PO_3^{3-}, PO_2^{3-}). Inorganic phosphite is abiotically produced in geothermal systems, but it can also derive from biological phosphonate degradation and anomalous phosphate reduction (Ewens et al., 2021). The pool of DOP is a complex mix of organophosphorus compounds, which include, among other molecules, phosphonates (Table 4.1 and Figure 4.2). Interestingly, in most oceanic habitats, the DOP pool is much larger (about 5 times on ocean average) than the phosphate pool (Karl & Björkman, 2002).

4.3.2 BIOLOGICAL CONSUMPTION OF P

The classical view assumed that P was biologically incorporated by phototrophic primary producers and microbes exclusively in the form of HPO_4^{2-}, which is the predominant phosphate anion (>90%) at the pH of seawater. Nevertheless, it is now well established that dissolved—inorganic and organic—reduced forms of P can also be biologically incorporated at high rates by diverse microbes. Among these reduced P compounds (i.e., with oxidation stage P^{3+}; Table 4.1), organic phosphonates and inorganic phosphite, and hypophosphite are of particular interest (Figure 4.2). Various heterotrophic bacteria, cyanobacteria, and archaea are known to contain the molecular machinery for transporting phosphonates and phosphite into the cell (Tyrrell, 1999; Dyhrman et al., 2006; Villarreal-Chiu et al., 2012; Metcalf et al., 2012; Van Mooy et al., 2015; Figueroa & Coates, 2017). The uptake of phosphonate and inorganic phosphite by microbes is mediated by ABC-type transporters, namely phosphonate-binding-protein (*phnCDE* genes) and ABC-type transporter, phosphite-binding-protein (*phnD1* gene), which are shared across different microbial phyla. Microbes possessing *phnD2* can transport both phosphonate and inorganic phosphite (Feingersch et al., 2012).

In summary, marine microorganisms are now known to be able to incorporate at least four forms of dissolved P: (i) inorganic oxyanions of multiple valence states and molecular form (i.e., Pi, PO_3^{3-}, PO_2^{3-}, inorganic poly-P); (ii) monoester- and diester-linked organophosphates (with C-O-P bonds); (iii) organic phosphoanhydrides (with C-O-P-O-P bonds); and (iv) organic phosphonates (with C-P bonds).

4.3.3 P ASSIMILATION

The P deriving from either phosphate directly taken into the microbial cell, or from intracellular degradation of other DIP or DOP forms previously incorporated, is finally assimilated (Figure 4.2). Through assimilation, the P atoms are transferred to new organic P compounds elaborated by the microbial cell (e.g., DNA, RNA, ATP, phospholipids, etc.).

Phosphonates are mostly used as organic sources of P for assimilation, although these molecules are characterized by a highly stable C-P bond that requires an enormous input of energy to break it. For this reason, it was initially assumed that only very few organisms would be able to process phosphonates. Nevertheless, it is now known (e.g., Villarreal-Chiu et al., 2012) that the C-P bond is cleaved by a variety of microbial enzymes C-P lyases (*phnG-P* genes) and C-P hydrolases (*phnX* genes), releasing phosphate for assimilation (Figure 4.2). Up to four different pathways for the degradation of phosphonates to release phosphate have been identified in different bacteria to date (White & Metcalf, 2007). Likewise, additional enzymes have been discovered for the oxidation (*htx* genes) of inorganic hypophosphite to phosphite (Figure 4.2). Phosphite can be further oxidized aerobically to produce phosphate for assimilation in a large variety of microbes (Figure 4.2). This pathway is known as assimilatory phosphite oxidation (APO). Two pathways have been characterized for phosphite assimilation: (i) through C-P lyases (*Phn* genes); and (ii) through bacterial alkaline phosphatase (BAP; *PhoA* genes).

Typically but not always, the expression of genes for phosphonate and phosphite incorporation into the microbial cells is upregulated by P-depleted situations. This pattern agrees with the general notion that the chronic environmental shortage of phosphates (typically << 1 μM in most ocean zones) negatively affects both phytoplankton productivity in oligotrophic areas of the modern photic ocean and the development of heterotrophic microbial communities in open waters away from coastal inputs of phosphates. Because reduced DOP (phosphonate) and DIP (phosphite, polyphosphate) are more abundant in seawater than phosphates, adaptions towards using them as alternative sources of bioavailable P may have been evolutionarily favored, even when these alternative biochemical pathways require notable energy investment.

Under oxic conditions and an abundance of intracellular P_i, inorganic poly-P is produced with ATP input, becoming a reserve of both P and energy. Poly-P is typically degraded under anoxic conditions, releasing P_i and producing ATP molecules (Figure 4.2). The reversible polymerization-depolymerization of poly-P is catalyzed by poly-P kinase enzymes (*ppk1*, *ppk2*, *ppx* genes).

4.3.4 P DISSIMILATION

Phosphite is not only used as a P source for assimilation, but it is also an energetically favorable electron donor that can be processed in anaerobic conditions through a microbial, dissimilatory phosphite oxidation (DPO) pathway (Figure 4.2). The biochemical reaction is coupled to the reduction of CO_2, sulfate, or nitrate, building ATP and releasing phosphate from the microbial cell (White & Metcalf, 2007; Ewens et al., 2021). The process is catalyzed by a phosphite dehydrogenase, encoded by the *ptxD* gene in association with a *ptxDE-ptdCFG2HI* gene cluster, which appears to be a characteristic marker for DPO microbes (Ewens et al., 2021).

4.3.5 P REMINERALIZATION

When the organisms die, the P incorporated into the organic molecules that make their bodies sink into the water column, as part of the particulate organic matter in degradation. Most of the assimilated P_i and DOP are remineralized back to dissolved P_i during both sinking and once at the sediment surface (Figure 4.2). Yet, the vast majority of remineralization appears to take place during the transit through the water column (Paytan & McLaughlin, 2007). The degrading microbes, both aerobic and anaerobic, possess alkaline phosphatase enzymes (*phoA*, *phoD*, *phoX* genes). Indeed, alkaline phosphatases are present in virtually every living organism. The microbial ones play a crucial role in the remineralization of organic P compounds, by hydrolyzing phosphates from many types of molecules such as nucleotides, proteins, alkaloids, phosphate esters, and anhydrides of phosphoric acid.

To discriminate and quantify the relative contribution of the various P fluxes deriving from the degradation of organic matter in terms of P remineralization is a complex process. For instance, under anoxic conditions in marine sediments or in seawater, bacterial remineralization of organic matter and also abiotic desorption/release of phosphate from the sediment typically cause the release of phosphate (Slomp et al., 2004; Paytan & McLaughlin, 2007; Lomnitz et al., 2016). However, in oxic sediments, the phosphate desorbed to the interstitial water is rapidly reabsorbed by iron oxyhydroxides or precipitated authigenically as apatite (*see* Section 4.3.6). Also, under oxic conditions, either in mesopelagic and deep-sea environments or in sediments, the prokaryotic microbial communities take up phosphate, because the oxygen consumption of aerobic respiration demands phosphate to generate adenosine triphosphate (ATP) molecules (Figure 4.2). Under oxic conditions, microbes may also polymerize phosphate to store energy and P into poly-P (Gächter & Meyer, 1993; Steenbergh et al., 2014). In general, these oxic reactions collectively lead to a net uptake of phosphate from the water by marine sediments. In contrast to this general pattern, most studied sponges on relatively oxic bottoms release phosphate. The mechanisms behind such an efflux from the sponges remain unclear (*see* Section 4.6).

4.3.6 PERMANENT P SINKS

It is estimated that only about 1% of the P carried to the seafloor by sinking organic matter is finally buried in the sediments (Paytan & McLaughlin, 2007). Major permanent sinks of phosphorous in the ocean are (i) burial in the sediments of organic P associated with organic matter sunk to the seafloor; (ii) burial of P adsorbed to Fe- and Mn-rich oxyhydroxide coatings of particles sinking through the water column; and (iii) burial of P associated to oxic authigenic carbonate fluorapatite formation in pore water of the sediments (Delaney, 1998; Slomp et al., 2004).

4.4 NET FLUX RATES OF SPONGES

About three decades ago, a pioneering study reported the release of dissolved inorganic nitrogen by marine sponges (Corredor et al., 1988) and unknowingly began a new research line at the time. It is now established that virtually every sponge releases and/or takes up a variety of dissolved inorganic nutrients, such as ammonium, nitrate, nitrite, and phosphates (reviews in Maldonado et al., 2012; Feng & Li, 2019; Folkers & Rombouts, 2020). Such nutrient fluxes result from a multifaceted combination of physiological processes that include not only the metabolism of the sponge cells but also that of complex microbial assemblages (i.e., microbiota) metabolically integrated within the sponges (reviews in Taylor et al., 2007; Webster & Taylor, 2012; Hentschel et al., 2012; Zhang et al., 2019a). In some cases, the microbes amount to up to 40% of the sponge body mass (Vacelet, 1971; Wilkinson, 1978; Taylor et al., 2007). The microbes can live extracellularly in the sponge mesohyl or within specialized cells called bacteriocytes (Figure 4.3) and, theoretically, such a difference in location may translate into major functional differences in terms of metabolic integration (*see* Section 4.5).

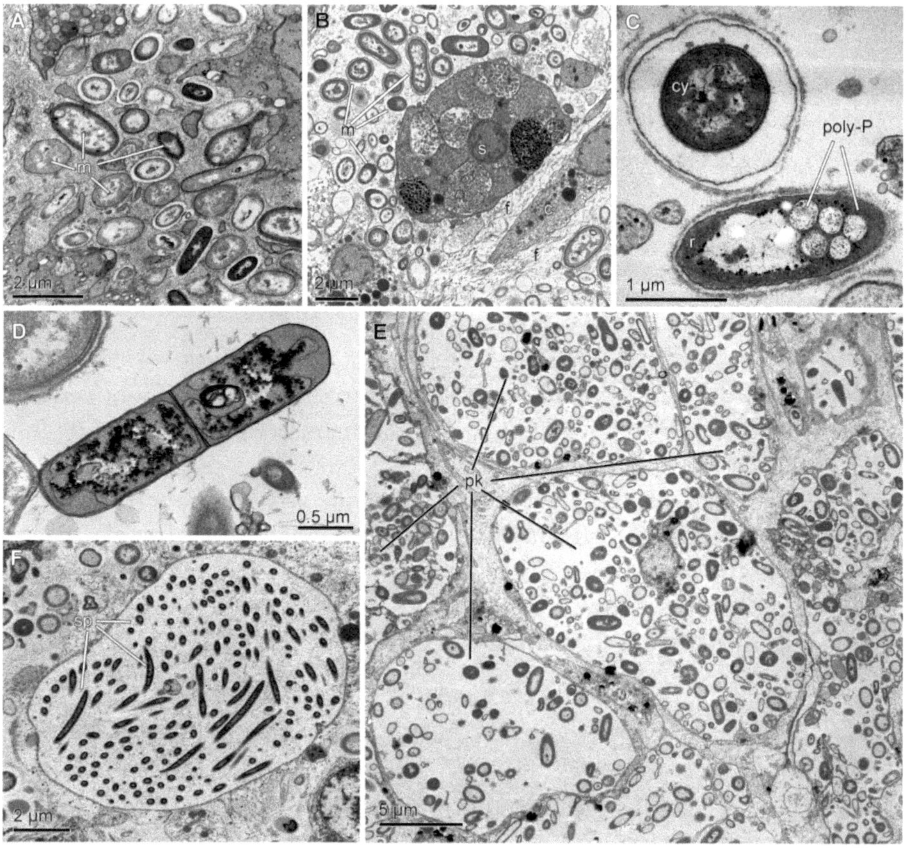

FIGURE 4.3 The abundance of microbes within sponges, either extracellularly or in bacteriocytes. (A) High abundance and diversity of microbes (m) living intercellularly in the mesohyl of the HMA, shallow-water sponge *Chondrosia reniformis*. (B) The abundance of extracellular microbes (m) in the mesohyl of the HMA, shallow-water sponge *Aplysina aerophoba*. The microbes are surrounding two sponge cells, a spherulous cell (s) and a collencyte (c). The latter one is involved in the production of the collagen fibrils (f) that make the intercellular medium of the mesohyl. (C) Detail of a cyanobacteria (cy) and a large bacterial rod (r) charged with intracytoplasmic granules (g) that could well be poly-phosphate accumulations (poly-P). These bacteria occurred intracellularly within a newly released parenchymella larva of the HMA, shallow-water sponge *Ircinia oros*, indicating that the microbial presence in the larva is the result of vertical transmission. (D) Detail of a dividing bacterial rod in the mesohyl of the sponge *A. aerophoba*. (E) General view of several specialized sponge cells (pocket bacteriocytes: pk) that collect and extracellularly maintain a diversity of microbes by folding the cell body to form a kind of pocket pouch. These cells are only known from the HMA, shallow-water sponge *Petrosia ficiformis*. (F) View of a true bacteriocyte in the HMA, shallow-water sponge *Aplysina cavernicola*. Note that the bacteriocyte contains intracellularly a single type of spirochete-like bacteria (sp).

Sponge species that host microbes in concentrations similar to those in the seawater of the habitat have been categorized as low microbial abundance (LMA) sponges and typically contain $<10^6$ microbial cells per g of sponge wet weight (Hentschel et al., 2003; Gloeckner et al., 2013). By comparison, sponges that host concentrations of 10^8 to 10^{10} microbial cells per g of sponge wet weight, which are typically two to four orders of magnitude higher than seawater, have been defined as high microbial abundance (HMA) sponges. HMA and LMA sponges are considered two different life strategies resulting from contrasting evolutionary pathways. However, not all sponges can be categorized as HMA or LMA (e.g., Maldonado et al., 2021b). Why and when differences in microbial abundance have or do not have effects on the capacity of the sponge species to release and/or incorporate dissolved nutrients is still under discussion. To better identify the biological patterns behind the nutrient fluxes of sponges, here we review and re-analyze quantitative data from sponge species for which both net flux rates and molecular analyses of their microbiomes are available (Tables 4.2–4.8, Supplementary Table 4.1), using all data available regardless of the number of studies carried out on each species.

TABLE 4.1 Main Forms of Nitrogen and Phosphorus in the Marine Environment, with Indication of Their Oxidation State*

Name	Formula	Oxidation State
Nitrate	NO_3^-	+5
Nitrogen dioxide	NO_2	+4
Nitrite	NO_2^-	+3
Nitric oxide	NO	+2
Nitrous oxide	N_2O	+1
Dinitrogen	N_2	0
Ammonia	NH_3	–3
Ammonium	NH_4^+	–3
Nitrogen (organic)	R-NH_3	–3
Hydrazine	N_2H_4	–2
Hydroxylamine	NH_2OH	–1
Phosphate	HPO_4^{2-}	+5
Polyphosphate	$n(P_2O_7)^{4-}$	+5
Phosphonates (organic)	R-CH_2-$PO(OR)_2$	+3
Phosphite	HPO_3^{2-}	+3
Hypophosphite	$H_2PO_2^-$	+1

*Compounds are inorganic unless their organic nature is indicated in brackets.

TABLE 4.2 Summary of Scientific Studies (N = 33) and Sponge Species (n = 92) Considered for This Review on Net Flux Rates of N and P Inorganic Compounds Through Sponges*

	Microbiome		Sponge Class				Depth zone		Method		
Reference	**HMA**	**LMA**	**De.**	**Ho.**	**He.**	**Ca.**	**sublittoral**	**deep sea**	**Lab incub.**	**In-situ incub.**	**In-Ex**
Corredor et al. (1988)	1	1	2	0	0	0	x			x	
Hatcher (1994)	1	1	2	0	0	0	x		x		
Diaz & Ward (1997)	4	0	3	1	0	0	x		x		
Davy et al. (2002)	0	1	1	0	0	0	x		x	x	
Yahel et al. (2003)	1	0	1	0	0	0	x				x
Bayer et al. (2007)	1	0	1	0	0	0	x		x		
Jimenez & Ribes (2007)	4	2	6	0	0	0	x		x		
Yahel et al. (2007)	0	2	0	0	2	0		x			x
Bayer et al. (2008)	1	0	1	0	0	0	x		x		
Southwell et al. (2008) a	5	1	6	0	0	0	x				x
Southwell et al. (2008) b	10	3	13	0	0	0	x		x		x
van Duyl et al. (2008)	2	0	1	0	1	0		x	x		
Hadas et al. (2009)	0	1	1	0	0	0	x		x		
Hoffmann et al. (2009)	1	0	1	0	0	0		x	x		
Schläppy et al. (2010)	1	1	2	0	0	0	x		x		
Perea-Blazquez et al. (2012)	1	9	7	1	0	2	x				x
Ribes et al. (2012)	2	1	3	0	0	0	x		x		
Radax et al. (2012)	1	3	4	0	0	0		x	x		
Fiore et al. (2013)	1	0	1	0	0	0	x				x
Keesing et al. (2013)	7	8	15	0	0	0	x			x	

TABLE 4.2 *(Continued)*

	Microbiome		Sponge Class				Depth zone		Method		
Reference	**HMA**	**LMA**	**De.**	**Ho.**	**He.**	**Ca.**	**sublittoral**	**deep sea**	**Lab incub.**	**In-situ incub.**	**In-Ex**
Archer et al. (2015)	0	1	1	0	0	0	x		x		
Morley et al. (2016)	0	4	4	0	0	0	x		x		
Archer et al. (2017)	1	0	1	0	0	0	x				x
Morganti et al. (2017)	3	2	5	0	0	0	x				x
Fang et al. (2018)	2	1	3	0	0	0		x	x		
Hoer et al. (2018)	3	8	11	0	0	0	x			x	
Leys et al. (2018)	1	0	1	0	0	0		x			x
Subina et al. (2018)	0	1	1	0	0	0	x		x		
López-Acosta et al. (2019)	0	1	1	0	0	0	x			x	
Valentine & Butler (2019)	7	3	10	0	0	0	x		x		
Rooks et al. (2020)	6	0	6	0	0	0		x	x		
Maldonado et al. (2021)	1	1	0	0	2	0		x		x	
Nemoy et al. (2021)	1	0	1	0	0	0	x				x
TOTAL	69	56	116	2	5	2	25	8	19	6	10

**Sponge classes are Demospongiae (De), Homoscleromorpha (Ho), Hexactinellida (He), and Calcarea (Ca). Microbiomes are either high microbial abundance (HMA) or low microbial abundance (LMA). Methods of flux rate determination are three*: (1) incubation in laboratory conditions (Lab incub.); (2) incubations in the field (*in-situ* incub.); and (3) in-ex determination (In-Ex), which typically happened in the field.

TABLE 4.3 Summary of Net Flux Rates of Ammonium Considered for This Review, Ordered by Microbial Abundance, Taxonomic Class, Species Name, and Year of Publication[#]

Microbial abundance	Taxonomic class	Sponge species	Method of rate determination	Geographical area	Depth range	Average rate	Associated error	Units	Reference
HMA	Demos.	*Agelas conifera*	InEx+pumping	Caribbean	sublittoral	0.049	0.061	μmol mL^{-1} sponge h^{-1}	Southwell et al. 2008 a
HMA	Demos.	*Agelas oroides*	InEx+pumping	Mediterranean	sublittoral	-0.053	0.044	μmol cm^{-2} sponge h^{-1}	Morganti et al. 2017
HMA	Demos.	*Aplysina aerophoba*	Lab incubation	Mediterranean	sublittoral	0.023	0.003	μmol g^{-1} WW h^{-1}	Bayer et al. 2007
HMA	Demos.	*Aplysina aerophoba*	Lab incubation	Mediterranean	sublittoral	0.302	0.337	μmol g^{-1} DW h^{-1}	Jimenez & Ribes 2007
HMA	Demos.	*Aplysina aerophoba*	Lab incubation	Mediterranean	sublittoral	0.012	0.003*	μmol g^{-1} DW h^{-1}	Bayer et al. 2008
HMA	Demos.	*Aplysina aerophoba*	Lab incubation	Mediterranean	sublittoral	-0.055	0.01*	μmol g^{-1} DW h^{-1}	Bayer et al. 2008
HMA	Demos.	*Aplysina aerophoba*	Lab incubation	Mediterranean	sublittoral	-0.574	0.087*	μmol g^{-1} DW h^{-1}	Bayer et al. 2008
HMA	Demos.	*Aplysina aerophoba*	Lab incubation	Mediterranean	sublittoral	-0.706	0.065*	μmol g^{-1} DW h^{-1}	Bayer et al. 2008
HMA	Demos.	*Aplysina aerophoba*	Lab incubation	Mediterranean	sublittoral	-1.571	0.235*	μmol g^{-1} DW h^{-1}	Bayer et al. 2008
HMA	Demos.	*Aplysina archeri*	InEx+pumping	Caribbean	sublittoral	-0.039	0.062	μmol mL^{-1} sponge h^{-1}	Southwell et al. 2008 a
HMA	Demos.	*Aplysina cauliformis*	Lab incubation	Caribbean	sublittoral	-0.50	0.50	μmol g^{-1} h^{-1}	Southwell et al. 2008 b
HMA	Demos.	*Aplysina lacunosa*	InEx+pumping	Caribbean	sublittoral	-0.020	0.068	μmol mL^{-1} sponge h^{-1}	Southwell et al. 2008 a
HMA	Demos.	*Aplysina* sp.	Lab incubation	W Australia	sublittoral	-0.87	N/A	μmol g^{-1} AFDW h^{-1}	Hatcher 1994
HMA	Demos.	*Chondrilla nucula*	Field incubation	Caribbean	sublittoral	-0.348	0.236*	μmol g^{-1} DW h^{-1}	Corredor et al. 1988
HMA	Demos.	*Chondrilla nucula*	Field incubation	Florida Keys	sublittoral	-0.001	0.004*	μmol mL^{-1} sponge h^{-1}	Hoer et al. 2018
HMA	Demos.	*Chondrilla nucula*	Lab incubation	Caribbean	sublittoral	-0.04	0.13	μmol g^{-1} DW h^{-1}	Diaz & Ward 1997
HMA	Demos.	*Chondrosia reniformis*	InEx+pumping	Mediterranean	sublittoral	0.051	0.034	μmol cm^{-2} sponge h^{-1}	Morganti et al. 2017
HMA	Demos.	*Chondrosia reniformis*	Lab incubation	Mediterranean	sublittoral	0.636	1.062	μmol g^{-1} DW h^{-1}	Jimenez & Ribes 2007
HMA	Demos.	*Chondrosia reniformis*	Lab incubation	Mediterranean	sublittoral	-0.196	N/A	μmol mL^{-1} sponge h^{-1}	Schläppy et al. 2010
HMA	Demos.	*Craniella* sp.	Field incubation	SW Australia	sublittoral	0.04	N/A	μmol g^{-1} DW h^{-1}	Keesing et al. 2013
HMA	Demos.	*Geodia barretti*	Lab incubation	Norwegian fjord	200	-0.006	0.002	μmol mL^{-1} sponge h^{-1}	Fang et al. 2018
HMA	Demos.	*Geodia barretti*	InEx+pumping	Norwegian fjord	160	0.0005	N/A	μmol mL^{-1} sponge h^{-1}	Leys et al. 2018

TABLE 4.3 *(Continued)*

Microbial	Taxonomic	Sponge	Method of	Geographical	Depth	Average	Associated	Units	Reference
HMA	Demos.	*Geodia* cf. sp.	Field incubation	SW Australia	sublittoral	0.02	N/A	μmol g^{-1} DW h^{-1}	Keesing et al. 2013
HMA	Demos.	*Geodia gibberosa*	Field incubation	Florida Keys	sublittoral	-0.009	0.008*	μmol mL^{-1} sponge h^{-1}	Hoer et al. 2018
HMA	Demos.	*Geodia* sp.	Field incubation	SW Australia	sublittoral	-0.02	N/A	μmol g^{-1} DW h^{-1}	Keesing et al. 2013
HMA	Demos.	*Higginsia thielei*	Lab incubation	NE Atlantic	570-785	-0.042	0.014	μmol mL^{-1} sponge h^{-1}	van Duyl et al. 2008
HMA	Demos.	*Hippospongia lachne*	Field incubation	Florida Keys	sublittoral	0.004	0.002*	μmol mL^{-1} sponge h^{-1}	Hoer et al. 2018
HMA	Demos.	*Hyrtios* sp.	Field incubation	SW Australia	sublittoral	0.02	N/A	μmol g^{-1} DW h^{-1}	Keesing et al. 2013
HMA	Demos.	*Hyrtios violaceus*	Lab incubation	Caribbean	sublittoral	-0.26	0.44	μmol g^{-1} DW h^{-1}	Diaz & Ward 1997
HMA	Demos.	*Ircinia campana*	Field incubation	Florida Keys	sublittoral	-0.010	0.006*	μmol mL^{-1} sponge h^{-1}	Hoer et al. 2018
HMA	Demos.	*Ircinia felix*	Field incubation	Florida Keys	sublittoral	-0.034	0.017*	μmol mL^{-1} sponge h^{-1}	Hoer et al. 2018
HMA	Demos.	*Ircinia strobilina*	InEx+pumping	Caribbean	sublittoral	-0.022	0.048	μmol mL^{-1} sponge h^{-1}	Southwell et al. 2008 a
HMA	Demos.	*Petrosia ficiformis*	InEx+pumping	Mediterranean	sublittoral	0.012	0.020	μmol cm^{-2} sponge h^{-1}	Morganti et al. 2017
HMA	Demos.	*Smenospongia aurea*	Lab incubation	Caribbean	sublittoral	0	0	μmol g^{-1} h^{-1}	Southwell et al. 2008 b
HMA	Demos.	*Spheciospongia vesparium*	Field incubation	Florida Keys	sublittoral	-0.009	0.002*	μmol mL^{-1} sponge h^{-1}	Hoer et al. 2018
HMA	Demos.	*Spongia* sp. morphotype 1	Field incubation	Florida Keys	sublittoral	-0.026	0.011*	μmol mL^{-1} sponge h^{-1}	Hoer et al. 2018
HMA	Demos.	*Spongia* sp. morphotype 2	Field incubation	Florida Keys	sublittoral	-0.006	0.006*	μmol mL^{-1} sponge h^{-1}	Hoer et al. 2018
HMA	Demos.	*Stryphnus fortis*	Lab incubation	Norwegian fjord	200	-0.003	0.001	μmol mL^{-1} sponge h^{-1}	Fang et al. 2018
HMA	Demos.	*Svenzea zeai*	Lab incubation	Caribbean	sublittoral	-0.19	0.30	μmol g^{-1} DW h^{-1}	Diaz & Ward 1997
HMA	Demos.	*Taonura* sp.	Field incubation	SW Australia	sublittoral	-0.01	0.01	μmol g^{-1} DW h^{-1}	Keesing et al. 2013
HMA	Demos.	*Thorectandra* cf. *choanoides*	Field incubation	SW Australia	sublittoral	-0.07	N/A	μmol g^{-1} DW h^{-1}	Keesing et al. 2013
HMA	Demos.	*Thorectandra* sp.	Field incubation	SW Australia	sublittoral	0.001	N/A	μmol g^{-1} DW h^{-1}	Keesing et al. 2013
HMA	Demos.	*Xestospongia muta*	InEx+pumping	Caribbean	sublittoral	0.002	0.059	μmol mL^{-1} sponge h^{-1}	Southwell et al. 2008 a

TABLE 4.3 *(Continued)*

Microbial	Taxonomic	Sponge	Method of	Geographical	Depth	Average	Associated	Units	Reference
HMA	Hexact.	*Nodastrella nodastrella*	Lab incubation	NE Atlantic	529-575	-0.002	0.001	µmol mL^{-1} sponge h^{-1}	van Duyl et al. 2008
HMA	Hexact.	*Vazella pourtalesii*	Field & Lab incubation	North Atlantic	700	0.006	0.014	µmol mL^{-1} sponge h^{-1}	Maldonado et al. 2021
HMA	Homo.	*Plakortis halichondrioides*	Lab incubation	Caribbean	sublittoral	-0.10	0.26	µmol g^{-1} DW h^{-1}	Diaz & Ward 1997
LMA	Demos.	*Acanthella* cf. *pulcherrima*	Field incubation	SW Australia	sublittoral	-0.05	N/A	µmol g^{-1} DW h^{-1}	Keesing et al. 2013
LMA	Demos.	*Antho dichotoma*	Lab incubation	Norwegian fjord	200-300	-0.174	0.017	µmol mL^{-1} sponge h^{-1}	Radax et al. 2012
LMA	Demos.	*Aplysinella* sp.	Field incubation	SW Australia	sublittoral	-0.06	0.13	µmol g^{-1} DW h^{-1}	Keesing et al. 2013
LMA	Demos.	*Cinachyrella* sp.	Field incubation	Florida Keys	sublittoral	0.005	0.002*	µmol mL^{-1} sponge h^{-1}	Hoer et al. 2018
LMA	Demos.	*Clathria nidificata*	Lab incubation	Antarctic	sublittoral	1.18	0.22	µmol g^{-1} AFDW h^{-1}	Morley et al. 2016
LMA	Demos.	*Clathria* sp.	Field incubation	SW Australia	sublittoral	-0.59	0.11	µmol g^{-1} DW h^{-1}	Keesing et al. 2013
LMA	Demos.	*Cliona varians*	Field incubation	Caribbean	sublittoral	-0.266	0.052*	µmol g^{-1} DW h^{-1}	Corredor et al. 1988
LMA	Demos.	*Crambe crambe*	InEx+pumping	Mediterranean	sublittoral	-0.019	0.020	µmol cm^{-2} sponge h^{-1}	Morganti et al. 2017
LMA	Demos.	*Dendrilla antarctica*	Lab incubation	Antarctic	sublittoral	0.25	0.01	µmol g^{-1} AFDW h^{-1}	Morley et al. 2016
LMA	Demos.	*Diplastrella* sp.	Field incubation	SW Australia	sublittoral	-0.02	N/A	µmol g^{-1} DW h^{-1}	Keesing et al. 2013
LMA	Demos.	*Dysidea avara*	InEx+pumping	Mediterranean	sublittoral	-0.081	0.098	µmol cm^{-2} sponge h^{-1}	Morganti et al. 2017
LMA	Demos.	*Dysidea avara*	Lab incubation	Mediterranean	sublittoral	-0.654	0.162	µmol g^{-1} DW h^{-1}	Jimenez & Ribes 2007
LMA	Demos.	*Dysidea avara*	Lab incubation	Mediterranean	sublittoral	-0.242	N/A	µmol mL^{-1} sponge h^{-1}	Schläppy et al. 2010
LMA	Demos.	*Halichondria melanadocia*	Lab incubation	Caribbean	sublittoral	-0.019	0.007	µmol mL^{-1} sponge h^{-1}	Archer et al. 2015
LMA	Demos.	*Halichondria melanodocia*	Field incubation	Florida Keys	sublittoral	-0.052	0.005*	µmol mL^{-1} sponge h^{-1}	Hoer et al. 2018
LMA	Demos.	*Haliclona* cf. sp.	Field incubation	SW Australia	sublittoral	-1.38	N/A	µmol g^{-1} DW h^{-1}	Keesing et al. 2013

TABLE 4.3 *(Continued)*

Microbial	Taxonomic	Sponge	Method of	Geographical	Depth	Average	Associated	Units	Reference
LMA	Demos.	*Haliclona cymaeformis*	Field & lab incubation	GBR, Australia	sublittoral	-4.6	N/A	µmol g^{-1} DW h^{-1}	Davy et al. 2002
LMA	Demos.	*Haliclona* sp.	Field incubation	Florida Keys	sublittoral	-0.052	0.007*	µmol mL^{-1} sponge h^{-1}	Hoer et al. 2018
LMA	Demos.	*Hexadella dedritifera*	Lab incubation	Norwegian fjord	200	0.000	N/A	µmol mL^{-1} sponge h^{-1}	Fang et al. 2018
LMA	Demos.	*Iotrochota baculifera*	Lab incubation	W Australia	sublittoral	-0.98	N/A	µmol g^{-1} AFDW h^{-1}	Hatcher 1994
LMA	Demos.	*Iotrochota* cf. *baculifera*	Field incubation	SW Australia	sublittoral	-0.07	0.08	µmol g^{-1} DW h^{-1}	Keesing et al. 2013
LMA	Demos.	*Monanchora* sp.	Field incubation	SW Australia	sublittoral	-0.4	N/A	µmol g^{-1} DW h^{-1}	Keesing et al. 2013
LMA	Demos.	*Mycale* cf. *mirabilis*	Field incubation	SW Australia	sublittoral	-0.42	0.28	µmol g^{-1} DW h^{-1}	Keesing et al. 2013
LMA	Demos.	*Negombata magnifica*	Lab incubation	Red Sea	sublittoral	-0.16	0.05	µmol g^{-1} WW h^{-1}	Hadas et al. 2009
LMA	Demos.	*Niphates digitalis*	InEx+pumping	Caribbean	sublittoral	-0.253	0.180	µmol mL^{-1} sponge h^{-1}	Southwell et al. 2008 a
LMA	Demos.	*Niphates erecta*	Lab incubation	Caribbean	sublittoral	1.50	0.30	µmol g^{-1} h^{-1}	Southwell et al. 2008 b
LMA	Demos.	*Phakellia ventilabrum*	Lab incubation	Norwegian fjord	200-300	-0.134	0.060	µmol mL^{-1} sponge h^{-1}	Radax et al. 2012
LMA	Demos.	*Phakellia ventilabrum*	Lab incubation	Norwegian fjord	200-300	-0.149	0.053	µmol mL^{-1} sponge h^{-1}	Radax et al. 2012
LMA	Demos.	*Phakellia ventilabrum*	Lab incubation	Norwegian fjord	200-300	-0.184	N/A	µmol mL^{-1} sponge h^{-1}	Radax et al. 2012
LMA	Demos.	*Phakellia ventilabrum*	Lab incubation	Norwegian fjord	200-300	-0.173	0.019	µmol mL^{-1} sponge h^{-1}	Radax et al. 2012
LMA	Demos.	*Sphaerotylus antarcticus*	Lab incubation	Antarctic	sublittoral	0.2	0.001	µmol g^{-1} AFDW h^{-1}	Morley et al. 2016
LMA	Demos.	*Suberites* sp.	Lab incubation	Antarctic	sublittoral	0.4	0.1	µmol g^{-1} AFDW h^{-1}	Morley et al. 2016
LMA	Demos.	*Tethya citrina*	Field incubation	NE Atlantic	sublittoral	-0.043	0.031	µmol mL^{-1} sponge h^{-1}	López-Acosta et al. 2019
LMA	Hexact.	*Schaudinnia rosea*	Field incubation	Arctic	600	-0.007	0.003	µmol mL^{-1} sponge h^{-1}	Maldonado et al. 2021

- flux = release **+ flux = uptake**

units converted to µmol, h, sponge mL or g rates

[#]Average (±SD) flux rates are normalized in diverse units of sponge biomass and hour. An asterisk (*) indicates specific cases in which the error associated with the average has not been calculated as ±SD but as ±SE. Negative values indicate nutrient release by the sponges and positive values nutrient incorporation. Sponge classes are Demospongiae, Hexactinellida, Homoscleromorpha, and Calcarea. Microbiomes are either high microbial abundance (HMA) or low microbial abundance (LMA).

TABLE 4.4 Summary of Net Flux Rates of Nitrite Considered for This Review, Ordered by Microbial Abundance, Taxonomic Class, Species Name, and Year of Publication[#]

Microbial abundance	Taxonomic class	Sponge species	Method of rate determination	Geographical area	Depth range	Average rate	Associated error	Units	Reference
HMA	Demos.	*Chondrilla nucula*	Lab incubation	Caribbean	sublittoral	–49.00	94.50	nmol g^{-1} DW h^{-1}	Diaz & Ward 1997
HMA	Demos.	*Chondrosia reniformis*	Lab incubation	Mediterranean	sublittoral	–0.8	N/A	nmol mL^{-1} sponge h^{-1}	Schläppy et al. 2010
HMA	Demos.	*Geodia barretti*	Lab incubation	Norwegian fjord	200	–0.0008	N/A	nmol mL^{-1} sponge h^{-1}	Fang et al. 2018
HMA	Demos.	*Geodia barretti*	InEx+pumping	Norwegian fjord	160	3.3	N/A	nmol mL^{-1} sponge h^{-1}	Leys et al. 2018
HMA	Demos.	*Higginsia thielei*	Lab incubation	NE Atlantic	570–785	–0.617	N/A	nmol mL^{-1} sponge h^{-1}	van Duyl et al. 2008
HMA	Demos.	*Hyrtios violaceus*	Lab incubation	Caribbean	sublittoral	–367.50	515.42	nmol g^{-1} DW h^{-1}	Diaz & Ward 1997
HMA	Demos.	*Stryphnus fortis*	Lab incubation	Norwegian fjord	200	–0.00035	0.00018	nmol mL^{-1} sponge h^{-1}	Fang et al. 2018
HMA	Demos.	*Svenzea zeai*	Lab incubation	Caribbean	sublittoral	–10.00	16.05	nmol g^{-1} DW h^{-1}	Diaz & Ward 1997
HMA	Hexact.	*Nodastrella nodastrella*	Lab incubation	NE Atlantic	529–575	–0.029	N/A	nmol mL^{-1} sponge h^{-1}	van Duyl et al. 2008
HMA	Hexact.	*Vazella pourtalesii*	Field incubation	North Atlantic	700	–24.3	9.5	nmol mL^{-1} sponge h^{-1}	Maldonado et al. 2021
HMA	Homos.	*Plakortis halichondrioides*	Lab incubation	Caribbean	sublittoral	–30.00	63.56	nmol g^{-1} DW h^{-1}	Diaz & Ward 1997
LMA	Demos.	*Antho dichotoma*	Lab incubation	Norwegian fjord	200–300	–12.50	1.25	nmol mL^{-1} sponge h^{-1}	Radax et al. 2012
LMA	Demos.	*Dysidea avara*	Lab incubation	Mediterranean	sublittoral	–0.8	N/A	nmol mL^{-1} sponge h^{-1}	Schläppy et al. 2010
LMA	Demos.	*Phakellia ventilabrum*	Lab incubation	Norwegian fjord	200–300	–5.83	2.92	nmol mL^{-1} sponge h^{-1}	Radax et al. 2012
LMA	Demos.	*Phakellia ventilabrum*	Lab incubation	Norwegian fjord	200–300	–0.63	1.25	nmol mL^{-1} sponge h^{-1}	Radax et al. 2012
LMA	Demos.	*Phakellia ventilabrum*	Lab incubation	Norwegian fjord	200–300	–6.67	2.92	nmol mL^{-1} sponge h^{-1}	Radax et al. 2012
LMA	Demos.	*Phakellia ventilabrum*	Lab incubation	Norwegian fjord	200–300	0.83	0.33	nmol mL^{-1} sponge h^{-1}	Radax et al. 2012
LMA	Demos.	*Tethya citrina*	Field incubation	NE Atlantic	sublittoral	–7.00	3.00	nmol mL^{-1} sponge h^{-1}	López-Acosta et al. 2019
LMA	Hexact.	*Schaudinnia rosea*	Field incubation	Arctic	600	0.0	0.1	nmol mL^{-1} sponge h^{-1}	Maldonado et al. 2021

- flux = release + flux = uptake

units converted to nmol, h, sponge mL or g rates

[#]Average (±SD) flux rates are normalized in diverse units of sponge biomass and hour. Negative values indicate nutrient release by the sponges and positive values nutrient incorporation. Sponge classes are Demospongiae, Hexactinellida, Homoscleromorpha, and Calcarea. Microbiomes are either high microbial abundance (HMA) or low microbial abundance (LMA).

TABLE 4.5 Summary of Net Flux Rates of Nitrate Considered for This Review, Ordered by Microbial Abundance, Taxonomic Class, Species Name, and Year of Publication#

Microbial abundance	Taxonomic class	Sponge species	Method of rate determination	Geographical area	Depth range	Average rate	Associated error	Units	Reference
HMA	Demos.	*Aplysina aerophoba*	Lab incubation	Mediterranean	Sublittoral	30.8	6.3	nmol g^{-1} WW h^{-1}	Bayer et al. 2007
HMA	Demos.	*Aplysina aerophoba*	Lab incubation	Mediterranean	Sublittoral	–326	117*	nmol g^{-1} DW h^{-1}	Bayer et al. 2008
HMA	Demos.	*Aplysina aerophoba*	Lab incubation	Mediterranean	Sublittoral	–180	46*	nmol g^{-1} DW h^{-1}	Bayer et al. 2008
HMA	Demos.	*Aplysina aerophoba*	Lab incubation	Mediterranean	Sublittoral	–344	189*	nmol g^{-1} DW h^{-1}	Bayer et al. 2008
HMA	Demos.	*Aplysina aerophoba*	Lab incubation	Mediterranean	Sublittoral	–291	40*	nmol g^{-1} DW h^{-1}	Bayer et al. 2008
HMA	Demos.	*Aplysina aerophoba*	Lab incubation	Mediterranean	Sublittoral	–89	52*	nmol g^{-1} DW h^{-1}	Bayer et al. 2008
HMA	Demos.	*Chondrilla nucula*	Lab incubation	Caribbean	Sublittoral	–1395.0	1616.3	nmol g^{-1} DW h^{-1}	Diaz & Ward 1997
HMA	Demos.	*Chondrosia reniformis*	Lab incubation	Mediterranean	Sublittoral	–8	N/A	nmol mL^{-1} sponge h^{-1}	Schläppy et al. 2010
HMA	Demos.	*Geodia atlantica*	Lab incubation	Norwegian fjord	200	0.89	N/A	nmol mL^{-1} sponge h^{-1}	Rooks et al. 2020
HMA	Demos.	*Geodia barretti*	Lab incubation	Norwegian fjord	200	–25.8	N/A	nmol mL^{-1} sponge h^{-1}	Fang et al. 2018
HMA	Demos.	*Geodia barretti*	InEx+pumping	Norwegian fjord	160	–40.9	N/A	nmol mL^{-1} sponge h^{-1}	Leys et al. 2018
HMA	Demos.	*Geodia barretti*	Lab incubation	Norwegian fjord	200	5.29	N/A	nmol mL^{-1} sponge h^{-1}	Rooks et al. 2020
HMA	Demos.	*Geodia hentscheli*	Lab incubation	Arctic	700	5.73	N/A	nmol mL^{-1} sponge h^{-1}	Rooks et al. 2020
HMA	Demos.	*Geodia parva*	Lab incubation	Arctic	700	7.16	N/A	nmol mL^{-1} sponge h^{-1}	Rooks et al. 2020
HMA	Demos.	*Hyrtios violaceus*	Lab incubation	Caribbean	Sublittoral	–157.5	261.1	nmol g^{-1} DW h^{-1}	Diaz & Ward 1997
HMA	Demos.	*Stelletta rhaphidiophora*	Lab incubation	Arctic	700	1.03	N/A	nmol mL^{-1} sponge h^{-1}	Rooks et al. 2020
HMA	Demos.	*Stryphnus fortis*	Lab incubation	Norwegian fjord	200	–19.7	1.90	nmol mL^{-1} sponge h^{-1}	Fang et al. 2018
HMA	Demos.	*Stryphnus fortis*	Lab incubation	Norwegian fjord	200	1.24	N/A	nmol mL^{-1} sponge h^{-1}	Rooks et al. 2020
HMA	Demos.	*Svenzea zeai*	Lab incubation	Caribbean	Sublittoral	–400.0	730.8	nmol g^{-1} DW h^{-1}	Diaz & Ward 1997
HMA	Hexact.	*Vazella pourtalesii*	Field incubation	North Atlantic	700	–2.1	5.8	nmol mL^{-1} sponge h^{-1}	Maldonado et al. 2021
HMA	Homos.	*Plakortis halichondrioides*	Lab incubation	Caribbean	Sublittoral	–150.0	226.5	nmol g^{-1} DW h^{-1}	Diaz & Ward 1997
LMA	Demos.	*Dysidea avara*	Lab incubation	Mediterranean	Sublittoral	–13.9	N/A	nmol mL^{-1} sponge h^{-1}	Schläppy et al. 2010

TABLE 4.5 *(Continued)*

Microbial	Taxonomic	Sponge	Method of	Geographical	Depth	Average	Associated	Units	Reference
LMA	Demos.	*Phakellia ventilabrum*	Lab incubation	Norwegian fjord	200-300	–14.0	13.8	nmol mL^{-1} sponge h^{-1}	Radax et al. 2012
LMA	Demos.	*Phakellia ventilabrum*	Lab incubation	Norwegian fjord	200-300	–5.0	5.8	nmol mL^{-1} sponge h^{-1}	Radax et al. 2012
LMA	Demos.	*Phakellia ventilabrum*	Lab incubation	Norwegian fjord	200-300	–67.5	12.5	nmol mL^{-1} sponge h^{-1}	Radax et al. 2012
LMA	Demos.	*Phakellia ventilabrum*	Lab incubation	Norwegian fjord	200-300	–15.8	2.1	nmol mL^{-1} sponge h^{-1}	Radax et al. 2012
LMA	Demos.	*Tethya citrina*	Field incubation	NE Atlantic	Sublittoral	–63	31	nmol mL^{-1} sponge h^{-1}	López-Acosta et al. 2019
LMA	Hexact.	*Schaudinnia rosea*	Field incubation	Arctic	600	–1	6.9	nmol mL^{-1} sponge h^{-1}	Maldonado et al. 2021

- flux = release + flux = uptake

units converted to nmol, h, sponge mL or g rates

#Average (±SD) flux rates are normalized in diverse units of sponge biomass and hour. An asterisk (*) indicates specific cases in which the error associated with the average has not been calculated as ±SD but as ±SE. Negative values indicate nutrient release by the sponges and positive values nutrient incorporation. Sponge classes are Demospongiae, Hexactinellida, Homoscleromorpha, and Calcarea. Microbiomes are either high microbial abundance (HMA) or low microbial abundance (LMA).

TABLE 4.6 Summary of Net Flux Rates of Phosphate Considered for This Review, Ordered by Microbial Abundance, Taxonomic Class, Species Name, and Year of Publication[#]

Microbial abundance	Taxonomic class	Sponge species	Method of rate determination	Geographical area	Depth range	Average rate	Associated error	Units	Reference
HMA	Demos.	*Agelas oroides*	Lab incubation	Mediterranean	Sublittoral	–58.8	12.6	nmol g^{-1} DW h^{-1}	Jimenez & Ribes 2007
HMA	Demos.	*Aplysina aerophoba*	Lab incubation	Mediterranean	Sublittoral	–26.4	10.8	nmol g^{-1} DW h^{-1}	Jimenez & Ribes 2007
HMA	Demos.	*Aplysina* sp.	Lab incubation	W Australia	Sublittoral	–6.76	N/A	nmol g^{-1} AFDW h^{-1}	Hatcher 1994
HMA	Hexact.	*Vazella pourtalesii*	Field incubation	North Atlantic	700	–0.8	0.5	nmol mL^{-1} sponge h^{-1}	Maldonado et al. 2021
LMA	Demos.	*Dysidea avara*	Lab incubation	Mediterranean	Sublittoral	–51	20.4	nmol g^{-1} DW h^{-1}	Jimenez & Ribes 2007
LMA	Demos.	*Iotrochota baculifera*	Lab incubation	W Australia	Sublittoral	–7.31	N/A	nmol g^{-1} AFDW h^{-1}	Hatcher 1994
LMA	Demos.	*Tethya citrina*	Field incubation	NE Atlantic	Sublittoral	–4	5	nmol mL^{-1} sponge h^{-1}	López-Acosta et al. 2019
LMA	Hexact.	*Schaudinnia rosea*	Field incubation	Arctic	600	–2.6	2.6	nmol mL^{-1} sponge h^{-1}	Maldonado et al. 2021

- flux = release + flux = uptake

units converted to nmol, h, sponge mL or g rates

[#]Average (±SD) flux rates are normalized in diverse units of sponge biomass and hour. Negative values indicate nutrient release by the sponges and positive values nutrient incorporation. Sponge classes are Demospongiae and Hexactinellida. Microbiomes are either high microbial abundance (HMA) or low microbial abundance (LMA).

TABLE 4.7 Summary of Microbes Possibly Involved in Nitrogen Cycle Within Sponge Hosts for Which Nitrogen Fluxes are Known, Ordered by Sponge Species*

Sponge Species	Microbes	Method	Function	Process	References
Agelas oroides	Archaea	PCR (*amoA*, 16S rDNA)	Ammonia oxidation	Nitrification	Ribes et al. (2012)
Antho dichotoma	Crenarchaeota	PCR (*amoA*, 16S rDNA)	Ammonia oxidation	Nitrification	Radax et al. (2012)
Aplysina aerophoba	Betaproteobacteria (*Nitrosospira* sp.)	PCR (*amoA*, 16S rDNA)	Ammonia oxidation	Nitrification	Bayer et al. (2008)
	Crenarcheaota				
	Alphaproteobacteria (*Rhodospirillaceae*)	Metagenomics (*narYJIG*, *napBD*, *GlnA*, *GltB*)	Nitrate reduction	Denitrification	Karimi et al. (2018)
			Nitrate and nitrite ammonification	Nitrogen utilization	
			Ammonia assimilation	Ammonia assimilation	
	Bacteria and archaea	Functional gene array (*amoA*, *hao*)	Ammonia oxidation	Nitrification	Bayer et al. (2014)
		Functional gene array (*narG*)	Nitrate reduction	Denitrification	
		Functional gene array (*nirS*)	Nitrite reduction		
		Functional gene array (*nosZ*)	Nitrous oxide reduction		
		Functional gene array (*nrfA*)	Nitrite reduction	DNRA	
		Functional gene array (*ureC*)	Urea utilization	Nitrogen utilization	
	Alphaproteobacteria (bins52, 36, 56, 65, 95, 129)	Metagenomics (*narG*)	Nitrate reduction	Denitrification	Slaby et al. (2017)
	Deltaproteobacteria (bin18); Gammaproteobacteria (bin55); Chloroflexi (bins22, 34)	Metagenomics (*norB*)	Nitric oxide reduction		
	Bacteriodetes (bin80)	Metagenomics (*nosZ*)	Nitrous oxide reduction		
	Nitrospinae (bin107); Delta-proteobacteria (bin18); Alpha-proteobacteria (bins52, 56, 65, 129); Bacteriodetes (bin80); Chloroflexi (bins22, 87, 127); Gammaproteobacteria (bin55)	Metagenomics (*nirB*)	–	DNRA	
	Chloroflexi (bins34, 22, 87); Proteobacteria (bins18, 52, 56, 131); Spirochaetae (bin103); Nitrospira (bins75, 107); Cyanobacteria (bin9)	Metagenomics (*ureABCHF*)	Urea utilization	Ammonification	

TABLE 4.7 *(Continued)*

Sponge Species	Microbes	Method	Function	Process	References
	Cyanobacteria (bin9)	Metagenomics (*nifH*, *nifD*)	Dinitrogen reduction	Nitrogen fixation	
	Nitrospira sp.	Metagenomics (*narGH*)	Nitrate reduction	Denitrification	De Mares et al. (2018)
	Pseudovibrio sp.	Metagenomics (*nirK*)	Nitrite reduction		
	Entotheonellaceae	Metagenomics (*norQ*)	Nitric oxide reduction		
Axinella polypoides	Thaumarchaeota (*Cenarchaeum* sp.)	PCR (amoA, 16S rDNA)	Ammonia oxidization	Nitrification	Steger et al. (2008)
Chondrilla nucula	Betaproteobacteria (*Nitrosomonas* spp.)	PCR (16S rDNA)	Ammonia oxidization	Nitrification	Diaz et al. (2004)
Chondrosia reniformis	Gammaproteobacteria (*Nitrosococcus* spp.)	PCR (*amoA*)	Ammonia oxidation	Nitrification	Ribes et al. (2012)
	Thaumarchaeota (*Cenarchaeum* sp.)	PCR (*amoA*)	Ammonia oxidation	Nitrification	Steger et al. (2008)
Cinachyrella sp.	Alphaproteobacteria	PCR/qPCR (*nirK*)	Nitrite reductase	Denitrification	Zhang et al. (2013)
	Gammaproteobacteria	PCR/qPCR (*nosZ*)	Nitrous oxide reduction		
Cliona varians	Thaumarchaeota (*Cenarchaeum* sp., *Nitrosopumilus* sp.)	PCR (*amoA*)	Ammonia oxidization	Nitrification	Steger et al. (2008)
Dysidea avara	Betaproteobacteria (*Nitrososspira* spp.)	PCR (*amoA*)	Ammonia oxidation	Nitrification	Bayer et al. (2008); Ribes et al. (2012)
–	Bacteria and archaea	Functional gene array (*amoA*, *hao*)	Ammonia oxidation	Nitrification	Bayer et al. (2014)
–		Functional gene array (*narG*)	Nitrate reductase	Denitrification	
–		Functional gene array (*nirS*)	Nitrite reductase		
–		Functional gene array (*nosZ*)	Nitrous oxide reductase		
–		Functional gene array (*nrfA*)	Cytochrome c nitrite reductase	DNRA	
–		Functional gene array (*ureC*)	Urea utilization	Ammonification	
Geodia atlantica		qPCR (*nirS/K*)	Nitrite reduction	Denitrification	Rooks et al. (2020)
Geodia barretti	Thaumarchaeota (*Cenarchaeum* sp., *Nitrosopumilus* sp.)	PCR, transcriptomics (*amoA*)	Ammonia oxidation	Nitrification	Hoffmann et al. (2009); Radax et al. (2012)

TABLE 4.7 *(Continued)*

Sponge Species	Microbes	Method	Function	Process	References
	Nitrospirae (*Nitrospira* spp.)	PCR, transcriptomics (16S rDNA)	Nitrite oxidation	–	–
	Planctomycetes (Ca. Scalindua *sp.*)	PCR (16S rDNA)	Anaerobe ammonia oxidation	Anammox Denitrification	–
	Betaproteobacteria (*Alcaligenes faecali*, *Thauera mechernichensis*)	PCR (*nirS*)	Nitrite reduction	Denitrification	–
		qPCR (*nirS/K*)	Nitrite reduction	Denitrification	Rooks et al. (2020)
Geodia hentscheli		qPCR (*nirS/K*)	Nitrite reduction	Denitrification	Rooks et al. (2020)
Geodia parva		qPCR (*nirS/K*)	Nitrite reduction	Denitrification	Rooks et al. (2020)
Haliclona cf. sp.	Thaumarchaeota (*Nitrosopumilus* ssp.)	PCR (16S rDNA)	Ammonia oxidization	Nitrification	Steger et al. (2008)
Hexadella dedritifera	*Ca.* Nitrosopumilus hexadellus	Metagenomics (*amoA/ure*)	Ammonia oxidization urea utilization	Nitrification ammonification	Zhang et al. (2019)
Hexadella cf. detritifera	*Ca.* Nitrosopumilus detritiferus	Metagenomics (*amoA/ure*)	Ammonia oxidization urea utilization	Nitrification ammonification	Zhang et al. (2019)
Ircinia felix		Predicted based on 16S rDNA Amplicon	–	PS and N-metabolism	Archer et al. (2017)
Ircinia oros	Thaumarchaeota (*Nitrosopumilus* ssp.)	PCR (*amoA*/16S rDNA)	Ammonia oxidation	Nitrification	Zhang et al. (2014)
Ircinia strobilina	*Nitrosospira* sp.	PCR (*amoA*)	Ammonia oxidation	Nitrification	Mohamed et al. (2010)
	Thaumarchaeota (*Nitrosopumilus* sp.)	PCR (*amoA*/16S rDNA)	Ammonia oxidation	Nitrification	Zhang et al. (2014)
	Gammaproteobacteria (*Azotobacterial* spp.) Alphaproteobacteria (*Methylocystis* sp.); Cyanobacteria (*Tolypothrix* sp.). Deltaproteobacteria (*Desulfovibrio* sp.)	PCR (*nifH*)	Dinitrogen reductase	Nitrogen fixation	Mohamed et al. (2008)
	Cyanobacteria (*Leptolyngbya* sp., *Myxosarcina* sp.)	PCR/RT-PCR (*nifH*)	Dinitrogen reductase	Nitrogen fixation	

TABLE 4.7 *(Continued)*

Sponge Species	Microbes	Method	Function	Process	References
	Cyanobacteria (*Leptolyngbya* sp., *Cyanothece* sp., *Anabena* sp., *Okeania plumata*, *Hydrocoleum lyngbyaceum*, *Microcoleus chthonoplastes Lyngbya wollei*)	PCR/RT-PCR (*nifH*)	Dinitrogen reductase	Nitrogen fixation	Zhang et al. (2014b)
	Alphaproteobacteria (*Xanthobacter* sp., *Sulfitobacter* sp.)				
	Gammaproteobacteria (*Klebsiella* spp., *Stenotrophomonas* sp., *Halorhodospira halophila*)				
	Deltaproteobacteria (*Desulfobulbus* sp.)				
Petrosia ficiformis	*Ca. Nitrosopumilus* sp. Pfa	Metagenomics (*amoA*/*ure*)	Ammonia oxidization	Nitrification	Haber et al. (2021)
–	–	–	Urea utilization	Ammonification	–
Phakellia ventilabrum	Crenarchaeota	PCR (*amoA*/16S rDNA)	Ammonia oxidation	Nitrification	Radax et al. (2012)
	Betaproteobacteria (*Nitrosospira* sp.)	–	Nitrite oxidation	–	–
Spheciospongia vesparium	Betaproteobacteria (*Nitrosospira* sp.)	PCR (*amoA*)	Ammonia oxidization	Nitrification	He et al. (2021)
	Shewanella sp.	PCR (*napA*)	Nitrate reduction	Denitrification	–
	Gammaproteobacteria (*Pseudomonas* sp.)	PCR (*narG*)	–		–
		PCR (*nirK*)	Nitrite reduction		–
	Alphaproteobacteria (*Rhodobacter* sp.)	PCR (*norB*)	Nitric oxide reduction		–
		PCR (*nosZ*)	Nitrous oxide reduction		–
	Alphaproteobacteria (*Methylo-bacterium* sp., *Oceaniovalibus* sp., *Roseobacter* sp.); Gamma-proteobacteria (*Psychrobacter* sp., *Oceanimonas* sp., *Pseudo-alteromonas* sp.); *Bacteroidetes* (*Bizionia* sp.), Betaproteobacteria (*Comamonas* sp.)	PCR (*ureC*)	Urea utilization	Ammonification	–

TABLE 4.7 *(Continued)*

Sponge Species	Microbes	Method	Function	Process	References
Spheciospongia vesparium	Gammaproteobacteria (*Oceanimonas* sp.)	PCR (*gltB*)	Ammonia assimilation	Ammonia assimilation	He et al. (2021)
	Epsilonbacteria (*Arcobacter* sp.)	PCR (*nifH*)	Dinitrogen reductase	Nitrogen fixation	
	Gammaproteobacteria (*Shewanella* sp.); Bacteroidetes (*Bizionia* sp.)	PCT (*nrfA*)	Nitrite reductase	DNRA	
	Alpha- and betaproteobacteria	PCR/qPCR (*nirK*)	Nitrite reductase	Denitrification	Zhang et al. (2013)
	Alphaproteobacteria (*Rhodospirillaceae*)	PCR/qPCR (*nosZ*)	Nitrous oxide reduction	–	–
Stelletta rhaphidiophora		qPCR (*nirS/K*)	Nitrite reduction	Denitrification	Rooks et al. (2020)
Stryphnus fortis		qPCR (*nirK*)	Nitrite reduction	Denitrification	Rooks et al. (2020)
Suberites sp.	*Ca.* Nitrosopumilus sp. Nsub	Metagenomics (*amoA/ure*)	Ammonia oxidization	Nitrification	Tian et al. (2017); Haber et al. (2021)
Tentorium semisuberites	Crenarchaeota	PCR (*amoA*/16S rDNA)	Ammonia oxididation	Nitrification	Radax et al. (2012)
Theonella swinhoei	Thaumarchaeon TS	Metagenomics (*amoA/ure*)	Ammonia oxidization urea utilization	Nitrification ammonification	Haber et al. (2021)
–	Betaproteobacteria (*Nitrosospira* sp.)	PCR (*amoA*/16S rDNA)	Ammonia oxidization	Nitrification	Bayer et al. (2008)
–	Thaumarchaeota	qPCR (*amoA*)	Ammonia oxidization	Nitrification	Feng et al. (2018)
–	Cyanobacteria; *Ca.* Poribacteria; alpha-, epsilon-, gamma-, and beta-proteobacteria	qPCR (*glnA*)	Glutamine synthetase	Ammonia assilimilation	Feng et al. (2018)
–		Metagenomics (*amo*, *hao*)	Ammonia oxidization	Nitrification	Li et al. (2015)
–		Metagenomics (*nor*, *nxr*)	Nitrite oxidation		
–		Metagenomics (*nir*)	Anaerobe ammonia oxidation	Anammox	
–		Metagenomics (*nar*, *napA*)	Nitrate reduction	Denitrification	
–		Metagenomics (*nirS/K*)	Nitrite reduction		

TABLE 4.7 *(Continued)*

Sponge Species	Microbes	Method	Function	Process	References
–		Metagenomics (*nor*)	Nitric oxide reduction		
–		Metagenomics (*nosZ*)	Nitrous oxide reduction		
–		Metagenomics (*nif*)	Dinitrogen reductase	Nitrogen fixation	
–		Metagenomics (*nrfA*)	Nitrite reductase	DNRA	
–		Metagenomics (*GDH2*, *Glt*)	Ammonia assimilation	Ammonia assimilation	
Vazella pourtalesii	Crenarchaeota (bins 74, 90, 131, 143); Betaproteobacteria (bin46)	Metagenomics (*amoA*)	Ammonia oxidation	Nitrification	Bayer et al. (2020); Maldonado et al. (2021b)
–		Metagenomics (*hao*)	Ammonia oxidation		
–	Alphaproteobacteria (bin70); Actinobacteriota (bin64); Planctomycetota (bin52)	Metagenomics (*narG*, *narJ*)	Nitrate reductase	Denitrification	–
–	Crenarchaeota (bins36, 90, 86, 143, 101, 143); Acidobacteriota (bin86)	Metagenomics (*nirK*, *nirS*)	Nitrite reductase		–
–		Metagenomics (*nosZ*)	Nitrous-oxide reductase	–	–
–	Alphaproteobacteria (bin70); Actinobacteriota (bin64); Planctomycetota (bin52)	Metagenomics (*napA*)	Nitrate reductase	DNRA	–
–		Metagenomics (*nrfA*, *nrfH*)	Nitrite reductase	–	–
–	Planctomycetota (bins52, 99); Actinobacteriota (bin 64), Gammaproteobacteria (bin89)	Metagenomics (*nirB*, *nirD*)	Nitrite reductase	–	–
–	Firmicutes (bins77, 88, 127), Planctomycetota (bin91); Bacteroidota (bin24); Alpha-proteobacteria (bin135)	Metagenomics (*nifD*, *nifK*, *nifH*, *anfG*)	Dinitrogen reductase	Nitrogen fixation	–
–	Alpha- and Gammaproteobacteria; Spirochaetota; Crenarchaeota; Planctomycetota; SAR324; Actino-bacteriota; Verrucomicrobiota; Chloroflexota; Firmicutes; Bacteroidota; Acidobacteriota, Thermoplasmatota; Marinisomatota	Metagenomics (*glnA*, *GLUL*)	Glutamine synthetase	Ammonia assimilation	–

TABLE 4.7 *(Continued)*

Sponge Species	Microbes	Method	Function	Process	References
Vazella pourtalesii	Chloroflexota (bin53); Alphaproteobacteria (bins70, 124, 135, 141); Crenarchaeota (bin131); Gammaproteobacteria (bins35, 46); Actinobacteriota (bin64); Acidobacteriota (bin86); Bacteroidota (bins24, 87); Marinisomatota (bin37); Verrucomicrobiota (bin23)	Metagenomics (ure)	Urea utilization	Ammonification	Bayer et al. (2020); Maldonado et al. (2021b)
Xestospongia muta	Crenarchaeota	qPCR (amoA)	Ammonia oxidation	Nitrification	López-Legentil et al. (2010)
–	Crenarchaeota	PCR (amoA)	–	–	Fiore et al. (2013, 2015)
–	Betaproteobacteria (Nitrosospira spp.)	PCR (amoA, 16S rDNA)	–	–	Bayer et al. (2008)
–	Thaumarchaeota; Gammaproteobacteria	Transcriptomics (amo)	Ammonia oxidation	Nitrification	Fiore et al. (2015)
–	Nitrospirae (Nitrospira fluvii)	Transcriptomics (nar)	Nitrate oxidoreductase	Denitrification	–
–	Alphaproteobacteria (*Sinorhizobium* sp.) Gammaproteobacteria (*Pseudomonas* sp.)	PCR (nirK)	Nitrite reduction	–	–
–	Thaumarchaeota; Gammaproteo-bacteria (*Oleispira antarctica*); Alphaproteobacteria (*Mesorhizobium amorphae*)	Transcriptomics (nirK)	–	–	–
–	Alpha-, Gammaproteobacteria, Euryarchaeota	Transcriptomics (nor)	Nitric oxide reduction	–	–
–	Betaproteobacteria (Alcaligenes-like)	PCR (qnorB)	Nitric oxide (NO)-reduction (quinol-oxidizing)	–	–
–	Cyanobacteria (*Xenococcus* sp., *Myxosarcina* sp., UCYN-A-like)	PCR (nifH)	Dinitrogen reductase	N-Fixation	–
–	Alphaproteobacteria (*Bradyrhizobium* sp.)	PCR (nifH)	–	–	–
–	Gammaproteobacteria (*Vibrio* spp., *Azotobacter* spp.)	PCR (nifH)	–	–	–

TABLE 4.7 *(Continued)*

Sponge Species	Microbes	Method	Function	Process	References
–	Deltaproteobacteria (*Desulfovibrio* spp.)	PCR (nifH)	–	–	–
–	Gammaproteobacteria (*Shewanella* spp.)	Transcriptomics (nrfA)	Nitrite reductase	DNRA	–
–	Alphaproteobacteria (*Nitratireductor aquibiodomus*)	Transcriptomics (formamidase)	Formamidase	Ammonia utilization	–
–	Poribacteria, Gamma-, Delta-, Alphaproteobacteria, Nitropinae, Eury-and Thaumarchaeota	Transcriptomics (glnA)	Glutamine synthase	Ammonia assimilation	–
–	Poribacteria, Chloroflexi, Alphaproteobacteria, NC10	Transcriptomics (gltB, GLT1, GLU)	Glutamate synthase	–	–
Xestospongia testudinaria	Alphaproteobacteria (*Ruegeria* sp., *Rhizobium* sp., *Magnetococcus* sp. *Rhodopseudomonas* sp., Meso-rhizobium, *Pseudomonas* sp.)	PCR/RT-PCR (ureC)	Urea utilization	Ammonification	Su et al. (2013)
–	Thaumarchaeota Cenarchaeum symbiosum A	PCR (ureC)	–	–	–
–	Cyanobacteria (*Synechococcus* sp., *Prochlorococcus* sp.)	–	–	–	–
–	Gammaproteobacteria (*Pseudomonas* sp.)	PCR/RT-PCR (ureC)	–	–	–
Xestospongia testudinaria	FCB group (*Hallella* sp.)	PCR/RT-PCR (ureC)	Urea utilization	Ammonification	Su et al (2013)
–	Terrabacteria (*Rhodococcus* sp.)	–	–	–	–
–	Alphaproteobacteria	PCR/qPCR (nosZ)	Nitrous oxide reduction	Denitrification	Zhang et al. (2013)

*It is indicated the phylogenetic affiliation of microbes if known; if unknown, it is indicated by "?." Indicated are also the method used for detection, the function in the pathway as well as the respective bibliographic references.

TABLE 4.8 Summary of Microbes Possibly Involved in P Transformations Within Sponge Hosts, Ordered by Studies[#]

Sponge Species	Microbes	Method	Function	References
Aplysina aerophoba	Alphaproteobacteria *(Rhodospirillaceae)*	Metagenomics (*ppk1-2*)	Poly-P synthesis	Karimi et al. (2018)
		Metagenomics (*ppx*)	Poly-P hydrolysis	
	Nitrospinae (bin107); Nitrospirae (bin75); Alphaproteobacteria (bins36, 52, 56, 65, 95, 129, 131); Gemmatimonadetes (bins43, 94); Gammaproteobacteria (bins40, 55); Deltaproteobacteria (bin18); Acidobacteria (bins61, 110, 134); Bacteriodetes (bin25, 80); Chloroflexi (bins5, 20, 34); Actinobacteria (bin76); Deinococcus-Thermus (bin119); SRB1093 (bin63); Cyanobacteria (bin9)	Metagenomics (*ppk*)	Poly-P synthesis	Slaby et al. (2017)
	Alphaproteobacteria (bins131, 36)	Metagenomics (*ppx1*)	Poly-P hydrolysis	
	All expect: bin98 (Alphaproteobacteria); bin40 (Gammaproteobacteria); bins25, 40 (Bacteriodetes)	Metagenomics (*pstSCAB*)	Phosphate ABC transporter	
	Nitrospinae (bin107), Alphaproteo-bacteria (bins36, 52, 56, 65, 95, 129, 131), Chloroflexi-SAR202 (bins16, 22, 87, 90), Deinicoccus-Thermus (bin119), Cynanobacteria (bin9)	metagenomics (*phnDCE*) *C* and *E* not found	phosphonate binding protein of the ABC-transporter	
	Nitrospinae (bin107); Alphaproteobacteria (bin36, 65, 131); Actinobacteria (bin76); Deinococcus-Thermus (bin119)	Metagenomics (*phnF*)	Transcriptional regulators	
	Alphaproteobacteria (bins36, 52, 95)	Metagenomics (*phnGHIJKL*)	C-P lyase	
	Alphaproteobacteria (bin36); Gemmatimonadetes (bins43, 94); Acidobacteria (bin110); Gammaproteobacteria (bin55); SRB1093 (bin63)	Metagenomics (*phoD*)	Phosphodiesterase/ alkaline phospha-tase D	

TABLE 4.8 *(Continued)*

Sponge Species	Microbes	Method	Function	References
	Nitrospinae (bin107); Alphaprcteobacteria (bin36, 52, 56, 65, 95, 129); Gemmatimondetes (bin34, 94); Deltaproteobacteria (bin18); Bacteriodetes (bin80); Chloroflexi (bin22, 87, 125); Actinobacteria (bin134); Ca. Poribacteria (bin44, 70); Gammaproteobacteria (bin55); Nitrospirae (bin75); Cyanobacteria (bin9)	Metagenomics (*phoX*)	Predicted phosphatase	
	Acidobacteria (bins61, 110); Ca. Poribacteria (bin70); Bacteriodetes (bin25)	Metagenomics (*phoA*)	alkaline phosphatase A	
Haliclona sp.	Cyanobacteria (*Synechococcales*)	PCR (*ppk*)	Poly-P synthesis	Ou et al. (2019)
Cladocroce sp.	Cyanobacteria (*Synechococcales*) Gammaproteobacteria (*Oceanospirillales*)			
Ircinia sp.	Alphaproteobacteria; Betaproteobacteria (*Nitrosomonadales*); Gammaproteobacteria (*Acidiferrobacterales*)			
Mycale sp.	Cyanobacteria (*Synechococcales*); Gammaproteobacteria (*Enterobacterales*, OMG group, *Xantomonadales*, *Thiotrichales*); Actinobacteria (*Streptomycetales*, *Frankiales*); Acidobacteria			
Tedania sp.	Cyanobacteria (*Synechococcales*); Acidobacteria; Gammaproteobacteria (*Enterobacterales*, *Xantomonadales*)			
Sigmaxinella sp.	Cyanobacteria (Synechococcales)			
Lissodendoryx sp.	Cyanobacteria (Synechococcales)			

TABLE 4.8 *(Continued)*

Sponge Species	Microbes	Method	Function	References
Vazella pourtalesii	Alphaproteobacteria bins 58, 65, 70, 71; Actinobacteriota bins 64, 80; Firmicutes (bins44, 55); Bacteroidota (bins3, 54, 60, 92); Spirochaetota (bin56); Planctomycetota (bins32, 52, 91, 113, 121, 134); Verrucomicrobiota (bins85, 112)	Metagenomics (*ppk1*)	Poly-P synthesis	Bayer et al. (2020); Maldonado et al. (2021b)
	Myxococcota (bin59); Thermoplasmatota (bins11, 15, 133); Gammaproteobacteria (bin40); Alphaproteobacteria (bin124); Actinobacteriota (bin64); Firmicutes (bin55); Planctomycetota (bins 52, 94, 119, 121); Verrucomicrobiota (bins23, 112)	Metagenomics (*ppk2*)	Poly-P synthesis	
	Alphaproteobacteria (bins19, 21, 27, 42, 50, 58, 65, 109, 124, 135, 141); Gammaproteobacteria (bins40, 51, 139); Planctomycetota (bins121, 134); Bacteroidota (bins28, 54, 60, 87, 138); Verrucomicrobiota (bin112); Acidobacteriota (bins72, 86); Actinobacteriota (bin64)	Metagenomics (*ppx*)	Poly-P hydrolysis	
	Alphaproteobacteria (bins19, 21, 42, 50, 65, 96, 104, 109, 114, 118, 124); Gammaproteobacteria (bins4, 20, 33, 139); Actinobacteriota (bin64); Firmicutes (bins55, 77, 127); Planctomycetota (bins52, 67, 121); Acidobacteriota (bin86); Crenarchaeota (bin90); Myxococcota (bin22); Chloroflexota (bins53, 100); SAR324 (bin126)	Metagenomics (*pstSCAB*)	High-affinity phosphate-specific ABC transport system.	
	Myxococcota (bin22); Crenarchaeota (bin 67, 143); Thermoplasmatota (bins15, 83, 107, 117, 130); Verrucomicrobiota (bin112); Actinobacteriota (bins17, 111); Bacteroidota (bin49); Alphaproteobacteria (124); Chloroflexota (bin100)	Metagenomics (*phnDCE*)	ABC transporter	
	Alphaproteobacteria (bin124)	Metagenomics (*phnGHIJKLM*)	C-P lyase	

TABLE 4.8 *(Continued)*

Sponge Species	Microbes	Method	Function	References
Vazella pourtalesii	Planctomycetota (bins52, 94, 116, 119); Crenarchaeota (bins74, 90, 143, 123, 131); Alphaproteobacteria (bins75, 124, 135, 141); Gammaproteobacteria (bins10, 9, 35, 63, 89); Chloroflexota (bins47, 53); Myxococcota (bin59); Firmicutes (bin44); Thermoplasmatota (bins15, 83); Acidobacteriota (bin86); Verrucomicrobiota (bins23, 84, 112); Marinisomatota (bins37, 82); Bacteroidota (bin24)	Metagenomics (*phnW*)	2AEP catabolism	Bayer et al. (2020); Maldonado et al. (2021b)
	Planctomycetota (bins52, 119)	Metagenomics (*phnX*)		
	Planctomycetota (bins75, 121); Myxococcota (bin59); Alphaproteobacteria (bins124, 135), Planctomycetota (bin75); Gammaproteobacteria (bin 63); Verrucomicrobiota (bins 23, 85, 112)	Metagenomics (*phnA*)	Phosphonoacetate hydrolase	
Ircinia strobilina	Cyanobacteria; Alphaproteobacteria; Gammaproteobacteria; Deltaproteobacteria	RT-PCR (*ppk*)	Poly-P synthesis	Zhang et al. (2015)
		PCR (*ppk*)	Poly-P synthesis	
Xestospongia muta		PCR/RT-PCR (*ppk*)	Poly-P synthesis	

[#]Indicated are the phylogenetic affiliation of microbes if known; if unknown, it is indicated by "?." Indicated are also the method used for detection, the function in the pathway as well as the respective bibliographic references.

4.4.1 DETERMINATION OF NUTRIENT FLUX RATES: METHODS AND ASSUMPTIONS

For quantitatively summarizing net flux rates, we have considered 33 studies, involving 92 sponge species (Table 4.2, Supplementary Table 4.1). About 40% of those studies were field, in-situ approaches, and about 60% were laboratory research. Studies are concentrated in the Atlantic-Mediterranean biogeographical region (Figure 4.4), initially focused on accessible shallow-water sponges. There is a prevalence of species from the Caribbean and the

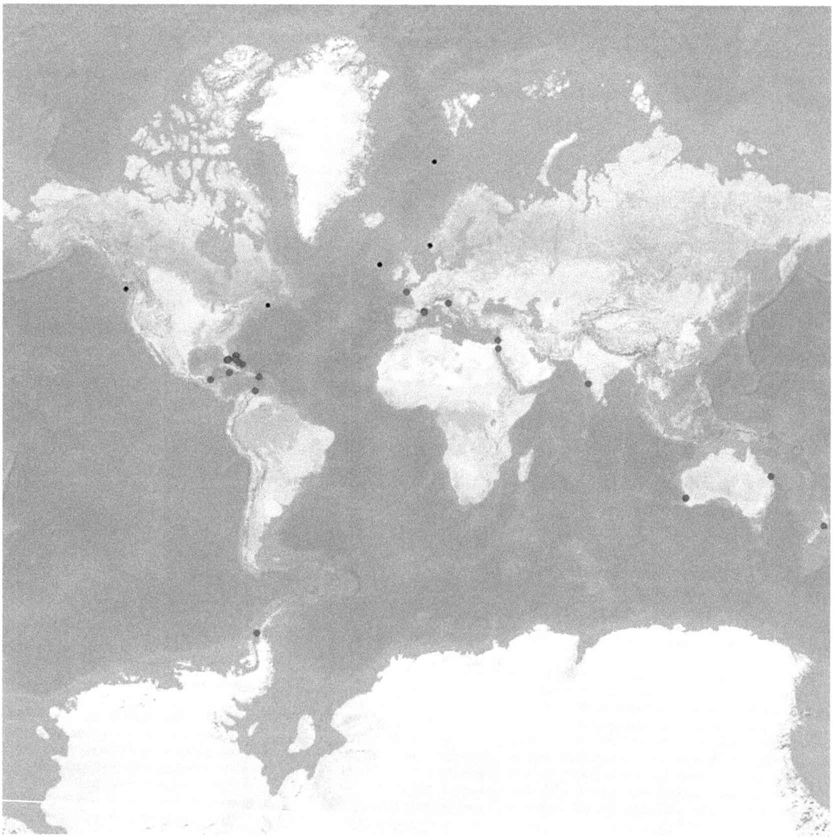

FIGURE 4.4 Summary of the geographical location of the 33 studies in which flux rates of inorganic nitrogen and phosphorus have been investigated for a total of 92 sponge species (see Table 4.2). Gray dots represent study sites (n = 36) for shallow-water sponges on continental shelves and black dots represent deep-sea sites (n = 11) for deep-water sponges. Note the important geographical and habitat bias in knowledge, with most studies investigating Atlantic-Mediterranean sponges, with a preference for Caribbean, shallow-water habitats. See also Supplementary Table 4.1.

Mediterranean Sea, but also species from the Great Barrier Reef. Thus, the bulk of available information bears a significant geographical and habitat bias. Yet, a small but growing number of studies are available for other areas, also for deep-sea species (Figure 4.4 and Supplementary Table 4.1). There is also bias regarding the nutrient type, with DIN flux rates having been quantified in some 92 sponge species to date, but phosphate flux rates being available for only 8 species (Supplementary Table 4.1). Finally, there is a bias regarding the sponge lineages studied, with, for instance, 83 out of the 92 species in which DIN has been quantified belonging to the class Demospongiae, but only 5 species in the Class Hexactinellida, and 2 species in each of the classes Homoscleromorpha and Calcarea (Table 4.2, Supplementary Table 4.1). Therefore, there is not enough information to carry out relevant between-class comparisons.

The N and P flux rates of sponges have been measured mainly through two different methods (Figure 4.5): (i) incubation (e.g., Jiménez & Ribes, 2007; de Goeij et al., 2008; López-Acosta et al., 2019; Maldonado et al., 2021b; Bart et al., 2021); and (ii) simultaneous sampling of the inhalant and exhalant water flows, an approach hereafter referred to as the "in-ex" method (e.g., Yahel et al., 2005, 2007; Pile & Young, 2006; Morganti et al., 2016). Both methods have pros and cons and provide only approximate estimates of the true nutrient flux rates. The difficulties identified in the in-ex method are several. First, the total volume of filtered water per unit of time and osculum during each sampling has to be known for calculating the flux rate. To measure such a volume with accuracy during a given time period is difficult because most sponges have several to many oscula and not all have similar levels of activity at the same time. Even for those sponges having a single osculum, accurately measuring the flow volume requires integration of measured values through the osculum diameter using equations developed to characterize flows at the outlet of flumes, because flow velocity is typically maximum at the center of the osculum and it lowers towards the periphery, depending on osculum morphology. The use of fluorescein dye *in situ* (e.g., Yahel et al., 2005; Leys et al., 2011; Morganti et al., 2019) and small velocimeters or thermistor flowmeters in the laboratory (Strehlow et al., 2016; Leys et al., 2018) are two approaches that may help but both bear some considerations. To determine during scuba dives the exact distance traveled in a few seconds by a diffusing cloud of dye may be tricky, even if the trajectory is recorded on video. Attention is to be paid to space distortions and loss of parallelism between the flow trajectory and the main axis of the spatial reference frame used. Use of velocimeters often means work has to be done in the laboratory, to place their sensor at a particular distance (5 cm) from the osculum, and ensure that suspended particles (e.g., microplastics from the manufacturer) are in the seawater for the particle counters to detect flow velocity in the aquarium. Because of these difficulties in determining the water volume pumped by each sponge during the in-ex sampling period, estimates of flux rates are expected to have variable repeatability across different studies of the same species. Thus, high replication is advised to produce average values reliable for extrapolation. Second, the in-ex method records empirical values for a period of only minutes, which has to be subsequently extrapolated to the daily physiology cycle. Often, the measured values are linearly extrapolated over 24 h periods, although there is evidence for many sponges that pumping is neither a continuous activity nor with unvarying intensity over time. Some species may maintain rather a constant pumping activity, but a transient cessation of pumping also occurs randomly through populations (Reiswig, 1971b; Bell et al., 1999; McMurray et al., 2014; Strehlow et al., 2016; Morganti et al., 2019). Therefore, it is advised to monitor a large enough number of individuals per species to reliably understand variability patterns prior to numerical extrapolations. Third, the in-ex approach assumes that the exchange of nutrients between the internal epithelia (i.e., choanocytes and pinacocytes) and the seawater reflects the metabolic activity of the entire sponge. This assumption disregards the exchange that may take place through the external epithelium (exopinacocytes). Although internal epithelia collectively represent a much greater exchange surface area than that of the external epithelium, in the case of sponges with phototrophic symbionts accumulated in peripheral, well-lighted mesohyl zones below the external epithelium, the in-ex approach may neglect a substantial component of the nutrient exchange taking place through the sponge skin (i.e., the ectoderm). Fourth, if during a single round of seawater circulation through the sponge, the rate of uptake or release for a given nutrient is low, the in-ex method may not detect it, leading to a false negative.

Incubations in chambers do not suffer from the considerations commented for the in-ex approach, but have others of similar magnitude. For incubation, sponges are enclosed within a chamber of known volume for a given period of time (typically hours rather than minutes), seawater being sampled from the chamber at the beginning and the end of incubation —sometimes also at intermediate times (Figure 4.5(C) and (F))—to determine the change in nutrient concentration with time. The sealing of the chamber causes oxygen concentration to decrease over time. Although a variety of experimental incubations indicate that the survival of most sponges is not compromised

FIGURE 4.5 Illustration of devices used for determining nutrient flux rates in sponges. (A) In-Ex device developed by Jasper de Goeij (the University of Amsterdam, The Netherlands) according to the VacuSip design (Morganti et al., 2016) to determine nutrient fluxes by scuba diving. The coral reef demosponge *Ircinia campana* is pictured while being sampled through capillary tips placed adjacent to the inhalant (in) and exhalant (ex) surfaces of the sponge at Curaçao reefs (Courtesy of Benjamin Mueller). (B) In-Ex device developed by Sally Leys (University of Alberta, Canada) and Gitai Yahel (Ruppin Academic Group, Israel) to determine nutrient fluxes in deep-water sponges using ROVs, pictured at British Columbia fjord wall

(Continued)

(Northeastern Pacific) during sampling of *Aphrocallistes vastus* through capillary tips placed adjacent to the inhalant (in) surface and in the exhalant (ex) opening of the sponge (Courtesy of Sally Leys). (C) Incubation chamber developed by Laurent Chauvaud (IUEM/CNRS, France) to be operated by scuba diving, pictured at the Bay of Brest (France, Eastern North Atlantic) during incubation of *Tethya citrina* by María López-Acosta (Courtesy of Thierry Le Bec). The device includes an incubation chamber (c), through which a water flow is generated by a pump fed by a battery (b) and can also incorporate a number of intercalated sampling units (f) if required. (D) The deployment of incubation chambers in the deep sea is a logistically complex process that requires manned submersibles or, as illustrated in the picture, high-class ROVs. Here ROPOS ROV is pictured ready to start a 180 m-deep dive aiming for the deployment of 2 manually operated incubation chambers (m) and a larger automatized chamber (a). Further chamber descriptions are given in pictures E and F. (E) Incubation chamber developed by Manuel Maldonado (CEAB-CSIC, Spain), pictured at Emerald Basin (Nova Scotian Shelf, Canada, Western North Atlantic) while incubating a large individual of the hexactinellid *Vazella pourtalesii* in complete isolation from the external environment. The chamber incorporates externally two sampling steel bottles (100 mL, Teflon folded inside, 2,000 m depth-rated). Sampling bottles are activated by the arms of ROVs or submersibles at the initiation (0) and end (1) of the incubation period for their respective capillaries collecting water from inside the chamber. (F) Incubation chamber developed by Furu Mienis (NIOZ, The Netherlands) and Jasper de Goeij (the University of Amsterdam, The Netherlands), pictured during deployment at the *Vazella* grounds of Emerald Basin (Nova Scotian Shelf, Canada, Western North Atlantic) to incubate the soft-bottom, benthic community (Courtesy of Jasper de Goeij). The chamber incorporates a battery connected to a programmable hardware unit (b) that activates a stirrer and rotatory sampling unit with six 60 mL syringes (s) that can be pre-programmed at different stirring speeds and sample intervals. Additionally, there is a pre-programmable port to inject a 30-mL sample (e.g., tracer injection) and a magnetic door (d) to ventilate the chamber, if required. Oxygen and temperature are continuously logged and the chamber is depth-rated at 6,000 m.

by maintaining them in decreasing oxygen conditions for a few hours to a day (Frøhlich & Barthel, 1997; Maldonado et al., 2011, 2020; López-Acosta et al., 2019; Micaroni et al., 2021), the oxygen decrease may favor a better performance of the anaerobic members of the sponge microbiome versus the aerobic ones, altering natural uptake and release rates of the P and N nutrients in which the various microbial populations are involved. In the closed environment of the incubation chamber, the pools of organic and inorganic compounds, as well as food, decrease over time and it might affect the sponge metabolism in a variety of ways, depending on the species and the duration of the incubations. The incubation process can also distort natural patterns of pumping (Yahel et al., 2005; Hadas et al., 2008), among other things, because many sponges might rely on horizontal flows passing over their oscula to enhance passively pumping as a result of the Bernoulli's principle (Vogel & Bretz, 1972; Vogel, 1977; Maldonado & Young, 1998). This deficiency is theoretically overcome in incubation chambers incorporating a battery-fed pump (Figure 4.5(C)), which generates a water flow over the oscula of the incubated sponges (e.g., Ribes et al., 2000; López-Acosta et al., 2019).

In some cases, sponges have been cut off into smaller pieces, and the resulting fragments are incubated in the laboratory to determine the uptake or release of given nutrients. Such an approach may be acceptable to reveal intermediate nutrient conversions that occur internally in the sponge tissues and that may never be reflected in the net global flux; it can also help to identify the microbial lineages responsible for such conversions (Radax et al., 2012a; Subina et al., 2018; Rooks et al., 2020). However, since pumping capacity is drastically disrupted in the dissected body fragments, and the homeostasis and the natural aerobic/anaerobic conditions of those tissues are also altered, the results of such fragment incubations could be unreliable when used to derive net nutrients fluxes that are extrapolated to the community or population level.

A common problem among the studies quantifying net nutrient fluxes in sponges is that rates are often expressed in a variety of non-comparable units. The results of in-ex approaches were initially presented as instantaneous changes in concentrations between the inhalant and exhalant flows rather than real flux rates (Supplementary Table 4.1), due to the above-commented difficulties to determine accurately the volume of water pumped by the sponges during the in-ex sampling. When flux rates have been measured, they appear reported in the literature under a variety of non-comparable reference units, expressed per sponge surface area, sponge wet weight, dry weight, sponge volume, or sponge ash-free dry weight. Frustratingly, interconversion factors for this disparity of parameters are rarely provided.

Sponge surface is a parameter hard to be quantified with accuracy even for encrusting sponges, whenever they have conulose or wrinkled surfaces, making between-species comparisons difficult. Furthermore, since the surface area of the internal aquiferous canal system can be orders of magnitude larger than the body surface, to refer fluxes to the latter one may therefore introduce an enormous artefactual bias in between-species comparisons, particularly between encrusting and massive morphotypes. To refer rates to wet weight is not advisable, even if associated conversions to dry weight or volume are provided. The reason is that the porosity of the sponge body may differ drastically from one species to another and variable amounts of seawater will be retained in the canals, making between-species comparisons based on wet weight problematic. Flux rates expressed per dry weight unit or volume unit are more accurate than wet weight, and reliable factors for conversion are easily obtainable and, therefore, often provided in the studies. Expressing rates by sponge volume facilitates subsequent extrapolation across sponge populations using merely image approaches and avoiding additional destructive samplings. Sponge volume is typically estimated by the water displacement method, that is, by collecting and measuring the volume of the seawater that overflows from a completely filled container after inserting the sponge. The main concerns are that highly porous species with large internal spaces (e.g., *Farrea* spp.) will release an important part of their internal seawater content during the transference from the storage container to the measuring container, while species with a more compact "endosome" (e.g., *Geodia* spp., *Aplysina* spp.) will not. Consequently, unless further correcting actions are implemented, the volume measured in highly porous species will mostly represent the volume of the sponge tissues excluding the internal voids, since the internal water drips out of the sponge body during the transfer from the maintenance container to the measuring container. In contrast, the approach will rather reflect global body volume (internal voids included) in sponges with low body porosity. In all cases, rates expressed by either volume or dry weight are not taking into account which proportion of the sponge body consists of active biological tissues and which is an inert mineral skeleton. Therefore, in sponges with a heavy mineral skeleton, both approaches will systematically underestimate the real metabolic effort carried out by the holobiont to generate the measured net flux rate. Likely, the most accurate way to express physiological rates to facilitate subsequent between-species comparisons (also between-phyla comparisons) is to refer the flux to as ash-free dry weight, which will consider only the amount of sponge organic components (i.e., ash-free dry weight) involved in the physiological activity, eliminating the bias introduced by the mineral skeleton (i.e., ash dry weight). Unfortunately, among the many studies on sponge flux rates compiled from the literature (Supplementary Table 4.1), few refer to rates as ash-free dry weight or provide conversion factors from volume or dry weight units. In some laboratories and field stations, suitable furnaces for combustion to ash-free dry weight may not even be available. Thus, retention of voucher pieces for further combustion and production of interconversion coefficients is encouraged as a routine procedure.

4.4.2 AMMONIUM

Net release of ammonium (negative flux) has been detected in a majority of sponges, irrespective of being studied in the field or the laboratory (Table 4.3 and Figure 4.6(A)). The LMA/HMA condition of the sponges does not appear to be related to the net flux being released or consumed. In the compiled literature, we have identified a total of 31 LMA species releasing ammonium, 5 taking it up, and 2 in which both the release and consumption of ammonium have been reported (Supplementary Table 4.1). Similarly, 21 HMA species released ammonium, and 17 incorporated it, with some species releasing or consuming ammonium depending on the environmental conditions during field or laboratory work (Bayer et al., 2008; Hoffmann et al., 2009; Ribes et al., 2012; Morganti et al., 2017; Leys et al., 2018; Hoer et al., 2018; Valentine & Butler, 2019). Among the releasers, the average rate (hereafter summarizing only studies in the two or three more frequent types of the unit; Table 4.3 and Figure 4.6(A)) of the HMA sponge was –0.34 ± 0.35 μmol g^{-1} sponge DW h^{-1} (n=13) or –0.03 ± 0.05 μmol mL^{-1} sponge h^{-1} (n=15). In LMA sponges, the average release rate was –0.77 ± 0.99 μmol g^{-1} sponge DW h^{-1} (n=11), –0.60 ± 0.43 μmol g^{-1} sponge AFDW h^{-1} (n=5), or –0.11 ± 0.10 μmol mL^{-1} sponge h^{-1} (n=13). Among the ammonium HMA consumers, the average rate was 0.17 ± 0.59 μmol g^{-1} sponge DW h^{-1} (n=6) or 0.01 ± 0.05 μmol mL^{-1} sponge h^{-1} (n=5). In LMA sponges, ammonium consumption rates have been quantified only in two Caribbean sponges (Table 4.3), *Niphates erecta* (1.50 ± 0.30 μmol g^{-1} sponge DW h^{-1}; Southwell et al., 2008) and *Cinachyrella* sp. (0.005, SD not available, SE = 0.002, μmol mL^{-1} sponge h^{-1}; Hoer et al., 2018). Note that flux rates of ammonium (typically expressed in micromol) are about three orders of magnitude more intense than those of nitrite and nitrate (typically given in nanomol; Figure 4.6).

4.4.3 NITRITE

Many studies have reported combined "nitrite + nitrate" (NO_x^-) fluxes, with sponges showing predominantly release (Supplementary Table 4.1). In the few cases quantifying nitrite separately, most species were also sources, irrespective of their LMA/HMA condition (Table 4.4 and Figure 4.6(B)). In HMA sponges, the average rate was –114.1 ± 272.5 nmol g^{-1} sponge DW h^{-1} (n=4) or –4.3 ± 12.5 nmol mL^{-1} sponge h^{-1} (n=6). In LMA sponges, the average rate was –5.6 ± 4.5 nmol mL^{-1} sponge h^{-1} (n=6). Pooling all species together, the global average rate of nitrite release was –114.1 ± 272.5 nmol g^{-1} sponge DW h^{-1} (n=4) or –4.9 ± 8.3 nmol mL^{-1} sponge h^{-1} (n=12). Net nitrite consumption has been detected in the species *Phakellia ventilabrum* and *Geodia barretti* (Supplementary Table 4.1), at the respective rates of 0.8 ± 0.3 nmol mL^{-1} sponge h^{-1} and 3.3 (±SD not available). However, in the former study, some specimens of *P. ventilabrum* showed nitrite release at

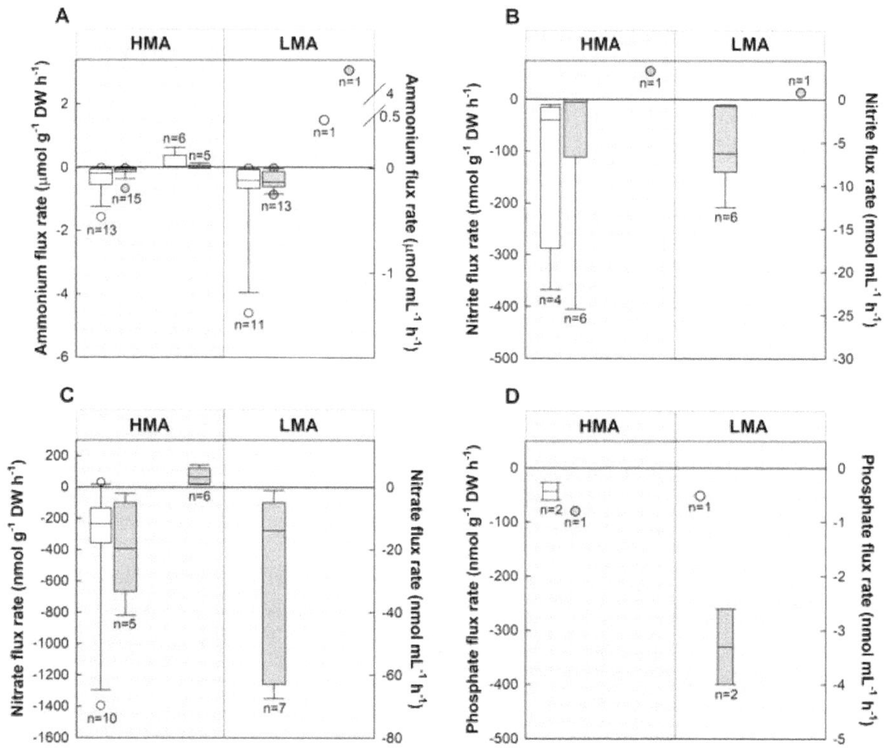

FIGURE 4.6 Visual comparison between HMA sponge species (cream background) and LMA species (blue background) for the flux rate of ammonium (data in Table 4.3), nitrite (data in Table 4.4), nitrate (data in Table 4.5), and phosphate (data in Table 4.6). Data are represented as either boxplot—when the number of sponge species (n) considered in the calculation of the statistic is >1—or as isolated data points—when n = 1. For each nutrient, rates are shown in the two most common types of comparable units (i.e., nmol g^{-1} DW h^{-1} = white boxplots; nmol mL^{-1} of sponge h^{-1} = grey boxplots). Positive flux values indicate nutrient incorporation by the sponges and negative values indicate nutrient release. Note that rates of ammonium (in micromole) are three orders of magnitude more intense than those of the rest of the nutrients (in nanomole).

rates ranging from -0.6 ± 1.3 to -6.7 ± 2.9 nmol mL^{-1} sponge h^{-1}, indicating important between-individual variability in the net nitrite flux.

4.4.4 *NITRATE*

In the cases in which nitrate has been measured separately from nitrite, sponges were predominantly nitrate releasers (Table 4.5 and Figure 4.6(C)). In HMA sponges, the average nitrate release rate was -330.2 ± 746.0 nmol

g^{-1} sponge DW h^{-1} (n=10) or –19.3 ± 13.6 nmol mL^{-1} sponge h^{-1} (n=5). In LMA sponges, the average rate of nitrate release was –25.7 ± 29.9 nmol mL^{-1} sponge h^{-1} (n=7). Nitrate consumption has been reported in only 6 HMA boreal species (Rooks et al., 2020), but the rates, resulting from laboratory incubations of sponge fragments rather than complete, living sponges, are debatable. Estimated nitrate consumption under those unnatural conditions averaged 3.6 ± 2.8 nmol mL^{-1} sponge h^{-1} (Figure 4.6(C)).

4.4.5 PHOSPHATE

Several in-ex studies have dealt with phosphate release/uptake but quantifying it as instant changes in P concentration that could not be translated into net flux rates. To date, only 5 studies have quantified rates of phosphate for a total of 8 species (Table 4.6 and Figure 4.6(D); Supplementary Table 4.1). Release of phosphate has been reported for nearly all sponge species examined in this regard, with an average release rate of –45.4 ± 19.7 nmol g^{-1} sponge DW h^{-1} (n=3), –7.0 ± 0.4 nmol g^{-1} sponge AFDW h^{-1} (n=2), or –2.5 ± 3.2 nmol mL^{-1} sponge h^{-1} (n=3). In the current stage of knowledge, differences between HMA and LMA species are not recognizable.

4.5 NITROGEN PATHWAYS IN SPONGES

4.5.1 SPONGE CELLS AND N COMPOUNDS

Sponge cells by themselves play a role in the production of inorganic and organic N forms. Sponges are animals in which specific organs to deal with the excretion of N products do not occur (or have not been identified yet). Thus, it is assumed that the cellular catabolism of proteins generates ammonia (NH_3) as waste, which is excreted from the cytoplasm of cells into the intercellular medium (the mesohyl) and/or released to the outgoing water flow. Ammonia is a toxic molecule, but it is very small, gets easily ionized as ammonium (NH_4^+), and readily diffuses downhill along a gradient across cell membranes. Another potential excretory N product of sponges might be urea. However, urea is the typical excretion product of higher animals (e.g., vertebrates), which combine ammonia and CO_2 in the liver to produce urea molecules. Unlike ammonia, urea has the advantage of not being toxic and its production recycles another metabolic "waste" (i.e., CO_2). Comparative physiological disadvantages are that urea is a relatively large molecule that does not go easily through cell membranes and specific excretion organs are needed for its elimination, such as those evolved in vertebrates and missing in sponges. Urease (*ureA*, *B*, *C* genes) has been identified in a variety of sponge symbionts (*see* Section 4.5.2.7). It might be that symbiotic microbes could help sponges to break down their urea if sponges produce it, an aspect that remains undetermined. Nevertheless, the most likely pathway is that the urea used for ammonification by the sponge microbiome is of external origin, incorporated from seawater, which includes the urea excreted by fish dwelling into the sponges. Interestingly, a deeper knowledge of the "N-side" of this triple organismal interaction among sponge host, the sponge microbiome, and sponge-dwelling fish, may help to better understand the functional connections among all 3 partners and explain why the association between sponges and fish is widespread (Tyler & Böhlke, 1972; Randall & Lobel, 2009; Pérez-Botello & Simões, 2021). To date, the net flux rate of urea uptake or release has not been quantified for any sponge. The magnitude and sign of such a flux would help to assess the ecological significance of urea processing by sponges, as well as to better understand the level of metabolic integration between the sponge cells and the urea-degrading microbes.

4.5.2 THE SPONGE MICROBIOME AND N CYCLING

Microbial functions in the sponge N-cycle are often detected by amplifying the genes encoding key enzymes (e.g., *amoA*, *nifH*) in given biochemical pathways, which is interpreted as evidence for one enzyme reaction that assumedly takes place in member/s of the sponge microbiome. Nonetheless, such studies might be somehow biased through the exclusion of (yet unknown) genomic information that may result from primer mismatches. Metagenomic studies are possibly more complete and less biased if no restriction through, for instance, primer usage occurs, whereas high GC-content DNA from, for instance, *Actinobacteria* might be difficult to sequence. A first milestone in metagenomics analysis was established by demonstrating that genomic information accounting for a closed nitrogen circuit was available in the shallow-water demosponge *Theonella swinhoei* and the deep-water demosponge *Neamphius huxleyi*, although assignment of the respective biochemical pathways to members of the microbiomes was not accomplished (Li et al., 2016). Several studies on the demosponge *Xestospongia muta* (Bayer et al., 2008; López-Legentil et al., 2010; Fiore et al., 2013, 2015) have also helped to shed light on the N cycling and the involved microbes (Table 4.7). Nitrogen utilization has also been investigated in depth in the demosponge *Aplysina aerophoba*, with 37 metagenome-assembled genomes (Slaby et al., 2017). More recently, pyrosequencing of the 16S rRNA genes in the Chinese population of the coral-reef sponge *Spheciospongia* cf. *vesparium* has revealed a microbiome dominated by Proteobacteria (*Shewanella* spp.) and Bacteroidetes (e.g., *Bizionia* spp.) containing up to 10 nitrogen-related genes which would account for a nearly complete N cycle (He et al., 2021). Unfortunately, the net N inputs and outputs of the sponge and how they could relate to the internal N transformations remained unaddressed. The puzzling absence of archaea in the microbiome of *S.* cf. *vesparium* also suggests that the real microbial diversity in this sponge could be underestimated due to primer bias. Recently, the net N fluxes, along with the putative microbial lineages involved and the biochemical pathways that would account for a closed N circuit have been reported for the hexactinellid *Vazella pourtalesii* (Maldonado et al., 2021b). This latter study was used as a base for the genomic searches of the 137 previous metagenome-assembled genomes (Bayer et al., 2020). As part of this review, we will also re-analyze the recovered metagenome bins from the previous studies on *A. aerophoba* (Slaby et al., 2017) and *V. pourtalesii* (Bayer et al., 2020).

4.5.2.1 NITRIFICATION

Within the N metabolism of sponges, ammonia oxidation is perhaps the best-studied pathway, with ammonia-oxidizing archaea (AOA; Thaumarchaeota) and bacteria (AOB; Beta- and Gammaproteobacteria) detected in a large variety of sponges (Table 4.7). AOA and AOB are commonly detected by amplifying the gene encoding the key enzyme *amoA* (Figure 4.1). In most studies, AOB associated with sponges belong to beta-proteobacterial *Nitrosospira* species (e.g., Bayer et al., 2008; Meyer & Kuever, 2008; Mohamed et al., 2010; Maldonado et al., 2012; He et al., 2021). Diaz et al. (2004) amplified *amoA* genes in the sponge *Lissodendoryx isodictyalis*, most likely belonging to the beta-proteobacterial *Nitrosomonas* genus. Gammaproteobacterial *amoA* genes were detected in *Chondrosia reniformis* (Ribes et al., 2012), and transcripts were found in *Xestospongia muta* (Fiore et al., 2015). Bacterial and/or archaeal *amoA* genes were also recovered from the microbiome of six other sponges, *Cymbastella coralliophila*, *Cymbastella concentrica*, *Rhopaloeides odorabile*, *Scopalina* sp., *Stylissa* sp., and *Tedania anhelans* (Fan et al., 2012). *Hao* genes and *amoA* genes (both bacterial and archaeal) have been detected as well, using a functional gene array in *Aplysina aerophoba*, *Dysidea avara*, *Xestospongia testudinaria*, and *Stylissa carteri* (Bayer et al., 2014). AOA (Thaumarcheota) and AOB (Betaproteobacteria) have been found in metagenome-assembled bins of *Vazella pourtalesii*, along with the gene *hao*. Nevertheless, the latter could not be assigned to a microbial symbiont (Maldonado et al., 2021b).

The emerging view is that ammonia-oxidizing Archaea —rather than bacteria – are the main drivers of nitrification in sponges. The first report of the presence of Archaea in sponges was given by Preston et al. (1996) in the genus *Axinella*. The isolation of the first AOA from a tropical marine aquarium, Ca. *Nitrosopumilus maritimus* (Könneke et al., 2005) represented a breakthrough in AOA characterization. It provided pioneering insight into the general physiology and specific metabolism of these organisms, demonstrating the ability of *N. maritimus* strain SCM1 to convert ammonia into nitrite under aerobic conditions. Soon after, the metabolic potential for ammonia oxidation was demonstrated for AOA of the sponge microbiome, through a fosmid library-based metagenomics analysis of the sponge-associated *Cenarchaeum symbiosium* genome (Hallam et al., 2006). Since then, the research on sponge AOAs has grown overwhelmingly and these microbes have been detected in numerous sponges, first using PCR approaches (Diaz et al., 2004; Bayer et al., 2007; Jiménez & Ribes, 2007; Mohamed et al., 2008; Steger et al., 2008; Hoffmann et al., 2009; Radax et al., 2012a; Zhang et al., 2014), then functional gene arrays (Bayer et al., 2014), and more recently, metagenomics (Fan et al., 2012; Fiore et al., 2015; Tian et al., 2017; Moitinho-Silva et al., 2017a; Zhang et al., 2019b; Moeller et al., 2019; Haber et al., 2021). In general, the mere detection of these microbes has led researchers to assume that they are driving part of the nitrogen cycle in the sponges. The recent analysis of 11 thaumarchaeal sponge symbiont genomes (including the new reconstruction of three) from different sponge species just confirmed the broad genomic potential for ammonia oxidation of Thaumarchaeota (Haber et al., 2021). Therefore, it may be that ammonia oxidation occurs even in more sponges than is thought since Thaumarchaeota are frequently detected in sponge microbiomes (Blanquer et al., 2013; Thomas et al., 2016; Turon & Uriz, 2020; Steinert et al., 2020; Busch et al., 2020a).

NOB is detected in sponges mainly by amplifying their 16S rRNA genes. The phylogenetic affiliation of sponge symbionts to the phylum Nitrospirae or proteobacterial genera such as *Nitrospira* has often led to the assumption that those microbes oxidize nitrite, completing the nitrification process in sponges. PCR- and amplicon-based detection of NOB is common in sponges and Nitrospirae typically emerge as members of sponge microbiomes (e.g., Schmitt et al., 2008; Hoffmann et al., 2009; Radax et al., 2012a; b; Blanquer et al., 2013; Tian et al., 2016; Thomas et al., 2016). The metagenome-assembled genome of a microbe associated with the sponge *Cymbastella concentrica* has confirmed the genomic potential of *Nitrospira* sp. (CCNi genome bin) for nitrite-oxidation (Moitinho-Silva et al., 2017a). In addition, recent studies have revealed that COMplete AMMonia OXidation (Commammox) can be carried out by some *Nitrospira* species (Daims et al., 2015; Fowler et al., 2018; Koch et al., 2019). Although this function has not yet been demonstrated in sponge microbiomes, the pathway is plausible since Nitrospirae microbes are common symbionts of sponges.

A general pattern appears to be that genes involved in the initial step of nitrification (oxidation of ammonium to nitrite) are found more often in the microbiome of sponges than genes responsible for the oxidation of nitrite to nitrate in the second nitrification step. On the one side, this might initially be interpreted as a feature leading to incomplete nitrification in many sponges (Webster & Taylor, 2012). However, the pattern might not be real but an artifact derived from unbalanced coverage of research efforts on these two microbial processes. In theory, incomplete nitrification should lead to the accumulation of nitrite within the sponge tissues for its subsequent release from the sponges (as detected in many sponges) and/or to be used as an N source to fuel other pathways of the sponge microbiome (Figure 4.7), as recently suggested for the hexactinellid sponge *Vazella pourtalesii* (Maldonado et al., 2021b). In any case, the causes behind the putative prevalence in sponge microbiomes of the first nitrification step remain unclear.

4.5.2.2 DENITRIFICATION

The functional genomic analysis of a variety of sponge microbiomes initially revealed ample representation of genes coding for enzymes involved in denitrification (Fan et al., 2012), including complete denitrification in six sponge species of diverse phylogenetic adscription (*Cymbastella coralliophila*, *Cymbastella concentrica*, *Rhopaloeides odorabile*, *Scopalina* sp., *Stylissa* sp., *Tedania anhelans*). Subsequently, the use of a functional gene array revealed genes (*narG*, *nirS*, and *nosZ*) coding for enzymes involved in denitrification in additional sponge species, such as *A. aerophoba*, *D. avara*, *X. testudinaria*, and *S. carteri* (Bayer et al., 2014). More recently, the entire set of genes necessary for a complete denitrification pathway has been

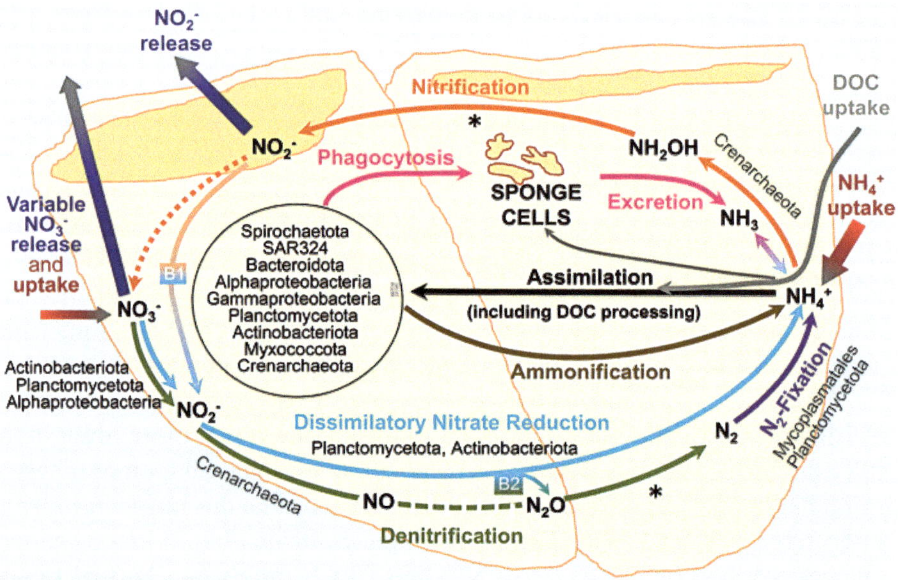

FIGURE 4.7 The hypothesis of the N cycle in the hexactinellid sponge *Vazella pourtalesii* indicates microbes possibly involved in the various biochemical pathways (adapted from Maldonado et al., 2021). Microbes able to participate in ammonification and assimilation are a large variety (*see* Table 4.6). For the other pathways only microbes that were enriched in the sponge compared to seawater are indicated. The dashed lines indicate enzymatic pathways for which genes were not detected. Asterisk (*) indicates genes that were present only in the unbinned dataset. "B1" and "B2" refer to putative bridges between major biochemical pathways, which provide alternative routes allowing an efficient closing of the N cycle in the holobiont. Uptake and release fluxes detected empirically in the sponges are also indicated. The holobiont N cycling appears to revolve around the incorporation of ammonium from the seawater and its subsequent internal recycling. This nutrient fosters, through N assimilation, the sponge microbiota, which also benefits from important amounts of DOC taken up from seawater by the sponge. It is proposed that a relevant part of the nourishment of *V. pourtalesii* derives from the phagocytosis of the microbiota farmed out of the N cycling, with farming occurring both intercellularly and within bacteriosyncytia.

identified after analyzing metagenomes in the sponges *Theonella swinhoei* and *Namphius huxleyi* (Li et al., 2016). Unfortunately, none of the three mentioned studies determined which members of the microbiomes were encoding the genes and displaying the functions.

It is suspected (Table 4.7) that Proteobacteria might be capable of denitrification in the sponges *Cinachyrella australiensis*, *Cinachyrella* sp., *Iotrochota* sp., *Spheciospongia vesparium*, and *X. testudinaria* (Zhang et al., 2013). The metagenomic assembly of alphaproteobacterial (family

Rhodospirillaceae) symbiont genomes from the sponges *A. aerophoba* and *Spongia officinalis* revealed *NarGHIJ* genes for nitrate reductases (Karimi et al., 2018). Our current re-analysis has revealed the occurrence of gene *narG* in six metagenome bins recovered from the sponge *A. aerophoba* and all belonging to the phylum Alphaproteobacteria (bin52, bin36, bin56, bin65, bin95, bin129). Also in the metagenome of *A. aerophoba*, De Mares et al. (2018) detected *narG* and *narH* genes to be highly homolog to that of *Nitrospira* sp. Using PCR and qPCR, Rooks et al. (2020) identified *nirS* and *nirK* genes in four *Geodia* spp. and in *Stelletta rhaphidiophora*, while only *nirK* was recovered from the sponge *Stryphnus fortis* (Table 4.7). *NirK* homolog to that of a *Pseudovibrio* sp. was detected in the *A. aerophoba* metagenome by De Mares et al. (2018). *NirS* genes detected in *G. barretti* were concluded to be a homolog of genes from Betaproteobacteria *Thauera mechernensis* and *Alcaligenes faecalis* (Hoffmann et al., 2009). In *Cymbastella concentrica*, the gene *nirK* was detected in the genome of CCThau bin, likely corresponding to Ca. *Nitrosopumilus cymbastellus* (Moitinho-Silva et al., 2017a). In the same sponge species, a bacterium belonging to the family *Phyllobacteriaceae* (Alphaproteobacteria) was detected to have *nar*-genes (Fan et al., 2012). In *A. aerophoba*, the gene *norB* (encoding nitric oxide reductase) was found by Slaby et al. (2017) in bin18 (Deltaproteobacteria), bin55 (Gammaproteobacteria), bin22, and bin34 (Chloroflexi), along with *nosZ* (N_2O reduction) in bin80 (Bacteriodetes). Using a combination of transcriptomic and PCR-based analysis, Fiore et al. (2015) suggested that Nitrospirae (*nar* genes), Alpha-, Gammaproteobacteria, and Thaumarchaeota (*nirK*), and Betaproteobacteria (*qnorB*) are possibly involved in denitrification in the sponge *X. muta* (Table 4.7). The gene *norQ* (encoding the nitric oxide reductase) was recovered from the metagenome of *A. aerophoba* and it is likely attributable to Ca. *Enthoteonella* sp. (De Mares et al., 2018). *NosZ* genes from Alphaproteobacteria were recovered by PCR and RT-PCR from the sponges *Spheciospongia vesparium*, *Cinachyrella australiensis* and *Iotrochota* sp., and gammaproteobacterial *nosZ* genes were recovered from the sponge *Cinachyrella* sp. (Zhang et al., 2013). Recently genes encoding for denitrification enzymes were found in the hexactinellid *V. pourtalesii* (Bayer et al., 2020; Maldonado et al., 2021b), suggesting that Proteobacteria, Nitrospirae, and Archaea may participate in the pathway (Table 4.7). Likewise, PCR-based analysis of denitrifying bacteria in the Chinese population of sponge *S.* cf. *vesparium* confirmed that at least some Proteobacteria carry genes for this pathway (He et al., 2021).

It is well established that there is no single prokaryote carrying the genomic potential for complete denitrification. Rather, completion of such a pathway must result from the coupled activity of various members of the sponge microbiome (Table 4.7), as proposed in several specific case studies (e.g., Hoffmann et al., 2009; Fiore et al., 2015; Maldonado et al., 2021b). Theoretically, incomplete denitrification should lead to the accumulation of nitric oxide (NO) (which at saturating concentrations will escape as gas from the body), unless this intermediate compound (which is also an important signaling molecule) is used to fuel pathways alternative to denitrification (Fan et al., 2012). The microbial transformation of NO to nitrogen and oxygen (Figure 4.1) has been suggested to be one of those alternative routes (Ettwig et al., 2010), which would also help to restore aerobic conditions and respiration in the sponge tissues (Webster & Thomas, 2016). Also, in the absence of NO reductase, denitrification could be completed, because nitrous oxide can potentially be supplied in anaerobic conditions from an intermediate stage of the dissimilatory nitrate reduction pathway (Figures 4.1 and 4.7), as suggested for *V. pourtalesii* (Maldonado et al., 2021b).

4.5.2.3 DISSIMILATORY NITRATE REDUCTION TO AMMONIUM (DNRA)

Genes encoding for enzymes involved in DNRA are found less often in sponge microbiomes than genes for nitrification and denitrification. However, such a pattern could reflect unbalanced research on the DNRA pathway, because its genes are frequently found in metagenomic studies. Following a metagenomics approach for six demosponge species (Fan et al., 2012), the gene *napA* (encoding for periplasmic nitrate reductase) was detected in abundance in only two (*Scopalina* sp. and *Tedania anhelans*), whereas the gene *narG* (encoding respiratory nitrate reductase) was more abundant in the four others (*Cymbastella concentrica*, *Cymbastella coralloides*, *Stylissa* sp., *Rhopaloeides odorrabile*). The genes *napA*, *napB*, and *napC* (encoding periplasmic nitrate reductase) and *nrfA* and *nrfD* (encoding respiratory nitrite reductase) have been found in the metagenome of the sponge *Hymeniacidon heliophila* (Weigel & Erwin, 2017). *NapA* and *nfrA* gene sequences are similar to those of *Shewanella* spp. were detected in the sponge *Spheciospongia vesparium* (He et al., 2021) as well as *nrfA* genes from Bacteriodetes bacteria (Table 4.7). GeoChip-based analysis (Table 4.7) had previously revealed the presence of *nrfA* genes in the microbiome of the sponges *Aplysina aerophoba*, *Disydea avara*, *Xestospongia testudinaria*, and *Stylissa carteri* (Bayer et al., 2014). Likewise, metatranscriptomics of sponges *Stylissa massa* and *Xestospongia testudinaria* had also revealed gene *nrfA* (de Voogd et al., 2015), which in the case of *Xetospongia muta*

(Table 4.7) was attributed to gammaproteobacterial *Shewanella* spp. (Fiore et al., 2015). Metagenomics has revealed genes *nrfA* and *nrfD* in the unbinned data of *V. pourtalesii* (Maldonado et al., 2021b), but their assignment to a microbial lineage remains unresolved. However, in the same study, *nirB* and *nirD* genes were recovered as possible belonging to Plantomycetota, Actinobacteriota, and Gammaproteobacteria, and gene *napA* belonging to Plantomycetota as well (Table 4.7).

4.5.2.4 ASSIMILATORY NITRATE REDUCTION (ANR) AND AMMONIA ASSIMILATION

To date, these N processes are in general scarcely investigated in sponges, but the situation might improve in the future with the contribution of metagenome analyses. The *nasA* gene (encoding for assimilatory nitrate reductase) and *nirA* genes (encoding ferredoxin-nitrite reductase in ANR process) were detected in the metagenome of the sponge *Hymeniacidon heliophila* (Weigel & Erwin, 2017). Our current re-analysis has found *nirB* genes encoded in several metagenome bins from *A. aerophoba* (bin107: Nitrospinae; bin18: Deltaproteobacteria; bins52, 56, 65, 129: Alphaproteobacteria; bin80: Bacteriodetes; bins 22, 87, 127: Chloroflexi; bin55: Gammaproteobacteria). Interestingly, no gene coding enzymes for the ANR pathway were found in 137 metagenome bins recovered from the hexactinellid *Vazella pourtalesii* (Maldonado et al., 2021b). A nitrate depletion in the habitat of the sponge cannot be blamed for such microbial absence, since the average nitrate concentration in bottom waters of the Central Scotian Shelf — where the 2,000 km^2 aggregation of *V. pourtalesii* is established— is not low, ranging from 9 to 20 μM over the annual cycle (Petrie et al., 1999).

As new data are obtained, the emerging view is that the assimilation of ammonia into amino acids may be a ubiquitous function in sponge microbiomes. Assimilation is essential for building crucial macromolecules for life and it also represents an effective recycling of waste produced by the sponge cells. The gene *glnA*, encoding the enzyme glutamine synthetase, is commonly used as an indicator of microbial ammonia assimilation. For example, *glnA* genes have been detected in all but four of the 37 high-quality metagenomic bins (representing 13 microbial phyla) from *A. aerophoba.* Yet, none of the bins was recovered completely (Slaby et al., 2017). *Gln* was detected in the metagenomes and transcriptomes of several sponges such as *Hymeniacidon heliophyla*, *Cymbastella concentrica*, *Theonella swinhoei*, *Neamphius huxleyi*, *Plakortis simplex*, *Phakellia fusca*, *Stylissa carteri,* and

Xetospongia muta (Thomas et al., 2010; Moitinho-Silva et al., 2014; Fiore et al., 2015; Li et al., 2016; Feng et al., 2018). A large proportion of members of the microbiome of *V. pourtalesii* (Table 4.7 and Figure 4.7) also appear able to assimilate ammonia/ammonium (Maldonado et al., 2021b). The widespread detection of ammonia assimilation in sponge microbiomes is suggestive that the bulk of the sponge microbes preferentially use ammonium for assimilation rather than for oxidation (i.e., respiration).

4.5.2.5 ANAEROBIC AMMONIUM OXIDATION (ANAMMOX)

Microbes that carry out this reaction are bacteria from the phylum *Planctomycetales*, characterized by distinct organelles, the anamoxomes. The first evidence of anammox activity in sponges came from *G. barretti*, whose symbiotic anammox bacteria were found to be closely related to the Planctomycetes *Scalindua* sp. (Hoffmann et al., 2009). To date, members of the Plancomycetales are known to be common in sponges (Mohamed et al., 2010; Thomas et al., 2016; Busch et al., 2020a). However, besides measured anammox rates (Hoffmann et al., 2009), functional evidence by amplifying genes or even transcripts of the key enzyme (*hzsA*) for this pathway is still lacking, a situation probably derived from unsolved primer mismatches in PCR-based studies. Interestingly, the gene *hzsA* was not found in any of the more than 170 metagenome bins collectively recovered for *A. aerophoba* and *V. pourtalesii*, despite including members of Planctomycetales (Slaby et al., 2017; Bayer et al., 2020).

The anammox process may occur only intermittently in the sponges since the chemical reaction can uniquely be conducted in a sub-oxic environment within the sponge tissue. For this reason, the anammox function is hypothetically favored if: (i) water pumping is arrested during long periods; and (ii) within bacteriocyte cells (Figures 4.3(E) and (F)), which charged with anammox bacteria, would be able to migrate to low oxygen zones of the sponge mesohyl (*see* Section 4.7) and/or provide suboxic conditions to the microbes enclosed in their intracellular vesicle.

4.5.2.6 NITROGEN FIXATION

Some microbes can provide nitrogen to the budget of sponges through fixation, an advantageous ability in oligotrophic environments with low nitrogen availability (Taylor et al., 2007). Nitrogen fixation was first reported in Red Sea sponges (*Siphonochalina tabernacula, Theonella swinhoei*) by measuring enzyme activity (Wilkinson & Fay, 1979) and it was further demonstrated using $^{15}N_2$ stable isotope incorporation (Wilkinson et al., 1999). Subsequently, phylogenetic assignment of this function to microbial linages became possible after characterizing the gene *nifH* (encoding the nitrogenase subunit). *NifH* genes have been detected in several sponges and here we highlight some examples. *NifH* genes and their transcribed counterparts were reported in the sponges *Ircinia strobilina* and *Mycale laxissima* (Mohamed et al., 2008; Zhang et al., 2014). These studies revealed that several Cyanobacteria and Proteobacteria lineages harbor the genetic potential for nitrogen fixation. Similarly, *nifH* genes from proteobacterial groups (Alpha-, Gamma-, Delta-), as well as from Cyanobacteria (*Xenococcus* sp., *Myxosarcina* sp., *UCYN-A*-like), were amplified in the sponge *Xestospongia muta* (Fiore et al., 2015). However, in this latter study, no *nifH*-transcript was recovered. Nitrogen fixation was recently corroborated in *T. swinhoei* (Li et al., 2016), but without assignment to microbes. In the metagenome of the sponge *V. pourtalesii*, several genes responsible for N fixation were found to be affiliated with microbes belonging to Firmicutes, Planctomycetota, Bacteroidota, and Alphaproteobacteria (Table 4.7). Our current re-analysis has revealed genes *nifH* and *nif D* in the genome bin9 recovered of *A. aerophoba* (Table 4.7), which is homolog to Ca. *Synechococcus spongarium* (Slaby et al., 2017).

Some cyanobacterial and several proteobacterial lineages, which are common in both LMA and HMA sponges, appear to have the capability for nitrogen fixation. Thus, this process might well take place in a large variety of sponge species.

4.5.2.7 AMMONIFICATION

Among the various organic N products that are degraded to produce ammonia/ammonium within the sponges, urea is of special interest. As indicated in Section 4.5.1, it remains unclear whether the urea is generated internally through the catabolism of sponge cells or is imported from the seawater, or both. Because urea is an abundant organic N compound in oligotrophic ocean areas (e.g., coral reefs) where other nitrogen sources are scarce (Bronk et al., 1994), it becomes a handy, alternative nitrogen source to ammonia, nitrate, and nitrite within the sponge holobiont (Webster & Thomas, 2016). Typically, microbes use the enzyme urease to hydrolyze urea to ammonia and carbamate, which are subsequently assimilated in making amino acids.

The capability of some sponge symbionts to process urea is well demonstrated. Proteobacterial urease genes and transcripts have been identified in the microbiome of the coral-reef sponge *Xestospongia testudinaria* (Su et al., 2013). GeoChip-based analysis of the microbiomes showed genomic evidence for urease in the sponges *A. aerophoba*, *D. avara*, *X. testudinaria*, and *S. carteri* (Bayer et al., 2014). Urease-encoding gene clusters and urea transporters are also reported in a few sponge symbionts from genomes assembled from metagenomic data, such as *Poribacteria* (Siegl et al., 2011) and Entotheonella spp. in the sponge *T. swinhoei* (Liu et al., 2016), and members of an unclassified microbial lineage referred as SAUL/PAUC34f (Astudillo-García et al., 2018). Several genome bins assembled from *A. aerophoba* microbiome, belonging to the phyla Chloroflexi, Proteobacteria, Spirochaetae, Nitrospira, and the genus *Synechoccocus* (Slaby et al., 2017), also encoded genes for urease (Table 4.7). Additionally, Moitinho-Silva et al. (2017a) found a *Nitrospira* sp. (CCNi bin) associated with the sponge *Cymbastella concentrica* that encodes urease. Genomic potential for urea utilization was also detected in several metagenome-assembled thaumarchaeal genomes (Moitinho-Silva et al., 2017a; Haber et al., 2021), such as in *Nitrosospongia ianthellae* from the sponge *Ianthella basta*, in *Nitrosopumilus* sp. (ESC bin) from the sponge *Hymedesmia methanophila*, in *Nitrosopumilus* sp. (Pfa bin) from *Petrosia ficiformis*, and in *Nitrosopumilus cymbastellus* (CCThau bin) from *C. concentrica*. *UreC* genes were detected by PCR in the microbiome of the sponge *Spheciospongia vesparium* and assigned to Proteobacteria and Bacteriodetes (He et al., 2021). Additionally, the urease gene cluster has been identified (Table 4.7) in metagenome bins from Chloroflexi, Proteobacteria, Spirochaetae, Nitrospirae, and Cyanobacteria in *A. aerophoba* microbiome (Slaby et al., 2017). Likewise, by re-analyzing data from *V. pourtalesii* (Bayer et al., 2020), genes encoding urease in several metagenome bins belonging to different phyla have been identified (Table 4.7).

Some of the microbial groups identified, such as Poribacteria and SAUL, are typical members of the microbiome of HMA sponges, whereas Thaumarchaeota, Proteobacteria, and Nitrospirae commonly occur in both LMA and HMA sponges (Moitinho-Silva et al., 2017b). Therefore, the ability to use urea as an N source is widespread in sponges, but, surprisingly, the net flux of this compound has not been measured for any sponge yet.

4.6 PHOSPHOROUS PATHWAYS IN SPONGES

Recent advances have made it evident that marine sponges are involved in the recycling of P in a variety of ways. It is expected that future metagenomic studies will shed new light on the subject. For this review, in addition to information compiled from the literature, we have reanalyzed genes involved in P cycling among the metagenome bins of *A. aerophoba* (n=37) reported in Slaby et al. (2017) and of *V. pourtalesii* (n=137) reported in Bayer et al. (2020), in order to gain insight into possible roles of the sponge microbiome in the biochemical processing of phosphorus.

4.6.1 POLYPHOSPHATE (POLY-P)

An activity recently discovered (Zhang et al., 2015) in members of the microbiome of some coral reef sponges is the consumption of phosphate to produce Poly-P macromolecules intracellularly, which can become poly-P deposits (granules) in the sponge mesohyl (Figure 4.8). Poly-P is found in a large variety of organisms, from bacteria to plants and animals. The main poly-P synthesis gene in prokaryotes is *polyphosphate kinase* (*ppk*). In sponges, the gene was first identified in *Amphimedon queenslandica*, described as a sequence without introns and with about 60% amino acid similarity to sequences from chemolithoautotrophic bacteria (Zhang et al., 2015). *Ppk* genes and transcripts were also amplified in the sponges *Mycale laxissima*, *Ircinia strobilina*, and *Xestospongia muta*, and the transcripts from *I. strobilina* were cloned and sequenced. Cyanobacteria and some Proteobacteria (Table 4.8) actively expressed this gene (Zhang et al., 2015). The gene *ppk* was also amplified by PCR from several sponges from the South China Sea (Table 4.8). In all those sponges, Cyanobacteria appeared to be involved in poly-P synthesis, along with Proteobacteria and Acidobacteria (Ou et al., 2020). The metagenomic approach has shown that the genome of a sponge-associated Alphaproteobacterium in *Spongia officinalis* has the potential for both poly-P synthesis and hydrolysis (Karimi et al., 2018). In the metagenome bins recovered from *Aplysina aerophoba* and *Vazella pourtalesii*, we detected *ppk* genes for poly-P-synthesis in several bins, suggesting genomic potential for that pathway in some of the symbionts (Table 4.8). On the other hand, in our re-analysis of metageomic data, poly-P hydrolysis (*ppx* genes) was restricted to only two alphaproteobacterial bins in *A. aerophoba*, while it was more represented in the microbiome of *V. pourtalesii* (Table 4.8). Whether this pattern is significant, and suggests that poly-P degradation could be more common in hexactinellids (*V. pourtalesii*) than in demosponges (*A. aerophoba*) or more common in sponges from deep waters (*V. pourtalesii*) than in those of shallow waters (*A. aerophoba*), remains to be assessed in further studies.

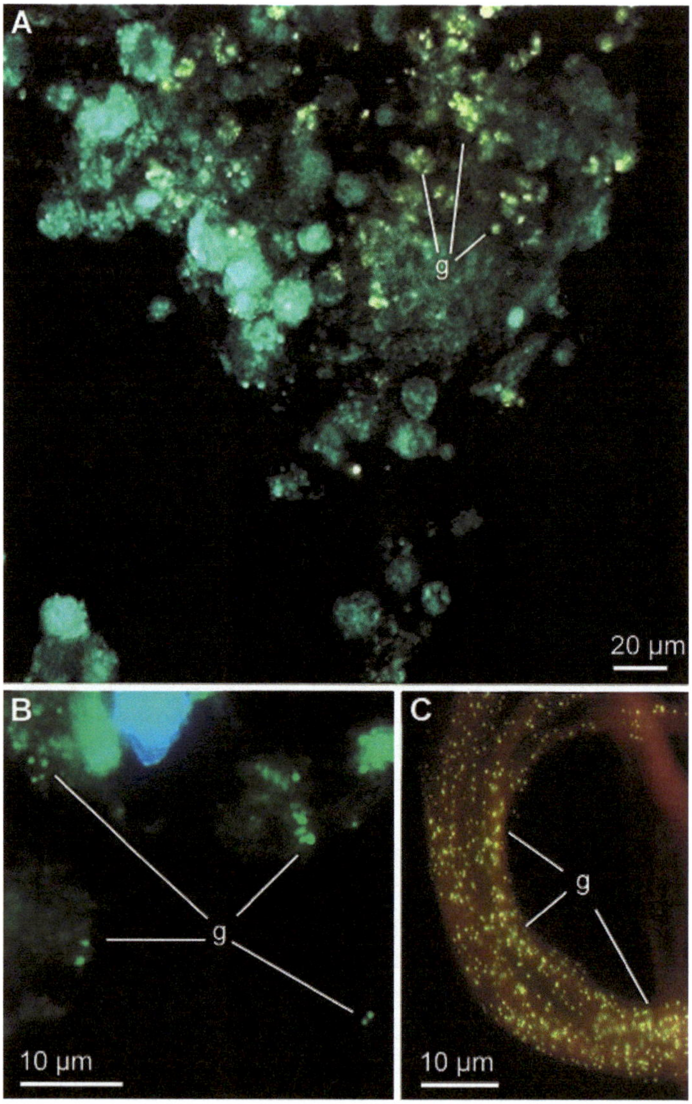

FIGURE 4.8 Visualization of poly-P granules in sponges. (A-B) Yellow fluorescent granules (g) were observed in the DAPI-stained tissue section of the sponge *Xestospongia muta* under a confocal fluorescent microscope. (C) Yellow fluorescent granules (g) were observed in the DAPI-stained culture of the filamentous cyanobacterium *Leptolyngbya* sp. isolated from the sponge *Ircinia strobilina*. All pictures by courtesy of Fan Zhang (Baylor College of Medicine, USA) and Russell Hill (University of Maryland, Center for Environmental Sciences, USA).

In prokaryotes, the main function that poly-P is thought to serve is as a handy storage of bioavailable phosphate as well as an energy reserve. The phosphate stored as poly-P by the sponge microbiome has been estimated to account for 25 to 40% of the total phosphate present in the holobiont (Zhang et al., 2015). The poly-P kinase enzyme reversibly polymerizes phosphate ions through phosphoanhydride bonds to form high-energy bond polymers. Such chemical bonds are analogous to those in ATP and allow ATP to recharge upon their breakage. Therefore, poly-P granules may serve not only as a reserve of P in times of deprivation but also as transient storage of energy for metabolism. It has been suggested that the Poly-P granules produced by the microbiomes may be used to protect the whole holobiont during periods of phosphate deprivation (Colman, 2015). As indicated by Webster & Taylor (2012), if such a hypothesis is validated, it would represent another unique mechanism of microbial metabolic integration to help sustain the sponge holobiont. However, given that most sponges studied show a net efflux of phosphate rather than consumption, it is difficult to believe that the metabolism of the holobiont is limited by phosphate shortage. Alternatively, if the primary function of poly-P in the holobiont would be transient storage of energy rather than that of P, the continuous mobilization of such an energy reserve by anoxic Poly-P degradation would cause a concomitant net release of phosphate. This interpretation (i.e., poly-P being produced mostly to transiently store energy) is more consistent with the phosphate efflux observed in most sponge holobionts. Yet many questions persist, since the poly-P initially accumulated within the cytoplasm of microbes forms relatively large oval granules (0.5–3 μm) in the sponge mesohyl (Figure 4.8), in which P is precipitated as apatite, combined with calcium into calcium phosphate, or converted into phosphite precursors. The interpretation of the significance of such large P granules is puzzling. It has been suggested that the formation of such large granules might lead the sponges to function as important P sinks (Zhang et al., 2015), a view in conflict with the experimental evidence that sponges are net sources of phosphate. A first quantification of P granule formation versus that of phosphate release is still needed before the role of sponges in the marine P cycle can be preliminarily assessed.

4.6.2 REDUCED P COMPOUNDS: PHOSPHONATE AND PHOSPHITE

It is now well established that a variety of microbes can obtain P for assimilation from reduced P compounds, such as phosphonates and phosphites, in addition to the conventional phosphate source. The use of phosphite is of particular interest since this ion can be oxidized to phosphate either aerobically (APO) or anaerobically (DPO). Whereas the phosphate produced by the aerobic oxidation of phosphite is thought to be used mostly for assimilation by microbial cells, the one processed by anaerobic oxidation acts as an electron donor for C fixation and, after its anaerobic oxidation with CO_2, nitrate, or sulfate, the resulting phosphate is released from the microbial cell (*see* Sections 4.3.3 and 4.3.4). Thus, the DPO pathway would provide a plausible mechanism that explains the phosphate efflux consistently noticed in the sponges. However, while the DPO pathway has empirically been confirmed in six bacterial groups to date (Ewens et al., 2021), it has never been reported from sponge microbiomes. For this reason, in the present study, we searched for the key gene responsible for DPO (*ptxD*) in two sponge species in which a phosphate efflux has been well established, the demosponge *Aplysina aerophoba* and the hexactinellid *Vazella pourtalesii* (Table 4.6). Furthermore, because phosphite could potentially be directly transported from seawater into the sponges but also obtained internally through the degradation of phosphonates or oxidation of hypophosphite, we also searched the metagenomes for a variety of genes coding for phosphonate transporters (*phnCDE* gene system), phosphite transporters (*ptxB* and *phnD* genes), hypophosphite transporters (*htxB* gene), and hypophosphite oxidation to phosphite (*htxA* gene). To our surprise, the gene systems involved in sourcing and processing phosphite by DPO were absent in the bins of both sponge species, and, consequently, the hypothesis of DPO as the origin of the phosphate efflux in these sponges was not proved. We found only one gene of the system: that encoded the ABC phosphonate transporter (*phnDCE*) in both *V. pourtalesii* and *A. aerophoba* (Table 4.8). Yet, none of the genomes encoded for the complete transporter, which might be due to the incompleteness of genomes.

The findings collectively suggest that DPO is an unlikely pathway in the microbiomes of the two sponge species inspected herein and that at least some members of the sponge microbiomes are able to use phosphonate as a source of P for assimilation. There is also the remarkable pattern that *phnG-P* genes, which encode a C-P lyase for enzymatic degradation of the C-P bonds of phosphonate, occurred exclusively in Alphaproteobacteria of both *V. pourtalesii* and *A. aerophoba* (Table 4.8). Therefore, it appears that a limited number of members of the sponge microbiomes can incorporate and degrade phosphonate for energy production and P assimilation. Alphaproteobacteria might be the only microbiome members that encode phosphonate lyase and complete the Pentose phosphate pathway in *V. pourtalesii* (Table 4.8).

Given the benefits of using phosphite both as a phosphorus source and as an electron donor, it cannot be discounted that phosphite-oxidizing microbes will be discovered sooner or later in the microbiome of sponges other than the two species herein examined. If so, these activities are predicted to occur in diverse groups of microorganisms that are not phylogenetically related, because there is preliminary evidence that DPO genes can be transmitted preferentially by lateral transfer (Figueroa & Coates, 2017).

4.6.3 PHOSPHATE

Phosphate, either obtained from seawater or produced internally from the degradation of DOP compounds, is assumed to be the conventional P source used by the bulk of the sponge microbiome. The gene *pstS*, encoding the high-affinity phosphate-binding protein used as a marker for phosphate processing in cells, has been detected in the sponges *Aplysina aerophoba*, *Dysidea avara*, *Xestospongia testudinaria*, and *Stylissa carteri* using a functional gene array (Bayer et al., 2014). We found *pstSCAB* genes encoding for the phosphate ABC transporter (Table 4.8) in nearly all metagenome bins recovered from *A. aerophoba* (Slaby et al., 2017) and *Vazella pourtalesii* (Bayer et al., 2020). There were also two interesting findings. First, genes encoding for alkaline phosphatases (*phoA*, *phoD*, *phoX*) that are involved in DOP degradation (i.e., dephosphorylation) to obtain phosphate for assimilation were detected only in several metagenome bins of *A. aerophoba*, but not in genomes of *V. pourtalesii* (Table 4.8). Second, genes involved in phosphonate degradation (*phnA*, *phnX*, *phnW*) were detected in the microbiome of *V. pourtalesii*, but not in that of *A. aerophoba* (Table 4.8). These preliminary patterns suggest that, beyond incompleteness of the metagenomes, potential differences between either "demosponges vs hexactinellids" or "shallow-water vs deep-water" sponges might exist relative to the way in which microbiomes obtain phosphate from organic P sources. Further research is necessary to clarify these aspects.

4.7 CYTOLOGICAL SPONGE ADAPTIONS FOR N AND P FLUXES

It has long been known that a large variety of symbiotic and merely opportunistic prokaryotes can coexist together in different numbers within sponge tissues (Figure 4.3). Although the initial finding of microbes in sponges lead to the interpretation that they could be pathogens (Dosse, 1939), their reiterative presence in healthy sponges, along with the subsequent discovery that those microbes were vertically transmitted across sponge generations by the oocytes (Duboscq & Tuzet, 1942; Lévi & Porte, 1962), led to the more correct assumption that some positive functional interactions (symbiosis, mutualism, opportunism) exist between the sponges and their microbes.

Most sponge microbes occur in the intercellular medium of the sponge mesohyl (Figure 4.3(A)–(D)). However, storage of the microbes in specialized cells, generically called bacteriocytes (Figures 4.3(E) and (F)), has also been reported in sponge species of the classes Demospongiae, Calcarea, and Hexactinellida. To our knowledge, bacteriocytes have not yet been described in Homoscleromorpha. To date, three main types of bacteriocytes have been found in sponges: (i) cells with one or several, large intracytoplasmic vesicles (hereafter referred to as vesicle bacteriocytes) in which microbes are enclosed intracellularly (Figure 4.3(F)); (ii) flattened cells that fold over themselves forming of bag or pocket (hereafter referred to as pocket bacteriocytes) where microbes are enclosed despite remaining extracellular (Figure 4.3(E); Maldonado, 2007); and (iii) giant, multinucleate bacteriocytes (hereafter referred to as bacteriosyncytia) with many small intracytoplasmic vesicles that enclose microbes (Maldonado et al., 2021b).

Within a single bacteriocyte, sponges can keep either different species of microbes or just a single one (Figure 4.3(E) and (F)). Importantly, at least some of the microbes, irrespective of being in the mesohyl or within the bacteriocytes, can be digested by the sponge cells. Therefore, digestion of part of the microbiome is believed to be a common contributor to sponge nourishment (Wilkinson & Garrone, 1980; Ilan & Abelson, 1995; Vacelet et al., 1996; Maldonado, 2007; Lanna & Klautau, 2010; Leys et al., 2018). The exploitation of this internal food source is not irrelevant from a physiological and ecological point of view, since by feeding on a microbiome mostly farmed out of the various fluxes of inorganic N and P (also C), sponges may largely reduce their dependence on the external availability of particulate food (*see* Section 4.8 for study cases).

To date, no study has addressed the implications of microbes occurring either free in the mesohyl or enclosed within bacteriocytes. Herein, we suggest that these different locations can be of great importance in functional terms for efficient co-existence and metabolic integration between the microbes and the host, but also between the aerobic and anaerobic members of the microbiome. Within their body, sponges need to provide oxic and anoxic (or suboxic) environments simultaneously in order to facilitate the co-existence and continuous growth of the diverse microbial lineages. Indeed, to attain efficient cycling of N and P compounds in the sponges, aerobic, and anaerobic biochemical pathways must be coupled internally in the holobionts. For instance, it has been commonly suggested that the nitrite and nitrate resulting, respectively from partial or complete oxic nitrification are used to feed anaerobic processes such as denitrification (Bayer et al., 2007; Hoffmann et al., 2009), dissimilatory nitrate reduction to ammonium (DNRA) (Maldonado et al., 2021b) and anammox (Hoffmann et al., 2009). Likewise, the co-existence of oxic and suboxic/anoxic conditions is required within the sponges for the respective polymerization and de-polymerization of poly-P molecules (Zhang et al., 2015). Both aerobic and anaerobic conditions would also be required to make assimilatory and DPO possible (Figueroa & Coates, 2017), if these pathways are ever demonstrated in sponges.

On the basis of experiments with *G. barretti* (Hoffmann et al., 2005a, b), it was initially proposed that when sponges are pumping water, the continuous water flow facilitates O_2 penetration in the mesohyl, creating well-oxygenated conditions favorable for the biochemical reactions mediated by aerobic microbes. When pumping ceases, partial O_2 concentrations drop quickly in some zones of the mesohyl, favoring the activity of anaerobic microbes. While this mechanism is possibly responsible for oxic and suboxic situations in many sponge species with massive bodies similar to those of *Geodia* spp., pumping activity may be less effective to trigger anoxia in other morphotypes, such as encrusting or pseudo-encrusting sponges (Schläppy et al., 2007, 2010), vase-shaped and fan-shaped forms with thin walls, etc. In these sponges, O_2 can well reach most of the mesohyl zones by mere diffusion from ambient water. More importantly, the "pumping vs. non-pumping" activity will provide only an alternation between oxic and anoxic conditions in some mesohyl areas where aerobic and anaerobic microbes are assumed to coexist. Transient oxic conditions may be tolerated by facultative anaerobes but are harmful to strict anaerobes, and vice-versa for aerobic microbes under oxic conditions.

We herein propose that to deal with that problem and provide simultaneously oxic and suboxic-anoxic conditions to the respective microbial groups, sponges have evolved specialized cells for holding microbes, the bacteriocytes. These cells allow the internal medium of the sponge to be compartmentalized, supplying the required oxic or suboxic conditions to the enclosed microbes, independently of the oxygenation level of the intercellular medium of the mesohyl. Furthermore, because the bacteriocytes are motile (i.e., crawling) cells, they can potentially move towards zones of the mesohyl where O_2 concentration is more suitable, depending on the nature of the enclosed microbes. Unfortunately, the phylogenetic and functional differences between the microbes enclosed in the bacteriocytes and those existing intercellularly in the mesohyl have never been addressed in any sponge to date. This challenging subject remains uninvestigated despite its critical importance in fully understanding the intricacies of the metabolic integration between sponge cells and microbes.

4.8 ECOLOGICAL EFFECTS OF THE N AND P FLUXES OF SPONGES

The ecological capabilities of the sponges are thought to be expanded through a functional (i.e., physiological) integration between the sponge and microbiome metabolisms. The ways in which this might happen are multiple and complex to study because the functionality emerges through particular co-evolution processes within each holobiont species. This includes a conundrum of microbial lineages that may change substantially from one sponge species to another and also between individuals of a given species (Zhang et al., 2014; Archer et al., 2017; Freeman et al., 2021), and that co-exist with other microbial lines faithfully maintained within sponge species through vertical transmission.

Sponge-mediated N and P fluxes have rarely been quantified in scenarios involving interactions between several sponge species (but see Valentine & Butler, 2019) or scaled up to the ecosystem level. Nevertheless, some relevant examples are available. One of the most studied processes of the sponge microbiome is ammonium consumption through incomplete or complete nitrification to yield nitrite or nitrate, respectively. While most sponges release ammonium (Figure 4.6), some consume it, releasing nitrite + nitrate (Scheffers et al., 2004; Jiménez & Ribes, 2007; Morganti et al., 2017; Leys et al., 2018; Feng & Li, 2019; Maldonado et al., 2021b), often with indications of incomplete nitrification. Theoretically, the ammonium that fuels nitrification is initially obtained from the excretion of the sponge cells, but typically it has to be supplemented with ammonium taken up from seawater (Table 4.3 and Figure 4.6), since the rates of ammonium utilization by the ammonium oxidizers (aerobic and anaerobic) of the microbiomes typically exceed those of ammonium excretion by the sponge cells, as demonstrated for some sponge species by Hofmann et al. (2009). Therefore, many nitrifying sponges are expected to function as net ammonium sinks in benthic communities. This consumption is not expected to be ecologically trivial, since ammonium flux rates of sponges are about three orders of magnitude greater than those of nitrite and nitrate (Figure 4.6). Ammonium occurs at low concentrations, particularly in the photic ocean, where it is quickly consumed by the bacterioplankton and phytoplankton communities that use it as a preferred N source.

A recent example of the ecological importance of ammonium consumption by sponges comes from the monospecific aggregations of the hexactinellid *V. pourtalesii* (Maldonado et al., 2021b). In that study, the net annual uptake of ammonium by the entire sponge aggregation, which extends over an area of 2,105 km^2 at 100–200 m depths on the Nova Scotia continental shelf (Canada), was estimated at 144.8×10^6 mol (some 2.6 million kg of ammonium per year). A comparison between the hourly average rate of ammonium consumption by *V. pourtalesii* (5.9 $\mu mol\ L^{-1}$ of sponge tissue h^{-1}) and that measured for the marine plankton community (bacterioplankton + phytoplankton) of a coastal area (47×10^{-3} $\mu mol\ L^{-1}$ of seawater h^{-1}) in the eutrophic Baltic Sea indicates that a *V. pourtalesii* individual of 1 liter in body volume consumes about three orders of magnitude more ammonium than the entire plankton community in a liter of eutrophic coastal seawater, which, in turn, is known to have higher rates than those of the plankton community in worldwide oceanic and estuarine systems (Klawonn et al., 2019). It has also been measured that the ammonium concentration at the habitat of *V. pourtalesii* is abnormally low, ranging from 0.04 to 0.56 μM. Thus, the effect of this sponge aggregation on the availability of ammonium in bottom waters is not insignificant at the ecosystem scale. In this context, it has been proposed that the massive ammonium consumption by *V. pourtalesii* decreases ammonium availability for the free-living microbial populations in the seawater around the large aggregations of this sponge (Maldonado et al., 2021b). Ammonium depletion driven by sponges has already been demonstrated in mesocosm experiments with Caribbean reef sponges (Valentine & Butler, 2019). A poor bacterioplankton community will decrease, in turn, the feeding chances of those sponge species that do not farm their microbes internally but instead rely on filter feeding on external bacterioplankton and picophytoplankton. In the case of *V. pourtalesii*, this view is consistent with the independent finding that the bulk energy for the aerobic metabolism of this sponge is not obtained through filter feeding of the free-living bacterioplankton (only about 10%), but rather through DOC incorporation (Bart et al., 2020) and digestion of the microbiome farmed in bacteriosyncytia (Maldonado et al., 2021b). In summary, the remarkable incorporation of ammonium by *V. pourtalesii* may result in a nutritional competitive exclusion of those sponge species that rely on the filter-feeding of the free-living bacterioplankton. This may be one of the reasons why the aggregations of *V. pourtalesii* are markedly monospecific. This hypothesis also appears to agree with patterns seen in other sponge aggregations. For instance, the hexactinellid sponges *Schaudinnia rosea*, which appears to be characterized by LMA (Busch et al., 2020b), and *Aphrocallistes vastus*, which has a poorly associated microbiota (Kahn et al., 2018), both release ammonium and both form multispecies sponge assemblages rather than monospecific aggregations.

Another example of the ecological importance of sponge-mediated nitrification and denitrification is provided by several species of the family Geodiidae. These sponges can form extensive multispecific aggregations (covering many thousands of km^2) in the North Atlantic Ocean at depths between 200 and 1,000 m, reaching densities of up to 30 kg sponge wet weight m^{-2} (Klitgaard & Tendal, 2004; Kutti et al., 2013; Pham et al., 2019). Extrapolating laboratory results, Hofmann et al. (2009) estimated that sponge-mediated nitrification would transform up to 16 mmol N m^{-2} day^{-1} on average and that sponge-mediated denitrification would remove 2.7 mmol N m^{-2} day^{-1} as N_2. These results suggest that the rate of N removal by the *Geodia* spp. aggregations, though much smaller than that of nitrified N, would still be 2 to 10 times higher than that occurring in continental shelf sediments at similar water depths, where denitrification rates are only 0.1–1 mmol N m^{-2} day^{-1} (Seitzinger & Giblin, 1996; Middelburg et al., 1996). Likewise, using data from Maldonado et al. (2021b), the rate of N removal from seawater through ammonium consumption for nitrification+ assimilation in *V. pourtalesii* aggregations is herein estimated as 68.8 mmol N m^{-2} day^{-1}. A comparison of the above sponge rates of ammonium consumption through nitrification with those of ammonium release or consumption by a variety of eutrophic sediments (–1.32 to 2.03 mmol m^{-2} day^{-1}) of the Croatian continental shelf (Adriatic Sea) reported by Baric et al. (2002) indicates that the N fluxes through the sponge communities (16 and 68.8 mmol N m^{-2} day^{-1}) are typically over one order of magnitude more intense than through bare sediment. The importance of sponge-mediated nitrification has also been demonstrated for coral reefs (Diaz & Ward, 1997; Southwell et al., 2008) and temperate rocky-bottom communities (Jiménez & Ribes, 2007). Altogether, the available studies exemplify well that, against the traditional view of the marine N cycle, sponge-dominated communities can be locally more influential for nitrogen mineralization processes than marine sediments.

Regarding carbon fixation through nitrification, it was long established that the ratio of carbon fixation to nitrification by bacteria oxidizing ammonia is approximately 1 mol of CO_2 for every 9 mol of ammonium oxidized to nitrate (Feliatra & Bianchi, 1993). This allowed Hofmann et al. (2009) to conclude that nitrification rates in the sponge *G. barretti* (582 nmol cm^{-3} day^{-1}, regarding nitrate lost by denitrification) allow the nitrifying microbial community in the sponges to fix up to 64 nmol CO_2 cm^{-3} sponge day^{-1}. By assuming that 1 mmol O_2 consumption per cm^3 sponge per day was due to the aerobic oxidation of ammonium and the remainder due to carbon oxidation, microbial nitrification would provide less than 1% of the carbon demand of *G. barretti*. On the basis of this limited C provision, it may be assumed that the main benefit of the complex nitrogen cycling in sponges would not be the fixation of new C but rather an efficient internal recycling of the sponge ammonia waste and, more importantly, the farming of its microbial food. The need for some sponges to incorporate ammonium from seawater (Table 4.3) suggests that the relevance of the symbiotic nitrifiers goes also beyond the recycling of toxic ammonia waste excreted by the sponge cells. An important contribution from the microbes to the holobiont, in addition to C fixation and ammonia recycling, is their consumption as food (e.g., Maldonado, 2007, 2021b; Leys et al., 2018; de Kluijver et al., 2021). For this feeding process to be efficient, the microbes have to be farmed out of a combination of energetically inexpensive, organic, and inorganic C, N, and P products, either generated internally by the sponges as a result of the cellular catabolism or easily available from the seawater. Such conditions particularly apply in holobionts in which the N transformations within the sponge make a closed circuit. By coupling diverse pathways, the N budget of the holobiont should go closer to neutral. An additional advantage of coupling various biochemical pathways involving different microbial capabilities to close an N cycle within the sponges is the avoidance of detrimental accumulation of compounds that may be harmful to the sponge cells, such as ammonia, nitrite, nitrate, nitrous, and NO, etc. Also importantly, the coexistence of aerobic and anaerobic microbial metabolisms facilitates sponges to keep functioning during transient low oxygen conditions that may be recurrent or prevailing in many deep-sea bottom communities, caves, mangroves, methane seeps, and other marine habitats known to have sponge-rich fauna (Maldonado et al., 2017). Therefore, the biochemical coupling and the microbial diversity detected in modern sponges may have been favored selectively through co-evolution (e.g., O'Brien et al., 2021) and maintained with fidelity through vertical transmission (Figure 4.3(C)) of specific microbial lineages (e.g., Maldonado, 2007; Steger et al., 2008; Schmitt et al., 2008).

Although denitrification and anammox, in contrast to nitrification, remove nitrate and nitrite from the environment, they appear to be less represented functions in the microbial metagenomes of sponges (Table 4.7). The need for anaerobic conditions for these pathways is probably also responsible for them being of smaller magnitude than aerobic nitrification (e.g., de Kluijver et al., 2021). Therefore, their effects in the net N sponge fluxes are assumed to be of lower magnitude than those of nitrification. However, high rates of denitrification have recently been reported from six deep-sea boreal species (Rooks et al., 2020). Unfortunately, the flux rates obtained arose from experiments with small fragments of sponges that lacked any pumping activity. Consequently, the flux rates cannot be reliably extrapolated as if they reflected the net fluxes of living sponges. The alternation between oxic and suboxic conditions within the sponge tissues, along with the occurrence of bacteriocytes, likely favors the coupling between aerobic and anaerobic pathways. For instance, denitrification and anammox pathways can theoretically be fed by final and intermediate products resulting from the nitrification pathway (Figure 4.7). Likewise, different anaerobic processes can bridge with each other to avoid N losses, such as denitrification and nitrogen fixation (Figure 4.1). This coupling, rather than leading to a loss of N_2 gas, would regenerate the ammonium supply internally for the nitrifying members of the microbiome, as proposed for the hexactinellid *V. pourtalesii* (Figure 4.7). Theoretically, the DNRA and the anammox pathways could also bridge with each other under suboxic conditions to regenerate ammonium (Figures 4.1 and 4.7). How the internally coupled biochemical pathways contribute to the net fluxes measured in live sponges remains little understood, with only one available study to date from the sponge *G. barrettii* (de Kluijver et al., 2021). In that study, a linear inverse model approach indicated that nitrification was essentially fueled by both ammonia internally produced by cellular excretion and DNRA, and that nitrification-associated chemoautotrophic production contributed 5.1–6.7 ± 3.0% to total sponge production.

A majority of sponges are known to release nitrite and phosphate (Tables 4.4 and 4.6). Nitrite is an intermediate product used as a source of nitrogen by the phytoplankton and, in a more important way, by the bacterioplankton. Because in seawater the nitrite concentration is typically low, the continuous nitrite release by some sponges (e.g., Valentine & Butler, 2019) is expected to favor the coexistence of different populations of nitrifiers, avoiding the typical competitive exclusions driven by low nitrite + low ammonium availability (Zakem et al., 2018). A similar favorable effect would apply to the release of phosphate detected in most investigated sponges (Table 4.6) since a phosphate shortage is blamed to limit the development of free-living phytoplankton and bacterioplankton communities in many marine oligotrophic environments. Using data from Maldonado et al. (2021b), we herein estimate that the aggregation of *V. pourtalesii* on the Scotian shelf releases annually about 20.5 x 10^6 mol of phosphate (some 1.9 million kg) to the local seawater, with a daily average flux rate of 9.7 mmol per m^2 and day. As in the case of ammonium, the flux of phosphate through the sponge communities is —either release of uptake— one to two orders of magnitude more intense than that, for instance, estimated (–0.031 to 0.164 mmol m^{-2} d^{-1}) for a variety of eutrophic sediments (Baric et al., 2002). Therefore, the phosphate efflux from the sponge holobionts is expected to have ecological relevance at the ecosystem scale. Furthermore, the N and P nutrient fluxes mediated by sponges appear to be of some ecological effect even in those benthic systems where sponges are not the dominant organism type. For instance, in the rich benthic community of maërl in the Bay of Brest (France, Eastern North Atlantic), the sponge *Tethya citrina* accounts for about 15% of the sessile animal biomass. The species contributes only about 1.5% of the ammonium benthic efflux from the bottom to the water column and 2.1% of the phosphate efflux, but it provides about 13.8% nitrate+nitrite efflux (López-Acosta et al., 2019).

The ecological facilitation derived from the metabolic integration between the sponges and their microbiomes may well account for the occurrence of sponges in some harsh environments. For instance, it has recently been discovered in Antarctica that sponges can establish simple benthic communities on boulders under the ice shelf, somewhere between 625 and 1,500 km in the direction of water flow from the nearest region of photosynthesis (Griffiths et al., 2021). In the absence of nearby primary production, the authors wondered what the sources of particulate food maintaining those benthic communities could be. We are herein proposing that the role of the microbiome could be critical to the survival of those sponges. A variety of organic and inorganic compounds that are common in seawater (DOC, CO_2, ammonium, nitrate, phosphate, phosphonate, etc.) might be easily incorporated by the sponges established far under the ice to farm their microbiomes, which would be subsequently digested as food. Importantly, if those sponges replace periodically their epithelial and mesohyl cells and/or discharge epithelial mucus, as many other sponges do (de Goeij & Van Duyl, 2007; Alexander et al., 2014; Maldonado, 2015), an output of particulate organic matter would be created that may provide particulate food to nearby detritivorous invertebrates unable to use dissolved compounds. These cascading processes, known as "the sponge loop" (de Goeij et al., 2013), would facilitate the onset of the animal food web at the incipient benthic community under the ice. Thus, a fine-tuned integration between microbial and physiological processes, similar to that seen in other sponges from other habitats, should probably be enough to sustain a small benthic community under the ice and far from the front of phototrophic primary production. Whether these mechanisms are actually in the sponges under the ice remains an issue to be experimentally corroborated. Indeed, food shortage may be a common situation not only for sponges under the ice but also for many other sponges that live in dense aggregations. For instance, it has been calculated that *G. barretti* filters bacterioplankton with a retention efficiency of 99%, but such a food supply provides only about 5% of the C needs of the sponges (de Kluijver et al., 2021). DOC uptake and digestion of the farmed microbiome —some 2.9×10^{11} microbes per mL of sponge— are estimated to provide the rest of the total carbon needs. A metabolic integration between microbes and the sponge could be behind the enormous ecological success of this species, which forms extensive aggregations at bathyal depths through the North Atlantic Ocean.

To our knowledge, the ecological importance of sponge-mediate nutrient fluxes has never been assessed through economic value, but at least the filtering capacity has been a case study (Pham et al., 2019). The bathyal sponge aggregations dominated by *G. barretti* at Grand Bank, Flemish Pass, and Flemish Cap (Canada) have an estimated extension of 1,35,000 km^2 and are calculated to filter some 56,000 million liters of seawater daily, with an associated respiration rate of 56.8 tons O_2 per day, and a daily consumption of 63.1±11.8 tons of organic carbon obtained from the bottom water. In addition, such sponge biomass was estimated to consume 0.02 tons of ammonium day^{-1} and about 0.41 tons of nitrite day^{-1}, releasing nitrate at 6.5 tons day^{-1}. All these are phenomenal fluxes. Pham et al. (2019) also estimated that deep-sea trawling activity in the area inflicted a sponge removal of 2,580 tons with a consequent average decrease of the pumping capacity in the sponge ground of 627 million liters day^{-1}. When the filtering capacity of the sponges was compared to that of the Ashbridges Bay Treatment Plant (ABTP) operating in the city of Toronto and serving a population of 1,524,000 people, the average treatment activity recorded between 2015 and 2017 was about 598 million liters per day^{-1}. Such a figure is similar to the estimated filtering capacity of the sponge biomass removed each day by bottom trawl fishing. As posed by the authors, the cost of building and maintaining annually a human-engineered system clearing bacteria from seawater at a rate similar to that of the *G. barretti* biomass daily removed by bottom fishing would rise to some USD 187 million. This is clear evidence that the sponge-mediated fluxes matter, both ecologically and economically.

4.9 MAJOR GAPS IN KNOWLEDGE AND FUTURE GUIDELINES

The advent of molecular techniques has contributed notably to understanding the astonishing phylogenetic diversity of sponge microbiomes and to identifying the genes that putatively encode the enzymes catalyzing transformations of key inorganic N and P compounds. Nevertheless, at this moment, there is far more descriptive information on the presence of genes and microbial lineages in the sponges than empirical demonstrations of their function in the context of the holobiont and its ecological benefits. As part of this review, a variety of significant gaps in knowledge have been identified relative to the fluxes of inorganic N and P compounds through sponges, as follows:

- There is poor knowledge about the qualitative and quantitative connections between the net flux rates of inorganic nutrients measured going in and out of the sponges and the internal biochemical networks identified for their microbiomes. Mass balance models aiming to couple net entries and outputs of inorganic nutrients to the internal transformations occurring within the sponge holobiont are urgently required.
- It has been repeatedly suggested that the various aerobic and anaerobic microbial pathways must couple and/or bridge with each other through either final or intermediate products to establish efficient, closed cycles in the holobionts. However, there is almost no empirical demonstration of such plausible (and necessary) coupling between pathways.
- Fluxes of inorganic N and P nutrients cannot be understood appropriately if consistently studied separately from fluxes of organic carbon and organic nitrogen, sulfur metabolism, oxygen utilization, and the energy balance of the sponge holobiont. Therefore, integrative empirical and modeling approaches are required for generating comprehensive metabolic knowledge.
- Molecular studies have demonstrated that the ability to process urea as a probable N source is widespread in sponge microbiomes. However, no study has attempted to estimate net urea fluxes in sponges, despite the foreseeable importance of this compound to understand in full the metabolic integration between sponges and microbes and to establish net N entries and outputs for future modeling.

- The phosphorus metabolism of sponge microbiomes is less studied and more poorly understood than nitrogen metabolism. The causes of the phosphate efflux detected in most investigated sponge species remain elusive. It is particularly puzzling the idea that poly-P, an energetically costly compound for organisms, is released from both microbial cells and the sponges themselves to form deposits in the marine sediments. There is currently a sort of contradictory interpretation of sponges, being simultaneously suggested as sources and sinks of inorganic P. Such a view probably derives from our still poor and fragmentary quantitative understanding of both phosphorus cycling within sponges and its ecological connections to the benthic ecosystems.
- Recent and ongoing research on the sponge holobiont is mostly approached from a microbiological perspective, overlooking the interest in cellular and physiological adaptions that sponges may have co-evolved to favor the rich functionality of their biodiverse microbiomes. Here we propose that bacteriocytes may play a crucial role in the co-existence and cooperative functioning of aerobic and anaerobic members of the microbiomes. However, no phylogenetic characterization of the microbes (or their functionalities) enclosed in bacteriocytes versus those living extracellular in the mesohyl is available yet for any sponge holobiont. Likewise, whether nutrient fluxes going in and out of the sponges are based exclusively on mere diffusion processes (unlikely) or on special transporters located in sponge cells —which cells?— remains uninvestigated. The latter scenario would facilitate the supply and transference of the various compounds through the internal sponge compartments. Sponge physiology experiments coupled with transcriptomes and differential gene expression (bulk and single-cell) are needed to solve the superlative puzzle of holobiont metabolic integration.
- Net nutrient fluxes measured in individual sponges have seldom been scaled up to the community or ecosystem level. Likewise, few attempts have been made to quantify the direct effects of the sponge net fluxes on the local nutrient concentrations and to characterize the responses of the bacterioplankton and phytoplankton. However, the plausible cascading effects in terms of primary productivity and scenarios of integrative ecosystem ecology may be important. Consequently, the foreseeable ecological and biogeochemical consequences of the sponge-derived fluxes of inorganic nutrients

remain to be unfolded even for benthic ecosystems as emblematic as coral reefs, mangroves, caves, a large variety of deep-sea sponge aggregations, etc. This missing quantitative knowledge is also necessary for management and conservation strategies in the context of an anthropogenic ocean for which major, rapid changes are predicted in the midterm.

ACKNOWLEDGMENTS

The authors thank Jasper de Goeij, Russell Hill, Thierry Le Bec, Sally Leys, Furu Mienis, Benjamin Muller, and Fan Zhang, who kindly provided unpublished pictures to illustrate this study. The authors also thank Beate Slaby for her help in searching for metagenome bins. This study has been partially supported by grant MICIU-PID2019-108627RB-I00 to MM.

KEYWORDS

- **deep-sea sponge aggregations**
- **dissimilatory phosphite oxidation**
- **genes coding**
- **holobionts**
- **microbiota**
- **non-comparable units**
- **phosphorus cycling**

SUPPLEMENTARY DATA AVAILABILITY

Supplementary data (Supplementary Table 4.1) is available at ERA: https://doi.org/10.7939/r3-1j24-6216

REFERENCES

Achlatis, M., Pernice, M., Green, K., De Goeij, J. M., Guagliardo, P., Kilburn, M. R., Hoegh-Guldberg, O., & Dove, S., (2019). Single-cell visualization indicates direct role of sponge host in uptake of dissolved organic matter. *Proc. R. Soc. B Biol. Sci., 286*. doi: 10.1098/rspb.2019.2153.

Alexander, B. E., Liebrand, K., Osinga, R., et al., (2014). Cell turnover and detritus production in marine sponges from tropical and temperate benthic ecosystems. *PLoS One, 9*, e109486.

Archer, S. K., Stevens, J. L., Rossi, R. E., Matterson, K. O., & Layman, A. C., (2017). Abiotic conditions drive significant variability in nutrient processing by a common Caribbean sponge, *Ircinia felix*. *Limnol. Oceanogr., 62,* 1783–1793. doi: 10.1002/lno.10533.

Arellano, S., Lee, O., Lafi, F., Yang, J., Wang, Y., Young, C., & Qian, & P, Y., (2013). Deep sequencing of *Myxilla (Ectyomyxilla) methanophila*, an epibiotic sponge on cold-seep tubeworms, reveals methylotrophic, thiotrophic, and putative hydrocarbon-degrading microbial associations. *Microb. Ecol., 65*, 450–461. doi: 10.1007/s00248-012-0130-y.

Astudillo-García, C., Slaby, B. M., Waite, D. W., Bayer, K., Hentschel, U., & Taylor, W. M., (2018). Phylogeny and genomics of SAUL, an enigmatic bacterial lineage frequently associated with marine sponges. *Environ. Microbiol., 20*, 561–576. doi: 10.1111/1462-2920.13965.

Baric, A., Kuspilic, G., & Matijevic, S., (2002). Nutrient (N, P, Si) fluxes between marine sediments and water column in coastal and open adriatic. In: *Nutrients and Eutrophication in Estuaries and Coastal Waters* (pp. 151–159). Springer Netherlands.

Bart, M. C., Hudspith, M., Rapp, H. T., Verdonschot, P. F. M., & De Goeij, J. M., (2021). A deep-sea sponge loop? sponges transfer dissolved and particulate organic carbon and nitrogen to associated fauna. *Front. Mar. Sci., 8*, 229. doi: 10.3389/fmars.2021.604879.

Bart, M. C., Mueller, B., Rombouts, T., et al., (2020). Dissolved organic carbon (DOC) is essential to balance the metabolic demands of four dominant North-Atlantic deep-sea sponges. *Limnol. Oceanogr*. doi: 10.1002/lno.11652.

Bayer, K., Busch, K., Kenchington, E., Beazley, L., Franzenburg, S., Michels, J., Hentschel, U., & Slaby, M. B., (2020). Microbial strategies for survival in the glass sponge *Vazella pourtalesii*. *mSystems, 5*, e00473–20. doi: 10.1128/mSystems.00473-20.

Bayer, K., Moitinho-Silva, L., Brümmer, F., Cannistraci, C. V., Ravasi, T., & Hentschel, U., (2014). GeoChip-based insights into the microbial functional gene repertoire of marine sponges (high microbial abundance, low microbial abundance) and seawater. *FEMS Microbiol. Ecol., 90*, 832–843. doi: 10.1111/1574-6941.12441.

Bayer, K., Schmitt, S., & Hentschel, U., (2007). Microbial nitrification in Mediterranean sponges: Possible involvement of ammonia-oxidizing betaproteobacteria. In: Custódio, M. C., Lôbo-Hajdu, G., Hajdu, E., & Muricy, G., (eds.), *Porifera Research: Biodiversity, Innovation and Sustainability* (pp. 165–171). Museu Nacional do Rio de Janeiro.

Bayer, K., Schmitt, S., & Hentschel, U., (2008). Physiology, phylogeny and *in situ* evidence for bacterial and archaeal nitrifiers in the marine sponge *Aplysina aerophoba*. *Environ. Microbiol., 10*, 2942–2955. doi: 10.1111/j.1462-2920.2008.01582.x.

Bell, A. H., Bergquist, P. R., & Battershill, N. C., (1999). Feeding biology of *Polymastia croceus*. *Mem. Queensl. Museum, 44*, 51–56.

Bell, J. J., (2008). The functional roles of marine sponges. *Estuar. Coast. Shelf Sci., 79*, 341–353. doi: http://dx.doi.org/10.1016/j.ecss.2008.05.002.

Blanquer, A., Uriz, M. J., & Galand, E. P., (2013). Removing environmental sources of variation to gain insight on symbionts vs. transient microbes in high and low microbial abundance sponges. *Environ. Microbiol., 15*, 3008–3019. doi: 10.1111/1462-2920.12261.

Bronk, D. A., Gilbert, P. M., & Ward, B. B., (1994). Nitrogen uptake, dissolved organic nitrogen release, and new production. *Science, 265*, 1843–1846. doi: 10.1126/science.265.5180.1843.

Busch, K., Beazley, L., Kenchington, E., Whoriskey, F., Slaby, B. M., & Hentschel, U., (2020a). Microbial diversity of the glass sponge *Vazella pourtalesii* in response to anthropogenic activities. *Conserv. Genet., 21*, 1001–1010. doi: 10.1007/s10592-020-01305-2.

Busch, K., Hanz, U., Mienis, F., Mueller, B., Franke, A., Martyn, R. E., Tore, R. H., & Hentschel, U., (2020b). On giant shoulders: How a seamount affects the microbial community composition of seawater and sponges. *Biogeosciences, 17*, 3471–3486. doi: 10.5194/bg-17-3471-2020.

Colman, A. S., (2015). Sponge symbionts and the marine P cycle. *Proc. Natl. Acad. Sci. U. S. A., 112*, 4191–4192. doi: 10.1073/pnas.1502763112.

Corredor, J. E., Wilkinson, C. R., Vicente, V. P., Morell, J. M., & Otero, E., (1988). Nitrate release by Caribbeaan reef sponges. *Limnol. Oceanogr., 33*, 114–120.

Daims, H., Lebedeva, E. V., Pjevac, P., et al., (2015). Complete nitrification by *Nitrospira* bacteria. *Nature, 528*, 504–509. doi: 10.1038/nature16461.

Dash, S., Das, S. K., Samal, J., & Thatoi, N. H., (2018). Epidermal mucus, a major determinant in fish health: A review. *Iran. J. Vet. Res., 19*, 72–81. doi: 10.22099/ijvr.2018.4849.

Dayton, P. K., (1989). Interdecadal variation in an Antarctic sponge and its predator from oceanographic climate shifts. *Science, 245*, 1484–1486. doi: 10.1126/science.245.4925.1484.

Dayton, P. K., Robilliard, G. A., Paine, R. T., & Dayton, B. L., (1974). Biological accommodation in the benthic community at the McMurdo sound, Antarctica. *Ecol. Monogr., 44*, 105–128.

De Goeij, J. M., & Van, D. F. C., (2007). Coral cavities are sinks of dissolved organic carbon (DOC). *Limnol. Oceanogr., 52*, 2608–2617.

De Goeij, J. M., De Kluijver, A., Van, D. F. C., Vacelet, J., Wijffels, R. H., De Goeij, A. F. P. M., Cleutjens, J. P. M., & Schutte, B., (2009). Cell kinetics of the marine sponge *Halisarca caerulea* reveal rapid cell turnover and shedding. *J. Exp. Biol., 212*, 3892–3900.

De Goeij, J. M., Lesser, M. P., & Pawlik, R. J., (2017). Nutrient fluxes and ecological functions of coral reef sponges in a changing ocean. In: *Climate Change, Ocean Acidification and Sponges: Impacts Across Multiple Levels of Organization* (pp. 373–410). Springer International Publishing AG.

De Goeij, J. M., Van De, B. H., Van, O. M., Epping, E. H. G., & Van, D. F. C., (2008). Major bulk dissolved organic carbon (DOC) removal by encrusting coral reef cavity sponges. *Mar. Ecol. Prog. Ser., 357*, 139–151.

De Goeij, J. M., Van, O. D., Vermeij, M. J. A., Osinga, R., Middelburg, J. J., De Goeij, A. F. P. M., & Admiraal, W., (2013). Surviving in a marine desert: The sponge loop retains resources within coral reefs. *Science, 342*, 108–110.

De Kluijver, A., Bart, M. C., Van, O. D., et al., (2021). An integrative model of carbon and nitrogen metabolism in a common deep-sea sponge (*Geodia barretti*). *Front. Mar. Sci., 7*, 1131. doi: 10.3389/fmars.2020.596251.

De Mares, M. C., Jiménez, D. J., Palladino, G., et al., (2018). Expressed protein profile of a *Tectomicrobium* and other microbial symbionts in the marine sponge *Aplysina aerophoba* as evidenced by metaproteomics. *Sci. Rep., 8*, 1–14. doi: 10.1038/s41598-018-30134-0.

De Voogd, N. J., Alvarez, B., Boury-Esnault, N., et al., (2021). *World Porifera Database.* doi: 10.14284/359.

De Voogd, N. J., Cleary, D. F. R., Polónia, A. R. M., & Gomes, N. C. M., (2015). Bacterial community composition and predicted functional ecology of sponges, sediment and seawater from the thousand islands reef complex, West Java, Indonesia. *FEMS Microbiol. Ecol., 91*, 19. doi: 10.1093/FEMSEC/FIV019.

Delaney, M. L., (1998). Phosphorus accumulation in marine sediments and the oceanic phosphorus cycle. *Global Biogeochem. Cycles, 12*, 563–572. doi: 10.1029/98GB02263.

Diaz, M. C., & Rützler, K., (2001). Sponges: An essential component of Caribbean coral reefs. *Bull. Mar. Sci., 69*, 535–546.

Diaz, M. C., Akob, D., & Cary, C. S., (2004). Denaturing gradient gel electrophoreis of nitrifying microbes associated with tropical sponges. *Boll. dei Musei e degli Ist. Biol., 68*, 279–289.

Diaz, M., & Ward, B., (1997). Sponge-mediated nitrification in tropical benthic communities. *Mar. Ecol. Prog. Ser., 156*, 97–107. doi: 10.3354/meps156097.

Dosse, G., (1939). Bacterial and fungal findings as well as pathological and putrefactive processes in marine and freshwater sponges. *Zeitschrift für Parasitenkd., 11*, 331–356.

Duboscq, O., & Tuzet, O., (1942). Recherches complementaires sur l'ovogenèse, la fécondation et les premiers stades du développement des éponges calcaires. *Arch. Zool. Expérimentale Générale, 81*, 395–466.

Dyhrman, S. T., Chappell, P. D., Haley, S. T., Moffett, J. W., Orchard, E. D., Waterbury, J. B., & Webb, A. E., (2006). Phosphonate utilization by the globally important marine diazotroph *Trichodesmium*. *Nature, 439*, 68–71. doi: 10.1038/nature04203.

Ettwig, K. F., Butler, M. K., D. Le Paslier, et al., (2010). Nitrite-driven anaerobic methane oxidation by oxygenic bacteria. *Nature, 464*, 543–548. doi: 10.1038/nature08883.

Ewens, S. D., Gomberg, A. F. S., Barnum, T. P., Borton, M. A., Carlson, H. K., Wrighton, K. C., & Coates, D. J., (2021). The diversity and evolution of microbial dissimilatory phosphite oxidation. *Proc. Natl. Acad. Sci. U. S. A., 118*. doi: 10.1073/pnas.2020024118.

Fan, L., Reynolds, D., Liu, M., Stark, M., Kjelleberg, S., Webster, N. S., & Thomas, T., (2012). Functional equivalence and evolutionary convergence in complex communities of microbial sponge symbionts. *Proc. Natl. Acad. Sci. U. S. A., 109*, E1878–E1887. doi: 10.1073/pnas.1203287109.

Feingersch, R., Philosof, A., Mejuch, T., Glaser, F., Alalouf, O., Shoham, Y., & Béjà, O., (2012). Potential for phosphite and phosphonate utilization by *Prochlorococcus*. *ISME J., 6*, 827–834. doi: 10.1038/ismej.2011.149.

Feliatra, F., & Bianchi, M., (1993). Rates of nitrification and carbon uptake in the Rhône River plume (northwestern Mediterranean Sea). *Microb. Ecol., 26*(1), 21–28. doi: 10.1007/BF00166026.

Feng, G., & Li, Z., (2019). Carbon and nitrogen metabolism of sponge microbiome. In: *Symbiotic Microbiomes of Coral Reefs Sponges and Corals* (pp. 145–169). Springer Netherlands.

Feng, G., Sun, W., Zhang, F., Orlić, S., & Li, Z., (2018). Functional transcripts indicate phylogenetically diverse active ammonia-scavenging microbiota in sympatric sponges. *Mar. Biotechnol., 20*, 131–143. doi: 10.1007/s10126-018-9797-5.

Figueroa, I. A., & Coates, D. J., (2017). Microbial phosphite oxidation and its potential role in the global phosphorus and carbon cycles. *Adv. Appl. Microbiol., 98*, 93–117. doi: 10.1016/bs.aambs.2016.09.004.

Finks, R. M., Reid, R. E. H., & Rigby, K. J., (2004). In: Moore, R. C., & Kaesler, R. L., (eds.), *Treatise on invertebrate paleontology, Part E: Porifera* (Vol. 3). The Geological Society of America and The University of Kansas.

Fiore, C. L., Baker, D. M., & Lesser, P. M., (2013). Nitrogen biogeochemistry in the Caribbean sponge, *Xestospongia muta*: A source or sink of dissolved inorganic nitrogen? *PLoS One, 8*. doi: 10.1371/journal.pone.0072961.

Fiore, C. L., Labrie, M., Jarett, J. K., & Lesser, P. M., (2015). Transcriptional activity of the giant barrel sponge, *Xestospongia muta* holobiont: Molecular evidence for metabolic interchange. *Front. Microbiol., 6*. doi: 10.3389/fmicb.2015.00364.

Folkers, M., & Rombouts, T., (2020). Sponges revealed: A synthesis of their overlooked ecological functions within aquatic ecosystems. In: *YOUMARES 9 – The Oceans: Our Research, Our Future* (pp. 181–193). Springer International Publishing.

Fowler, S. J., Palomo, A., Dechesne, A., Mines, P. D., & Smets, F. B., (2018). Comammox *Nitrospira* are abundant ammonia oxidizers in diverse groundwater-fed rapid sand filter communities. *Environ. Microbiol., 20*, 1002–1015. doi: 10.1111/1462–2920.14033.

Freeman, C. J., Easson, C. G., Fiore, C. L., & Thacker, W. R., (2021). Sponge–microbe interactions on coral reefs: Multiple evolutionary solutions to a complex environment. *Front. Mar. Sci., 8*, 917. doi: 10.3389/FMARS.2021.705053.

Frøhlich, H., & Barthel, D., (1997). Silica uptake on the marine sponge *Halichondria panicea* in Kiel bight. *Mar. Biol., 128*, 115–125.

Gächter, R., & Meyer, S. J., (1993). The role of microorganisms in mobilization and fixation of phosphorus in sediments. *Hydrobiologia, 253*, 103–121. doi: 10.1007/BF00050731.

Gloeckner, V., Hentschel, U., Ereskovsky, A. V., & Schmitt, S., (2013). Unique and species-specific microbial communities in *Oscarella lobularis* and other Mediterranean *Oscarella* species (Porifera: Homoscleromorpha). *Mar. Biol., 160*, 781–791. doi: 10.1007/s00227-012-2133-0.

Griffiths, H. J., Anker, P., Linse, K., Maxwell, J., Post, A. L., Stevens, C., Tulaczyk, S., & Smith, A. J., (2021). Breaking all the rules: The first recorded hard substrate sessile benthic community far beneath an Antarctic ice shelf. *Front. Mar. Sci., 8*, 76. doi: 10.3389/fmars.2021.642040.

Haber, M., Burgsdorf, I., Handley, K. M., Rubin-Blum, M., & Steindler, L., (2021). Genomic insights into the lifestyles of *Thaumarchaeota* inside sponges. *Front. Microbiol., 11*, 3441. doi: 10.3389/fmicb.2020.622824.

Hadas, E., Ilan, M., & Shpigel, M., (2008). Oxygen consumption by a coral reef sponge. *J. Exp. Biol., 211*, 2185–2190.

Hallam, S. J., Konstantinidis, K. T., Putnam, N., et al., (2006). Genomic analysis of the uncultivated marine crenarchaeote *Cenarchaeum symbiosum*. *Proc. Natl. Acad. Sci. U. S. A., 103*, 18296–18301. doi: 10.1073/pnas.0608549103.

He, L., Karlep, L., & Li, Z., (2021). The nitrogen-cycling network of bacterial symbionts in the sponge *Spheciospongia vesparium*. *J. Ocean Univ. China., 20*(4), 999–1012. doi: 10.1007/S11802-021-4530-9.

Hentschel, U., Fieseler, L., Wehrl, M., Gernert, C., Steinert, M., Hacker, J., & Horn, M., (2003). Microbial diversity of marine sponges. In: Mueller, W. E. G., (ed.), *Molecular Marine Biology of Sponges* (p. 60–88). Springer Verlag.

Hentschel, U., Piel, J., Degnan, S. M., & Taylor, W. M., (2012). Genomic insights into the marine sponge microbiome. *Nat. Rev. Microbiol., 10*(9), 641–654. doi: 10.1038/nrmicro2839.

Hoer, D. R., Tommerdahl, J. P., Lindquist, N. L., & Martens, S. C., (2018). Dissolved inorganic nitrogen fluxes from common Florida Bay (U.S.A.) sponges. *Limnol. Oceanogr., 63*, 2563–2578. doi: 10.1002/lno.10960.

Hoffmann, F., Larsen, O., Rapp, H. T., & Osinga, R., (2005a). Oxygen dynamics in choanosomal sponge explants. *Mar. Biol. Res., 1*, 160–163. doi: 10.1080/17451000510019006.

Hoffmann, F., Larsen, O., Thiel, V., Rapp, H. T., Pape, T., Michaelis, W., & Reitner, J., (2005b). An anaerobic world in sponges. *Geomicrobiol. J., 22*, 1–10. doi: 10.1080/01490450590922505.

Hoffmann, F., Radax, R., Woebken, D., et al., (2009). Complex nitrogen cycling in the sponge *Geodia barretti*. *Environ. Microbiol., 11*, 2228–2243. doi: 10.1111/j.1462-2920.2009.01944.x.

Ilan, M., & Abelson, A., (1995). The life of a sponge in a sandy lagoon. *Biol. Bull. Mar. Biol. Lab. Woods Hole, 189*, 363–369.

Jiménez, E., & Ribes, M., (2007). Sponges as a source of dissolved inorganic nitrogen: Nitrification mediated by temperate sponges. *Limnol. Oceanogr., 52*, 948–958. doi: 10.4319/lo.2007.52.3.0948.

Kahn, A. S., Chu, J. W. F., & Leys, P. S., (2018). Trophic ecology of glass sponge reefs in the strait of Georgia, British Columbia. *Sci. Rep., 8*, 756. doi: 10.1038/s41598-017-19107-x.

Kamp, A., Høgslund, S., Risgaard-Petersen, N., & Stief, P., (2015). Nitrate storage and dissimilatory nitrate reduction by eukaryotic microbes. *Front. Microbiol., 6*, 1492. doi: 10.3389/fmicb.2015.01492.

Karimi, E., Slaby, B. M., Soares, A. R., Blom, J., Hentschel, U., & Costa, R., (2018). Metagenomic binning reveals versatile nutrient cycling and distinct adaptive features in alphaproteobacterial symbionts of marine sponges. *FEMS Microbiol. Ecol., 94*. doi: 10.1093/femsec/fiy074.

Karl, D. M., & Björkman, M. K., (2002). Dynamics of DOP. In: Hansell, D. A., & Carlson, C. A., (eds.), *Biogeochemistry of Marine Dissolved Organic Matter* (pp. 249–366). Elsevier.

Klawonn, I., Bonaglia, S., Whitehouse, M. J., Littmann, S., Tienken, D., Kuypers, M. M. M., Brüchert, V., & Ploug, H., (2019). Untangling hidden nutrient dynamics: Rapid ammonium cycling and single-cell ammonium assimilation in marine plankton communities. *ISME J., 13*, 1960–1974. doi: 10.1038/s41396-019-0386-z.

Klitgaard, A. B., & Tendal, S. O., (2004). Distribution and species composition of mass occurrences of large-sized sponges in the northeast Atlantic. *Prog. Oceanogr., 61*, 57–98. doi: http://dx.doi.org/10.1016/j.pocean.2004.06.002.

Klitgaard, A. B., (1995). The fauna associated with outer shelf and upper slope sponges (Porifera, Demospongiae) at the faroe islands, northeastern *Atlantic. Sarsia, 80*, 1–22. doi: 10.1080/00364827.1995.10413574.

Koch, H., Van, K. M. A. H. J., & Lücker, S., (2019). Complete nitrification: Insights into the ecophysiology of comammox *Nitrospira*. *Appl. Microbiol. Biotechnol., 103*, 177–189. doi: 10.1007/s00253-018-9486-3.

Könneke, M., Bernhard, A. E., De La Torre, J. R., Walker, C. B., Waterbury, J. B., & Stahl, A. D., (2005). Isolation of an autotrophic ammonia-oxidizing marine archaeon. *Nature, 437*, 543–546. doi: 10.1038/nature03911.

Kutti, T., Bannister, R. J., & Fosså, H. J., (2013). Community structure and ecological function of deep-water sponge grounds in the traenadypet MPA—Northern Norwegian continental shelf. *Cont. Shelf Res., 69*, 21–30. doi: http://dx.doi.org/10.1016/j.csr.2013.09.011.

Lanna, E., & Klautau, M., (2010). Oogenesis and spermatogenesis in *Paraleucilla magna* (*Porifera, Calcarea*). *Zoomorphology, 129*, 249–261. doi: 10.1007/s00435-010-0117-5.

Laubenfels, M. W. D., (1955). In: Moore, R. C., (ed.), *Porifera*. Geological Society of America and University of Kansas Press.

Lesser, M. P., (2006). Benthic-pelagic coupling on coral reefs: Feeding and growth of Caribbean sponges. *J. Exp. Mar. Bio. Ecol., 328*, 277–288.

Lévi, C., & Porte, A., (1962). Étude au microscope électronique de l'éponge *Oscarella lobularis* Schmidt et de sa larve amphiblastula. *Cah. Biol. Mar., 3*, 307–315.

Leys, S. P., Kahn, A. S., Fang, J. K. H., Kutti, T., & Bannister, J. R., (2018). Phagocytosis of microbial symbionts balances the carbon and nitrogen budget for the deep-water boreal sponge *Geodia barretti. Limnol. Oceanogr., 63*, 187–202. doi: 10.1002/lno.10623.

Leys, S. P., Yahel, G., Reidenbach, M. A., Tunnicliffe, V., Shavit, U., & Reiswig, M. H., (2011). The sponge pump: The role of current induced flow in the design of the sponge body plan. *PLoS One, 6*, e27787.

Li, Y., Wang, Y., Fu, L., Gao, Y., Zhao, H., & Zhou, W., (2017). Aerobic-heterotrophic nitrogen removal through nitrate reduction and ammonium assimilation by marine bacterium *Vibrio* sp. Y1-5. *Bioresour. Technol., 230*, 103–111. doi: 10.1016/j.biortech.2017.01.049.

Li, Z., Wang, Y., Li, J., Liu, F., He, L., He, Y., & Wang, S., (2016). Metagenomic analysis of genes encoding nutrient cycling pathways in the microbiota of deep-sea and shallow-water sponges. *Mar. Biotechnol., 18*, 659–671. doi: 10.1007/s10126-016-9725-5.

Liu, F., Li, J., Feng, G., & Li, Z., (2016). New genomic insights into "Entotheonella" symbionts in *Theonella swinhoei*: Mixotrophy, anaerobic adaptation, resilience, and interaction. *Front. Microbiol., 7*, 1333. doi: 10.3389/fmicb.2016.01333.

Lomnitz, U., Sommer, S., Dale, A. W., Löscher, C. R., Noffke, A., Wallmann, K., & Hensen, C., (2016). Benthic phosphorus cycling in the Peruvian oxygen minimum zone. *Biogeosciences, 13*, 1367–1386. doi: 10.5194/bg-13–1367–2016.

López-Acosta, M., Leynaert, A., Chavaud, L., Amice, E., Bihannic, I., Le Bec, T., & Maldonado, M., (2019). *In situ* determination of Si, N, and P utilization by the demosponge *Tethya citrina*: A benthic-chamber approach. *PLoS One, 14*. doi: 10.1371/journal.pone.0218787.

López-Legentil, S., Erwin, P. M., Pawlik, J. R., & Song, B., (2010). Effects of sponge bleaching on ammonia-oxidizing archaea: Distribution and relative expression of ammonia monooxygenase genes associated with the barrel sponge *Xestospongia muta. Microb. Ecol., 60*, 561–571. doi: 10.1007/s00248-010-9662-1.

Maldonado, M., & Young, M. C., (1998). Limits on the bathymetric distribution of keratose sponges: A field test in deep water. *Mar. Ecol. Prog. Ser., 174*, 123–139. doi: 10.3354/meps174123.

Maldonado, M., (2007). Intergenerational transmission of symbiotic bacteria in oviparous and viviparous demosponges, with emphasis on intracytoplasmically-compartmented bacterial types. *J. Mar. Biol. Assoc. United Kingdom, 87*, 1701–1713. doi: 10.1017/S0025315407058080.

Maldonado, M., (2015). Sponge waste that fuels marine oligotrophic food webs: A re-assessment of its origin and nature. *Mar. Ecol., 37*, 1–15. doi: 10.1111/maec.12256.

Maldonado, M., Aguilar, R., Bannister, R. J., et al., (2017). Sponge grounds as key marine habitats: A synthetic review of types, structure, functional roles, and conservation concerns. In: Sergio, R., Bramanti, L., Gori, A., & Orejas, C., (eds.), *Marine Animal Forests: The Ecology of Benthic Biodiversity Hotspots* (pp. 145–183). Springer International Publishing.

Maldonado, M., Beazley, L., López-Acosta, M., Kenchington, E., Casault, B., Hanz, U., & Mienis, F., (2021a). Massive silicon utilization facilitated by a benthic-pelagic coupled feedback sustains deep-sea sponge aggregations. *Limnol. Oceanogr., 66*, 366–391. doi: 10.1002/lno.11610.

Maldonado, M., López-Acosta, M., Beazley, L., Kenchington, E., Koutsouveli, V., & Riesgo, A., (2020). Cooperation between passive and active silicon transporters clarifies the

ecophysiology and evolution of biosilicification in sponges. *Sci. Adv., 6*, eaba9322. doi: 10.1126/sciadv.aba9322.
Maldonado, M., López-Acosta, M., Busch, K., et al., (2021b). A microbial nitrogen engine modulated by bacteriosyncytia in hexactinellid sponges: Ecological implications for deep-sea communities. *Front. Mar. Sci., 8*, 638505. doi: 10.3389/fmars.2021.638505.
Maldonado, M., López-Acosta, M., Sitjà, C., García-Puig, M., Galobart, C., Ercilla, G., & Leynaert, A., (2019). Sponge skeletons as an important sink of silicon in the global oceans. *Nat. Geosci., 12*, 815–822. doi: 10.1038/s41561-019-0430-7.
Maldonado, M., Navarro, L., Grasa, A., Gonzalez, A., & Vaquerizo, I., (2011). Silicon uptake by sponges: A twist to understanding nutrient cycling on continental margins. *Sci. Rep., 1*. doi: 10.1038/srep00030.
Maldonado, M., Ribes, M., & Van, D. F. C., (2012). Nutrient fluxes through sponges. Biology, budgets, and ecological implications. *Adv. Mar. Biol., 62*, 113–182. doi: 10.1016/B978-0-12-394283-8.00003-5.
McMurray, S. E., Pawlik, J. R., & Finelli, M. C., (2014). Trait-mediated ecosystem impacts: How morphology and size affect pumping rates of the Caribbean giant barrel sponge. *Aquat. Biol., 23*, 1–13. doi: 10.3354/AB00612.
Metcalf, W. W., Griffin, B. M., Cicchillo, R. M., et al., (2012). Synthesis of methylphosphonic acid by marine microbes: A source for methane in the aerobic ocean. *Science, 337*, 1104–1107. doi: 10.1126/science.1219875.
Meyer, B., & Kuever, J., (2008). Homology modeling of dissimilatory APS reductases (AprBA) of sulfur-oxidizing and sulfate-reducing prokaryotes. *PLoS One, 3*, e1514. doi: 10.1371/journal.pone.0001514.
Micaroni, V., Strano, F., Mcallen, R., Woods, L., Turner, J., Harman, L., & Bell, J. J., (2021). Adaptive strategies of sponges to deoxygenated oceans. *Glob. Chang. Biol.* doi: 10.1111/GCB.16013.
Middelburg, J. J., Soetaert, K., Herman, P. M. J., & Heip, C. H. R., (1996). Denitrification in marine sediments: A model study. *Global Biogeochem. Cycles, 10*, 661–673. doi: 10.1029/96GB02562.
Moeller, F. U., Webster, N. S., Herbold, C. W., et al., (2019). Characterization of a thaumarchaeal symbiont that drives incomplete nitrification in the tropical sponge *Ianthella basta*. *Environ. Microbiol., 21*, 3831–3854. doi: 10.1111/1462-2920.14732.
Mohamed, N. M., Colman, A. S., Tal, Y., & Hill, T. R., (2008). Diversity and expression of nitrogen fixation genes in bacterial symbionts of marine sponges. *Environ. Microbiol., 10*, 2910–2921. doi: 10.1111/j.1462-2920.2008.01704.x.
Mohamed, N. M., Saito, K., Tal, Y., & Hill, T. R., (2010). Diversity of aerobic and anaerobic ammonia-oxidizing bacteria in marine sponges. *ISME J., 4*, 38–48. doi: 10.1038/ismej.2009.84.
Moitinho-Silva, L., Díez-Vives, C., Batani, G., Esteves, A. I. S., Jahn, M. T., & Thomas, T., (2017a). Integrated metabolism in sponge-microbe symbiosis revealed by genome-centered metatranscriptomics. *ISME J., 11*, 1651–1666. doi: 10.1038/ismej.2017.25.
Moitinho-Silva, L., Seridi, L., Ryu, T., Voolstra, C. R., Ravasi, T., & Hentschel, U., (2014). Revealing microbial functional activities in the Red Sea sponge *Stylissa carteri* by metatranscriptomics. *Environ. Microbiol., 16*, 3683–3698. doi: 10.1111/1462-2920.12533.
Moitinho-Silva, L., Steinert, G., Nielsen, S., et al., (2017b). Predicting the HMA-LMA status in marine sponges by machine learning. *Front. Microbiol., 8*, 752. doi: 10.3389/fmicb.2017.00752.

Morganti, T. M., Ribes, M., Yahel, G., & Coma, R., (2019). Size is the major determinant of pumping rates in marine sponges. *Front. Physiol., 10*. doi: 10.3389/fphys.2019.01474.

Morganti, T., Coma, R., Yahel, G., & Ribes, M., (2017). Trophic niche separation that facilitates co-existence of high and low microbial abundance sponges is revealed by *in situ* study of carbon and nitrogen fluxes. *Limnol. Oceanogr., 62*, 1963–1983. doi: 10.1002/lno.10546.

Morganti, T., Yahel, G., Ribes, M., & Coma, R., (2016). VacuSIP, an improved in-ex method for *in situ* measurement of particulate and dissolved compounds processed by active suspension feeders. *J. Vis. Exp., 114*, e54221. doi:doi: 10.3791/54221.

Nelson, C. E., Alldredge, A. L., McCliment, E. A., Amaral-Zettler, L. A., & Carlson, A. C., (2011). Depleted dissolved organic carbon and distinct bacterial communities in the water column of a rapid-flushing coral reef ecosystem. *ISME J., 5*, 1374–1387.

Ni, S. Q., & Zhang, J., (2013). Anaerobic ammonium oxidation: From laboratory to full-scale application. *Biomed Res. Int., 2013*. doi: 10.1155/2013/469360.

O'Brien, P. A., Andreakis, N., Tan, S., Miller, D. J., Webster, N. S., Zhang, G., & Bourne, G. D., (2021). Testing cophylogeny between coral reef invertebrates and their bacterial and archaeal symbionts. *Mol. Ecol.* mec.16006. doi: 10.1111/mec.16006.

Ou, H., Li, M., Wu, S., Jia, L., Hill, R. T., & Zhao, J., (2020). Characteristic microbiomes correlate with polyphosphate accumulation of marine sponges in South China sea areas. *Microorganisms, 8*, 63. doi: 10.3390/microorganisms8010063.

Paytan, A., & McLaughlin, K., (2007). The oceanic phosphorus cycle. *Chem. Rev., 107*, 563–576. doi: 10.1021/cr0503613.

Perea-Blázquez, A., Davy, S., & Bell, J., (2012). Nutrient utilization by shallow water temperate sponges in New Zealand. *Hydrobiologia, 687*, 237–250. doi: 10.1007/s10750-011-0798-x.

Pérez-Botello, A. M., & Simões, N., (2021). Sponge-dwelling fauna: A review of known species from the Northwest Tropical Atlantic coral reefs. *Biodivers. Data J., 9*, 1–18. doi: 10.3897/BDJ.9.e63372.

Petrie, B., Yeats, P., & Strain, P., (1999). Nitrate, Silicate and Phosphate Atlas for the Scotian Shelfand the Gulf of Maine. Fisheries and Oceans Canada. *Canadian Technical Report of Hydrography and Ocean Sciences, 342*.

Pham, C. K., Murillo, F. J., Lirette, C., Maldonado, M., Colaço, A., Ottaviani, D., & Kenchington, E., (2019). Removal of deep-sea sponges by bottom trawling in the Flemish cap area: Conservation, ecology and economic assessment. *Sci. Rep., 9*. doi: 10.1038/s41598-019-52250-1.

Pile, A. J., & Young, M. C., (2006). The natural diet of a hexactinellid sponge: Benthic-pelagic coupling in a deep-sea microbial food web. *Deep. Res. Part I Oceanogr. Res. Pap., 53*, 1148–1156. doi: https://doi.org/10.1016/j.dsr.2006.03.008.

Preston, C. M., Wu, K. Y., Molinski, T. F., & DeLong, F. E., (1996). A psychrophilic crenarchaeon inhabits a marine sponge: *Cenarchaeum symbiosum* gen. nov., sp. nov. *Proc. Natl. Acad. Sci. USA, 93*, 6241–6246.

Radax, R., Hoffmann, F., Rapp, H. T., Leininger, S., & Schleper, C., (2012a). Ammonia-oxidizing archaea as main drivers of nitrification in cold-water sponges. *Environ. Microbiol., 14*, 909–923. doi: 10.1111/j.1462-2920.2011.02661.x.

Radax, R., Rattei, T., Lanzen, A., Bayer, C., Rapp, H. T., Urich, T., & Schleper, C., (2012b). Metatranscriptomics of the marine sponge *Geodia barretti:* Tackling phylogeny and function of its microbial community. *Environ. Microbiol., 14*, 1308–1324. doi: 10.1111/j.1462-2920.2012.02714.x.

Randall, J. E., & Lobel, S. P., (2009). A literature review of the sponge-dwelling gobiid fishes of the genus *Elacatinus* from the western Atlantic, with description of two new Caribbean species. *Zootaxa, 2133*, 1–19. doi: 10.11646/zootaxa.2133.1.1.
Reiswig, H. M., (1971a). Particle feeding in natural populations of three marine demosponges. *Biol. Bull. Mar. Biol. Lab. Woods Hole, 141*, 568–591.
Reiswig, H. M., (1971b). *In situ* pumping activities of tropical *Demospongiae*. *Mar. Biol., 9*, 38–50.
Reiswig, H. M., (1981). Partial carbon and energy budgets of the bacteriosponge *Verongia fistularis* (Porifera: Demopongiae) in Barbados. *Pubbl. della Stn. Zool. di Napoli Mar. Ecol., 2*, 273–293.
Ribes, M., Coma, R., Gili, J. M., Svoboda, A., Julia, A., & Parera, J., (2000). A "semi-closed" recirculating system for the *in-situ* study of feeding and respiration of benthic suspension feeders. *Scirntia Mar., 64*, 265–275.
Ribes, M., Jiménez, E., Yahel, G., López-Sendino, P., Diez, B., Massana, R., Sharp, J. H., & Coma, R., (2012). Functional convergence of microbes associated with temperate marine sponges. *Environ. Microbiol., 14*, 1224–1239. doi: 10.1111/j.1462-2920.2012.02701.x.
Richter, C., & Wunsch, M., (1999). Cavity-dwelling suspension feeders in coral reefs - a new link in reef trophodynamics. *Mar. Ecol. Progess Ser., 188*, 105–116.
Robertson, G. P., & Groffman, M. P., (2015). Nitrogen transformations. In: Paul, E. A., (ed.), *Soil Microbiology, Ecology and Biochemistry* (pp. 421–446). Elsevier.
Romillac, N., (2018). Ammonification. In: *Encyclopedia of Ecology* (pp. 256–263). Elsevier.
Rooks, C., Kar-Hei, F. J., Mørkved, P. T., Zhao, R., Tore, R. H., Xavier, J. J., & Hoffmann, F., (2020). Deep-sea sponge grounds as nutrient sinks: Denitrification is common in boreo-arctic sponges. *Biogeosciences, 17*, 1231–1245. doi: 10.5194/bg-17-1231-2020.
Scheffers, S. R., Nieuwland, G., Bak, R. P. M., & Van, D. F. C., (2004). Removal of bacteria and nutrient dynamics within the coral reef framework of Curaçao (Netherlands Antilles). *Coral Reefs, 23*, 413–422. doi: 10.1007/s00338-004-0400-3.
Schläppy, M. L., Hoffmann, F., Roy, H., Wijffels, R. H., Mendola, D., Sidri, M., & De Beer, D., (2007). Oxygen dynamics and flow patterns of *Dysidea avara* (Porifera: Demospongiae). *J. Mar. Biol. Assoc. United Kingdom, 87*, 1677–1682.
Schläppy, M. L., Weber, M., Mendola, D., Hoffmann, F., & De Beer, D., (2010). Heterogeneous oxygenation resulting from active and passive flow in two Mediterranean sponges, *Dysidea avara* and *Chondrosia reniformis*. *Limnol. Oceanogr., 55*, 1289–1300. doi: 10.4319/lo.2010.55.3.1289.
Schmitt, S., Angermeier, H., Schiller, R., Lindquist, N., & Hentschel, U., (2008). Molecular microbial diversity survey of sponge reproductive stages and mechanistic insights into vertical transmission of microbial symbionts. *Appl. Environ. Microbiol., 74*, 7694–7708. doi: 10.1128/AEM.00878-08.
Seitzinger, S. P., & Giblin, E. A., (1996). Estimating denitrification in North Atlantic continental shelf sediments. *Biogeochem., 35*(1), 235–260. doi: 10.1007/BF02179829.
Siegl, A., Kamke, J., Hochmuth, T., Piel, J., Richter, M., Liang, C., Dandekar, T., & Hentschel, U., (2011). Single-cell genomics reveals the lifestyle of poribacteria, a candidate phylum symbiotically associated with marine sponges. *ISME J., 5*, 61–70.
Simpson, T. L., & Volcani, E. B., (1981). *Silicon and Siliceous Structures in Biological Systems*. Springer-Verlag.

Slaby, B. M., Hackl, T., Horn, H., Bayer, K., & Hentschel, U., (2017). Metagenomic binning of a marine sponge microbiome reveals unity in defense but metabolic specialization. *ISME J.* doi: 10.1038/ismej.2017.101.

Slomp, C. P., Thomson, J., & De Lange, G. J., (2004). Controls on phosphorus regeneration and burial during formation of eastern Mediterranean sapropels. *Mar. Geol., 203*, 141–159. doi: 10.1016/S0025-3227(03)00335-9.

Southwell, M. W., Popp, B. N., & Martens, S. C., (2008). Nitrification controls on fluxes and isotopic composition of nitrate from Florida keys sponges. *Mar. Chem., 108*, 96–108. doi: 10.1016/j.marchem.2007.10.005.

Steenbergh, A. K., Bodelier, P. L. E., Slomp, C. P., & Laanbroek, J. H., (2014). Effect of redox conditions on bacterial community structure in Baltic Sea sediments with contrasting phosphorus fluxes. *PLoS One, 9*, 92401. doi: 10.1371/journal.pone.0092401.

Steger, D., Ettinger-Epstein, P., Whalan, S., Hentschel, U., De Nys, R., Wagner, M., & Taylor, W. M., (2008). Diversity and mode of transmission of ammonia-oxidizing archaea in marine sponges. *Environ. Microbiol., 10*, 1087–1094.

Steinert, G., Busch, K., Bayer, K., et al., (2020). Compositional and quantitative insights into bacterial and archaeal communities of South Pacific deep-sea sponges (Demospongiae and Hexactinellida). *Front. Microbiol., 11*. doi: 10.3389/fmicb.2020.00716.

Strehlow, B. W., Jorgensen, D., Webster, N. S., Pineda, M. C., & Duckworth, A., (2016). Using a thermistor flowmeter with attached video camera for monitoring sponge excurrent speed and oscular behavior. *PeerJ, 2016*, 1–13. doi: 10.7717/peerj.2761.

Su, J., Jin, L., Jiang, Q., Sun, W., Zhang, F., & Li, Z., (2013). Phylogenetically diverse ureC genes and their expression suggest the urea utilization by bacterial symbionts in marine sponge *Xestospongia testudinaria*. *PLoS One, 8*, 64848. doi: 10.1371/journal.pone.0064848.

Subina, N. S., Thorat, B. R., & Gonsalves, J. M., (2018). Nitrification in intertidal sponge *Cinachyrella cavernosa*. *Aquat. Ecol., 52*, 155–164. doi: 10.1007/s10452-018-9651-x.

Taylor, M. W., Radax, R., Steger, D., & Wagner, M., (2007). Sponge-associated microorganisms: Evolution, ecology, and biotechnological potential. *Microbiol. Mol. Biol. Rev., 71*, 295–347. doi: 10.1128/mmbr.00040-06.

Thomas, T., Moitinho-Silva, L., Lurgi, M., et al., (2016). Diversity, structure and convergent evolution of the global sponge microbiome. *Nat. Commun., 7*, 11870. doi: 10.1038/ncomms11870.

Thomas, T., Rusch, D., DeMaere, M. Z., et al., (2010). Functional genomic signatures of sponge bacteria reveal unique and shared features of symbiosis. *ISME J., 4*, 1557–1567. doi: 10.1038/ismej.2010.74.

Tian, R. M., Sun, J., Cai, L., Zhang, W. P., Zhou, G. W., Qiu, J. W., & Qian, Y. P., (2016). The deep-sea glass sponge *Lophophysema eversa* harbors potential symbionts responsible for the nutrient conversions of carbon, nitrogen and sulfur. *Environ. Microbiol., 18*, 2481–2494. doi: 10.1111/1462-2920.13161.

Tian, R. M., Zhang, W., Cai, L., Wong, Y. H., Ding, W., & Qian, P. Y., (2017). Genome reduction and microbe-host interactions drive adaptation of a sulfur-oxidizing bacterium associated with a cold seep sponge. *mSystems, 2*. doi: 10.1128/msystems.00184-16.

Turon, M., & Uriz, J. M., (2020). New insights into the archaeal consortium of tropical sponges. *Front. Mar. Sci., 6*, 789. doi: 10.3389/fmars.2019.00789.

Tyler, J. C., & Böhlke, E. J., (1972). Records of sponge-dwelling fishes, primarily of the Caribbean. *Bull. Mar. Sci., 22*, 601–642.

Tyrrell, T., (1999). The relative influences of nitrogen and phosphorus on oceanic primary production. *Nat., 400*, 525–531. doi: 10.1038/22941.

Uriz, M. J., Turon, X., Becerro, M. A., & Agell, G., (2003). Siliceous spicules and skeleton frameworks in sponges: Origin, diversity, ultrastructural patterns, and biological functions. *Microsc. Res. Tech., 62*, 279–299. doi: 10.1002/jemt.10395.

Vacelet, J., (1971). Etude en microscopie électronique de l'association entre une cyanophycée chroococcale et une éponge du genre *Verongia. J. Microsc.* (Paris), *12*, 363–380.

Vacelet, J., Boury-Esnault, N., Fiala-Medioni, A., & Fisher, R. C., (1995). A methanotrophic carnivorous sponge. *Nature, 377*, 296.

Vacelet, J., Fiala-Medioni, A., Fisher, C. R., & Boury-Esnault, N., (1996). Symbiosis between a methane-oxidizing bacteria and a deep-sea carnivoroous cladorhizid sponge. *Mar. Ecol. Prog. Ser., 145*, 77–85.

Valentine, M. M., & Butler, J. M., (2019). Sponges structure water-column characteristics in shallow tropical coastal ecosystems. *Mar. Ecol. Prog. Ser., 608*, 133–147. doi: 10.3354/meps12758.

Van, D. F. C., Lengger, S. K., Schouten, S., Lundälv, T., Van, O. D., & Müller, E. C., (2020). Dark CO2 fixation into phospholipid-derived fatty acids by the cold-water coral associated sponge *Hymedesmia (Stylopus) coriacea* (Tisler Reef, NE Skagerrak). *Mar. Biol. Res.*, 1–17. doi: 10.1080/17451000.2019.1704019.

Van, K. M. A. H. J., Speth, D. R., Albertsen, M., Nielsen, P. H., Op Den, C. H. J. M., Kartal, B., et al., (2015). Complete nitrification by a single microorganism. *Nature, 528*, 555–559. doi: 10.1038/nature16459.

Van, M. B. A. S., Krupke, A., Dyhrman, S. T., et al., (2015). Major role of planktonic phosphate reduction in the marine phosphorus redox cycle. *Science, 348*, 783–785. doi: 10.1126/science.aaa8181.

Villarreal-Chiu, J. F., Quinn, J. P., & McGrath, W. J., (2012). The genes and enzymes of phosphonate metabolism by bacteria, and their distribution in the marine environment. *Front. Microbiol., 3*. doi: 10.3389/fmicb.2012.00019.

Vogel, S., & Bretz, L. W., (1972). Interfacial organisms: Passive ventilation in the velocity gradients near surfaces. *Science, 175*, 210–211.

Vogel, S., (1977). Current-induced flow through living sponges in nature. *Proc. Natl. Acad. Sci. USA, 74*, 2069–2071.

Voultsiadou-Koukoura, H. E., Koukouras, A., & Eleftheriou, A., (1987). Macrofauna associated with the sponge *Verongia aerophoba* in the north Aegean Sea. *Estuar. Coast. Shelf Sci., 24*, 265–278. doi: 10.1016/0272-7714(87)90069-2.

Webster, N. S., & Taylor, W. M., (2012). Marine sponges and their microbial symbionts: Love and other relationships. *Environ. Microbiol., 14*, 335–346. doi: 10.1111/j.1462-2920.2011.02460.x.

Webster, N. S., & Thomas, T., (2016). The sponge hologenome. *MBio, 7*. doi: 10.1128/mBio.00135-16.

Weigel, B. L., & Erwin, M. P., (2017). Effects of reciprocal transplantation on the microbiome and putative nitrogen cycling functions of the intertidal sponge, *Hymeniacidon heliophila. Sci. Rep., 7*, 1–12. doi: 10.1038/srep43247.

Weisz, J. B., Massaro, A. J., Ramsby, B. D., & Hill, S. M., (2010). Zooxanthellar symbionts shape host sponge trophic status through translocation of carbon. *Biol. Bull. Mar. Biol. Lab. Woods Hole, 219*, 189–197.

White, A. K., & Metcalf, W. W., (2007). Microbial metabolism of reduced phosphorus compounds. *Annu. Rev. Microbiol., 61*, 379–400. doi: 10.1146/annurev.micro.61.080706.093357.

Wiedenmayer, F., (1994). Contributions to the knowledge of post-Paleozoic neritic and archibental sponges (Porifera). *Schweizerische Paläontologidche Abhandlungen, 116*, 1–147.

Wild, C., Hüttel, M., Klüter, A., Kremb, S. G., Rasheed, M. Y. M., & Jørgensen, B. B., (2004). Coral mucus functions as an energy carrier and particle trap in the reef ecosystem. *Nature, 428*, 66–70.

Wild, C., Mayr, C., Wehrmann, L., Schottner, S. I., Naumann, M., Hoffmann, F., & Rapp, T. H., (2008). Organic matter release by cold water corals and its implication for fauna-microbe interaction. *Mar. Ecol. Prog. Ser., 372*, 67–75.

Wilkinson, C. R., & Fay, P., (1979). Nitrogen fixation in coral reef sponges with symbiotic cyanobacteria. *Nature, 279*, 527–529.

Wilkinson, C. R., & Garrone, R., (1980). Ultrastructure of siliceous spicules and microsclerocytes in the marine sponge *Neofibularia irata* N. sp. *J. Morphol., 166*, 51–63. doi: 10.1002/jmor.1051660105.

Wilkinson, C. R., (1978). Microbial associations in sponges. II. Numerical analysis of sponge and water bacterial populations. *Mar. Biol., 49*, 169–176.

Wilkinson, C. R., (1987). Interocean differences in size and nutrition of coral reefs sponge populations. *Science, 236*, 1654–1657.

Wilkinson, C. R., Summons, R. E., & Evans, E., (1999). Nitrogen fixation in symbiotic marine sponges: Ecological significance and difficulties in detection. *Mem. Queensl. Museum, 44*, 667–673.

Wulff, J. L., (2001). Assessing and monitoring coral reef sponges: Why and how? *Bull. Mar. Sci., 69*, 831–846.

Yahel, G., Marie, D., & Genin, A., (2005). InEx-a direct *in situ* method to measure filtration rates, nutrition and metabolism of active suspension feeders. *Limnol. Oceanogr. Methods, 3*, 46–58.

Yahel, G., Whitney, F., Reiswig, H. M., Eerkes-Medrano, D. I., & Leys, S. P., (2007). *In situ* feeding and metabolism of glass sponges (Hexactinellida, Porifera) studied in a deep temperate fjord with a remotely operated submersible. *Limnol. Oceanogr., 52*, 428–440.

Zakem, E. J., Al-Haj, A., Church, M. J., et al., (2018). Ecological control of nitrite in the upper ocean. *Nat. Commun., 9*, 1–13. doi: 10.1038/s41467-018-03553-w.

Zhang, F., Blasiak, L. C., Karolin, J. O., Powell, R. J., Geddes, C. D., Hill, R. T., & Karl, M. D., (2015). Phosphorus sequestration in the form of polyphosphate by microbial symbionts in marine sponges. *Proc. Natl. Acad. Sci. U. S. A., 112*, 4381–4386. doi: 10.1073/pnas.1423768112.

Zhang, F., Jonas, L., Lin, H., & Hill, R. T., (2019a). Microbially mediated nutrient cycles in marine sponges. *FEMS Microbiol. Ecol., 95*, 155. doi: 10.1093/femsec/fiz155.

Zhang, F., Vicente, J., & Hill, T. R., (2014). Temporal changes in the diazotrophic bacterial communities associated with Caribbean sponges *Ircinia stroblina* and *Mycale laxissima*. *Front. Microbiol., 5*, 561. doi: 10.3389/fmicb.2014.00561.

Zhang, S., Song, W., Wemheuer, B., Reveillaud, J., Webster, N., & Thomas, T., (2019b). Comparative genomics reveals ecological and evolutionary insights into sponge-associated *Thaumarchaeota*. *mSystems, 4*. doi: 10.1128/msystems.00288-19.

Zhang, X., He, L., Zhang, F., Sun, W., & Li, Z., (2013). The different potential of sponge bacterial symbionts in N_2 release indicated by the phylogenetic diversity and abundance analyses of denitrification genes, *nirK* and *nosZ*. *PLoS One, 8*, 65142. doi: 10.1371/journal.pone.0065142.

CHAPTER 5

Nerve Nets and Centralized Nervous Systems in Cnidarians

RICHARD A. SATTERLIE

Department of Biology and Marine Biology, University of North Carolina Wilmington, Wilmington, NC, USA

ABSTRACT

The most noticeable characteristic of the cnidarian nervous system, the nerve net, has become a symbol of the nervous complexity of the group. This greatly underestimates the morphological and physiological capabilities of cnidarians. A nerve net, in isolation, has significant integrative capabilities. But the cnidarian nervous systems consist of multiple interacting conducting systems (some diffuse nerve nets) and are capable of controlling rich behavior repertoires. In many cnidarians, the compression of nerve nets gives the appearance and function of directional nerves, while the collection of multiple, interacting, conducting systems in the nerve rings (NRs) of hydromedusae form a centralization that functions like the central nervous systems of other animals. In scyphozoan and cubozoan medusae, rhopalia combine sensory structures with complex neuropils to form integration centers that function much like ganglia. The two forms of centralization in medusae perform the functions of central nervous systems but under the constraints of radial symmetry.

INTRODUCTION

Cnidarians occupy an interesting position when it comes to nervous system organization and function. They are among the earlier metazoans, with an initial appearance 600 million years ago (Park et al., 2012; Takahashi, 2020). Along

Frontiers in Invertebrate Physiology: A Collection of Reviews, Volume 1: Non-Bilaterian Phyla.
Saber Saleuddin, Sally P. Leys, Robert D. Roer, and Iain C. Wilkie, (Eds.)

with Ctenophores, they represent the earliest animals with distinct, multicellular nervous systems (Lichneckert & Reichert, 2007; Watanabe et al., 2009).

Multicellularity was a major evolutionary advance that allowed cellular specialization but also required developmental and regulatory cellular communication. Interestingly, many genes that are now associated with the multicellular condition appear to have had a unicellular origin (Sebe-Pedros et al., 2017). In particular, it was suggested that electrogenesis, sensation, and neurosecretion may have co-evolved (Westfall, 1973; Westfall & Kinnamon, 1978). Comparative genomic data further confirm that genes related to nervous system form and function, as well as neurogenesis, neuronal development and specification, and network organization that is found in other animals, are present in cnidarians (Marlow et al., 2009; Watanabe et al., 2009). Early speculation on the independent origins of neurons and muscle cells (Hertwig & Hertwig, 1878; Claus, 1878; Chun, 1880) has been advanced to suggest that neurons evolved multiple times (Moroz, 2009; Moroz & Kohn, 2016) and that neurons may have developed from secretory cells (Grundfest, 1959, 1965; Moroz, 2009, 2018).

The cnidarian nervous systems control relatively rich behavioral repertoires and yet show what is perceived as a relatively simple level of anatomical organization. Due to this, and their phylogenetic position, the cnidarian nervous system organization has been labeled as "primitive" in comparison to that of other animals. Efforts have been made to construct a step-wise pathway for the evolution of nervous systems, with the cnidarians representing a basal organization.

The most apparent feature of cnidarian nervous systems is the nerve net, typically a two-dimensional array of neurons that is capable of non-polarized conduction. The diffuse organization, frequently with neurons running in many directions, has become a widely recognized generalization for the cnidarian nervous system, as well as for some other basal phyla (Figure 5.1). This is so prevalent that these groups are known as "Nerve Net Animals." This well-accepted view greatly underestimates the actual nervous system complexity of many members of the phylum. And due to the development of cephalization in the Bilateria, the cnidarians are frequently assumed to lack any significant form of centralization.

Generalizations aside, the prevailing view is of a nervous system comprised of "simple" (diffuse) networks of neurons with some degree of regional condensation and sometimes with distinct polarization (Koizumi et al., 2004; Watanabe et al., 2009). The concept of the central nervous system is reserved for later lineages, with the possibility that complex brains and

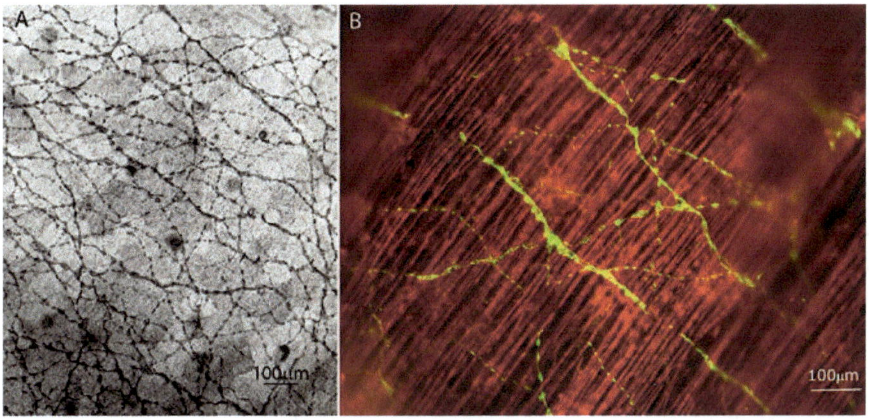

FIGURE 5.1 Subumbrellar nerve nets of scyphozoan (A) and cubozoan (B) medusae. (A) the motor nerve net of *Aurelia*, tubulin immunolabel; (B) the motor nerve net of *Carybdea*, yellow: tubulin immunolabel; red: actin immunolabel.

nervous systems had multiple origins in these lineages (Moroz, 2009). Here, it is necessary to distinguish between a Central Nervous System and centralization within nervous systems that lie outside of the bilaterian construct. This distinction does not negate the common functional value of these forms of centralization, namely an enhanced integrative ability.

To gain an understanding of the organization and function of cnidarian nervous systems, constraints on the systems will be considered, some of which contributed to the selective pressures that gave rise to the nervous systems of extant cnidarians. For the most part, general differences from the cnidarian groups will be emphasized, while species-specific differences, which result from ecological and environmental specializations, will not be detailed except for specific examples (*see* Satterlie, 2019 for a class-by-class chapter on cnidarian nervous systems).

5.1 BODY FORM

Cnidarians have been considered diploblastic organisms with a predominant radial symmetry. Both have influences on the nervous organization of the group. However, neither aspects of body organization are as clear-cut as many textbooks state.

The generalized diploblastic scheme includes two tissue layers – the epidermis (derived from the ectoderm) and the gastrodermis (derived from the endoderm), typically separated by a mesogleal connective tissue. This is in contrast to the majority of metazoan phyla, which are triploblastic, with a mesodermal layer that, most notably, gives rise to striated muscle. Recent molecular and genetic studies suggest that triploblasty preceded the emergence of the bilateria (Martindale et al., 2004; Boero et al., 2005, 2007; Seipel & Schmid, 2005, 2006; Burton, 2007; Wijesena et al., 2017). For example, the mesenteric retractor muscles of some anthozoan polyps and tentacle retractor muscles of some jellyfish show characteristics of mesodermal origin. In hydrozoan medusae, a specialized tissue called the endocodon produces tissue that is independent of the ectoderm or the endoderm and gives rise to the subumbrellar striated muscle (Boero et al., 1998; Ball et al., 2004). The endocodon may be homologous with the mesoderm of bilaterians (Boero et al., 1998, 2007; Spring et al., 2000, 2002; Muller et al., 2003; Seipel & Schmid, 2005; Burton, 2007; Wijesena et al., 2017). The concentration of parallel muscle cells into muscle bands is interesting from the viewpoint of neuronal organization because where muscle goes, neurons typically follow, except where muscle cells show intercellular conduction via gap junctions (discussed later).

Body symmetry has a significant influence on behavior, particularly locomotion. For this reason, directed locomotion is considered an important contributor to the evolution of bilateral symmetry (*see* Finnerty, 2003, 2005). Body symmetry differences (such as radial versus bilateral) will influence the location of sensory cells as well as locomotory musculature. Logic would dictate that a concentration of sensory cells in the leading end of a bilateral organism will require a concentration of neural tissue to integrate the sensory information, thus contributing to cephalization. In the radial condition, the circumferential location of sensory cells requires a very different organization of integrative neural tissue, although this does not rule out the centralization of neurons. But body symmetry in cnidarians is not strictly radial. There are internal structures (such as mesenteries in polyps) that have a bilateral arrangement, and planula larvae tend to be bilateral in their organization (*see* Malakhov, 2016). Current descriptors of cnidarian symmetry include bilateral, biradial, tetraradial, and radial. Once again, genetic and molecular markers of bilateral body axes are present in cnidarians, suggesting that a common ancestor to both cnidarians and bilaterians had molecular machinery for the bilateral body plan (*see* Martindale et al., 2002; Finnerty, 2003). While locomotion strategies likely contributed to the evolution of the bilateral body plan, other factors also likely contributed (Finnerty, 2005). Directed locomotion in cnidarians requires unique coordination within a radial locomotory system, so it is interesting that many cnidarians are sessile or fully planktonic.

5.2 THE NERVE NET

The keen focus on the nerve net as the primary component of the cnidarian nervous system is like describing the human nervous system by focusing solely on the motor output of the spinal nerves since many diffuse nerve nets function as efferent output to muscular systems. But if we consider a broad, two-dimensional (one cell thickness) muscle sheet that can be activated by sensory cells (e.g., mechanoreceptors or chemoreceptors), sensory organs (ocelli-eyes or statocysts), or rhythmic pacemakers that are located around the periphery of the muscle sheet, then a diffuse, non-polarized nerve net may be the most efficient means of activation of that muscle sheet (Satterlie, 2015). So, three additional factors that influence the nature of a conducting system in cnidarians are the location of sensory inputs to the system the location of pacemakers (if the motor system is rhythmic), and the architecture of the musculature that is being activated by the neuronal system. At the opposite extreme of the broad muscle sheet, such as in polyp retractor muscles and radial muscle bands in medusae, the motor networks tend to be highly compressed and polarized in their conduction patterns.

The classical experiments demonstrating the conducting properties of a nerve net were performed on scyphomedusae using a series of clever cutting experiments (Romanes, 1885). Three primary characteristics emerged from this work. First, conduction was diffuse. It could survive multiple interdigitating cuts to suggest that a contraction wave could take multiple paths through a muscle sheet. Second, conduction was non-polarized. Conduction in one direction through a bridge of tissue could also take place in the opposite direction through that same bridge. Third, conduction was through conducting. Point stimulation of a contraction wave in a muscle sheet resulted in a wave that was conducted through the entire muscle sheet.

A unique feature of cnidarian (and ctenophore) nervous systems is the symmetrical chemical synapse. Morphologically, they are similar to traditional synapses; however, synaptic vesicles are found on both sides of the synaptic densification (Horridge & Mackay, 1962; Horridge et al., 1962; Westfall, 1973, 1996; Anderson & Schwab, 1981; Anderson & Grunert, 1988). Symmetrical synapses are found in all cnidarian classes (Jha & Mackie, 1967; Westfall et al., 1971; Peteya, 1973; Satterlie, 1979; Gray et al., 2009). Anderson (1985) recorded from both neurons that formed a symmetrical synapse in the scyphomedusae *Cyanea* and confirmed two-way synaptic conduction in which an action potential in the pre-synaptic neuron produced a synaptic potential in the post-synaptic neuron with a synaptic delay of 1 ms. Switching stimulation of the two neurons produced similar results in the reverse direction. The possibility of electrical coupling between synaptically connected neurons was discounted (Anderson & Schwab, 1981; Anderson, 1985). Symmetrical synapses contribute to the non-polarized nature of nerve nets. Two-way electrical conduction via gap junctions is found in hydrozoans and will be discussed later.

Despite the apparent simplicity of a Romanes-type nerve net, it can produce variability in the contractile activity of the muscle sheet. A simple neuromuscular system activated by a nerve net can thus serve as an integrative unit to produce behavioral variability. The primary mechanism for producing this variability is frequency-dependent neuromuscular facilitation (Pantin, 1935a, b, 1965; Ross, 1952, 1955; Bullock, 1943) (Figure 5.2). This type of peripheral integrative mechanism allows very different muscular outputs based on the specific temporal characteristics of nerve net impulses, with shorter interpulse intervals producing large facilitation rates and greater overall contractions and longer intervals producing lesser facilitation rates and lower amplitude contractions. This allows graded muscle output from these relatively simple neuromuscular systems.

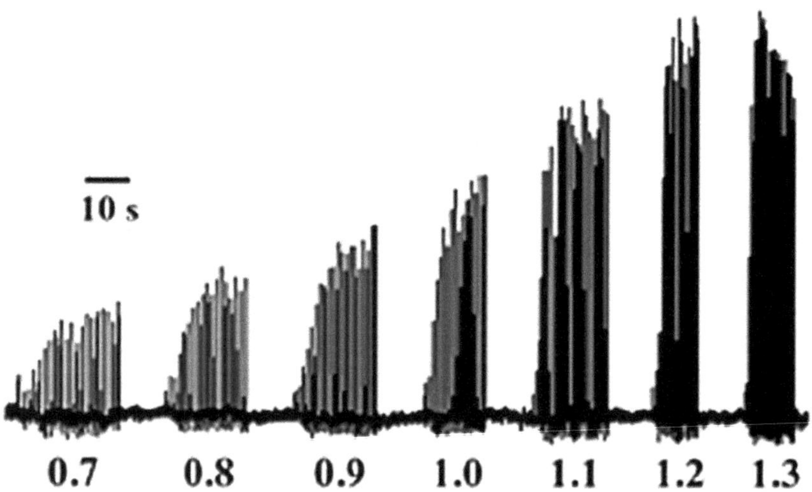

FIGURE 5.2 Frequency-dependent facilitation – facilitation profiles of subumbrellar swim musculature of the cubomedusae *Tamoya*. Uncalibrated force records with electrical stimulation at different frequencies (horizontal labels). Note the different rates of facilitation and the overall differences in the amplitude of contractions.

An additional type of behavioral variability is produced by nerve net-type conducting systems and is achieved by modification of one of the three primary properties of nerve net conduction. This is the concept of through conduction. In many colonial cnidarians, primarily of the Class Anthozoa, a colonial nerve net coordinates protective polyp withdrawal when a noxious stimulus is delivered to the organism. In many species, the colonial response is through-conducted, while in other species, incremental conduction is observed (Horridge, 1957; Anderson & Case, 1975; Shelton, 1975a–c; Anderson, 1976a, b; Satterlie et al., 1976, 1980; Shelton & McFarlane, 1976; Satterlie & Case, 1979, 1980). In incremental conduction, the initial stimulus produces polyp withdrawal in a limited area of the colony surrounding the stimulus, but with each subsequent stimulus, the area of polyp withdrawal increases. Some species exhibit increasing incremental conduction, in which through-conduction can eventually be achieved. Other species show decreasing incremental conduction (smaller area of polyp withdrawal with each stimulus), so the area of polyp withdrawal is limited within the colony. Even with colonial through-conduction, conduction velocities with successive stimuli initially increase and then decrease to contribute to the behavioral variability of the colony (Satterlie & Case, 1980; Satterlie et al., 1980). These examples represent plasticity within the nerve nets and, when coupled with neuromuscular facilitation, produce significant variability in motor responses from nerve net-controlled motor systems.

5.3 MULTIPLE CONDUCTING SYSTEMS AND CONTROL OF BEHAVIOR

The interaction between multiple, diffuse conducting systems increases the integrative capacity of cnidarian nervous systems. Members of the Class Anthozoa present the best example. Four conduction systems were identified electrophysiologically in sea anemones (McFarlane, 1969a, 1973), and subsequent work clarified the behavior output of these systems (Table 5.1). Of the four conducting systems described, the DIS is believed to be the output of shifting pacemaker sites (Jackson & McFarlane, 1976), so the overt behavior of anemones is likely the interplay between diffuse conducting systems and pacemaker-like activity within the nervous system. All three diffuse conducting systems (NN, SS1, SS2) are involved in feeding behavior in *Calliactis*.

TABLE 5.1 Summary of Conducting Systems in Anemones and Their Behavioral Roles

Through-conducting nerve net (NN)	Contraction of tentacles, oral disc, and mesenteries	Pantin, (1935a,b); Josephson, (1966); Robson and Josephson, (1969); Pickens, (1969); McFarlane (1969a, 1973)
Slow System 1 (SS1)	Expansion of the oral disc and column, body swaying; Activated by chemosensory Input	cFarlane (1969a, 1970); McFarlane and Lawn (1972)
	Pedal disc detachment in shell-climbing behavior in *Calliactis*	McFarlane (1969a, 1973)
	Pedal detachment and initial stages of swimming behavior in *Stomphia* following contact by a predatory sea star.	Lawn (1976)
Slow System 2 (SS2)	Reciprocal activity with NN is associated with burst-like contractions of parietal muscles in *Calliactis*; SS2 inhibits endodermal parietal and circular muscles	McFarlane (1972, 1973, 1974)
Delayed Initiation System (DIS)	Delays onset of SS1 activity following column stimulation In *Calliactis*	In *Calliactis* Jackson and McFarlane (1976)

Colonial anthozoans have colony-wide conducting systems in addition to polyp systems. A colonial nerve net is present in a variety of species and controls coordinated polyp withdrawal and colony-wide bioluminescence. In addition, a second, slower colonial conduction system has been identified (Anderson & Case, 1975; Shelton, 1975a–c; Anderson, 1976a, b; Satterlie et al., 1976, 1980; Shelton & McFarlane, 1976; Satterlie & Case, 1978, 1979, 1980).

As mentioned previously, the work on colonial anthozoans has added two integrative variables to nerve net function, both involving repetitive activity in the nerve nets: changes in conduction velocity, and changes in the degree of spread in non-through-conducting nerve nets.

The subumbrellar musculature of scyphozoan jellyfish is innervated by two separate diffuse conduction systems represented by nerve nets (Figure 5.3). The motor nerve net (MNN) activates the circular, striated muscles that produce swim movements (Shafer, 1878; Horridge, 1954a; Patton & Passano, 1972; Passano, 1973; Schwab & Anderson, 1980, 1981; Anderson & Schawb, 1981, 1983, 1984; Gremmelikhuijzen, 1983; Anderson et al., 1992; Carlberg et al., 1995; Moosler et al., 1997; Satterlie, & Eichinger, 2014). A second diffuse nerve net carries sensory information but also innervates the swim musculature to either produce or modulate contractions (Horridge, 1956a; Passano, 1965, 1982, 2004). Scyphomedusae possesses eight or more marginal neural structures, called rhopalia, which contain sensory cells and structures (including ocelli and statoliths), and the pacemakers for initiation of swim contractions. The MNN represents the motor output to the muscles and is activated by the pacemakers while the DNN serves more as a sensory/motor net (Horridge, 1956b; Yamasu & Yoshida, 1973; Hundgen & Biela, 1982; Nakanishi et al., 2009; Satterlie & Eichinger, 2014). Activity in the DNN can modulate the pacemakers in a rhopalium, allowing the modification of the swim rhythm by sensory input, and this can be localized or widespread.

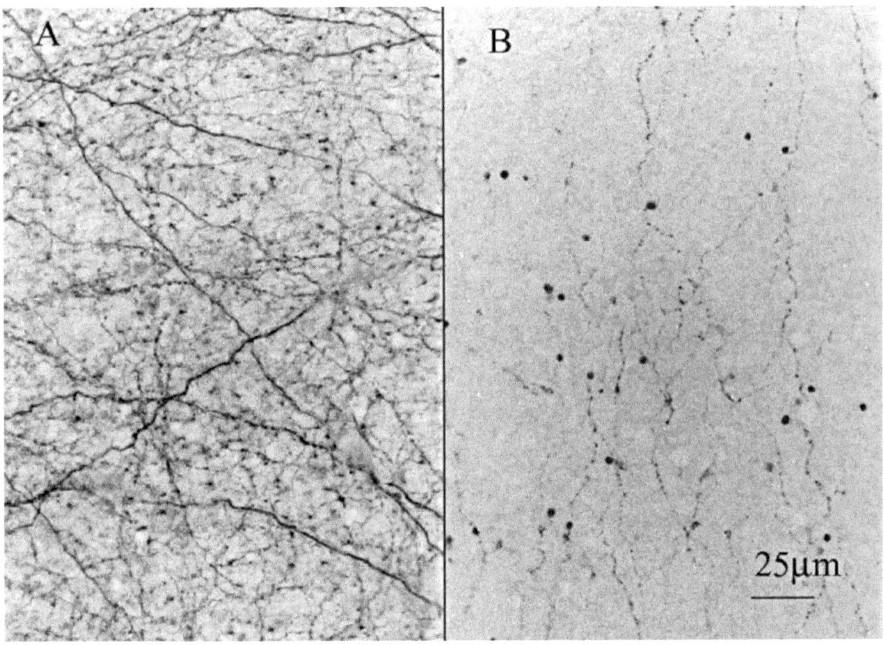

FIGURE 5.3 Immunolabeling of the MNN (A) and the DNN (B) of *Aurelia*. Both images are from the same area, but with different antibodies (tubulin for MNN and FMRFamide for DNN).

A third example of parallel conducting systems is found in the tentacles of cubozoan medusae. One network shows FMRFamide immunoreactivity, and it is superficial to the second, which is labeled with tubulin antibodies (Eighinger & Satterlie, 2014). The FMRFamide immunoreactive network includes surface sensory cells, while the tubulin-immunoreactive network is believed to be associated with the longitudinal muscle of the tentacles. The two systems allow local reflex activity. In addition, the sensory system conveys information to the rest of the animal, and the motor system distributes commands from the nervous system.

5.4 DIRECTIONAL AND COMPRESSED NETWORKS

Anatomical data show that not all nerve nets are diffuse and non-polarized. Various degrees of compression run from diffuse but directional nerve nets to nerve-like compressions. In all cases, the arrangement follows the organization of the innervated musculature. A good example is found in the subumbrellar musculature of the medusa stage of the hydrozoan *Aequorea*. The subumbrellar epidermis includes circular, striated muscle, which produces swim contractions, and radial, smooth muscle, which is involved in inward curling the subumbrella during a protective crumple response (Satterlie, 1985a, b, 2008). The circular muscle cells form a sheet that abuts the mesoglea and is overlain by a sheet-like layer of radial muscle (Figure 5.4). Two nerve nets are present in the subumbrella, shown with immunohistochemical staining using tubulin and FMRFamide antibodies (Satterlie, 2008). The tubulin network is associated with the circular muscle and is both diffuse and non-polarized (based on the orientation of individual neurons). The FMRFamide-immunoreactive network is more superficial, and while it forms a diffuse nerve net, there is a primary orientation of neurites in a radial direction (Figure 5.5).

Further nerve net compression is found in all cnidarian groups, where bundles of parallel muscle fibers are found. Examples include the mesenteric retractor muscles of anthozoans and radial muscles of hydrozoan, cubozoan, and some scyphozoan medusae. In these cases, compression produces polarization of the network that mirrors the orientation of the muscle bundles. In extreme cases, the compression produces a nerve-like structure. A specific example is found in cubomedusae, where four perradial bundles of smooth muscle run from the margin to the bell apex within the subumbrellar epithelium (Satterlie et al., 2005). The muscle bundles are involved in the protective

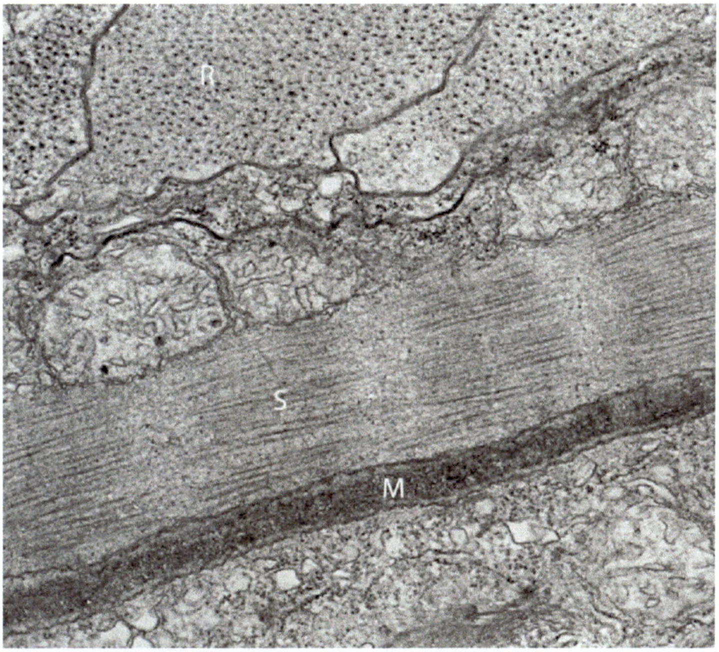

FIGURE 5.4 Subumbrellar musculature of the hydromedusa *Aequorea*. Circular, striated muscle cells (S) abut the mesoglea (M). Superficial to the circular muscle is a layer of radial muscle cells (R), which run perpendicular to the circular cells.

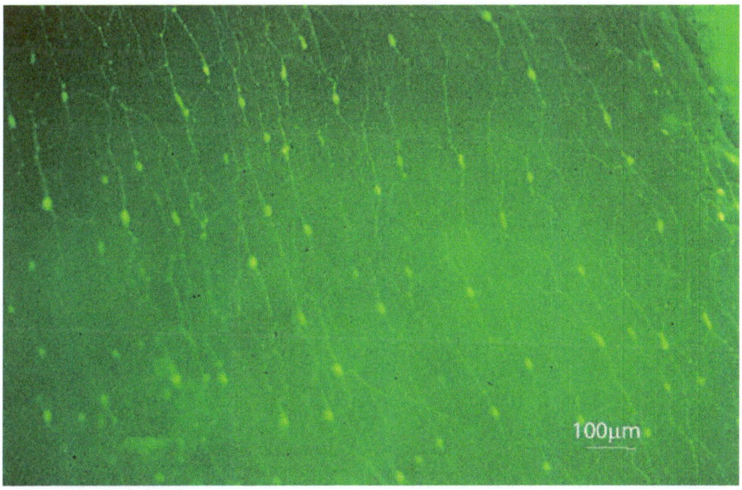

FIGURE 5.5 Subumbrellar wholemount from the hydromedusa *Aequorea* showing a nerve net with a direction (polarized) orientation. FMRFamide immunolabeling stained a nerve net associated with the radial muscle, and the nerve net has a preferred radial orientation of neurites. The bell margin is to the bottom right, and the bell apex is to the top left.

crumpling response, which involves the contraction and inward bending of all four pedalia, found at the base of the tentacles (Satterlie, 2014). Contraction of the radial muscles and the tentacular/pedalial muscles curl the margin inward and bring the contracted tentacles into the bell opening. Feeding responses are similar but involve the inward bending of a single tentacle (Larson, 1976). Two compressed networks of neurons are associated with the radial musculature, one immunoreactive to FMRFamide and the other to tubulin antibodies (Eichinger & Satterlie, 2014).

Similar compressed nets are found associated with radial muscle bundles in hydromedusae that overlie the radial canals of the digestive system (Singla, 1978a; Satterlie, 2002, 2011; Mackie, 2004b) (Figure 5.6). Activity in the radial muscle bands is involved in crumpling, feeding, and conduction within the compressed radial nerves, allowing two-way communication between the marginal neuronal structures and the manubrium (Mackie & Singla, 1975; Spencer, 1975, 1978; Mackie, 2004b).

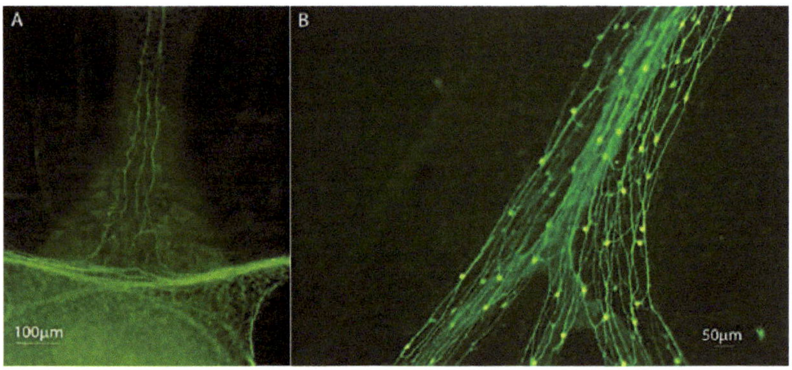

FIGURE 5.6 Compressed nerve nets in hydromedusae follow the path of muscle bundles. Both (A) and (B) are labeled with FMRFamide immunohistochemistry. (A) a radial nerve from *Aegina*; and (B) the compressed network following the radial muscle overlying the branching radial canal.

5.5 CENTRALIZATION THROUGH COMPRESSION OF MULTIPLE CONDUCTING SYSTEMS: HYDROMEDUSAE

A unique feature of the nervous systems of hydromedusae is the presence of inner and outer NRs, found at the bell margin, near the junction of the bell proper and the velum. The velum is a shelf of muscular tissue that narrows the bell opening for more efficient swimming. Within the inner NR is a compressed network of oversized neurons that serve as both the primary motor network and the pacemaker network for swim contractions (Anderson

& Mackie, 1977; Singla, 1978a; Spencer & Satterlie, 1980; Spencer, 1981, 1982; Satterlie & Spencer, 1983; Satterlie, 1985a, b; Meech & Mackie, 1995; Lin et al., 2001) (Figure 5.7). The neurons within this network exhibit another unique property of the Class Hydrozoa – electrical coupling via gap junctions (Spencer & Satterlie, 1980; Satterlie, 1985a, b). Thus far, up to 19 innexins have been identified through genomic analyses, while only one has been found in anthozoans and none have been found in scyphozoans (Moroz & Kohn, 2016). The confirmed use of gap junctions for neuron-to-neuron electrical conduction, epithelial conduction, and myoid conduction is limited to this class and is used extensively for the control of behavior. Myoid conduction within muscle sheets or groups is common in hydromedusae (Mackie & Singla, 1975; Keogh & Summers, 1976; Singla, 1978a,

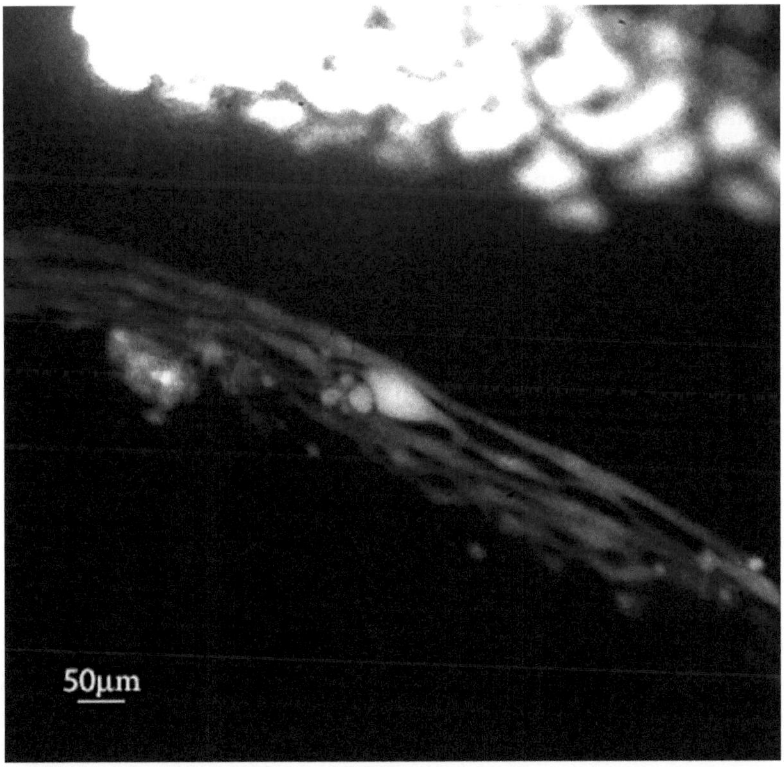

FIGURE 5.7 Lucifer Yellow fill of the motor network in the inner nerve ring of *Aequorea*. The swollen cell is the cell that was iotophoretically injected. Note the cells are dye-coupled (electrical coupling was verified with dual recordings). The bright cells to the top are photocytes that fluoresce with the filters used to visualize Lucifer Yellow (the autofluorescence is due to Green Fluorescent Protein).

b; Spencer, 1979; Satterlie, 1985a, 2002, 2008; Satterlie & Spencer, 1983; Kerfoot et al., 1985; Mackie, 2004a). In some hydromedusae, the circular, striated muscle sheet of the subumbrella (swim musculature) is aneural with innervation at the periphery of the sheet and myoid conduction throughout the sheet (for example, in *Polyorchis*: Spencer, 1978, 1979; Satterlie & Spencer, 1983). In other hydromedusae, myoid conduction is aided by input from a subumbrellar nerve net (for example, *Aequorea*: Satterlie, 1985a, 2002, 2008). In addition, excitable epithelia conduct electrical information that interacts with neural and muscular conduction systems to contribute to behavioral variability (Mackie, 1964, 1965, 1970, 1976, 1978; Josephson & Schwab, 1979; Spencer, 1978).

Around 14 conduction systems have been described for the hydromedusa *Aglantha*, eight of which are found within the NRs (*see* Mackie, 2004b). This allows significant interactions between the conduction systems to control behavioral output. The NRs thus represent a centralization of nervous/conduction system function.

While the inner NR contains circuitry for the control of the swim musculature, the outer NR contains sensory networks that interact with the inner NR. As with the inner NR motor network, the multiple sensory networks each are made up of electrically-coupled neurons (Spencer & Arkett, 1984; Arkett & Spencer, 1986a, b). The sensory systems include ocelli (Singla, 1974), statocysts (Singla, 1975), and mechanoreceptors, the best example of which are the tactile combs of *Aglantha* (Arkett & Mackie, 1988; Mackie, 2004b).

Giant neurons are incorporated into the nervous systems of some hydromedusae, as originally described for the stem of siphonophores (Mackie, 1973, 1978, 1989). These neurons control zooid retraction and fast escape (backward) swimming of the colony. A more extensive use of giant neurons is found in *Aglantha*, where they control ballistic escape swimming (Donaldson et al., 1980; Roberts & Mackie, 1980). Stimulation of the mechanoreceptive tactile combs activates a Ring Giant neuron in the outer NR (Roberts & Mackie, 1980). The Ring Giant activates eight Motor Giant neurons that extend from the margin toward the bell apex in the neural tracts that overlie the radial canals (Meech & Mackie, 1995). The Motor Giants have processes that enter the subumbrella muscle sheets and innervate the swim musculature (Singla, 1978a; Weber et al., 1982). The Motor Giants are unique in producing two types of action potentials: slow calcium action potentials for muscle activation characteristic of normal swimming and fast sodium action potentials for strong escape contractions (Mackie & Meech, 1985; Meech & Mackie, 1995).

5.6 CENTRALIZATION THROUGH MARGINAL GANGLION-LIKE STRUCTURES: SCYPHOMEDUSAE AND CUBOMEDUSAE

Scyphomedusae and Cubomedusae possess marginal sensory/ganglion-like structures called rhopalia. Each rhopalium bears at least one ocellus, a statolith, and other sensory cells, as well as neural tissue that includes pacemakers that control swim contractions. Scyphomedusae have at least eight rhopalia, while cubomedusae have four. Removal of the rhopalia stops swimming activity (Figure 5.8).

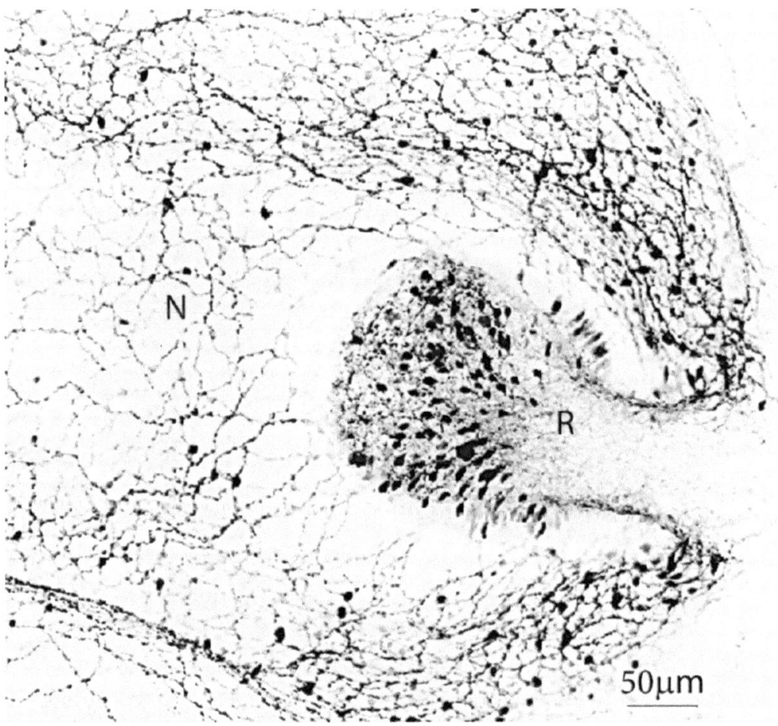

FIGURE 5.8 Rhopalium and rhopalial niche of the scyphomedusae *Cassiopea*, labeled with FMRFamide antibody. The network of the niche (N) is part of the diffuse nerve net. The Rhopalium has numerous surface-projecting sensory cells.

In scyphomedusae, the DNN (described earlier) provides incoming sensory information, which is sorted with that of the rhopalial sensory structures, to influence pacemaker activity. The pacemaker output is transmitted to the swim musculature via the MNN. The physically separated rhopalia interact with one another through a resetting hierarchy where the rhopalium with the fastest rhythm drives swim contractions. Due to sensory inputs to the individual rhopalia, the pacemaker dominance shifts between rhopalia. In this way, each rhopalium independently sorts local sensory information with more global input (from the DNN) to determine its activity, and the full set of rhopalia are tied together through their MNN output. This loose form of centralization allows whole-animal control of swimming while also producing localized responses for asymmetrical behaviors like turning.

Rhopalial development reaches a zenith in the cubomedusae, where each of the four rhopalia has two lens-bearing eyes, two slit ocelli, and two pit ocelli (Figure 5.9). Cubomedusae are extremely agile swimmers that can turn quickly and avoid obstacles. They exhibit a variety of visually guided responses. As with scyphomedusae, rhopalial pacemakers activate swim musculature via an MNN-like subumbrellar nerve net. However, a greater degree of centralization is achieved through the connections of the rhopalia, and tentacle bases, by an NR (Garm et al., 2006) (Figure 5.10). The NR shows regional anatomical specializations and does more than just distribute motor impulses to the swim musculature (Garm et al., 2007c). Cubomedusae lack a diffuse equivalent of the scyphozoan DNN; however, sensory information is passed through the rhopalial network in the NR. This, in effect, represents a compression of the DNN into discrete nerve-like pathways.

Each rhopaliam has a bilateral organization with the two lensed eyes on the midline, flanked by left and right pairs of slit and spot ocelli. The bilateral organization also extends to the neural organization of the neuropil, with symmetric tracts and neuron clusters (Skogh et al., 2006; Garm & Ekstrom, 2010). Each complex eye has a cellular lens, retina, and pigment layer, and is underlain by a neural layer that connects to the neuropil of the rhopalium (Berger, 1898; Yamasu & Yoshida, 1973; Laska & Hundgen, 1984; Martin, 2004; O'Connor et al., 2009; Gray et al., 2009). The lensed eyes are camera-type eyes capable of forming images but with limited resolution (Nilsson et al., 2005; Garm et al., 2007a; O'Connor et al., 2009). Immunohistochemical evidence suggests that a single opsin is present in the lensed eyes, and their spectral sensitivity suggests they do not see colors (Martin, 2004; Garm et al., 2007a; O'Connor et al., 2010; Ekstrom et al., 2008).

Obstacle avoidance behavior speaks to the integrative abilities of the cubomedusan centralized nervous system. It is a visually-guided behavior (Garm et al., 2007b) that involves contrast-dependent spatial resolution (Garm et al., 2013). To achieve turning, the swim frequency is increased, and an asymmetrical contraction is initiated in the velarium, a marginal shelf

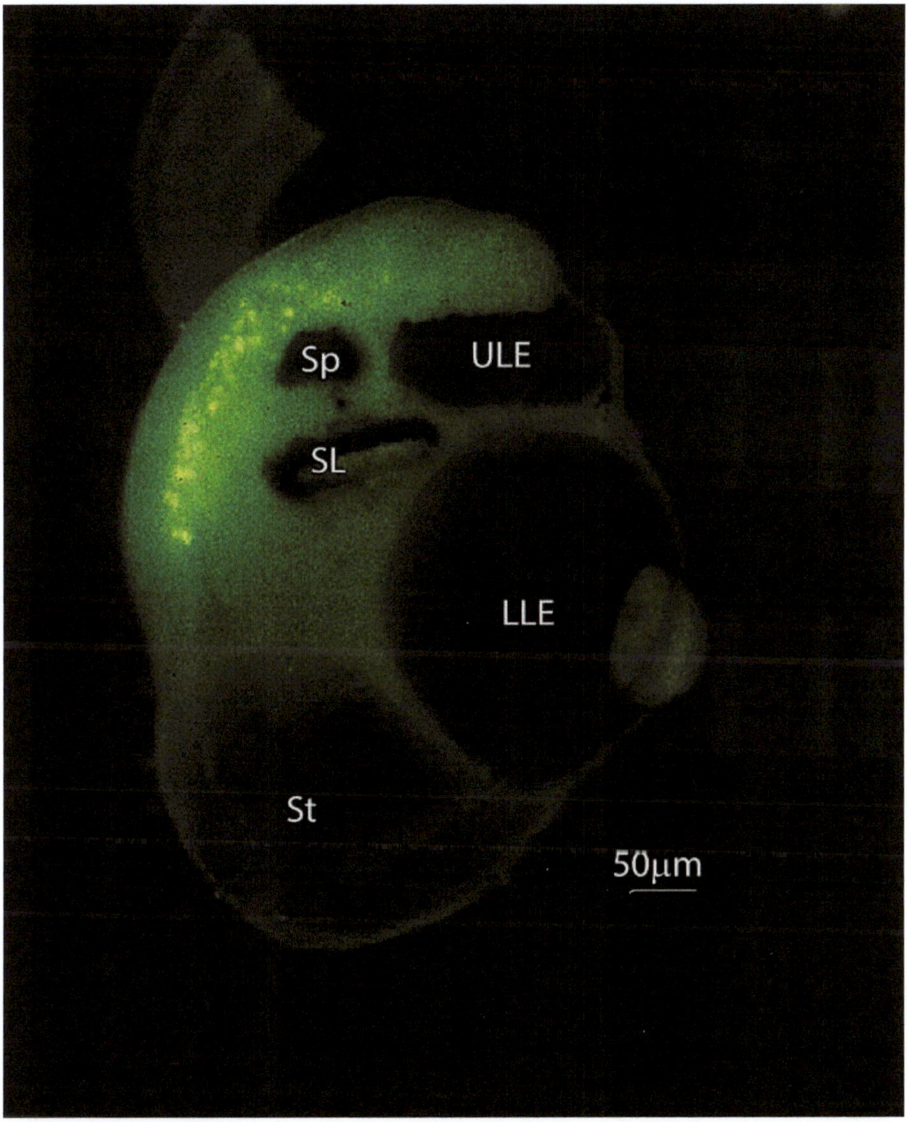

FIGURE 5.9 FMRFamide immunolabel of the rhopalium of the cubomedusa *Carybdea*. A compressed network of neurons is labeled along the left top of the figure (bright staining). Sp: spot ocellus; SL: slit ocellus; St: statolith; ULE: upper lensed eye; and LLE: lower lensed eye.

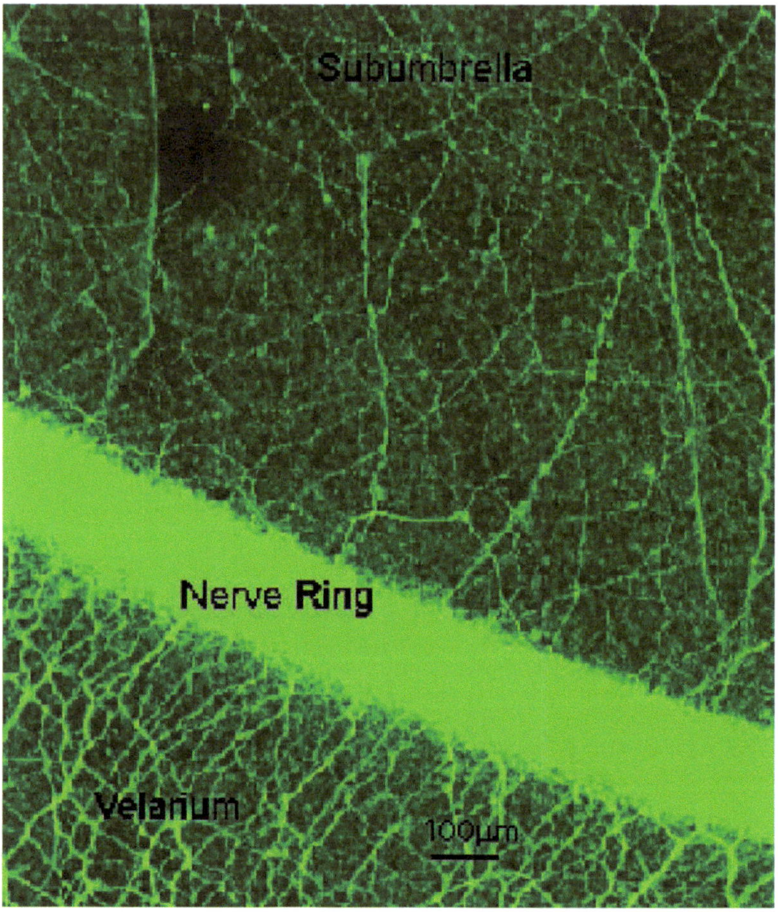

FIGURE 5.10 The nerve ring (NR) contributes neurons to the subumbrellar and velar nerve nets. Tubulin immunohistochemistry of the cubomedusa *Carybdea.*

of tissue that is analogous to the velum of hydromedusae (Petie et al., 2011, 2013a, b).

Changes in light intensity alter the frequency of pacemaker activity (Garm & Bielecki, 2008), and the light responses are transmitted to the other rhopalia in the NR (Satterlie, 2014). The different eyes and ocelli each have unique effects on pacemaker activity, further illustrating the complexity and sophistication of this centralized sensory/ganglionic structure (Garm & Mori, 2009). For example, in *Tripedalia*, the lower lensed eye helps keep the medusa in light shafts among the mangrove roots to allow feeding on photopositive prey (Garm & Bielecki, 2008), while evidence suggests the upper lensed eye is used for orientation to terrestrial cues (Garm et al., 2011).

The rhopalial neuropil is well developed, particularly in the area opposite the complex eyes, and includes tracts associated with the eyes and ocelli as well as commissural tracts and compressed networks (Martin, 2002, 2004; Satterlie, 2002; Parkefelt et al., 2005; Skogh et al., 2006; Parkefelt & Ekstrom, 2009; Eichinger & Satterlie, 2014). Recent immunohistochemical work has demonstrated distinct localizations of neurons expressing three neuropeptides in the rhopalia, NR, and other structures (Sofie et al., 2021). This reinforces the complex neural architecture of the ganglion-like rhopalia. It is also important since genomic evidence suggests that early nervous systems were peptidergic and classical transmitters appeared later (Moroz, 2009).

5.7 CENTRALIZED VERSUS CENTRAL NERVOUS SYSTEMS

The central nervous system is largely a bilaterian construct consisting of a brain and nerve cord(s), frequently with ganglionic swellings. Interestingly, genomic data suggests that brains and nervous systems had multiple origins (up to seven: Moroz, 2009). The cnidarian nervous systems do not fit this construct. But if we look at the location and accumulation of sensory structures and their relation to the distribution of associated neural tissue, we can envision the integrative value of cephalization in bilateral animals but also similarly appreciate the integrative value of NRs in hydromedusae rhopalia in scyphomedusae, the rhopalial-nerve ring system in cubomedusae, and even the multiple interacting nerve nets in anthozoans.

KEYWORDS

- **anthozoans**
- **central nervous systems**
- **Ganglion**
- **rhopalia**
- **rhopalial-nerve ring system**
- **scyphomedusae**

REFERENCES

Anderson, P. A. V., & Case, J. F., (1975). Electrical activity associated with luminescence and other colonial behavior in the pennatulid *Renilla kollikeri*. *Biol. Bull., 149*, 80–95.
Anderson, P. A. V., & Grunert, U., (1988). Three-dimensional structure of bidirectional, excitatory chemical synapses in the jellyfish *Cyanea capillata*. *Synapse, 2*, 606–613.
Anderson, P. A. V., & Mackie, G. O., (1977). Electrically coupled, photosensitive neurons control swimming in a jellyfish. *Science, 197*, 186–188.
Anderson, P. A. V., & Schwab, W. E., (1981). The organization and structure of nerve and muscle in the jellyfish *Cyanea capillata*. *J. Morph., 170*, 383–399.
Anderson, P. A. V., & Schwab, W. E., (1983). The action potential in neurons of the motor nerve net of *Cyanea* (Coelenterata). *J. Neurophysiol., 50*, 671–683.
Anderson, P. A. V., & Schwab, W. E., (1984). An epithelial cell-free preparation of the motor nerve net of *Cyanea* (Coelenterata; Scyphozoa). *Biol. Bull., 166*, 396–408.
Anderson, P. A. V., (1976a). An electrophysiological study of mechanisms controlling polyp retraction in colonies of the scleractinian coral *Goniopora lobata*. *J. Exp. Biol., 65*, 381–393.
Anderson, P. A. V., (1976b). The electrophysiology of the organ-pipe coral, *Tubipora musica*. *Biol. Bull., 150*, 337–347.
Anderson, P. A. V., (1985). Physiology of a bidirectional, excitatory, chemical synapse. *J. Neurophysiol., 53*, 821–835.
Anderson, P. A. V., Moosler, A., & Grimmelikhuijzen, C. J. P., (1992). The presence and distribution of Antho-RFamide-like material in schyphomedusae. *Cell Tiss. Res., 267*, 67–74.
Arkett, S. A., & Mackie, G. O., (1988). Hair cell mechanoreception in the jellyfish *Aglantha digitale*. *J. Exp. Biol., 135*, 329–342.
Arkett, S. A., & Spencer, A. N., (1986a). Neuronal mechanisms of a hydromedusan shadow reflex. I. Identified reflex components and sequence of events. *J. Comp. Physiol. A, 159*, 201–213.
Arkett, S. A., & Spencer, A. N., (1986b). Neuronal mechanisms of a hydromedusan shadow reflex. II. Graded response of reflex components, possible mechanisms of photic integration, and functional significance. *J. Comp. Physiol. A, 159*, 215–225.
Ball, E. E., Hayward, D. C., Saint, R., & Miller, D. J., (2004). A simple plan-cnidarians and the origins of developmental mechanisms. *Nat. Rev. Genet., 5*, 567–577.
Berger, E. W., (1898). The histological structure of the eyes of cubomedusae. *J. Comp. Neurol., 8*, 223–230.
Boero, F., Bouillon, J., & Piraino, S., (2005). The role of Cnidaria in evolution and ecology. *Ital. J. Zool., 72*, 65–71.
Boero, F., Gravili, C., Pagliara, P., Piraino, S., Bouillon, J., & Schmid, V., (1998). The cnidarian presises of metazoan evolution: From triploblasty to coelom formation, to metamery. *Ital. J. Zool., 65*, 5–9.
Boero, F., Schierwater, B., & Piraino, S., (2007). Cnidarian milestones in metazoan evolution. *Integ. Comp. Biol., 47*, 693–700.
Bullock, T. H., (1943). Neuromuscular facilitation in scyphomedusae. *J. Cell. Comp. Physiol., 22*, 251–272.
Burton, P. M., (2007). Insights from diploblasts; the evolution of mesoderm and muscle. *J. Exp. Zool., 308*, 1–10.

Carlberg, M., Alfredsson, K., Nielsen, S. O., & Anderson, P. A. V., (1995). Taurine-like immunoreactivity in the motor nerve net of the jellyfish *Cyanea capillata*. *Biol. Bull., 188*, 78–82.

Chun, C., (1880). *Die Ctenophoren des Golfes von Neapel und der agrenzenden Meeres-Abschnitte*. Engelmann, Leipzig.

Claus, C., (1878). Uber *Charybdea marsupialis*. *Arb. Zool. Inst. Univ. Wien., 1*, 221–276.

Donaldson, S., Mackie, G. O., & Roberts, A., (1980). Preliminary observations on escape swimming and giant neurons in *Aglantha digitale* (Hydromedusae: Trachylina). *Can. J. Zool., 58*, 549–552.

Eichinger, J. M., & Satterlie, R. A., (2014). Organization of the ectodermal nervous structures in medusae: Cubomedusae. *Biol. Bull., 226*, 41–55.

Ekstrom, P., Garm, A., Palsson, J., Vihtelic, T. S., & Nilsson, D. E., (2008). Immunohistochemical evidence for multiple photosystems in box jellyfish. *Cell Tiss. Res., 333*, 115–124.

Finnerty, J. R., (2003). The origins of axial patterning in the metazoan: How old is bilateral symmetry? *Int. J. Dev. Biol., 47*, 523–529.

Finnerty, J. R., (2005). Did internal transport, rather than directed locomotion, favor the evolution of bilateral symmetry in animals? *BioEssays, 27*, 1174–1180.

Garm, A., & Bielecki, H., (2008). The swim pacemakers of box jellyfish are modified by visual input. *J. Comp. Physiol. A, 194*, 641–651.

Garm, A., & Ekstrom, P., (2010). Evidence for multiple photosystems in jellyfish. *Int. Rev. Cell Molec. Biol., 280*, 41–78.

Garm, A., & Mori, S., (2009). Multiple photoreceptor systems control the swim pacemaker activity in box jellyfish. *J. Exp. Biol., 212*, 3951–3960.

Garm, A., Coates, M. M., Gad, R., Seymour, R., & Nilsson, D. E., (2007a). The lens eyes of the box jellyfish *Tripedalia cystophora* and *Chiropsalmus sp.* are slow and color-blind. *J. Comp. Physiol. A, 193*, 547–557.

Garm, A., Ekstrom, P., Boudes, M., & Nilsson, D. E., (2006). Rhopalia are integrated parts of the central nervous system of box jellyfish. *Cell Tiss. Res., 325*, 333–343.

Garm, A., Hedal, I., Majken, I., & Gurska, D., (2013). Pattern- and contrast-dependent visual response in the box jellyfish *Tripedalia cystophora*. *J. Exp. Biol., 216*, 4520–4529.

Garm, A., O'Connor, M., Parkefelt, L., & Nilsson, D. E., (2007b). Visually guided obstacle avoidance in the box jellyfish *Tripedalia cystophora* and *Chiropsella bronzie*. *J. Exp. Biol., 210*, 3616–3623.

Garm, A., Oskarsson, M., & Milsson, D. E., (2011). Box jellyfish use terrestrial visual cues for navigation. *Curr. Biol., 21*, 798–803.

Garm, A., Poussart, Y., Parkefelt, L., Edstrom, P., & Nilsson, D. E., (2007c). The ring nerve of the box jellyfish *Tripedalia cystophora*. *Cell Tiss. Res., 329*, 147–157.

Gray, G. C., Martin, V. J., & Satterlie, R. A., (2009). Ultrastructure of the retinal synapses in cubozoans. *Biol. Bull., 217*, 35–49.

Grimmelikhuijzen, C. J. P., (1983). FMRFamide immunoreactivity is generally occurring in the nervous systems of coelenterates. *Histochemistry, 78*, 361–381.

Grundfest, H., (1959). Evolution of conduction in the nervous system. In: Bass, A. D., (ed.), *Evolution of Nervous Control from Primitive Organisms to Man* (pp. 43–86). AAAS, Washington DC.

Grundfest, H., (1965). Evolution of electrophysiological properties among sensory receptor systems. In: Pringle, J. W. S., (ed.), *Essays on Physiological Evolution* (pp. 107–138). Pergammon Press, Oxford, UK.

Hertwig, O., & Hertwig, R., (1878). *Das Nervensystem und die Sinnesorgane der Medusen.* Vogel, Leipzig.
Horridge, G. A., & MacKay, B., (1962). Naked axons and symmetrical synapses in coelenterates. *Quart. J. Microsc. Sci., 103*, 531–541.
Horridge, G. A., (1954a). Observations on the nerve fibers of *Aurelia aurita. Quart. J. Microsc. Sci., 95*, 85–92.
Horridge, G. A., (1954b). The nerves and muscles of medusae. I. Conduction in the nervous system of *Aurelia aurita* Lamarck. *J. Exp. Biol., 31*, 594–600.
Horridge, G. A., (1956a). The nerves and muscles of medusae. V. Double innervation in Scyphozoa. *J. Exp. Biol., 33*, 366–383.
Horridge, G. A., (1956b). The nervous system of the ephyra larva of *Aurelia aurita. Quart. J. Microsc. Sci., 97*, 59–74.
Horridge, G. A., (1957). The co-ordination of the protective retraction of coral polyps. *Phil. Trans. Roy. Soc. London, 240*, 495–528.
Horridge, G. A., Chapman, D. M., & MacKay, B., (1962). Naked axons and symmetrical synapses in an elementary nervous system. *Nature, 193*, 899, 900.
Hundgen, L. H., & Biela, C., (1982). Fine structure of touch plates in the scyphomedusan *Aurelia aurita. J. Ultrastruct. Res., 80*, 178–184.
Jackson, A. J., & McFarlane, I. D., (1976). Delayed initiation of SS1 pulses in the sea anemone *Calliactis parasitica*: Evidence for a fourth conducting system. *J. Exp. Biol., 65*, 539–552.
Jha, R. K., & Mackie, G. O., (1967). The recognition, distribution, and ultrastructure of hydrozoan nerve elements. *J. Morph., 123*, 43–61.
Josephson, R. K., & Schwab, W. E., (1979). Electrical properties of an excitable epithelium. *J. Gen. Physiol., 74*, 213–236.
Josephson, R. K., (1966). Neuromuscular transmission in a sea anemone. *J. Exp. Biol., 45*, 305–319.
Keough, E. M., & Summers, R. G., (1976). An ultrastructural investigation of the striated subumbrellar musculature of the anthomedusan, *Pennaria tiarella. J. Morph., 149*, 507–526.
Kerfoot, P. A. H., Mackie, G. O., Meech, R. W., Roberts, A., & Singla, C. L., (1985). Neuromuscular transmission in the jellyfish *Aglantha digitale. J. Exp. Biol., 116*, 1–25.
Koizumi, O., Sato, N., & Goto, C., (2004). Chemical anatomy of hydra nervous system using antibodies against hydra neuropeptides: A review. *Hydrobiologia, 530, 531*, 41–47.
Larson, R. J., (1976). Cubomedusae: Feeding, functional morphology, behavior, and phylogenetic position. In: Mackie, G. O., (ed.), *Coelenterate Ecology and Behavior* (pp. 237–245). Plenum Press, New York.
Laska, G., & Hundgen, M., (1984). Die Ultrastruktur des neuromuscularen systems der medusen von *Tripedalia cystophora* und *Carybdea marsupialis* (Coelenterata, Cubozoa). *Zoomorphology, 104*, 163–170.
Lawn, I. D., (1976). Swimming in the sea anemone *Stomphia coccinea* triggered by a slow conduction system. *Nature, 262*, 708–709.
Lictneckert, R., & Reichert, H., (2007). Origin and evolution of the first nervous system. In: Kaas, J. H., (ed.), *Evolution of Nervous Systems* (pp. 289–315). Academic Press, Amsterdam.
Lin, Y. C., Gallin, W. J., & Spencer, A. N., (2001). The anatomy of the nervous system of the *hydrozoa*n jellyfish *Polyorchis penicillatus*, as revealed by a monoclonal antibody. *Invert. Neurosci., 4*, 65–75.

Mackie, (1973). Report on the giant nerve fibers in *Nanomia. Pub. Seto Mar. Biol. Lab., 20*, 12–19.
Mackie, G. L., & Singla, C. L., (1975). Neurobiology of *Stomotoca*. I. Action systems. *J. Neurbiol., 6*, 339–356.
Mackie, G. O., & Meech, R. W., (1985). Separate sodium and calcium spikes in the same axon. *Nature, 313*, 791–793.
Mackie, G. O., (1964). Analysis of locomotion in a siphonophores colony. *Proc. Roy. Soc. Lond. B, 159*, 366–391.
Mackie, G. O., (1965). Conduction in the nerve-free epithelia of siphonophores. *Am. Zool., 5*, 439–453.
Mackie, G. O., (1970). Neuroid conduction and the evolution of conducting tissues. *Quart. Rev. Biol., 45*, 319–332.
Mackie, G. O., (1976). Propagated spikes and secretion in a coelenterate glandular epithelium. *J. Gen. Physiol., 68*, 313–325.
Mackie, G. O., (1978). Coordination in physonectid siphonophores. *Mar. Behav. Physiol., 5*, 325–346.
Mackie, G. O., (1989). Evolution of cnidarian giant axons. In: Anderson, P. A. V., (ed.) *Evolution of the First Nervous Systems* (pp. 395–407). Plenum Press, New York.
Mackie, G. O., (2004a). Epithelial conduction: Recent findings, old questions, and where do we go from here? *Hydrobiologia, 530, 531*, 73–80.
Mackie, G. O., (2004b). Central neural circuitry in the jellyfish *Aglantha. Neurosignals, 13*, 5–19.
Malakhov, V. V., (2016). Symmetry and the tentacular apparatus in *Cnidaria, Russ. J. Mar. Biol., 42*, 287–298.
Marlow, H. Q., Srivastava, M., Matus, D. Q., Rokhsar, D., & Martindale, M. Q., (2009). Anatomy and development of the nervous system of *Nematostella vectensis*, an anthozoan cnidarian. *Dev. Neurobiol., 69*, 235–254.
Martin, V. J., (2002). Photoreceptors in cnidarians. *Can. J. Zool., 80*, 1703–1722.
Martin, V. J., (2004). Photoreceptors of cubozoan jellyfish. *Hydrobiologia, 530, 531*, 135–144.
Martindale, M. Q., Finnerty, J. R., & Henry, J. Q., (2002). The radiata and the evolutionary origins of the bilaterian body plan. *Molec. Phylogenet. Evol., 24*, 358–365.
Martindale, M. Q., Pang, K., & Finnerty, J. R., (2004). Investigating the origins of triploblasty: "Mesodermal" gene expression in a diploblastic animal, the sea anemone *Nematostella vectensis* (phylum Cnidaria; class Anthozoa). *Development, 131*, 2463–2474.
McFarlane, I. D., & Lawn, I. D., (1972). Expansion and contraction of the oral disc in the sea anemone *Tealia felina. J. Exp. Biol., 57*, 633–649.
McFarlane, I. D., (1969a). Two slow conduction systems in the sea anemone *Calliactis parasitica. J. Exp. Biol., 51*, 377–385.
McFarlane, I. D., (1969b). Coordination of pedal-disk detachment in the sea anemone *Calliactis parasitca. J. Exp. Biol., 51*, 387–396.
McFarlane, I. D., (1972). Spontaneous electrical activity in the sea anemone *Calliactis parasitica. J. Exp. Biol., 58*, 77–90.
McFarlane, I. D., (1973). Multiple conduction systems and the behavior of sea anemones. *Publ. Seto Mar. Biol. Lab., 20*, 513–523.
McFarlane, I. D., (1974). Excitatory and inhibitory control of inherent contractions in the sea anemone *Calliactis parasitica. J. Exp. Biol., 60*, 397–422.

Meech, R. W., & Mackie, G. O., (1995). Synaptic potentials underlying spike production in motor giant axons of *Aglantha digitale*. *J. Neurophysiol., 74*, 1662–1670.

Moosler, A., Rinehart, K. L., & Gremmelikhuijzen, C. J. P., (1997). Isolation of three novel neuropeptides, the Cyanea-RFamides I-III, from scyphomedusae. *Biochem. Biophys. Res. Comm., 236*, 743–749.

Moroz, L. L., & Kohn, A. B., (2016). Independent origins of neurons and synapses: Insights from ctenophores. *Phil. Trans. R. Soc. B, 371*, 20150041. http://dx.doi.org/10.1098/rstb.2015.0041.

Moroz, L. L., (2009). On the independent origins of complex brain and neurons. *Brain Behav. Evol., 74*, 177–190.

Moroz, L. L., (2018). NeuroSystematics and periodic system of neurons: Model vs reference species at single-cell resolution. *ACS Chem. Neurosci., 9*, 1884–1903.

Muller, P., Seipel, K., Yanze, N., Reber-Muller, S., Steifwolf-Engel, R., Stierwald, M., Spring, J., & Schmid, V., (2003). Evolutionary aspects of developmentally regulated helix-loop-helix transcription factors in striated muscle of jellyfish. *Dev. Biol., 255*, 216–229.

Nakanishi, N., Hartenstein, B., & Jacobs, D. K., (2009). Development of the rhopalial nervous system *in Aurelia sp.1* (Cnidaria, Scyphozoa). *Dev. Genes Evol., 219*, 301–317.

Nielsen, S. K. D., Koch, T. L., Wiisbye, S. H., Grimmelikhuijzen, J. P., & Garm, A., (2021). Neuropeptide expression in the box jellyfish *Tripedalia cystophora* – new insights into the complexity of a "simple" nervous system. *J. Comp. Neurol., 529*, 2865–2882.

Nilsson, D. E., Coates, M. M., Gislen, I., Skogh, C., & Garm, A., (2005). Advanced optics in a jellyfish eye. *Nature, 435*, 201–205.

O'Connor, M., Garm, A., & Nilsson, D. E., (2009). Structure and optics of the eyes of the box jellyfish *Chiropsella bronzie*. *J. Comp. Physiol. A, 195*, 557–569.

O'Connor, M., Nilsson, D. E., & Garm, A., (2010). Temporal properties of the lens eyes of the box jellyfish *Tripedalia cystophora*. *J. Comp. Physiol. A, 196*, 213–220.

Pantin, C. F. A., (1935a). The nerve net of the actinozoa. I. Facilitation. *J. Exp. Biol., 12*, 119–138.

Pantin, C. F. A., (1935b). The nerve net of the actinozoa. II. Plan of the nerve net. *J. Exp. Biol., 12*, 139–155.

Pantin, C. F. A., (1965). Capabilities of the coelenterate behavior machine. *Amer. Zool., 5*, 581–589.

Park, E., Hwang, D. S., Lee, J. S., Song, J. I., Seo, T. K., & Won, Y. J., (2012). Estimation of divergence times in cnidarian evolution based on mitochondrial protein-coding genes in the fossil record. *Molec. Phylogenet. Evol., 62*, 329–345.

Parkefelt, L., & Ekstrom, P., (2009). Prominent system of RFamide immunoreactive neurons in the rhopalia of box jellyfish (Cnidaria: Cubozoa). *J. Comp. Neurol., 516*, 157–165.

Parkefelt, L., Skogh, C., Nilsson, D. E., & Ekstrom, P., (2005). Bilateral symmetric organization of neuronal elements in the visual system of a coelenterate, *Tripedalia cystophora*. *J. Comp. Neurol., 492*, 251–262.

Passano, L. M., (1965). Pacemakers and activity patterns in medusae: Homage to Romanes. *Amer. Zool., 5*, 465–489.

Passano, L. M., (1973). Behavioral control systems in medusae: A comparison between hydro- and scyphomedusae. *Publ Seto Mar. Biol. Lab., 20*, 615–645.

Passano, L. M., (1982). Scyphozoa and cubozoa. In: Shelton, G. A. B., (ed.) *Electrical Conduction and Behavior in "Simple" Invertebrates* (pp. 149–202). Clarendon Press, Oxford.

Passano, L. M., (2004). Spasm behavior and the diffuse nerve-net in *Cassiopea xamachana* (Scyphozoa: Coelenterata). *Hydrobiologia, 530, 531*, 91–96.

Peteya, D. J., (1973). A light and electron microscope study of the nervous system of *Ceriantheopsis americanus* (Cnidaria, Ceriantharia). *Z. Zellforsch. 141*, 301–317.

Petie, R., Garm, A., & Nilsson, D. E., (2011). Visual control of steering in the box jellyfish *Tripedalia cystophora. J. Exp. Biol., 214*, 2809–2815.

Petie, R., Garm, A., & Nilsson, D. E., (2013a). Velarium control and visual steering in box jellyfish. *J. Comp. Physiol A, 199*, 315–324.

Petie, R., Garm, A., & Nilsson, D. E., (2013b). Contrast and rate of light intensity decrease control directional swimming in the box jellyfish *Tripedalia cystophora* (Cnidaria, Cubomedusae). *Hydrobiologia, 703*, 69–77.

Pickens, P. E., (1969). Rapid contractions and associated potentials in a sand-dwelling sea anemone. *J. Exp. Biol., 51*, 513–528.

Roberts, A., & Mackie, G. O., (1980). The giant axon escape system of a hydrozoan medusa, *Aglantha digitale. J. Exp. Biol., 84*, 303–318.

Robson, E. A., & Josephson, R. K., (1969). Neuromuscular properties of mesenteries from the sea-anemone *Metridium. J. Exp. Biol., 50*, 151–168.

Romanes, G. J., (1885). *Jellyfish, Starfish and Sea Urchins* (p. 323). Kegan Paul, Trench and Co., London.

Ross, D. M., (1952). Facilitation in sea anemones. *J. Exp. Biol., 29*, 235–254.

Ross, D. M., (1955). Facilitation in sea anemones. IV. The quick response of *Calliactis parasitica* at high temperatures. *J. Exp. Biol., 32*, 815–821.

Satterlie, R. A., & Case, J. F., (1978). Neurobiology of the Gorgonian coelenterates, *Muricia californica*, and *Lophogorgia chilensis*. II. morphology. *Cell Tiss. Res., 187,* 379–396.

Satterlie, R. A., & Case, J. F., (1979). Neurobiology of the Gorgonian coelenterates, *Muricia californica*, and *Lophogorgia chilensis*. I. Behavioral physiology. *J. Exp. Biol., 79*, 191–204.

Satterlie, R. A., & Case, J. F., (1980). Neurobiology of the stoloniferan octocoral *Clavularia sp. J. Exp. Zool., 212*, 87–99.

Satterlie, R. A., & Eichinger, J. M., (2014). Organization of the ectodermal nervous structures in jellyfish: Scyphomedusae. *Biol. Bull., 226*, 29–40.

Satterlie, R. A., & Spencer, A. N., (1983). Neural control of locomotion in hydrozoan medusae: A comparative story. *J. Comp. Physiol., 150*, 195–206.

Satterlie, R. A., (1979). Central control of swimming in the cubomedusan jellyfish *Carybdea rastonii. J. Comp. Physiol. A, 133*, 357–367.

Satterlie, R. A., (1985a). Control of swimming in the hydrozoan jellyfish *Aequorea aequorea*: Direct activation of the subumbrella. *J. Neurobiol., 16*, 211–226.

Satterlie, R. A., (1985b). Central generation of swimming activity in the hydrozoan jellyfish *Aequorea aequorea. J. Neurobiol., 16*, 41–55.

Satterlie, R. A., (2002). Neural control of swimming in jellyfish: A comparative story. *Can. J. Zool., 80*, 1654–1669.

Satterlie, R. A., (2008). Control of swimming in the hydrozoan jellyfish *Aequorea victoria*: Subumbrellar organization and local inhibition. *J. Exp. Biol., 211*, 3467–3477.

Satterlie, R. A., (2011). Commentary: Do jellyfish have central nervous systems? *J. Exp. Biol., 214*, 1215–1223.

Satterlie, R. A., (2014). Multiple conducting systems in the cubomedusa *Carybdea marsupialis. Biol. Bull., 227*, 274–284.

Satterlie, R. A., (2015b). Cnidarian nerve nets and neuromuscular efficiency. *Int. Comp. Biol., 55*, 1050–1057.
Satterlie, R. A., (2019). Cnidarian neurobiology. In: Byrne, J. H., (ed.), *The Oxford Handbook of Invertebrate Neurobiology* (pp. 185–218). Oxford Univ. Press. New York, NY.
Satterlie, R. A., Anderson, P. A. V., & Case, J. F., (1976). Morphology and electrophysiology of the through-conducting systems in pennatulid coelenterates. In: Mackie, G. O., (ed.), *Coelenterate Ecology and Behavior* (pp. 619–627). Plenum Press, New York.
Satterlie, R. A., Anderson, P. A. V., & Case, J. F., (1980). Colonial coordination in anthozoans: Pennatulacea. *Mar. Behav. Physiol., 7*, 25–46.
Satterlie, R. A., Thomas, K. S., & Gray, G. C., (2005). Muscle organization of the cubozoan jellyfish *Tripedalia cystophora* (*Conant*, 1897). *Biol. Bull., 209*, 154–163.
Schwab, W. E., & Anderson, P. A. V., (1980). Intracellular recordings of spontaneous and evoked electrical events in the motor neurons of *Cyanea capillata. Amer. Zool., 20*, 941.
Schwab, W. E., & Anderson, P. A. V., (1981). Symmetrical chemical synapses. *Amer. Zool., 21*, 942.
Sebe-Pedros, A., Degnan, B. M., & Ruiz-Trillo, I., (2017). The origin of metazoa: A unicellular perspective. *Nature Rev. Genet., 18*, 498–512.
Seipel, K., & Schmid, V., (2005) Evolution of striated muscle: Jellyfish and the origin of triploblasty. *Dev. Biol., 282*, 14–26.
Seipel, K., & Schmid, V., (2006). Mesodermal anatomies in cnidarian polyps and medusae. *Int. J. Dev. Biol., 50*, 589–599.
Shafer, E. A., (1878). Observations on the nervous system of *Aurelia aurita. Phil. Trans. B, 168*, 563–575.
Shelton, G. A. B., & McFarlane, I. D., (1976). Electrophysiology of two parallel conducting systems in the colonial hexacorallia. *Proc. Roy. Soc. London B, 193*, 77–87.
Shelton, G. A. B., (1975a). Electrical activity and colonial behavior in anthozoans hard corals. *Nature, 253*, 558–560.
Shelton, G. A. B., (1975b). Colonial conduction systems in the Anthozoa: Octocorallia. *J. Exp. Biol., 62*, 571–578.
Shelton, G. A. B., (1975c). Colonial behavior and electrical activity in the hexacorallia. *Proc. Roy. Soc. London B, 190*, 239–256.
Singla, C. L., (1974). Ocelli of hydromedusae. *Cell Tiss. Res., 149*, 413–429.
Singla, C. L., (1975). Statocysts of hydromedusae. *Cell Tiss. Res., 158*, 391–407.
Singla, C. L., (1978a). Locomotion and neuromuscular system of *Aglantha digitale. Cell Tiss. Res., 188*, 317–327.
Singla, C. L., (1978b). Fine structure of the neuromuscular system of *Polyorchis penicillatus* (Hydromedusae, Cnidaria). *Cell Tiss. Res., 193*, 163–174.
Skogh, C., Garm, A., Nilsson, D. E., & Ekstrom, P., (2006). Bilaterally symmetrical rhopalial nervous system of the box jellyfish *Tripedalia cystophora. J. Morph., 267*, 1391–1405.
Spencer, A. N., & Arkett, S. A., (1984). Radial symmetry and the organization of central neurons in a *hydrozoan* jellyfish. *J. Exp. Biol., 110*, 69–90.
Spencer, A. N., & Satterlie, R. A., (1980). Electrical and dye-coupling in an identified group of neurons in a coelenterate. *J. Neurobiol., 11*, 13–19.
Spencer, A. N., (1975). Behavior and electrical activity in the hydrozoan *Proboscidactyla flavicirrata* (Brandt). II. The medusa. *Biol. Bull., 149*, 236–250.
Spencer, A. N., (1978). Neurobiology of *Polyorchis*. I. Function of effector systems. *J. Neurobiol., 9*, 143–157.

Spencer, A. N., (1979). Neurobiology of *Polyorchis*. II. Structure of effector systems. *J. Neurobiol., 10,* 95–117.

Spencer, A. N., (1981). The parameters and properties of a group of electrically coupled neurons in the central nervous system of a hydrozoan jellyfish. *J. Exp. Biol., 93,* 33–50.

Spencer, A. N., (1982). The physiology of a coelenterate neuromuscular synapse. *J. Comp. Physiol., 148,* 353–363.

Spring, J., Yanze, N., Josch, C., Middel, A. M., Winninger, B., & Schmid, V., (2002). Conservation of *Brachyury, Mef2,* and *Snail* in the myogenic lineage of jellyfish: A connection to the mesoderm of bilateria. *Dev. Biol., 244,* 372–384.

Spring, J., Yanze, N., Middel, A. M., Stierwald, M., Groger, H., & Schmid, V., (2000). The mesoderm specification factor *twist* in the life cycle of jellyfish. *Dev. Biol., 228,* 363–375.

Takahashi, T., (2020). Comparative aspects of structure and function of cnidarian neuropeptides. *Front. Endocrin.* Doi: 10.3389/fendo.202000339.

Watanabe, H., Fujisawa, T., & Holstein, T. W., (2009). Cnidarians and the evolutionary origin of the nervous system. *Dev. Growth Differ., 51,* 167–183.

Weber, C., Singla, C. L., & Kerfoot, P. A. H., (1982). Microanatomy of the subumbrellar motor innervation in *Aglantha digitale* (Hydromedusae: Trachylina). *Cell Tiss. Res., 223,* 305–213.

Westfall, J. A., & Kinnamon, J. C., (1978). A second sensory-motor interneuron with neurosecretory granules in *Hydra. J. Neurocytol., 7,* 365–379.

Westfall, J. A., (1973). Ultrastructural evidence for neuromuscular systems in coelenterates. *Am. Zool., 13,* 237–246.

Westfall, J. A., (1996). Ultrastructure of synapses in the first-evolved nervous systems. *J. Neurocytol., 25,* 735–746.

Westfall, J. A., Yamataka, S., & Enos, P. D., (1971). Ultrastructural evidence for polarized synapses in the nerve net of *Hydra. J. Cell Biol., 51,* 318–323.

Wijesena, N., Simmons, D. K., & Marindale, M. Q., (2017). Antagonistic BMP-cWNT signaling in the cnidarian *Nemostella vectensis* reveals insight into the evolution of mesoderm. *Proc. Nat. Acad. Sci. U.S., 114,* E5608-E5615.

Yamasu, T., & Yoshida, M., (1973). Electron microscopy on the photoreceptors of an anthomedusa and a scyphomedusa. *Publ. Seto Mar. Biol. Lab., 20,* 757–778.

CHAPTER 6

Neuropeptides as Potentiators of Coral Polyp Contraction

SHINYA SHIKINA[1,2] and CHING-FONG CHANG[2,3]

[1]*Institute of Marine Environment and Ecology, National Taiwan Ocean University, Keelung, Taiwan*

[2]*Center of Excellence for the Oceans, National Taiwan Ocean University, Keelung, Taiwan*

[3]*Department of Aquaculture, Taiwan National Taiwan Ocean University, Keelung, Taiwan*

ABSTRACT

Scleractinian corals are the keystone animals that create coral reef ecosystems, which possess the highest marine biodiversity on the planet. They comprise one of the major groups of cnidarians, accounting for more than 10% of the phylum Cnidaria. Despite their ecological importance, our understanding of the biological characteristics of corals remains weak. In particular, their neurobiology has been largely unexplored; however, recently, some important advances have been made in that regard. The existence of neurons and neuropeptides in corals was first demonstrated in 2019. In addition, neuropeptides were shown to be involved in polyp contraction, a physiological response in corals. This chapter presents recent advances in coral neurobiology, with a focus on the glycine-leucine-tryptophan-amide family of neuropeptides (GLWamides). This chapter also covers current knowledge of GLWamides from adult corals, including the full-length cDNA of the preprohormone, predicted sequences of peptides, distributions of mRNA and peptides (GLWamide-positive [GLWamide$^+$] neurons), and

Frontiers in Invertebrate Physiology: A Collection of Reviews, Volume 1: Non-Bilaterian Phyla.
Saber Saleuddin, Sally P. Leys, Robert D. Roer, and Iain C. Wilkie, (Eds.)

their possible physiological functions. Finally, we discuss future prospects in coral neurobiology.

6.1 INTRODUCTION

Neuropeptides are short linear polypeptides secreted by neurons (Holmgren & Jensen, 2001). In general, neuropeptides are derived from larger precursor molecules called preprohormones (also referred to as prepropeptides). Preprohormones are post-translationally processed by proteolytic cleavage and modified to yield mature, small peptides (Holmgren & Jensen, 2001; Grimmelikhuijzen et al., 2002). Neuropeptides act either directly or indirectly as modulators and/or hormones, and they regulate various physiological processes (e.g., growth, reproduction, and metabolism) and even behavior (Holmgren & Jensen, 2011; Klavdieva, 1995; Jékely, 2013; Elphick et al., 2018). The phylum Cnidaria is characterized by the possession of cnidocytes. It includes about 10,000 species, mostly marine (corals, anemones, jellyfish, etc.) and a limited number of freshwater species (*Hydra*, etc.) (Brusca & Brusca, 2003; Pechenik, 2010). More than 50 types of cnidarian neuropeptides have been identified to date, and their significance has been demonstrated in various life stages (Takahashi & Takeda, 2015; Grimmelikhuijzen et al., 2002; Fujisawa & Hayakawa, 2012; Takahashi, 2020). It is not an exaggeration to say that neuropeptides are indispensable for understanding the physiological functions of cnidarians. A detailed discussion of cnidarian neuropeptides is beyond the scope of this chapter, but several reviews have addressed this subject (Grimmelikhuijzen et al., 2002; Takahashi & Hatta, 2011; Takahashi & Takeda, 2015; Kelava et al., 2015; Takahashi, 2020; Wiliams, 2020).

Scleractinian corals (hereinafter, corals) are one of the major groups of cnidarians, accounting for more than 10% of the phylum (over 1,300 species, according to the World List of Scleractinians: http://www.marinespecies.org/scleractinia/). They are also the main organisms that construct coral reef ecosystems, which support biological productivity in tropical and subtropical regions (Odum & Odum, 1955; Moberg & Folke, 1999). In spite of their importance, however, our knowledge of the biological characteristics of corals is still limited. Basic coral biology, including coral neurobiology, has been slow to develop.

The existence of neurons and neuropeptides was first demonstrated in 2019 (Attenborough et al., 2019). Soon thereafter, neuropeptides were shown to be involved in polyp contraction (Shikina et al., 2020). Although pertinent literature is very limited, with only a few reports (Attenborough et al., 2019; Shikina et al., 2020), these studies represented a significant step in coral neurobiology. This chapter reviews recent advances in this field, with a focus on neuropeptides of the glycine-leucine-tryptophan-amide family (GLWamides). First, the fundamental biological characteristics of corals and their nervous systems are described. Then, various aspects of GLWamides, including the full-length cDNA of the preprohormone, predicted sequences of peptides, distributions of the mRNA and peptides (GLWamide$^+$ neurons), and their possible physiological functions are described. Finally, future prospects in coral neurobiology are discussed.

6.1.1 FUNDAMENTAL BIOLOGICAL CHARACTERISTICS OF CORALS

The typical life cycle of corals starts with the development of fertilized eggs into free-swimming larvae called planulae. In general, planulae metamorphose into primary polyps after settling on appropriate substrates and initiate asexual reproduction (e.g., budding) (Harrison & Wallace, 1990). Polyps produced by asexual reproduction are genetically identical, and in most corals, they are connected to one another by tissue called the coenosarc (Figure 6.1). Most corals are colonial, and formed by long-term asexual reproduction. When polyps (or colonies) reach certain ages and sizes, they begin to reproduce sexually to create genetically diverse individuals. There is no medusa phase in the life history of corals, and they live exclusively as polyps (Harrison & Wallace, 1990; Technau & Steele, 2011; Jones et al., 2015; Shikina & Chang, 2018).

The structure of a coral body, like that of other cnidarians, is simple and basically comprises two layers, the epidermis and the gastrodermis (Technau & Steele, 2011). Mesoglea, a layer of extracellular matrix, exists between the epidermis and the gastrodermis (Figure 6.1). There are no distinct organs or organ systems, as found in vertebrates, but there are several functionally specialized tissues in corals. For example, tentacles extending from the top of the polyps capture prey and defend against enemies (Figure 6.1). The mouth and pharynx, located at the center of the tentacles, ingest seawater and prey into the gastrovascular cavity and also excrete wastes. Mesenterial filaments in the gastrodermis cavity are lamella-like extensions of mesenteries that are mainly responsible for digesting captured prey (Goldberg, 2002; Raz-Bahat et al., 2017) (Figure 6.1). The gonads are located in specific regions within the mesenteries and form distinct swellings during the reproductive season due to active gametogenesis. Most stony corals contain symbiotic algae in the gastrodermis (Figure 6.1). In addition, corals form exoskeletons composed primarily of calcium carbonate and an organic matrix (Figure 6.1).

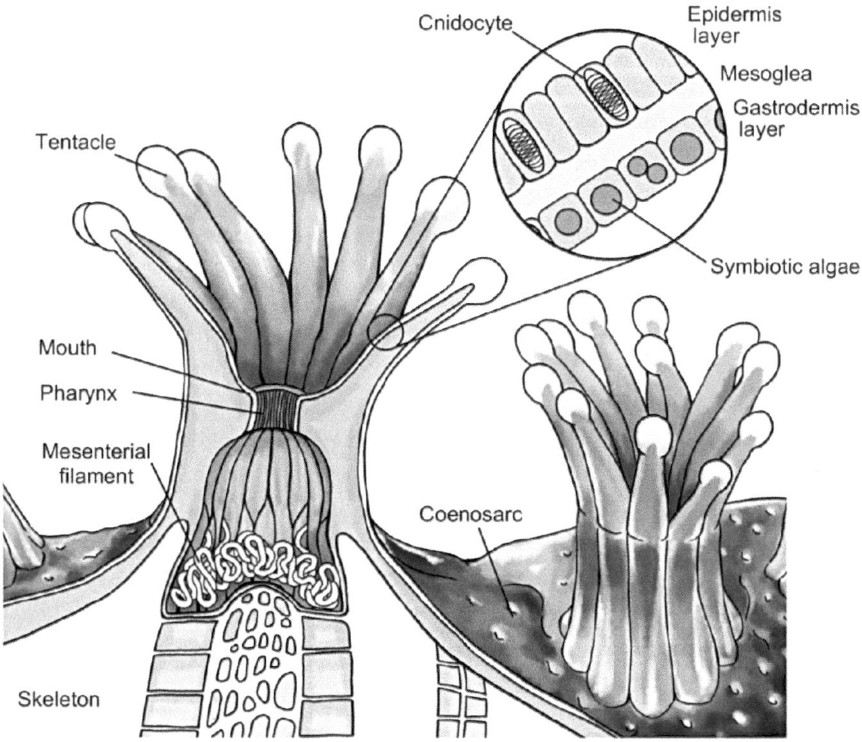

FIGURE 6.1 Gross morphology and cross-section of a coral polyp. Polyps are interconnected by the coenosarc. The calcium carbonate skeleton subtends the polyps and coenosarc.

6.1.2 CORAL NERVOUS SYSTEMS

In general, cnidarians lack a central nervous system, i.e., a brain, and have a simple, diffuse nervous system (Technau & Steele, 2011; Kelava et al., 2015; Takahashi & Takeda, 2015). In particular, in the freshwater cnidarian, *Hydra*, colonial marine hydroids, hydrozoan jellyfish, and sea anemones, distributional patterns of multiple types of neurons have been well demonstrated (Schmich et al., 1998; Takahashi & Hatta, 2011; Koizumi et al., 2015; Kelava et al., 2015; Havrilak et al., 2017). In corals, somewhat surprisingly, cytohistological evidence for the presence of neurons was not available until recently (Attenborough et al., 2019; Shikina et al., 2020). From a phylogenetic point of view, it had long been assumed that corals have a nervous system like that of other cnidarians. The presence of coral neurons and nervous systems had been inferred primarily from coral behavior. For instance, corals have the ability to expand and contract polyps (Figure 6.2). When expanding, they open their tentacles like flowers and extend their polyps (Figure 6.2(A)). The polyps take in seawater from the environment and distribute it to the tentacle tips, thereby maintaining hydrostatic pressure. When exposed to stresses such as desiccation, strong currents, or attacks by predators, entire polyps immediately contract into the skeleton (Figure 6.2(B) and (C)). Polyps return to their original extended state soon after the threats have passed. Some corals extend their polyps in response to insolation (Abe, 1939; Sweeney, 1976; Sebens & Deriemer, 1977; Lasker, 1979; Hoadley et al., 2011).

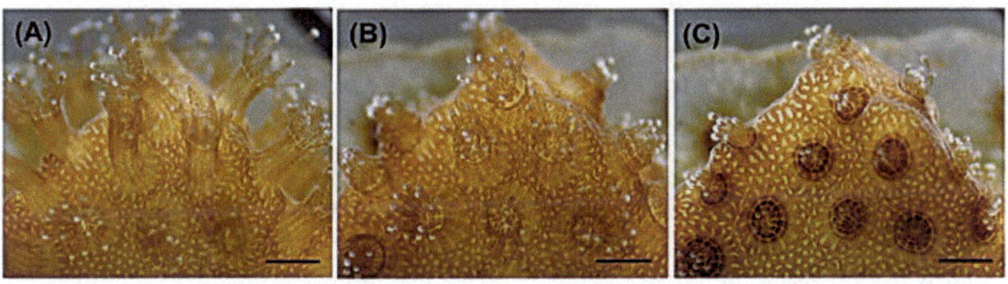

FIGURE 6.2 Contraction of polyps of the stony coral, *Pocillopora damicornis*, under strong water flow. (A) before; (B) during; and (C) after contraction. Scale bars = 1 mm.

Because many coral species are sessile, they necessarily have sensitive nervous systems that allow them to perceive biotic and abiotic factors in the environment and to respond physiologically for defense and predation. Neurons were first demonstrated cytohistologically in *Acropora millepora* in 2019. The distribution of a neuronal subpopulation, RFamide neurons, in larvae and adult polyps was shown using *in situ* hybridization (ISH) and immunostaining with an anti-FMRFamide antibody (Attenborough et al., 2019). In adult polyps, numerous FMRFamide-immunoreactive cells are widely distributed in the tentacles, oral discs, mouth, pharynx, and mesenterial filaments. Notably, immunoreactive cells form a neural ring around the mouth. There are also scattered immunoreactive cells in the coenosarc between polyps (Attenborough et al., 2019). In planula larvae, neurons are distributed in the epidermis. Moreover, Attenborough and colleagues also demonstrated another subpopulation of neurons, GLWamide^{+} neurons, in larvae (Attenborough et al., 2019). At present, very little is known about coral nervous systems. Much remains to be learned about other types of neurons, the overall integration of the nervous systems, and the physiological functions in which they are involved.

6.1.3 GLWAMIDES: NEUROPEPTIDES WITH MULTIPLE PHYSIOLOGICAL FUNCTIONS

Glycine-leucine-tryptophan-amide family neuropeptides (GLWamides, also referred to as LWamides) are important neuropeptides in cnidarians. The first cnidarian GLWamide, isolated using high-performance liquid chromatography (HPLC) in 1994 from the sea anemone, *Anthopleura elegantissima*, triggers larval metamorphosis of a colonial marine hydroid, *Hydractinia echinata* (Leitz et al., 1994). During the next three years, seven types of GLWamides were isolated from *Hydra magnipapillata* using a combination of HPLC, differential display-PCR, and bioassays (Takahashi et al., 1997). Some of these GLWamides are involved in a wide range of physiological functions in cnidarians, such as larval metamorphosis in *H. echinata* (Schmich et al., 1998) and the sea anemone, *Nematostella vectensis* (Nakanishi & Martindale, 2018), muscle contraction (myoactivity) in *H. magnipapillata* and the sea anemone, *A. fuscoviridis* (Takahashi et al., 2003), polyp detachment in *H. magnipapillata* (Takahashi et al., 1997, 2003), and oocyte maturation and ovulation in the hydrozoan jellyfish, *Cytaeis uchidae* (Takeda et al., 2013).

Like most other neuropeptides, GLWamides are derived from larger inactive preprohormones, which are incorporated into the endoplasmic reticulum (ER) and converted into prohormones. Subsequently, they are transported to the Golgi apparatus to be post-translationally processed by a series of proteolytic cleavages and C-terminal amidation to yield mature, active peptides (Takahashi & Takeda, 2015; Attenborough et al., 2019). Mature cnidarian GLWamides identified to date range from 5 to 14 amino acids. Some conserved structural features are found in their N- and C-terminal regions. For example, in the N-terminal region, the 7 *H. magnipapillata* GLWamides identified to date possess either a proline residue in the second position (X-Pro) or paired prolines in the second and third positions (X-Pro-Pro). The C-terminus has the residues GLW with C-terminal amidation (-Gly-Leu-Trp-NH_2) (Takahashi et al., 1997; Takahashi & Hatta, 2011; Takahashi & Takeda, 2015). The proline(s) near the N-terminus block degradation by nonspecific aminopeptidases (Carstensen et al., 1992).

The function of GLWamides in coral physiology was first reported by Iwao et al. (2002). They found that treatment with a *Hydra*-derived GLWamide peptide (Hym-248, EPLPIGLWamide) effectively induced metamorphosis in cultured planulae of nine Pacific *Acropora* species (Iwao et al., 2002). Although the mechanism of action has not yet been clarified, the discovery was sensational. Thereafter, it was shown that Hym-248 induces metamorphosis in planulae of various Pacific *Acropora* corals (Hatta & Iwao, 2003), a Caribbean coral *A. palmata* (Erwin & Szmant, 2010), and Pacific *A. millepora* (Grasso et al., 2011; Meyer et al., 2011). Since Hym-248 treatment is a highly reproducible method of inducing coral larval metamorphosis under laboratory conditions, it has been used in coral developmental studies, especially in research on morphological changes during metamorphosis from planulae into primary polyps and the genes involved (Hirose et al., 2008; Grasso et al., 2011; Meyer et al., 2011). Interestingly, Hym-248 has not been shown to induce metamorphosis in other coral genera (Iwao et al., 2002).

6.2 STRUCTURE AND EXPRESSION OF THE GLWAMIDE PREPROHORMONE IN CORALS

6.2.1 *SEQUENCE DISCOVERY*

Although the involvement of GLWamide in coral metamorphosis was widely recognized among coral biologists after the report by Iwao et al. (2002), the gene (cDNA) encoding the GLWamide preprohormone, GLWamide$^+$ neurons, and/or native GLWamides remained undiscovered until 2019, when the full sequence of GLWamide preprohormone was reported in *Acropora millepora* (Attenborough et al., 2019). That was followed by a similar report from the coral *Euphyllia ancora* (Shikina et al., 2020). In both studies, sequences were identified by *in silico* analyses using BLAST searches of the coral genome and/or transcriptomes and were further confirmed by cDNA cloning (Attenborough et al., 2019; Shikina et al., 2020).

6.2.2 *CDNA STRUCTURE, PREPROHORMONE STRUCTURE, AND POTENTIAL PEPTIDES*

The cDNA structure encoding GLWamide preprohormone is illustrated using an isolated cDNA sequence from *E. ancora* (hereinafter, EaGLWamide preprohormone) (Figure 6.3) (Shikina et al., 2020). The full-length cDNA for EaGLWamide preprohormone, determined by RACE-PCR, consists of

1,231 nucleotides, comprising 108 bases in the 5′ untranslated region, 780 bases in the open reading frame, 321 bases in the 3′ untranslated region, and a stretch of 19 adenosine residues (Figure 6.3(A)). A typical polyadenylation motif (AATAAA) preceding the poly-A tail is present in the 3′ untranslated region.

The cDNA predicts an open reading frame encoding a 260-amino acid preprohormone with a molecular mass of 29.5 kDa (Figure 6.3(B)) (Shikina et al., 2020). The predicted preprohormone comprises a signal peptide of 20 amino acids for protein translocation to the ER, 7 GLW motifs, 6 potential dibasic cleavage sites (*KR/RR*), and 7 C-terminal amidations with monobasic cleavage sites (*GRs*) (Figure 6.3(B)). A software program Peptraq, which was developed for *in silico* analyses of "-omic" data (Zatylny-Gaudin et al., 2016, https://peptraq.greyc.fr), predicts a single GIWamide peptide and 6 potential GLWamide peptides (Figure 6.3(C)). Although the lengths of those 6 potential GLWamide peptides range from 6 to 51 amino acids, all potential peptides bear tandem proline residues (PP) followed by GLW motifs at their C-terminal ends (Figure 6.3(C)).

6.2.3 SEQUENCE COMPARISONS AMONG STONY CORALS

Interspecific differences in GLWamide preprohormone sequences and the presence of conserved regions have been investigated by comparing sequences from multiple coral species (Figure 6.4(A)) (Shikina et al., 2020). Sequences of GLWamide preprohormones from *E. ancora*, *Stylophora pistillata*, *Pocillopora damicornis*, and *Acropora digitifera* (Figure 6.4(A)) are used for comparison. Sequence lengths and numbers of GLW motifs in the sequences vary among species. Their lengths range from 210 to 314 amino acids, and GLW motifs vary from 5 to 10 (Figure 6.4(B)). Overall identities among GLWamide preprohormones are low, ranging from 33 to 36%. However, the sequence "PPGLWGR," followed by an amidation-monobasic cleavage signal (GR), is highly conserved. Other sequence segments are highly variable (Figure 6.4(C)), but the conservation of the PPGLWGR sequence suggests its functional significance in coral physiology.

6.2.4 SEQUENCE COMPARISONS AMONG CNIDARIANS

Sequence comparisons between corals and other cnidarians have also been reported (Shikina et al., 2020). Sequence lengths and GLW motif numbers

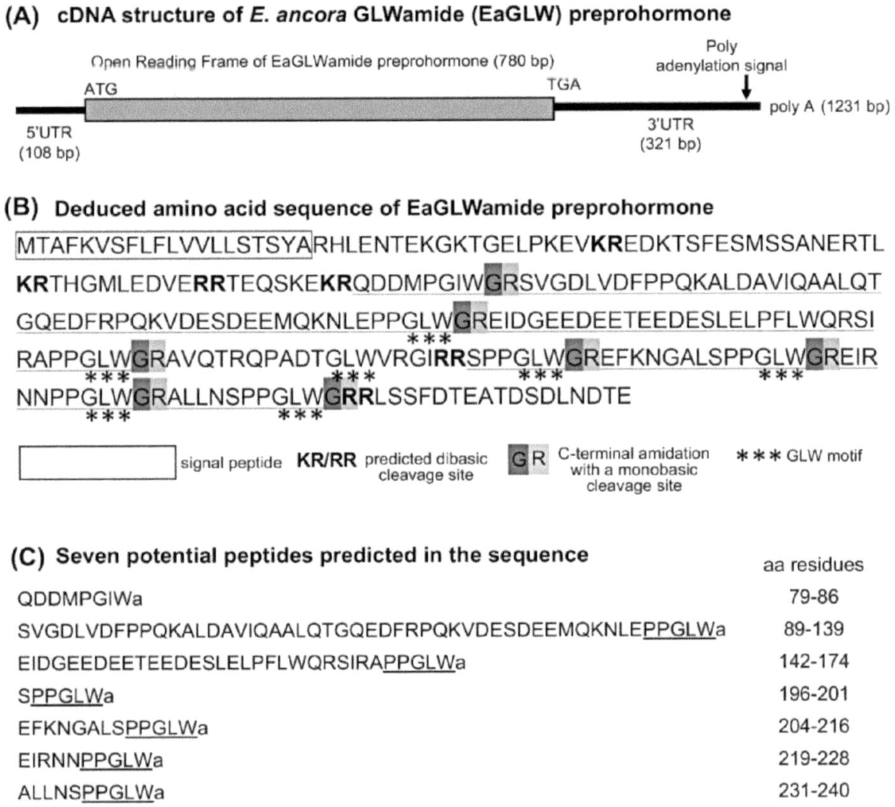

FIGURE 6.3 (A) Schematic representation of the mRNA and open reading frame of *Euphyllia ancora* GLWamide preprohormone. The 5′ and 3′ untranslated regions are shown as a heavy black line (5′UTR and 3′UTR). The open reading frame is shown as a shaded box. (B) The deduced amino acid sequence of the EaGLWamide preprohormone. Positions of the signal peptide, dibasic cleavage sites, C-terminal amidation with a monobasic cleavage site, and the GLW motif are shown in the figure. Locations of potential GLWamide peptide sequences are underlined. (C) Seven potential peptides are predicted from the EaGLWamide preprohormone using Peptraq software (Zatylny-Gaudin et al., 2016). Positions of potential peptides in the EaGLWamide preprohormone are shown on the right (amino acids). Note that all potential GLWamide peptides contain tandem proline residues followed by a GLW motif (PPGLW) at the C-terminus, as underlined.

Source: Modified and reproduced from: Shikina et al. (2020), with permission.

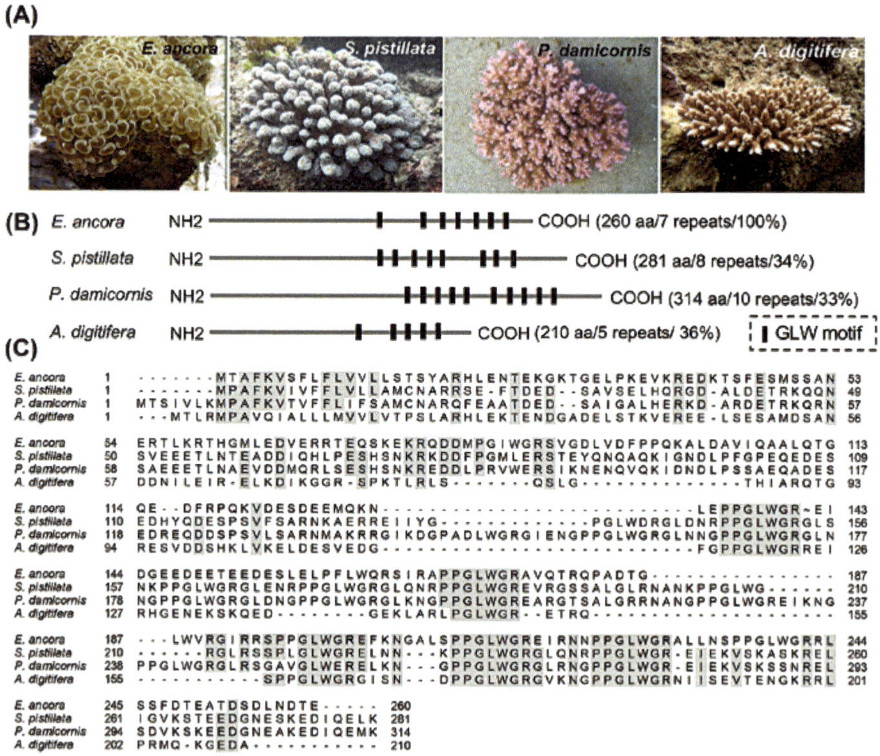

FIGURE 6.4 (A) Representative pictures of the stony corals *Euphyllia ancora*, *Stylophora pistillata*. *Pocillopora damicornis*, and *Acropora digitifera*. (B) Comparison of GLWamide preprohormones from these corals. Lengths, numbers of GLW motifs, and sequence identities of the amino acid sequence with EaGLW preprohormone are shown on the left. (C) Comparison of amino acid sequences of GLWamide preprohormones among *E. ancora* (Genbank ID: QAU55050), *S. pistillata* (Genbank ID: XP_022779355), *P. damicornis* (Genbank ID: XP_027060554), and *A digitifera* (Genbank ID: XP_015772272). The multiple sequence alignment was created with BioEdit software. Identical residues among the sequences are shaded in gray. The identity threshold is set at 75%. Gaps introduced in the sequences to maximize homology are indicated by hyphens.

Source: Photograph courtesy of Mr. Tzu-Chieh Lin, Ms. Yi-Ling Chiu, Ms. Yin-Chu Cheng, and Dr. Chuya Shinzato; Modified and reproduced from: Shikina et al. (2020), with permission.

vary widely among species (Figure 6.5). For instance, lengths range from 217 residues in the sea anemone, *Exaiptasia pallida* to 514 residues in the sea anemone *A. elegantissima*. GLW motif numbers range from 7 (*E. ancora*) to 38 (*A. elegantissima*). Sequence identities between EaGLWamide preprohormone and GLWamide preprohormones of other cnidarians are low, ranging from 20 to 27%.

6.2.5 EXPRESSION OF MRNA FOR GLWAMIDE PREPROHORMONE IN CORAL POLYPS

The mRNA distribution of *GLWamide preprohormone* within coral polyps has been investigated by quantitative reverse transcription (qRT)-PCR analysis in *E. ancora* (Shikina et al., 2020). Because *E. ancora* has large polyps, different parts of polyp tissues can be microscopically isolated (Shikina et al., 2012, 2013). Tentacles, mouths, including the pharynx, gonads (testis/ovary), and mesenterial filaments of *E. ancora* (Figure 6.5(A) and (B)) were isolated from polyps of male and female colonies, and mRNA levels of *EaGLW preprohormone* among these tissues were compared. The mRNA is highly expressed in the mouth, including the pharynx, compared with its expression in other tissues, in both male (Figure 6.5(C)) and female (Figure 6.5(D)) polyps. Tentacles also have high but non-significant levels of transcripts in polyps of both sexes (Figure 6.5(C) and (D)).

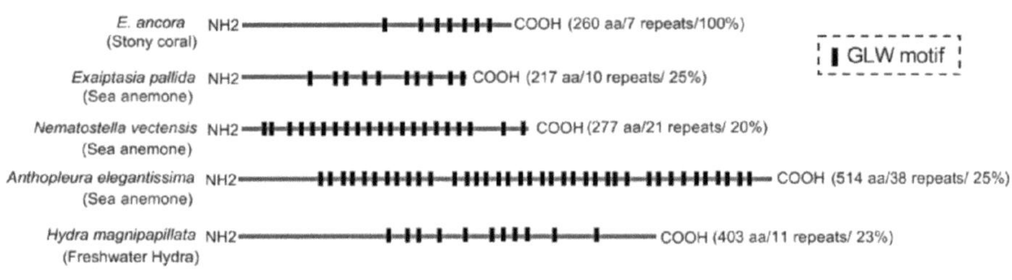

FIGURE 6.5 Comparison of GLWamide preprohormones among cnidarians. Lengths, numbers of GLW motifs, and percent sequence identities of the amino acid sequence with EaGLW preprohormone are shown at the right. Sequences from the sea anemones, *Exapitasia pallida* (Genbank ID: KXJ28092), *Nematostella vectensis* (Genbank ID: XM_001626406), *Anthopleura elegantissima* (Genbank ID: Q16992), the freshwater Hydra, *Hydra magnipapillata* (Genbank ID: XP_002164748) and the marine hydroid, *Hydractinia XXXchinate* (Genbank ID: X89734), were used for comparison.

Source: Modified and reproduced from: Shikina et al. (2020), with permission.

6.3 DISTRIBUTION OF GLWAMIDE⁺ NEURONS IN CORAL POLYPS

Identifying types of neurons and their spatial distribution patterns within polyps will help to better understand their functions in corals. Coral neurons have been detected either by immunodetection with antibodies specific for the target neuropeptides or ISH with target mRNA for the neuropeptide preprohormone. Distribution patterns of GLWamide$^+$ neurons in larvae and adult corals are discussed in subsections.

6.3.1 LARVA AND PRIMARY POLYPS

Transcript expression patterns of GLWamide preprohormone from larvae to post-settlement have been determined in *A. millepora* using mRNA ISH (Attenborough et al., 2019). Expressing cells (named AmLWamide-expressing cells) appear in the ectoderm in early planulae and persist for most of the planula stage. They are mainly distributed in the central portions of the cylindrical larvae and less in the oral and posterior regions. The cell bodies lie close to the mesoglea, extending long projections to the larval surface. Notably, AmLWamide-expressing cells start to disappear just before settlement and completely disappear in newly settled primary polyps. Thereafter, in post-settlement polyps, scattered AmLWamide-expressing cells begin to reappear on the oral side (Attenborough et al., 2019).

6.3.2 ADULT POLYPS

The distribution of GLWamide$^+$ neurons in adult polyps has been investigated in six Pacific coral species from five families, mainly using immunohistochemistry with a monoclonal antibody against GLWamide (Table 6.1; Shikina et al., 2020). In *E. ancora*, GLWamide$^+$ neurons are mainly detected in the mouth, pharynx, and tentacles (Figure 6.7(A)–(F)). This accords with distributional patterns of GLWamide preprohormone mRNA (Figure 6.6(C) and (D)). In the mouth and pharynx, thin fibers of GLWamide$^+$ neurons are distributed as a neural ring in the epidermis along the mesoglea, extending long projections to the surface (Figure 6.7(A)–(C)). In tentacles, GLWamide$^+$ neurons are distributed in the column but not in the apical region, which has a number of cnidoblasts (Figure 6.7(D)–(F)). By contrast, in the testis, ovary, and mesenterial filaments, only a few GLWamide$^+$ neurons (immunoreactive) are detected (Figure 6.7(G)–(I)). In other stony corals, including *S. pistillata*, *P. damicornis*, and *Acropora hyacinthus*, GLWamide$^+$ neurons are distributed mainly in the epidermis of the mouth and pharynx. In particular, GLWamide$^+$ neurons are found in a linear pattern along the mesoglea, and cell bodies are observed in the epidermis (Figure 6.7(A)–(I)). Numbers of GLWamide$^+$ neurons in other tissues, such as tentacles, vary among species (Table 6.1).

These data indicate that GLWamide⁺ neurons are commonly and predominantly distributed in the mouth and pharynx of coral polyps. Additionally, given their broad taxonomic distribution, GLWamides are probably involved in various physiological functions in corals.

TABLE 6.1 Distribution of GLWamide Neurons in Stony Corals as Assessed by Immunohistochemical Analysis

Order	Family	Species Name	Immunoreactivity			
			Mouth with Pharynx	Tentacle	Mesenterial Filament	Gonad
Scleractinia	Euphyllidae	*Euphyllia ancora*	+++	++	+	+
	Pocilloporidae	*Stylophora pistillata*	+++	+	+	unknown
	Pocilloporidae	*Pocillopora damicornis*	+++	+	+	unknown
	Acroporidae	*Acropora hyacinthus*	+++	+	+	unknown
	Merulinidae	*Favites pentagona*	+++	++	+	unknown
	Fungiidae	*Cycloseris* sp.	+++	++	+	unknown

Note: The levels of the number of cells exhibiting immunoreactivity were defined as four grades: many (+++); moderate (++); a few (+); and negative (–).

6.4 THE ROLE OF GLWAMIDES IN POLYP CONTRACTION

So far, the involvement of GLWamide peptides in polyp contraction has been demonstrated in *E. ancora* (Shikina et al., 2020) and *P. damicornis* (Shikina et al., unpublished). Their involvement in polyp contraction was examined by treating coral colony fragments or isolated tissues with GLWamide/related peptides and observing their responses. In treatment experiments, mainly synthetic PPGLWamide peptide (Pro-Pro-Gly-Leu-Trp-NH_2) was used. This sequence was selected because all GLWamides predicted from *E. ancora* GLWamide preprohormone contain a tandem proline residue followed by a GLW motif (PPGLW) at the C-terminus (Figure 6.3(C)) and because this sequence is highly conserved among corals (Figure 6.4(C)).

6.4.1 RESULTS OF TREATMENT EXPERIMENTS IN E. ANCORA

A preliminary treatment experiment was performed using tissues microscopically isolated from polyps. Treatments of tentacles, pieces of mouth with pharynx, and mesenterial filaments, with PPGLWamide induced contraction

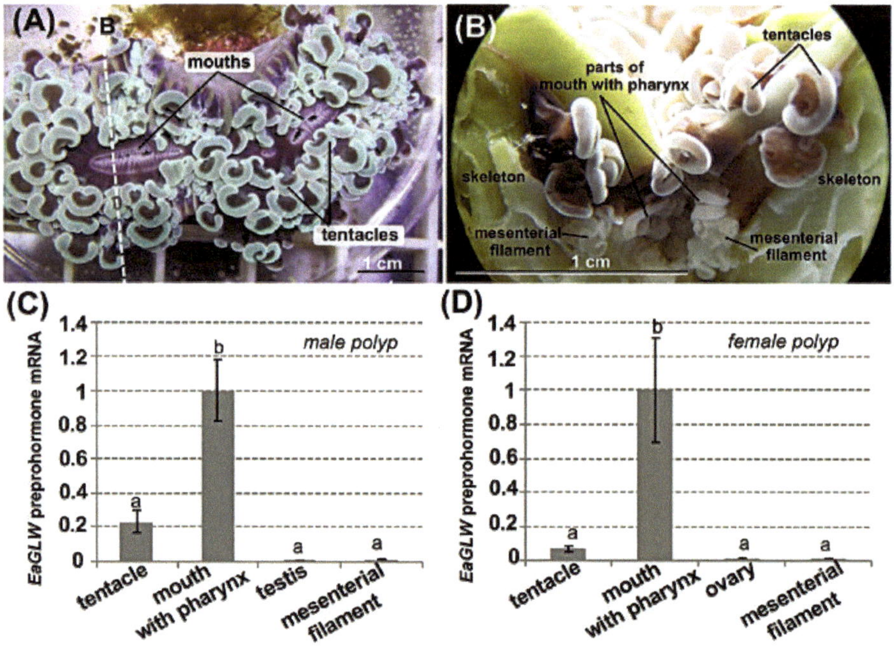

FIGURE 6.6 (A) Representative picture of a colony of the stony coral, *Euphyllia ancora*, in an aquarium (viewed from above). Bar, 1 cm. (B) Representative picture of a dissected polyp of *E. ancora*. The polyp was dissected through the mouth at the position of the broken line shown in (A). Three major polyp tissues, tentacles, mouth with the pharynx, and mesenterial filaments, are observed. (C) and (D) Comparison of expression levels of EaGLW preprohormone transcripts in male (C) and female polyp tissues (D). Data are shown as means ± SE (4 male or 3 female colonies) of the relative values of the mouth group. *E. ancora β-actin* was used as the reference gene. Groups with different letters are significantly different ($P < 0.05$).

Source: Reproduced from: Shikina et al. (2020), with permission.

of tentacles and pieces of mouth with pharynx (Figure 6.9(A) and (B)). No apparent response was observed in mesenterial filaments. Dose-dependent responses were also observed in more than 90% of tentacles and mouth with pharynx samples, showing contraction at 10 μM or higher (Figure 6.9(C)). Interestingly, this response is dependent on the sequence and modification state of the peptide. Responses were rarely observed with PPGLLamide (Pro-Pro-Gly-Leu-Leu-NH_2) or PPGLW-OH (Figure 6.9(D)).

Subsequently, treatment experiments were performed using colony fragments that contained a few polyps. As with the above experiments using isolated tissues, treatments of colony fragments with control vehicle (H_2O, 0 μM), 10 μM PPGLWamide, 10 μM PPGLW-OH, or 10 μM PPGLLamide

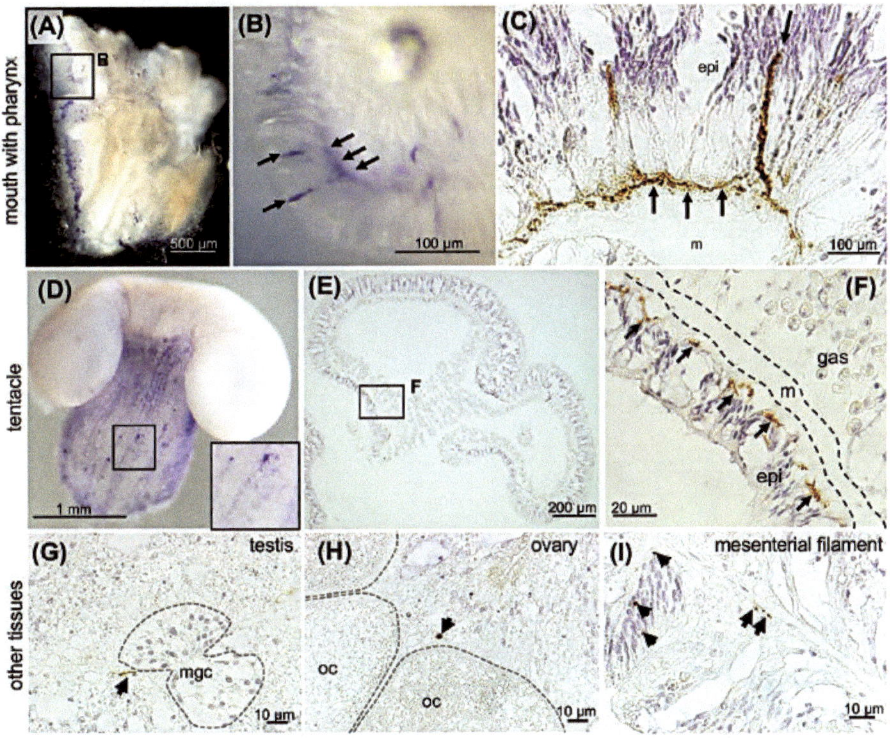

FIGURE 6.7 Distribution of GLWamide⁺ neurons in an adult *Euphyllia ancora* polyp. (A) Whole-mount immunodetection of GLWamide⁺ neurons in an isolated piece of mouth, including the pharynx. (B) Higher-magnification view of the inset in (A). (C) Immunohistochemical detection of GLWamide⁺ neurons in the mouth. (D) Whole-mount immunodetection of GLWamide⁺ neurons in an isolated tentacle. The inset shows a higher magnification of the area indicated by the arrows. (E) Immunohistochemical detection of GLWamide⁺ neurons in the tentacle. (F) Higher-magnification view of the inset in (E). Immunohistochemical detection of GLWamides in testis (G), ovary (H), and mesenterial filament (I). Note that only a few immunoreactive cells (arrow) were observed. Arrows indicate GLWamide⁺ neurons exhibiting immunoreactivity. M: mesoglea; epi: epidermis; gas: gastrodermis; mgc: male germ cells; and oc: oocyte.

Source: Modified and reproduced from: Shikina et al. (2020), with permission.

induced polyp contraction only when PPGLWamide was applied (Figure 6.10(A) and (B)).

Functional elements in the PPGLWamide peptide have also been investigated. In the treatment of polyps with five peptides of different lengths, i.e., with PPGLWamide, PGLWamide, GLWamide, LWamide, or Wamide peptide, no contraction resulted with Wamide peptide (Figure 6.10(C)).

From the results of these experiments, it was concluded that the sequence, LW, is essential for the biological activity of GLWamide in corals. It was further suggested that specific receptors for GLW (or LW) amides exist in the mouth region (including the pharynx) and tentacles of corals and that signal transduction for muscle contraction is induced by this treatment (Figures 6.8–6.10).

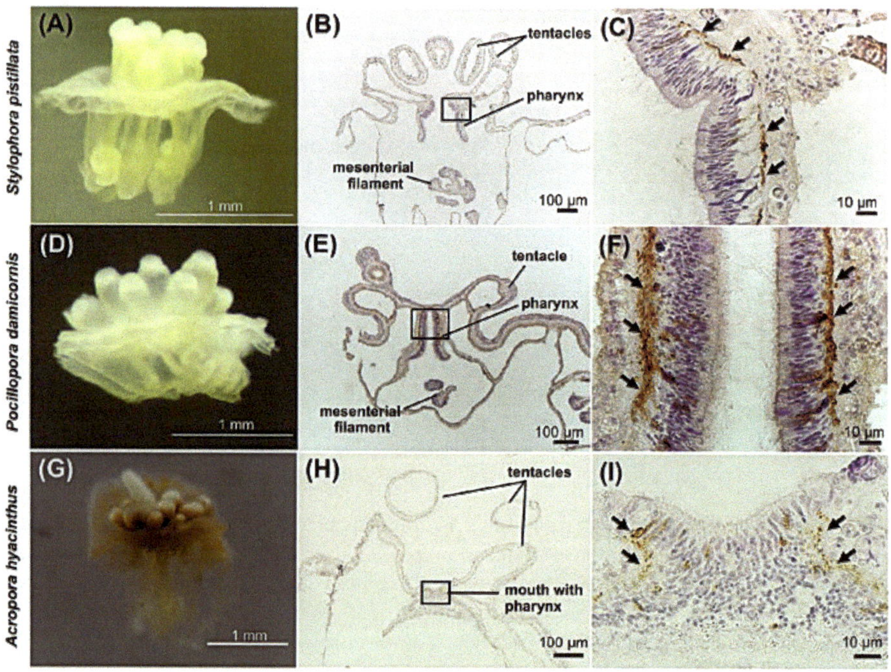

FIGURE 6.8 GLWamide$^+$ neurons in adult polyps of *Stylophora pistillata* (A-C), *Pocillopora damicornis* (D-F), and *Acropora hyacinthus* (G-I). (A), (D), and (G) Appearances of isolated polyps from each species. Skeletons were removed from the colonies with EDTA treatment. (B), (E), and (H) Immunohistochemical detection of GLWamide$^+$ neurons in coral polyps. (C), (F), and (I) Higher-magnification views of the insets shown in B, E, and H. Arrows indicate GLWamide$^+$ neurons exhibiting immunoreactivity.

Source: Reproduced from: Shikina et al. (2020), with permission.

6.4.2 *RESULTS OF TREATMENT EXPERIMENTS IN P. DAMICORNIS*

The involvement of GLWamide in polyp contraction has been investigated by observing microscopic changes in polyps during treatment. In the treatment of *P. damicornis* fragments (5 x 5 mm) with vehicle (H_2O), 10 µM

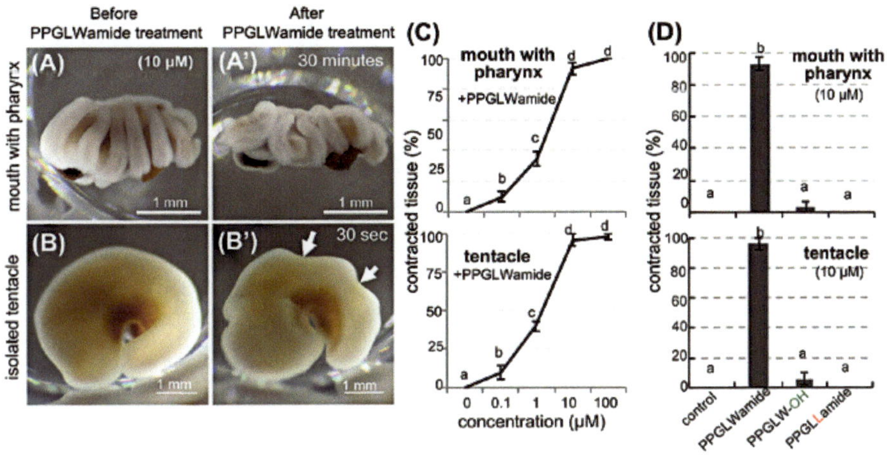

FIGURE 6.9 PPGLWamide peptide treatment induces contraction of isolated *Euphyllia ancora* polyp tissues. (A) Representative pictures of an isolated piece of mouth with the pharynx before treatment with 10 μM PPGLWamide peptide. (A') The same tissue was 30 min after treatment. (B) Representative pictures of an isolated tentacle before treatment. (B') The identical tissue was 30 sec after treatment. Arrows indicate dents observed in the tentacle induced by the treatment. (C) Effects of PPGLWamide peptide treatment (0–100 μM) on the contraction of the isolated piece of mouth with pharynx (top) and tentacles (bottom). (D) Effects of treatment of the isolated piece of mouth with control (vehicle, H_2O) and tentacles with 10 μM PPGLWamide, 10 μM PPGLW-OH, or 10 μM PPGLLamide peptide. In (C) and (D), percentages of isolated tissues exhibiting distinct contractions after treatment (contracted tissues) were determined. Data are shown as means ± SE (n=5 experiments). Groups with different lower-case letters are significantly different ($P < 0.05$).

Source: Reproduced from: Shikina et al. (2020), with permission.

PPGLWamide, or 10 μM PPGLW-OH, only PPGLWamide induced polyp contraction. Polyp responses are rapid, with most polyps showing distinct contraction within 10 s (Figure 6.11), and the effect lasts more than 10 min (Shikina et al., unpublished data). Because *P. damicornis* polyps are very small (~1 mm), treatment experiments using isolated tissues were not performed. These results indicate that GLWamides participate in polyp contraction in at least two coral species from different families. Because induction of muscle contraction (myoactivity) by GLWamides has also been demonstrated in other cnidarians, such as the sea anemone, *A. fuscoviridis,* and *Hydra* (Takahashi et al., 2003), this feature may be common to many cnidarians.

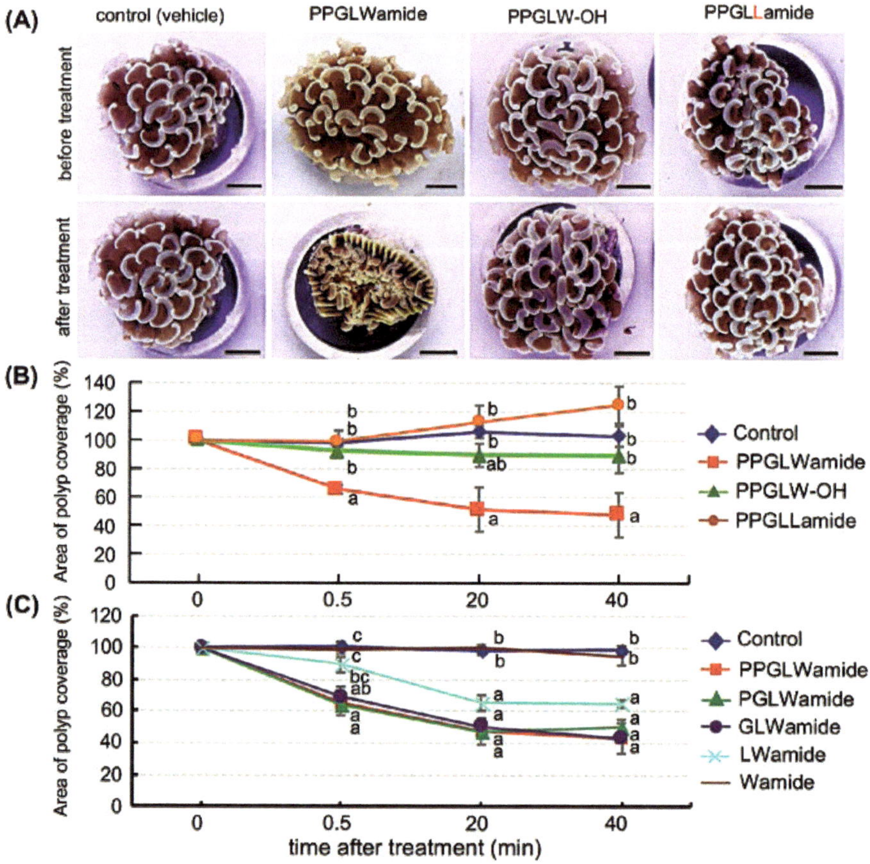

FIGURE 6.10 Treatment with GLWamide peptides induces the contraction of *Euphyllia ancora* polyps. (A) Representative pictures of *E. ancora* before treatment (top panels) and 40 min after treatment (bottom panels) with control vehicle H_2O, 10 μM PPGLWamide peptide, 10 μM PPGLW-OH peptide, or 10 μM PPGLLamide peptide. Polyp contraction was recorded by taking photographs with a digital camera that was placed on a height-fixing device. Only the polyp treated with PPGLWamide peptide exhibited distinct contraction. Scale bars = 1 cm. (B) Quantified effects with control (vehicle, H_2O), 10 μM PPGLWamide peptide, 10 μM PPGLW-OH peptide, or 10 μM PPGLLamide peptide on polyp contraction. (C) Quantified effects of the treatment with control (vehicle, H_2O), 10 μM PPGLWamide, 10 μM PGLWamide, 10 μM GLWamide, 10 μM LWamide, or 10 μM Wamide peptides on polyp contraction. The area of polyp coverage (%) was quantified using the formula (%) = 100 × [Tx/T0], where T0 is the area of tentacle coverage before treatment and Tx is that after treatment (0.5, 10, 20, 40, and 60 min). Image J64 software (National Institutes of Health, Bethesda, MD) was used for the quantification. Each data point represents a mean ± SE (n = 3 different colonies). Different lowercase letters above the symbols denote significant differences ($P < 0.05$) among the groups at each time point.

Source: Modified and reproduced from: Shikina et al. (2020), with permission.

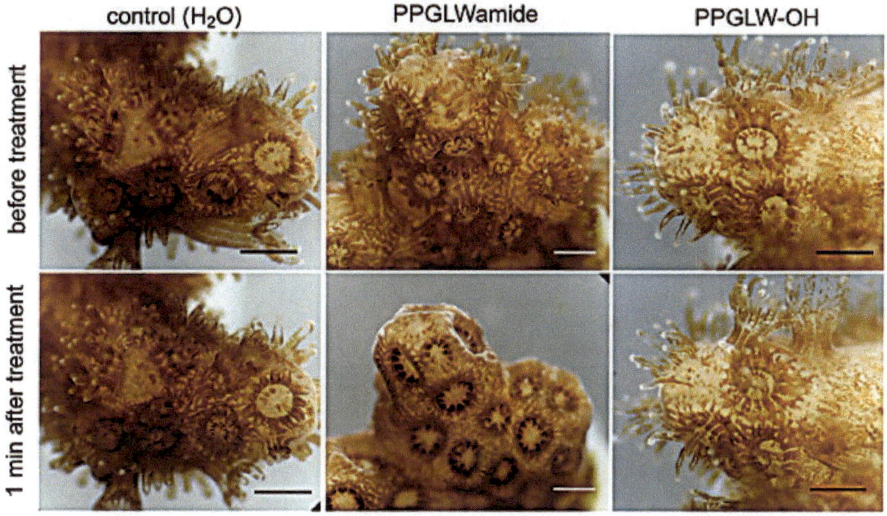

FIGURE 6.11 PPGLWamide peptide treatment induces the contraction of a *Pocillopora damicornis* polyp. Polyps were treated with vehicle H_2O (control), 10 μM PPGLWamide, or 10 μM PPGLW-OH and photographed before and 1 min after treatment. Note that only polyps treated with PPGLWamide peptide exhibited distinct contraction. Scale bars =1 mm.

6.5 CONCLUSION AND CHALLENGES

Despite the subtle species differences highlighted above, it is apparent that GLWamides are involved in polyp (muscle) contraction in corals. GLWamides may participate in defense and/or may coordinate daily polyp extension/contraction. To the best of our knowledge, at present, GLWamides are the only neuropeptides for which distribution and function in adult corals have been investigated. Native GLWamide peptides and receptors for GLW (or LW) amides that act in corals have not yet been identified. Although potential GLWamide peptides have been predicted from GLW preprohormones by sequence analysis (Attenborough et al., 2019; Shikina et al., 2020) because cnidarian preprohormones contain unusual or novel processing sites (such as S or N) (Leviev et al., 1997), native peptide lengths may differ from those of predicted peptides. Several attempts are currently underway to isolate native peptides from extracts of *E. ancora* tissue using liquid chromatography-tandem mass spectrometry (LC-MS/MS) (Figure 6.12). To identify the receptor, advanced mass spectrometry technology, such as affinity-trapping methods (e.g., biotin-tagged GLWamides) with LC-matrix-assisted laser desorption/

ionization (MALDI)-MS/MS (Shinya et al., 2010), or recent machine learning-associated research strategies (Shiraishi et al., 2019; Satake et al., 2021) may be applicable, using coral mouth with pharynx and tentacles as source materials. Identification of native GLWamides and their receptors will provide a more complete picture of GLWamide functions and mechanisms.

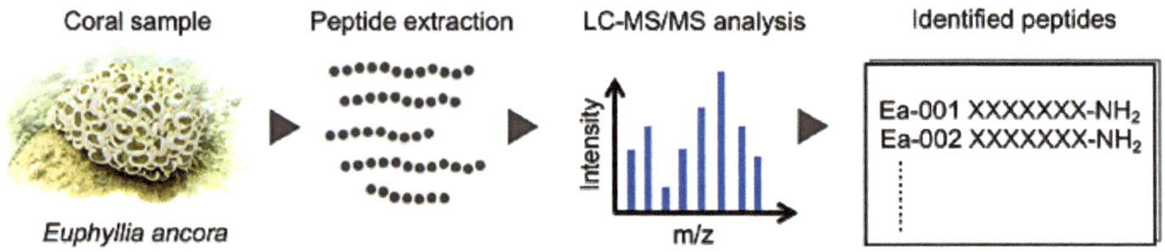

FIGURE 6.12 Schematic of the experimental procedure for identification of native GLWamides in *Euphyllia ancora*. LC-MS/MS, liquid chromatography-tandem mass spectrometry.

Research on coral neuropeptides has only recently begun. As with other cnidarians, many neuropeptides are assumed to be involved in the physiological functions of corals, but they remain largely unknown. For further development of coral neurobiology, the first questions that need to be addressed are the following. What types of neurons occur in corals? What kinds of neuropeptides do those neurons produce? In recent years, coral genomic and transcriptomic databases have been increasing in number (Grasso et al., 2011; Meyer et al., 2011; Traylor-Knowles et al., 2011; Shinzato et al., 2011, 2014, 2021; Anderson et al., 2016; Voolstra et al., 2017; Cunning et al., 2018; Ying et al., 2019; Chiu et al., 2020), and gene functions have been established for some species (Yasuoka et al., 2016; Cleves et al., 2018). In addition, with advances in mass spectrometry instrumentation and software, much more information can be obtained from extremely small amounts of samples (Kawashima & Ohara, 2018). By combining multi-omics data and functional analyses, we will be able to perform systematic identification and functional characterization of neuropeptides. Neuropeptides will be an important focus of future research in coral physiology. Further research on coral nervous systems and neuropeptides may lead to a better understanding of the physiological mechanisms of corals, ultimately deepening our understanding of coral biology.

ACKNOWLEDGMENTS

This work was supported by grants from the Ministry of Science and Technology, Taiwan (MOST) 104-2923-B-019-001-MY4 and MOST 108–2628-B-019-002. We thank colleagues and divers who helped us to collect coral samples in Kenting National Park (Nanwan, Pingtung, Taiwan). We especially thank Dr. Toshio Takahashi from the Suntory Foundation for Life Sciences, Bioorganic Research Institute (Kyoto, Japan) for kindly providing the monoclonal antibody against GLWamide.

KEYWORDS

- **functional analyses**
- **multi-omics data**
- **neurons**
- **neuropeptides**
- **scleractinian corals**
- **transcriptomic databases**

REFERENCES

Abe, N., (1939). On the expansion and contraction of the polyp of a reef coral *Caulastrea furcata* dana. *Palao Tropical Biological Station Studies, 1*, 651–670.

Anderson, D. A., Walz, M. E., Weil, E., Tonellato, P., & Smith, M. C., (2016). RNA-Seq of the Caribbean reef-building coral *Orbicella faveolata* (*Scleractinia-Merulinidae*) under bleaching and disease stress expands models of coral innate immunity. *PeerJ., 4*, e1616.

Attenborough, R. M. F., Hayward, D. C., Wiedemann, U., Forêt, S., Miller, D. J., & Ball, E. E., (2019). Expression of the neuropeptides RFamide and LWamide during development of the coral *Acropora millepora* in relation to settlement and metamorphosis. *Dev. Biol., 446*, 56–67.

Brusca, R. C., & Brusca, G. J., (2003). *Invertebrates* (2nd edn.). Sinauer Associates; Sunderland: Massachusetts.

Carstensen, K., Rinehart, K. L., McFarlane, I. D., & Grimmelikhuijzen, C. J. P., (1992). Isolation of Leu-Pro-Pro-Gly-Pro-Leu-Pro-Arg-Pro-NH2 (Antho-RPamide), an N-terminally protected, biologically active neuropeptide from sea anemones. *Peptides, 13*, 851–857.

Chiu, Y. L., Shikina, S., Yoshioka, Y., Shinzato, C., & Chang, C. F., (2020). De novo transcriptome assembly from the gonads of a scleractinian coral, *Euphyllia ancora*: Molecular mechanisms underlying scleractinian gametogenesis. *BMC Genomics, 21*, 732.

Cleves, P. A., Strader, M. E., Bay, L. K., Pringle, J. R., & Matz, M. V., (2018). CRISPR/Cas9-mediated genome editing in a reef-building coral. *Proc. Natl. Acad. Sci. U.S.A., 115*, 5235–5240.

Cunning, R., Bay, R. A., Gillette, P., Baker, A. C., & Traylor-Knowles, N., (2018). Comparative analysis of the *Pocillopora damicornis* genome highlights the role of the immune system in coral evolution. *Sci. Rep., 8*, 16134.

Elphick, M. R., Mirabeau, O., & Larhammar, D., (2018). Evolution of neuropeptide signaling systems. *J. Exp. Biol., 221*, jeb151092.

Erwin, P. M., & Szmant, A. M., (2010). Settlement induction of *Aropora palamata* planulae by a GLW-amide neuropeptide. *Coral Reefs., 29*, 929–939.

Fujisawa, T., & Hayakawa, E., (2012). Peptide signaling in Hydra. *Int. J. Dev. Biol., 56*, 543–550.

Goldberg, W., (2002). Gastrodermal structure and feeding responses in the scleractinian *Mycetophylla reesim*, a coral with novel digestive filaments. *Tissue Cell, 34*, 246–261.

Grasso, L. C., Negri, A. P., Fôret, S., Saint, R., Hayward, D. C., Miller, D. J., & Ball, E. E., (2011). The biology of coral metamorphosis: Molecular responses of larvae to inducers of settlement and metamorphosis. *Dev. Biol., 353*, 411–419.

Grimmelikhuijzen, C. J. P., Williamson, M., & Hansen, G. N., (2002). Neuropeptides in cnidarians. *Can. J. Zool., 80*, 1690–1702.

Harrison, P. L., & Wallace, C. C., (1990). Reproduction, dispersal, and recruitment of scleractinian corals. In: Dubinsky, Z., (ed.), *Ecosystems of the World 25, Coral Reefs* (pp. 133–207). Elsevier: Amsterdam.

Hatta, M., & Iwao, K., (2003). Metamorphosis induction and its possible application to coral seedlings production. In: Sexena, N., (ed.), *Recent Advances in Marine Science and Technology* (pp. 465–470). Japan International Science and Technology Federation Akaskak: Tokyo.

Havrilak, J. A., Faltine-Gonzalez, D., Wen, Y., Fodera, D., Simpson, A. C., Magie, C. R., & Layden, M. J., (2017) Characterization of NvLWamide-like neurons reveals stereotypy in *Nematostella* nerve net development. *Dev. Biol., 43*, 336–346.

Hirose, M., Yamamoto, H., & Nonaka, M., (2008). Metamorphosis and acquisition of symbiotic algae in planula larvae and primary polyps of *Acropora* spp. *Coral Reefs., 27*, 247–254.

Hoadley, K. D., Szmant, A. M., & Pyott, S. J., (2011). Circadian clock gene expression in the coral *Favia fragum* over diel and lunar reproductive cycles. *PLoS One, 6*, e19755.

Holmgren, S., & Jensen, J., (2001). Evolution of vertebrate neuropeptides. *Brain Res. Bull., 55*, 723–735.

Iwao, K., Fujisawa, T., & Hatta, M. A., (2002). Cnidarian neuropeptide of the GLWamide family induces metamorphosis of reef-building corals in the genus *Acropora. Coral Reefs, 21*, 127–129.

Jékely, G., (2013). Global view of the evolution and diversity of metazoan neuropeptide signaling. *Proc. Natl. Acad. Sci. U. S. A., 110*, 8702–8707.

Jones, R., Ricardo, G. F., & Negri, A. P., (2015). Effects of sediments on the reproductive cycle of corals. *Mar. Poll. Bull., 100*, 13–33.

Kawashima, Y., & Ohara, O., (2018). Development of a NanoLC-MS/MS system using a nonporous reverse phase column for ultrasensitive proteome analysis. *Anal. Chem., 90*, 12334–12338.

Kelava, I., Rentzsch, F., & Technau, U., (2015). Evolution of eumetazoan nervous systems: Insights from cnidarians. *Phil. Trans. R. Soc. B, 370*, 20150065.

Klavdieva, M. M., (1995). The history of neuropeptides 1. *Front Neuroendocrinol., 16*, 293–321.

Koizumi, O., Hamada, S., Minobe, S., Hamaguchi-Hamada, K., Kurumata-Shigeto, M., Nakamura, M., & Namikawa, H., (2015). The nerve ring in cnidarians: Its presence and structure in hydrozoan medusae. *Zoology, 118*, 79–88.

Lasker, H. R., (1979). Light-dependent activity patterns among reef corals: *Montastrea cavernosa. Biol. Bull., 156*, 196–211.

Leitz, T., Morand, K., & Mann, M., (1994). Metamorphosin A: A novel peptide controlling development of the lower metazoan *Hydractinia echinata* (Coelenterata, Hydrozoa). *Dev. Biol., 163*, 440–446.

Leviev, I., Williamson, M., & Grimmelikhuijzen, C. J. P., (1997). Molecular cloning of a preprohormone from *Hydra magnipapillata* containing multiple copies of hydra-LWamide (Leu-Trp-NH_2) neuropeptides: Evidence for processing at Ser and Asn residues. *J. Neurochem., 68*, 1319–1325.

Meyer, E., Aglyamova, G. V., & Matz, M. V., (2011). Profiling gene expression responses of coral larvae (*Acropora millepora*) to elevated temperature and settlement inducers using a novel RNA-Seq procedure. *Mol. Ecol., 20*, 3599–3616.

Moberg, F., & Folke, C., (1999). Ecological goods and services of coral reef ecosystems. *Ecol. Econ., 29*, 215–233.

Nakanishi, N., & Martindale, M. Q., (2018). CRISPR knockouts reveal an endogenous role for ancient neuropeptides in regulating developmental timing in a sea anemone. *eLIFE., 7*, e39742.

Odum, H. T., & Odum, E. P., (1955). Trophic structure and productivity of a windward coral reef community on Eniwetok Atoll. *Ecol. Monog., 25*, 291–320.

Pechenik, J. A., (2010). *Biology of the Invertebrates* (6th edn.). McGraw-Hill, Higher education: New York.

Raz-Bahat, M., Douek, J., Moiseeva, E., Peters, E. C., & Rinkevich, B., (2017). The digestive system of the stony coral *Stylophora pistillata. Cell Tissue Res., 368*, 311–323.

Satake, H., Osugi, T., & Shiraishi, A., (2021). Impact of machine learning-associated research strategies on the identification of peptide-receptor interactions in the post-omics era. *Neuroendocrinology*. doi: 10.1159/000518572.

Schmich, J., Rudolf, R., Trepel, S., & Leitz, T., (1998). Immunohistochemical studies of GLWamides in *Cnidaria. Cell Tissue Res., 294*, 169–177.

Schmich, J., Trepel, S., & Leitz, T., (1998). The role of GLWamides in the metamorphosis of *Hydractinia echinata. Dev. Genes. Evol., 208*, 267–273.

Sebens, K. P., & Deriemer, K., (1977). Diel cycles of expansion and contraction in coral reef anthozoans. *Mar. Biol., 43*, 247–256.

Shikina, S., & Chang, C. F., (2018). *Cnidaria*. In: Skinner, M., (ed.), *Encyclopedia of Reproduction* (2nd edn., Vol. 6, pp. 491–497). Academic Press, Elsevier.

Shikina, S., Chen, C. J., Chung, Y. J., Shao, Z. F., Liou, J. Y., Tseng, H. P., Lee, Y. H., & Chang, C. F., (2013). Yolk formation in a stony coral *Euphyllia ancora* (Cnidaria, Anthozoa):

Insight into the evolution of vitellogenesis in non-bilaterian animals. *Endocrinology, 154*, 3447–3459.

Shikina, S., Chen, C. J., Liou, J. Y., Shao, Z. F., Chung, Y. J., Lee, Y. H., & Chang, C. F., (2012). Germ cell development in the scleractinian coral *Euphyllia ancora* (Cnidaria, Anthozoa). *PloS One, 7*, e41569.

Shikina, S., Chiu, Y. L., Zhang, Y., Yao, Y. C., Liu, T. Y., Tsai, P. H., Zatylny-Gaudin, C., & Chang, C. F., (2020). Involvement of GLWamide neuropeptides in polyp contraction of the adult stony coral *Euphyllia ancora. Sci. Rep., 10*, 9427.

Shinya, T., Osada, T., Desaki, Y., Hatamoto, M., Yamanaka, Y., Hirano, H., Takai, R., et al., (2010). Characterization of receptor proteins using affinity cross-linking with biotinylated ligands. *Plant Cell Physiol., 51*, 262–270.

Shinzato, C., Inoue, M., & Kusakabe, M., (2014). A snapshot of a coral "holobiont": A transcriptome assembly of the scleractinian coral porites captures a wide variety of genes from both the host and symbiotic zooxanthellae. *PLoS One, 9*, e85182.

Shinzato, C., Khalturin, K., Inoue, J., Zayasu, Y., Kanda, M., Kawamitsu, M., Yoshioka, Y., et al., (2021). Eighteen coral genomes reveal the evolutionary origin of *Acropora* strategies to accommodate environmental changes. *Mol. Biol. Evol., 38*, 16–30.

Shinzato, C., Shoguchi, E., Kawashima, T., Hamada, M., Hisata, K., Tanaka, M., Fujie, M., et al., (2011). Using the *Acropora digitifera* genome to understand coral responses to environmental change. *Nature, 476*, 320–323.

Shiraishi, A., Okuda, T., Miyasaka, N., Osugi, T., Okuno, Y., Inoue, J., & Satake, H., (2019). Repertoires of G protein-coupled receptors for Ciona-specific neuropeptides. *Proc Natl Acad Sci U S A., 116*, 7847–7856.

Sweeney, B. M., (1976). Circadian rhythms in corals, particularly Fungiidae. *Biol. Bull., 151*, 236–246.

Takahashi, T., & Hatta, M., (2011). The importance of GLWamide neuropeptide in cnidarian development and physiology. *J. Amino Acids*, 424501.

Takahashi, T., & Takeda, N., (2015). Insight into the molecular and functional diversity of cnidarian neuropeptides. *Int. J. Mol. Sci., 16*, 2610–2625.

Takahashi, T., (2020). Comparative aspects of structure and function of cnidarian neuropeptides. *Front, Endocrinol. (Lausanne), 11*, 339.

Takahashi, T., Kobayakawa, Y., Muneoka, Y., Fujisawa, Y., Mohri, S., Hatta, M., Shimizu, H., et al., (2003). Identification of a new member of the GLWamide peptide family: Physiological activity and cellular localization in cnidarian polyps. *Comp. Biochem. Physiol. B. Biochem. Mol. Biol., 135*, 309–324.

Takahashi, T., Muneoka, Y., Lohmann, J., Lopez De, H. M. S., Solleder, G., Bosch, T. C., David, C. N., et al., (1997). Systematic isolation of peptide signal molecules regulating development in hydra: LWamide and PW families. *Proc. Natl. Acad. Sci. U. S. A., 94*, 1241–1246.

Takeda, N., Nakajima, Y., Koizumi, O., Fujisawa, T., Takahashi, T., Matsumoto, M., & Deguchi, R., (2013). Neuropeptides trigger oocyte maturation and subsequent spawning in the hydrozoan jellyfish *Cytaeis uchidae. Mol. Rep. Dev., 80*, 223–232.

Technau, U., & Steele, R. E., (2011). Evolutionary crossroads in developmental biology: *Cnidaria. Development, 138*, 1447–1458.

Traylor-Knowles, N., Granger, B. R., Lubinski, T. J., Parikh, J. R., Garamszegi, S., Xia, Y., Marto, J. A., et al., (2011). Production of a reference transcriptome and transcriptomic database (PocilloporaBase) for the cauliflower coral, *Pocillopora damicornis. BMC Genomics, 12*, 585.

Voolstra, C. R., Li, Y., Liew, Y. J., Baumgarten, S., Zoccola, D., Flot, J. F., Tambutté, S., et al., (2017). Comparative analysis of the genomes of *Stylophora pistillata* and *Acropora digitifera* provides evidence for extensive differences between species of corals. *Sci. Rep., 7*, 17583.

Williams, E., (2020). Function and distribution of the wamide neuropeptide superfamily in metazoans. *Front. Endocrinol. (Lausanne), 11*, 344.

Yasuoka, Y., Shinzato, C., & Satoh, N., (2016). The mesoderm-forming gene brachyury regulates ectoderm-endoderm demarcation in the coral *Acropora digitifera. Curr. Biol., 26*, 2885–2892.

Ying, H., Hayward, D. C., Cooke, I., Wang, W. W., Moya, A., Siemering, K. R., Sprungala, S., et al., (2019). The whole-genome sequence of the coral *Acropora millepora. Genome. Biol. Evol., 11*, 1374–1379.

Zatylny-Gaudin, C., Cornet, V., Leduc, A., Zanuttini, B., Corre, E., Le Corguillé, G., Bernay, B., et al., (2016). Neuropeptidome of the cephalopod *Sepia officinalis*: Identification, tissue mapping, and expression pattern of neuropeptides and neurohormones during egg laying. *J. Proteome. Res., 15*, 48–67.

CHAPTER 7

Coral Calcification at the Cellular Scale: Insight Through the 'Window' of the Growing Edge

A. A. VENN, E. TAMBUTTÉ, D. ZOCCOLA, L. CAPASSO, D. ALLEMAND, N. CAMINITI-SEGONDS, N. TECHER, P. GANOT, AND S. TAMBUTTÉ

Centre Scientifique de Monaco, Department of Marine Biology, Monaco

7.1 INTRODUCTION

Corals form the structural basis of coral reefs, the world's largest bioconstructions, which cover an area of about 284,300 km^2 globally (Spalding et al., 2001). Although this represents a relatively small portion of the ocean's surface area, coral reefs rank among the world's most valuable ecosystems in ecological and socio-economic terms. About a third of all marine species associate with coral reef habitats (Reaka-Kudla, 2005) and an estimated 500 million people depend on coral reef ecosystem goods and services (Hoegh-Guldberg et al., 2019; Wilkinson, 2008). Despite such profound importance to both marine biodiversity and humanity, much remains to be learned about the fundamental processes underlying the production of the building blocks that form coral reef architecture: coral skeletons.

Coral skeletons are biominerals, and in the case of scleractinian corals (the "stony" corals), they are composed of calcium carbonate ($CaCO_3$) predominantly in the form of aragonite embedded with organic material. The presence of calcium in the coral skeleton leads us to use the term 'calcification' when referring to the process of coral biomineralization, a term that we employ in this review. Coral calcification can be studied at multiple scales, ranging from nano-sized building blocks of skeletons to whole coral colonies, to the growth of entire coral reefs. Traditionally speaking, coral calcification has

Frontiers in Invertebrate Physiology: A Collection of Reviews, Volume 1: Non-Bilaterian Phyla.
Saber Saleuddin, Sally P. Leys, Robert D. Roer, and Iain C. Wilkie, (Eds.)

received far greater attention from the research community at larger scales (e.g., whole colonies) than at smaller scales (e.g., cells) (Weis et al., 2008). This is partly because investigations into coral biology at cellular scales are technically challenging due to the demands of culturing coral colonies in the laboratory and the constraints of working with complex morphology and a massive exoskeleton. Indeed, the convoluted interface of the tissues with the ornate morphology of most coral skeletons is an obstacle to the analysis of the calcifying cells themselves, particularly *in vivo*. Furthermore, coral cell cultures have proved challenging to maintain due to problems of contamination, the suitability of culturing media, and limited duration of viability (Nowotny et al., 2021; Rinkevich, 2011). To date, specific cultures of coral calcifying cells have never been achieved, although there are studies that have investigated mineral precipitation in cultures of re-aggregations of multiple corals cell types (e.g., Mass et al., 2012, 2017a).

In this review, we focus on recent advances in our understanding of coral calcification at the cellular scale, made possible by a different approach to working with whole colonies or cell cultures. Here, we describe research conducted in a specific zone at the periphery of corals that are in the process of laterally extending across a substrate by encrusting growth. This type of lateral growth is observed both in juvenile corals following settlement and metamorphosis, and in mature corals, following the breakage of a larger colony and adherence of fragments to the substrate (Gleason & Hofmann, 2011; Highsmith, 1982). The peripheral zone at the margin of laterally growing corals has been referred to previously as the "horizontal distal tissue edge" (Raz-Bahat et al., 2006), the "distal margin" (Venn et al., 2011) or the "growing edge" (Sevilgen et al., 2019). Here we employ the term growing edge "GE" and our aim is to bring together research on this zone that has spanned a diverse range of topics pertinent to calcification, including the ultra-structure of cells and calcium carbonate crystals, the mineralogy of crystals, and the cell physiology of the calcifying cell layer and other epithelia (Akiva et al., 2018; Mass et al., 2017b; Neder et al., 2019; Shapiro et al., 2016; Tambutté et al., 2007a; Venn et al., 2011, 2020). As is often typical of research on biomineralization, work on GE has encompassed many different scientific disciplines, including physiology, molecular biology, crystallography, structural biology, and geochemistry. The success of research conducted at the GE lies partly in the simplicity of its histology because tissues are aligned as neat parallel cell layers in this zone (reminiscent of diagrams of coral histology often found in reviews or textbooks). Additionally, the tissues at the GE are transparent and the skeleton is at its earliest stages of formation in this zone.

These characteristics allow unprecedented access to the calcifying cell layer and the primary steps of calcification by microscopy and other techniques in corals grown on glass slides, Petri-dishes, or coverslips, providing a unique 'window' into coral calcification at the cellular scale.

In order to provide a consistent and clear picture of what we learn about calcification through the GE, the current review mostly focuses on the widely used model coral species, *Stylophora pistillata*, for which the GE is best described and for which data on calcification are quite abundant. After presenting general information about this species, we discuss the current knowledge of the GE that has been accumulated by studies on the GE of both mature and juvenile corals. We review what we have learned about the mechanism of calcification from the GE, and review what the GE has told us about how calcification is affected by environmental parameters.

7.2 PRESENTATION OF *STYLOPHORA PISTILLATA*

7.2.1 GENERALITIES

Stylophora pistillata is a hexacoral that belongs to the order Scleractinia and the family Pocilloporidae (Figure 7.1). Among the Scleractinia, it belongs to the robust clade of corals that evolutionary diverged from the complex clade at least 245 mya ago (Voolstra et al., 2017). *S. pistillata* is widely distributed across the Indo-Pacific region (Veron, 1993), and is considered to be the "best-studied" species in the Red Sea where it can be found in a wide range of environments including shallow and mesophotic reefs, and reefs that are protected or exposed to wave action in both clear and turbid waters (Sawall et al., 2011; Scucchia et al., 2020). DNA barcoding indicates that the *Stylophora pistillata* taxon comprises four deeply divergent clades corresponding to different regions comprising of the Pacific and Western Australia; Chagos, Madagascar, and South Africa; the Gulf of Aden, Zanzibar, and Madagascar and the Red Sea; and the Persian/Arabian Gulf and Kenya. On the basis of the fossil record of *Stylophora*, these four clades diverged from one another 51.5–29.6 Mya, long before the closure of the Tethyan connection between the tropical Indo-West Pacific and Atlantic in the early Miocene (16–24 Mya). It has been argued that these clades should be recognized as four distinct species (Keshavmurthy et al., 2013). *S. pistillata* from the Northern Red Sea is regarded as thermally tolerant and may have been evolutionarily selected for heat tolerance during migration through the warmer Southern Red Sea after the last ice age (Krueger et al., 2017). It is a branching coral (Figure 7.2(A)), found as several morphotypes that vary according to habitats, depths, and geographic regions (Stefani et al., 2011). Like many tropical scleractinians, *S. pistillata* hosts endosymbiotic dinoflagellate algae (family Symbiodinaceae) (LaJeunesse et al., 2018), commonly called zooxanthellae, that contribute to its nutrition through the provision of photosynthetic products, giving it both autotrophic and heterotrophic capacity (Ezzat et al., 2017). Previous studies indicate that *S. pistillata* associates with several members of the Symbiodinaceae family depending on geographic location and depth. In the Gulf of Eilat, shallow water colonies associate with the genus *Symbiodinium* (formerly referred to as clade A1), while congeners sampled in deep-water harbor symbionts from the genus *Cladocopium* (formerly clade C) (Byler et al., 2013; Mieog et al., 2007). The presence of these endosymbiotic algae gives *S. pistillata* a predominantly brown coloration, although its coloration can vary among morphotypes due to fluorescent and non-fluorescent animal pigments that give it pink or green hues. It is a brooder (internal fertilization and release of offspring as larvae), with gonads observed almost throughout the year, and a larval release during several months of the first half of the year (Shlesinger & Loya, 1991). In the Gulf of Eilat, its reproductive state varies from hermaphrodite to male

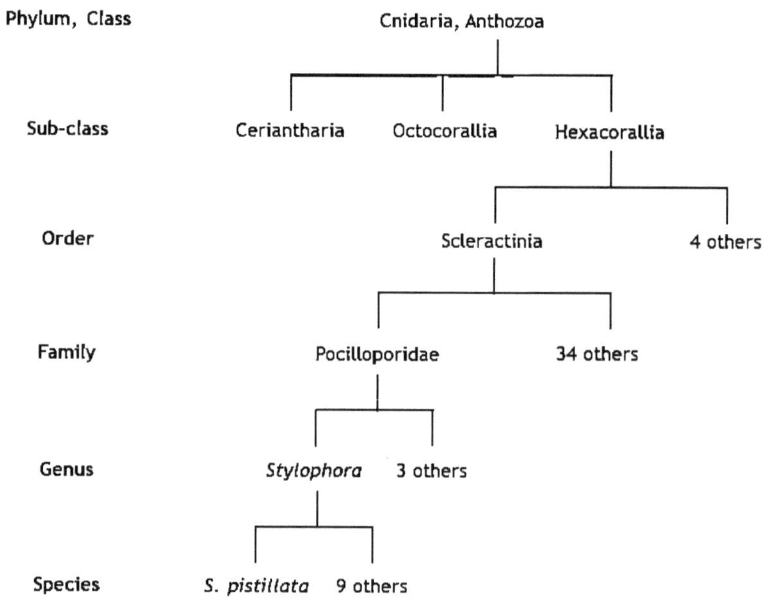

FIGURE 7.1 Phylogeny of the coral *Stylophora pistillata.*

or sterile modes depending on the year and stress conditions (Rinkevich & Loya, 1987). It has been regarded as a r-strategist or a major pioneer species in the succession of coral communities (Loya, 1976) and it is widely considered as a model species for field investigation as well as for laboratory-based experiments addressing various aspects of coral biology.

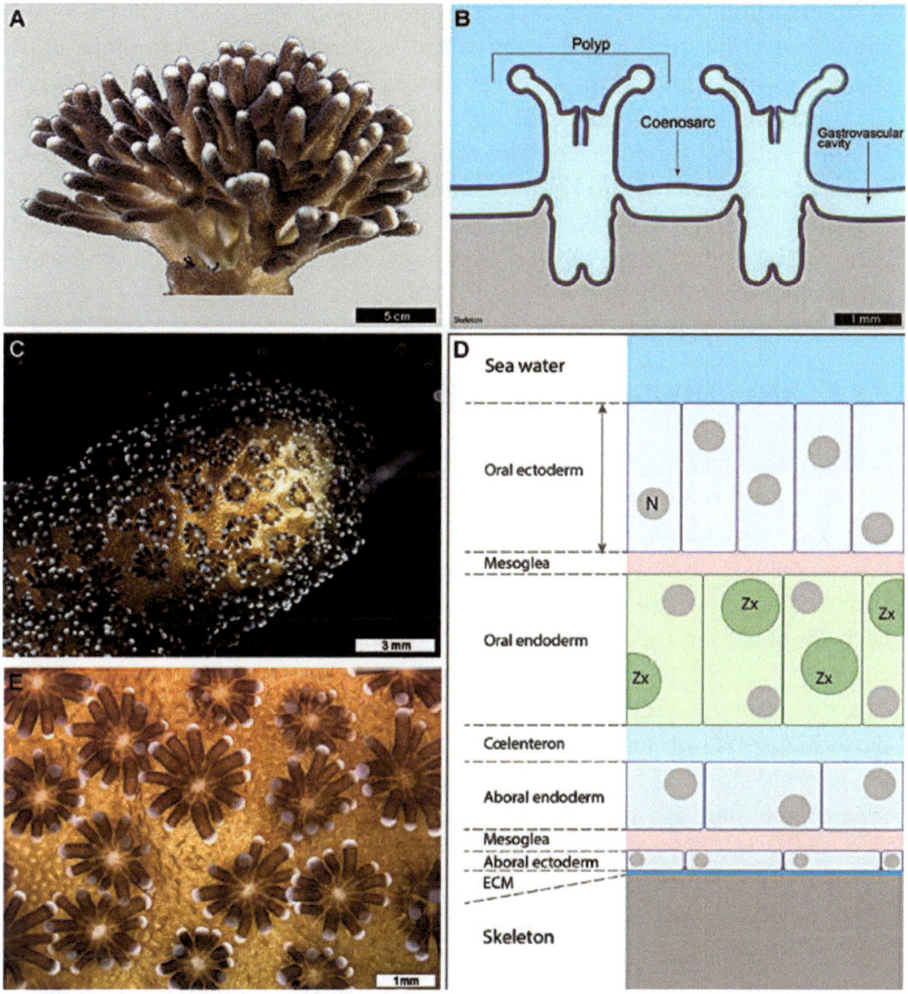

FIGURE 7.2 Anatomy and histology of *Stylophora pistillata*. (A) whole colony; (B) a diagram of a section through the polyps, coenosarc, and skeleton; (C) a colony branch; and (D) a diagram showing the histology of the coenosarc. (E) zoom on polyps. [N: Nucleus; Zx: Zooxanthella (=Symbiodinaceae); ECM: Extracellular calcifying medium].

A Web of Science search performed in 2020 found the species name *Stylophora pistillata* in the title of about 1,000 peer-reviewed articles. When using the search term "*Stylophora pistillata* + calcification" (or biomineralization), a Web of Science search in 2020 found about 270 peer-reviewed articles compared to a total of about 2000 publications for "coral + calcification" (or biomineralization). Considering that there are about 1,600 species of scleractinian corals, the fact that more than 10% of studies on coral calcification involve *Stylophora pistillata* underscores the importance of this species in this topic. Data for *S. pistillata* are available at different scales of molecular and biochemical organization, with "omics" data including genome data published in Voolstra et al., (2017), transcriptome data in Bhattacharya et al., (2016) and Karako-Lampert et al., (2014) (Zoccola) and skeletal proteome data in Drake et al., (2013). The literature also contains much data obtained at cellular, tissular, and physiological levels, e.g., reviewed in (Tambutté et al., 2011).

7.2.2 ANATOMY AND HISTOLOGY OF STYLOPHORA PISTILLATA

Stylophora pistillata is a colonial coral, with polyps connected by a common tissue usually referred to as the cœnosarc or cœnenchyme (Galloway et al., 2007). *S. pistillata* anatomy and histology are presented in Figure 7.2. The polyp consists of a column and a single opening called the mouth, which is surrounded by a crown of tentacles (Figure 7.2(B)–(E)). The mouth serves both for ingesting food and rejecting waste and leads into the gastrovascular cavity or cœlenteron, which is a simple sac-like stomach, where food is digested (Figure 7.2(B)). The cœlenteron is divided into a number of chambers by a series of septa (vertically-orientated plates). These septa support the internal stomach folds, known as mesenteries, which increase the surface area of the stomach thereby aiding the digestion of food. The gonads are attached to the mesenteries (Rinkevich & Loya, 1979).

The tissue in contact with seawater is the oral tissue whereas the tissue adjacent to the skeleton is the aboral tissue (Figure 7.2(D)). The oral tissue is thicker than the aboral tissue in the cœnosarc region (by a ratio of about 3:1) (Tambutté et al., 2007a). The tissues consist of two epithelial layers named epidermis and gastrodermis, referring to the adult epithelia, or ectoderm and endoderm, referring to their embryological origins (Fautin Gail & Mariscal, 1991; Galloway et al., 2007). Despite a long-standing debate on this topic, we will use the latter terms in this review. The two epithelial layers are separated by an essentially acellular connective layer of collagen-rich matrix called mesoglea (Young, 1973). The oral tissues comprise the oral ectoderm and endoderm, and the aboral tissues, the aboral ectoderm, and endoderm. However, in some areas such as the cœnosteal spines, there are only the oral and aboral ectodermal layers separated by the mesoglea, with no endodermal layer in between (Tambutté et al., 2007a). The cell layers contain several cell subtypes. For example, oral tissues contain cells specialized for food digestion and reproduction (gonads), cnidocytes, sensory cells, motor neurons, interneurons, and neurosecretory cells. Both endoderm layers have the specificity of containing the endosymbiotic dinoflagellate algae. The aboral ectoderm is a specialized tissue that is in contact with the skeleton and is responsible for its formation. This calcium carbonate-producing cell layer is often referred to as the skeletogenic tissue, the calicoblastic ectoderm, or simply the calicodermis (Galloway et al., 2007). We will use "calicoblastic ectoderm" in this review. The calicoblastic ectoderm contains calcifying cells called "calicoblasts" which owe their name to their calyx shape. The calicoblastic ectoderm also contains the "desmocytes" which are the cells that anchor the tissue to the skeleton/substrate (Muscatine et al., 1997). The calicoblasts are responsible for the formation of the skeleton and are thus of major interest when studying calcification. The extracellular calcifying medium (ECM, see below) is at the interface between them and the skeleton. In all cell layers, the basal side of the cells faces the mesoglea and the apical side faces the cœlenteron in the case of the endoderm, the seawater for the oral ectoderm, and the ECM for the calicoblastic ectoderm.

7.2.3 SKELETON OF *STYLOPHORA PISTILLATA*

The skeleton is extracellular, underlying the calicoblastic ectoderm of the polyp or the cœnosarc. The corallites and the cœnosteum (Figure 7.3) are the skeletal counterparts of the polyps and cœnosarc, respectively. Corallites are immersed, conical, or hooded (Veron, 1993). They have a solid style-like columella, six primary septa which may be short or fused with the columella (Figure 7.3(C)), and sometimes six short secondary septa. The cœnosteum is covered by fine spines (Figure 7.3(C)). The major mineral composition of the mature skeleton of *Stylophora pistillata* is calcium carbonate crystallized under the aragonitic polymorph. The aragonitic structures include centers of calcification (CoCs) (also referred to as rapid accretion deposits) and elongated acicular crystals termed fibers, radiating out of CoCs and forming plumose spherulites (Sun et al., 2017). Amorphous calcium carbonate (ACC), disordered aragonite phases, poorly crystalline aragonite (pAra) (Mass et al., 2017b), and high magnesium content have been observed in the CoCs, whereas low amounts of magnesium and strontium were detected in the aragonitic fibers (Neder et al., 2019). In addition to calcium carbonate, the skeleton contains organic molecules called the organic matrix (OM) (Tambutté et al., 2007c). The OM is composed of proteins, lipids, and sugars. A full description of OM characteristics falls outside of the focus of this review, but it is important to note that *S. pistillata* is one of the two scleractinian coral species together with *Acropora sp.* for which molecular data on the organic matrix are available (Drake et al., 2013; Ramos-Silva et al., 2014). At least 60 skeleton proteins have been identified in *S. pistillata* (Peled et al., 2020) (the first sequenced being Galaxin identified in *Galaxea fascicularis* (Fukuda et al., 2003)) and their putative functions have been proposed mostly based on results from *in vitro* experiments (Drake et al., 2013, 2014; Mass et al., 2013).

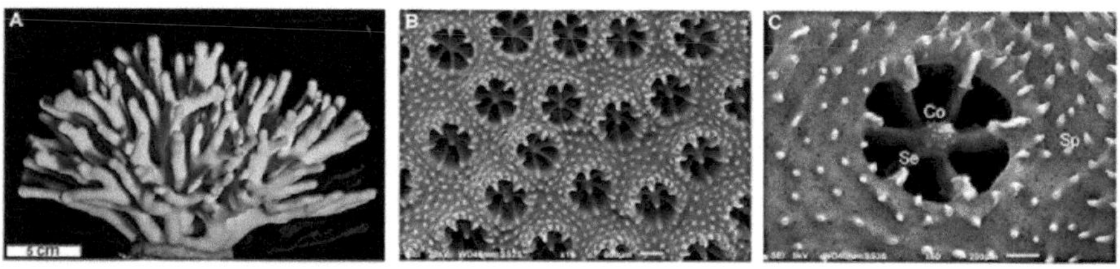

FIGURE 7.3 The skeleton of *Stylophora pistillata.* (A) the skeleton of a whole colony; (B) the surface of the skeleton showing corallites and coenosteum; and (C) corallite with columella and septae, surrounded by coenosteal spines. [Co: Columella; Se: Septa; Sp: Spine].

7.3 PRESENTATION OF THE GROWING EDGE (GE) OF *STYLOPHORA PISTILLATA*

7.3.1 PROPAGATION OF CORAL SAMPLES WITH A GROWING EDGE (GE)

Stylophora pistillata is grown in aquariums worldwide for display and scientific purposes. At the Centre Scientifique de Monaco, colonies collected in 1990 in the Red Sea (Gulf of Aqaba) have been cultured and propagated by fragmentation to provide a sustainable source of experimental material. Fragments, referred to as microcolonies or nubbins are grown on nylon nets (Tambutté et al., 2012), suspended via nylons lines (Tambutté et al., 1995, 1996) or can be fixed and grown on various substrates.

The technique of growing microcolonies of *S. pistillata* on glass slides from fragments of mature colonies was first developed by (Tambutté, 1996) and (Muscatine et al., 1997) and then was later used in many studies (e.g., Bockel & Rinkevich, 2019; Reynaud-Vaganay et al., 1999; Shafir et al., 2001; Venn et al., 2011). This method of preparation was referred to as the 'lateral skeleton preparative (LSP) Assay' by Raz-Bahat et al. (2006). In this method, fragments are cut from mature coral colonies and glued with epoxy on either coverslips or slides. New tissue develops at the edge of the sample within a few days and extends rapidly, doubling the surface area of the sample within about three weeks, firmly attaching the sample to the substrate (Figure 7.4(A)–(C)). If left to colonize the surface of glass slides or coverslips laterally for a period of months, they start to grow vertically by developing apexes and eventually form large branching colonies (Figure 7.4(D) and (E)).

In microcolonies fixed and grown on glass substrates by the LSP assay, mature polyps can be observed across the sample except at the lateral, peripheral margin where tissues are transparent and where there are only juvenile polyps or no polyps (Figure 7.5(A)–(C)). Raz-Bahat et al. (2006) identified different stages of tissue structure and skeletal formation that can be observed from the distal lateral margin of the preparation to the zone where the microcolony was originally glued to the glass slide. The typical histology consists of both oral and aboral endoderms and ectoderms, and cœlenteron, and can be observed everywhere except at the extreme outer edge of the sample, which is only composed of a single cell layer formed by the coalescence of the two ectoderms (Raz-Bahat et al., 2006). Symbiotic microalgae can be observed to reside in the endoderm cell layers and free algae are seen moving rapidly within the cœlenteron (Venn et al., 2011). Peripheral polyp mesenterial filaments at the outer edge are observed to actively "sweep" and remove detritus within the perimeter surrounding the microcolony (Pers. Obs). This has also been observed on *Acropora pulchra* growing on glass slides after tissue lesion and it has been proposed that the continuous removal of detrital and algal material is both a source of heterotrophic food for the growing polyps and also a way to prepare the substrate for calcified coral growth (Roff et al., 2008). Regarding the skeleton, there is a spatial-temporal progression in its formation (Raz-Bahat et al., 2006). When the tissue advances, it deposits individual crystals (Figure 7.6(A) and (B)). At progressively greater distances from the advancing distal edge of the colony, these isolated crystals and crystal bundles of various morphologies, join to form a continuous sheet of skeleton between 100 and 400 μm from the sample margin (Figure 7.6(C)) and finally form a mature skeleton (Figure 7.6(D)). It is this zone of lateral growth (that includes both mineral and tissue components), lying between the sample margin and the complete mature skeleton, that we term the "growing edge" (GE).

The GE can also be studied in juvenile corals cultivated in the laboratory (Figure 7.7). Planulae larvae of *S. pistillata* and other species including *Acropora digitifera* and *Pocillopora damicornis* obtained from colonies in the field or maintained in the laboratory are collected and allowed to settle in closed-circuit aquarium conditions on glass-based Petri-dishes, slides or coverslips that were previously placed in the aquarium (Gilis et al., 2014; Neder et al., 2019; Ohno et al., 2017a, b) (Figure 7.7). Following the settlement, the larvae metamorphose and form juvenile corals that grow out laterally across the glass coverslip, displaying a zone of isolated crystals at the distal margin that joins to form continuous skeletons further into the sample (Figure 7.7). Observations of both the tissue and minerals of the GE have been made in the early stages of single primary polyps to larger multi-polyp colonies in several investigations (Akiva et al., 2018; Gilis et al., 2014; Neder et al., 2019; Ohno et al., 2017a, b) (Figure 7.7).

7.3.2 MORPHOLOGY AND ULTRASTRUCTURE OF THE CALICOBLASTIC ECTODERM AT THE GROWING EDGE (GE)

Historically, the coral calicoblastic ectoderm has been difficult to access in most coral colonies as it lies between the overlying tissues and the underlying skeleton. Observations made at the GE allow inspection of the calicoblasts via inverted microscopy by looking up through the coverslip, between the gaps in the newly forming skeleton and crystals (Venn et al., 2011). Furthermore, at the GE, the oral and aboral tissues are arranged in neat parallel layers, more or less as they are frequently depicted in illustrations of coral histology (Figure 7.2; Raz-Bahat et al., 2006). This means that in fixed samples, the skeleton can be removed by decalcification allowing observation of the calicoblastic ectoderm as a relatively flat sheet (Tambutté et al., 2007a; Venn et al., 2011). Scanning electron microscopy (SEM) on decalcified samples has shown that two types of cells are observable, the desmocytes and the calicoblasts. Desmocytes that anchor the tissue to the substrate, i.e., the coverslip or the glass slide, are also observed at the GE (Figure 7.5(D) and (E)) (Tambutté et al., 2007a)). Regarding the calicoblasts, when observed with transmission electron microscopy (TEM, transversal tissue sections) at the GE, they appear long and thin on the zones where crystals are not yet formed (Tambutté et al., 2007a). When observed in SEM

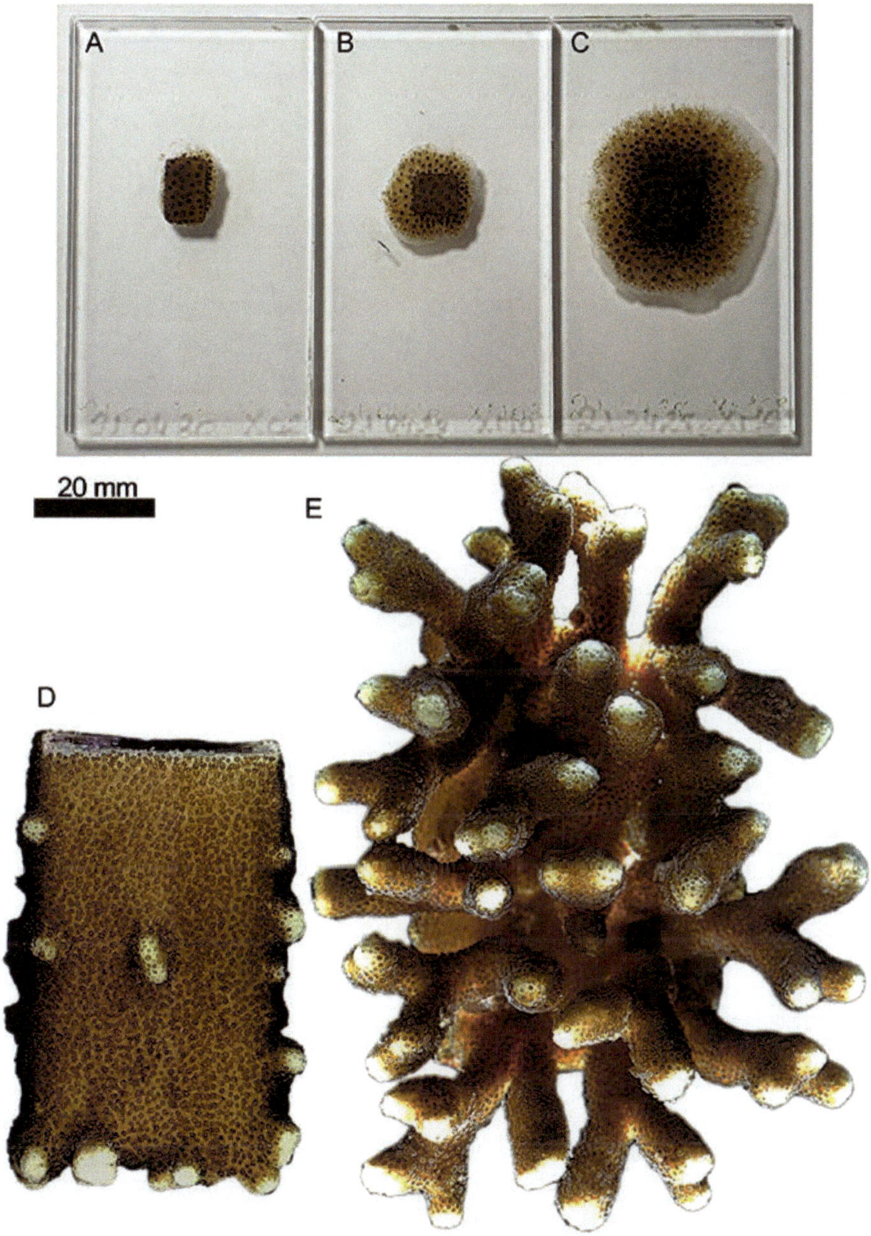

FIGURE 7.4 Fragments of corals cut from mature colonies and glued with epoxy on glass slides. (A), (B), (C) new tissue develops at the edge of the sample within a few days and extends rapidly, firmly attaching the sample to the substrate. (D) and (E) samples left to colonize the surface of the glass slides laterally for a period of months, start to grow vertically by developing apexes, and eventually forms a branching colony.

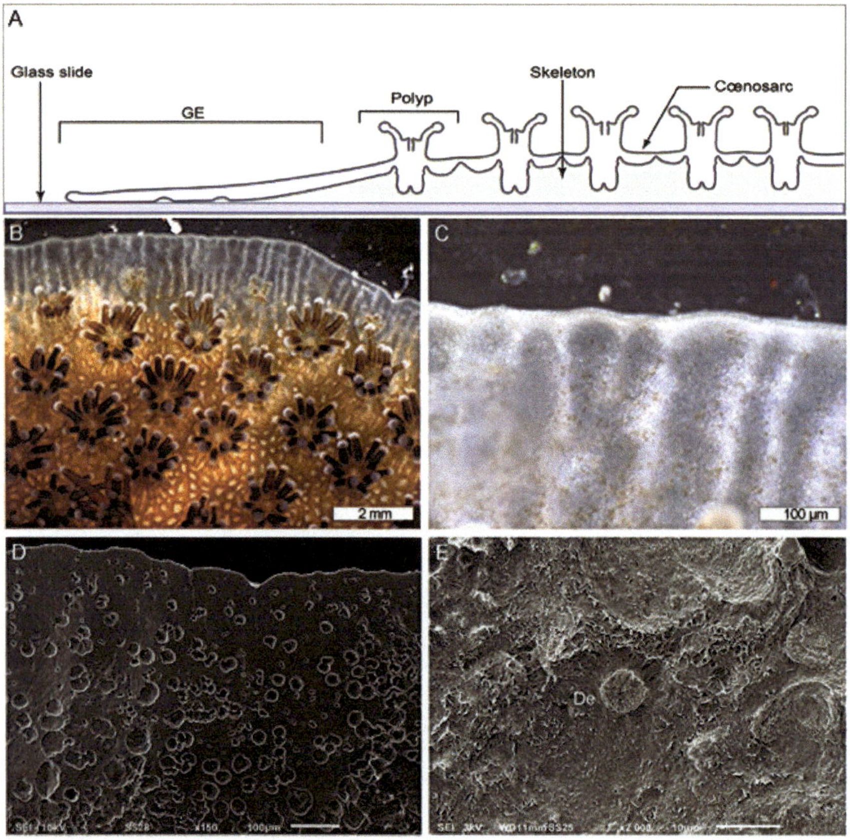

FIGURE 7.5 The growing edge (GE) of a microcolony of *Stylophora pistillata* grown on a glass slide. (A) schematic representation of a longitudinal section of the microcolony growing on the slide; and (B) a corresponding light microscope image taken from above the microcolony looking down on sample (C) Zoom on B showing the growing edge. (D) Scanning Electron Microscopy (SEM) image of the underside of the growing edge showing the calicoblastic epithelium with the imprints of newly formed, isolated crystals that were removed by decalcification during sample preparation (E) SEM image of the calicoblastic epithelium showing the presence of a desmocyte (De) in the calicoblastic epithelium.

from their apical side (not on sections but on whole cells), some appear flat with a hexagonal shape and others have a calyx (cup-like) shape (Tambutté et al., 2007a)). At the GE, the imprint of crystals appears in the calicoblastic epithelium as impressions in the shape of the crystals that they form (Figure 7.5(D); Venn et al., 2011).

At the GE, a monocilium is present on each calicoblastic cell that is in the process of forming a crystal (Tambutté et al., 2021). Despite differences

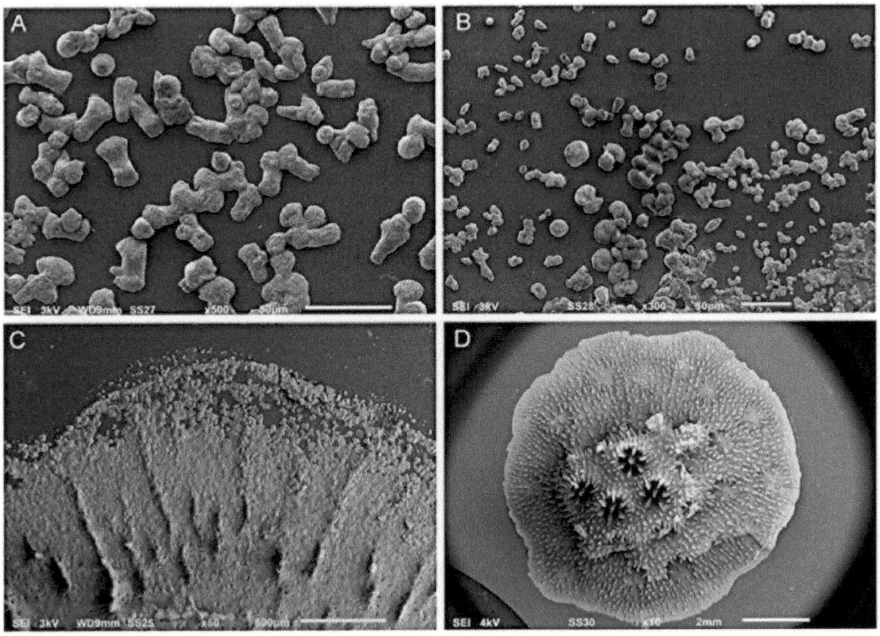

FIGURE 7.6 Scanning electron microscope images of the skeleton of a microcolony of *Stylophora pistillata* prepared by the lateral sample preparative (LSP) assay and grown on a glass slide. (A) (B) isolated $CaCO_3$ crystals at the growing edge (C) crystals and crystal bundles of various morphologies, join to form a continuous sheet of the skeleton at between 100 and 400 µm from the sample margin (D) View of the whole specimen (isolated crystals are visible in (C) around its perimeter).

in cell shape and density, there are no differences in cilium structure regardless of the zone of calcifying cells (i.e., over a newly forming skeleton, or different areas of a mature skeleton (Tambutté et al., 2021). Septate junctions between calicoblastic cells are always observed on the apico-lateral side of the cells both at the GE and in other zones of the calicoblastic ectoderm (Tambutté et al., 2007a). In TEM cross-sections, septate junctions display characteristic electron-dense ladder-like structures of 20 nm width called septa (Ganot et al., 2015; Tambutté et al., 2007b).

7.3.3 CRYSTAL AT THE GROWING EDGE (GE)

A defining characteristic of the GE is that the early steps of calcification can be observed in the form of isolated $CaCO_3$ crystals formed by the overlying calicoblastic ectoderm. Mineral deposits were described by light microscopy in *S. pistillata* prepared via the LSP assay from fragments of mature colonies (Raz-Bahat et al., 2006). Images of crystals at the GE of *S. pistillata* prepared by the LSP assay have also been captured by confocal and SEM (Venn et al., 2011, 2013, 2020). These studies show crystalline structures ranging in size from the sub-micrometric to tens of micrometers in length. They occur in several morphologies, from round or fusiform small crystals to larger 'dog-bone' or 'dumb-bell' shaped crystals (Figure 7.6(A) and (B)). The arrangement of the isolated crystals at the GE appears random, but in certain areas, crystals are arranged linearly to form the beginning of the cœnosteal spines (Figure 7.6(C)). Between 100 and 400 μm from the extremity of the GE, the mineral deposits merge to form a continuous skeleton ((Venn et al., 2011) Figure 7.6(C) and (D))).

The crystals at the GE of recently-settled juvenile corals show many similarities to mineral deposits found in LSP samples made from mature colonies. In primary polyps of *S. pistillata* (Neder et al., 2019) also observed a range of isolated crystal morphologies including pristine dumbbells and dumbbell-like structures that grow by the deposition of needle-shaped fibers radiating from the pristine dumbbells. Like in samples prepared by the LSP assay, in certain areas crystals are arranged linearly to form the beginning of the cœnosteal spines (Figure 7.7).

Isolated crystals at the GE of various morphologies have also been observed in a number of other coral species. For example, Neder et al. (2019) also characterized the isolated crystals found in primary polyps of *Pocillopora acuta* in a comparative study with *S. pistillata*. Before that Gilis et al (2014) made an extensive characterization of crystals and the steps in their development at the GE of another juvenile Pocilloporid, *P. damicornis*. Both studies indicate that there are interspecies differences in crystal morphology at the GE, but also many similarities, which is not surprising given that *S. pistillata* and *Pocillopora* sp. belong to the same family (Pocilloporidae). GE crystals have also been observed in more distantly related corals, including members of the complex clade such as *Acropora digitifera* (Ohno et al., 2017a, b). A detailed morphological characterization of crystals at the GE of acroporid corals has not yet been carried out, however.

Research on juvenile corals at the primary polyp stage has addressed crystal mineralogy as well as morphology. In *S. pistillata* (Neder et al., 2019) determined that the mineral phase of the pristine dumbbell crystals was Mg-calcite. The same authors also determined that Mg-calcite was present in rod-like crystals in *P. acuta*. Also (Gilis et al., 2014) identified calcite in early forming rod-like structures in *P. damicornis*. From Neder et al's work on *S. pistillata* and *P. acuta*, it appears that the Mg-calcite can be detected in the early stages of calcification, but in older samples (4 days after polyp

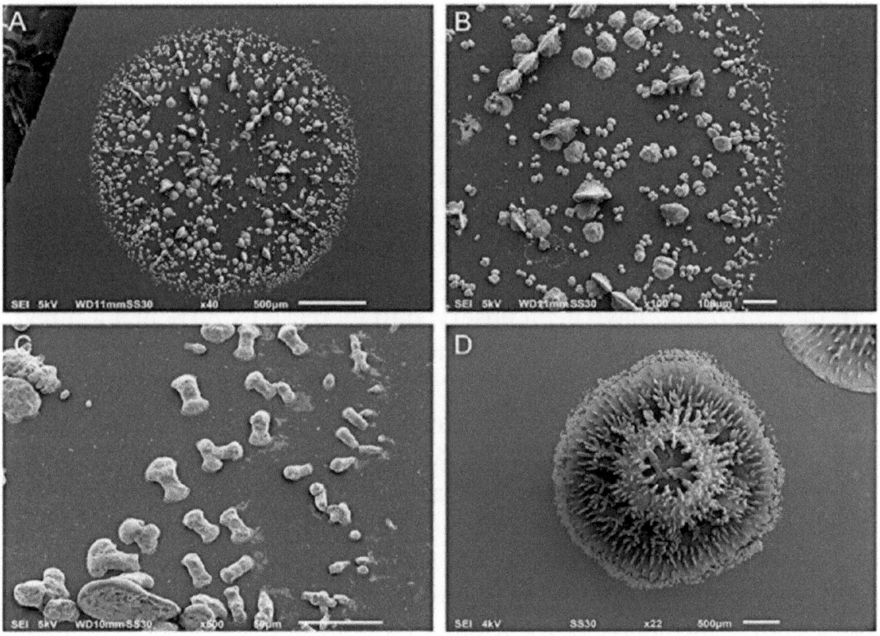

FIGURE 7.7 Scanning electron microscope images of the skeleton of juvenile specimens of *Stylophora pistillata* that have previously settled and metamorphosed on a glass slide. (A), (B), (C) crystals deposited observed at different scales in specimens approximately 24 h after settlement. (D) the entire skeleton of the primary polyp approximately three days after settlement (isolated crystals are visible around its perimeter).

settlement instead of 2 days), aragonite was more abundant. These authors proposed that aragonite crystals form on top of Mg-calcite dumbells and suggest that the same process occurs in the centers of the calcification of mature coral skeletons. In the case of other coral species for which the GE has been analyzed such as *Acropora digitifera* (Ohno et al., 2017a, b), it has not yet been possible to discriminate between aragonite or calcite crystals. Furthermore, the mineral phases of GE crystals of samples prepared from fragments of mature corals by the LSP assay have not been determined. This is an area for future research.

7.3.4 CALCIFICATION AND GROWTH RATES AT THE GROWING EDGE (GE)

It is noteworthy that the GE is a rapidly calcifying zone, and as such, crystals forming at the GE can be used for performing qualitative and quantitative measurements of calcification rates. Fluorescent dyes, such as calcein and alizarin, widely used in the field of biomineralization, are incorporated into crystals at the GE and have been demonstrated to be non-toxic to corals (Holcomb et al., 2013). Rates of calcein incorporation into the skeleton as an indicator for $CaCO_3$ formation agree with traditional methods of measuring calcification such as the alkalinity anomaly (Tambutté et al., 2012). In experiments over time scales of a few hours, calcein can be used to mark time points in crystal growth (Venn et al., 2013) allowing changes in crystal cross-sectional area to be calculated as a percentage of their original size. By this approach, crystals of approximately 20 μm length at the GE of *S. pistillata* have been observed to increase by 50% in a 4 h period. This relatively short time period within which appreciable crystal growth can be measured makes it feasible to compare crystal growth rates between corals in different levels of environmental stress under experimental conditions (Venn, 2013; Section 7.5.2). Another approach has been to measure calcein incorporation into all crystals within a certain area at the GE by increases in fluorescence in order to assess the impact of the inhibition of certain cellular processes on the calcification rate (Barott et al., 2020). Both approaches have exploited the accessibility and rapid growth rates at the GE to investigate calcification at the crystal scale, a scale that can be easily related to cell biology in this zone.

In addition to quantitative measurements of crystal growth, qualitative observations of their development can also shed light on calcification processes in real time. Time-lapse *in vivo* imaging using high-resolution forms of microscopy such as confocal, allows the first appearance of crystals in the ECM to be documented on the submicrometric scale and their subsequent growth recorded (Figure 7.8).

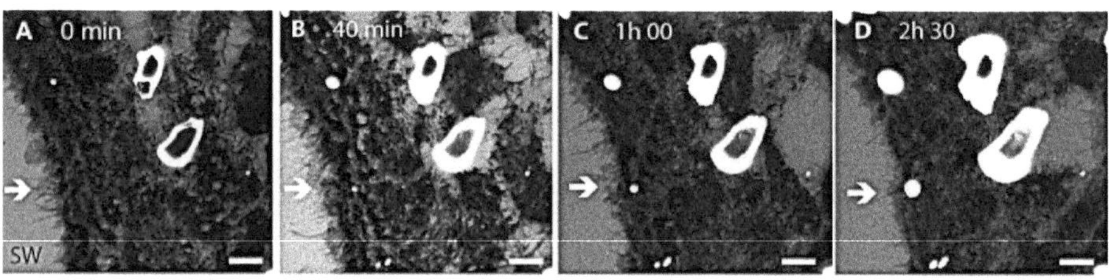

FIGURE 7.8 Time-lapse sequence of confocal images of isolated crystal growth at the growing edge of *Stylophora pistillata* incubated in calcein dye. SW = the seawater surrounding the coral. Calicoblastic epithelium is shown in black. Crystals are shown in bright white (note the increase in size). Extracellular calcifying medium (ECM) is shown in grey. The arrow points to the position of a new crystal that forms during the time-lapse.

7.3.5 STRUCTURE OF THE INTERFACE BETWEEN THE CALICOBLASTIC ECTODERM AND SKELETON

Growth of the skeleton occurs at the interface between the calicoblastic ectoderm and the existing skeleton, in an extracellular medium. Multiple terms have been employed to describe both the interface and the medium. In studies on the GE, these include subcalicoblastic medium (Ohno et al., 2017a, b; Tambutté et al., 2011, 2015; Venn et al., 2011, 2013), calcifying fluid (Comeau et al., 2017; Sun et al., 2020), ECM (Sevilgen et al., 2019; Venn et al., 2019, 2020) or site of calcification (Sevilgen et al., 2019). In studies on fully developed coral skeletons terms include "extracytoplasmic calcifying fluid" (Gattuso et al., 1999; Meibom et al., 2006), "calcifying mother liquor" (Adkins et al., 2003), "extracellular calcifying compartment" (Cohen et al., 2009), "subcalicoblastic space" (Tambutté et al., 2007a), "peri-crystalline fluid" (Marubini et al., 2008), "sub-calicoblastic extracellular calcifying medium" (Allemand et al., 2011), "calcifying fluid" (McCulloch et al., 2012b) or the more general term of "privileged space" (Marchitto et al., 2018). In the present review, we use the term extracellular calcifying medium (ECM).

Observations of the ECM in corals with mature skeletons are tricky since this medium is sandwiched between the skeleton and the tissues, which impedes its direct observation. Estimates of its thickness have been derived from electron microscopy observations on fixed samples. Sample preparation for electron microscopy includes steps of dehydration that can cause shrinkage of the samples, making it difficult to have an exact value of the size of the ECM. The estimates available in the literature vary from the order of a few nanometers (Tambutté et al., 2007a, 2011) to more than 1 µm in width (Isa, 1986; Johnston, 1980; Le Tissier, 1990). "Pockets" or "cavities" where the calicodermis is lifted away from the skeletal surface have also been described by electron microscopy in corals with mature skeletons (Barnes, 1970, 1972; Tambutté et al., 2011).

The application of confocal microscopy on the GE allowed direct *in vivo* imaging of the ECM for the first time (Venn et al., 2011). For these observations, corals prepared by the LSP assay and grown on coverslips were fitted into a cell-cultivation perfusion chamber and mounted on the confocal microscope (Figure 7.9(A)). Analysis of z stacks of optical sections taken by confocal microscopy at the interface between the coral tissue and the glass coverslip in *S. pistillata* showed that the cell impermeant forms of the dyes calcein and carboxyseminaphthorhodafluor-1

(SNARF-1) stain the ECM under the calicoblastic ectoderm (Figure 7.9(B) and (D)). Simultaneous imaging of the crystals and tissues revealed that the calicoblastic ectoderm is generally in close contact with the crystals and with the coverslip, but is lifted away from the glass coverslip in places creating spaces or "pockets" under the calicoblastic tissue ((Venn et al., 2011), Figure 7.9(B) and (C)). Although crystals were observed to be in close contact with the calicoblastic ectoderm, they are frequently found at the periphery of spaces containing the ECM (Figure 7.9(B)). It is important to note that although the ECM is most easily observable in the pockets of lifted calicoblastic ectoderm, close inspection of the interface between the calicoblastic ectoderm and the underlying substrate (the coverslip or skeletal elements) shows that a thin, nanometric layer of ECM stained with SNARF-1 is visible across the whole extent of the GE. This includes the contours of the newly forming cœnosteal spines (Figure 7.9(C)). This indicates that the ECM is not restricted to the pockets of lifted tissue, but is a continuous layer under most of the calicoblastic ectoderm at the GE.

Although almost all *in vivo* work on the ECM and the interface of the calicoblastic ectoderm and skeleton has been conducted on *S. pistillata*, imaging of the ECM has also been conducted in *P. damicornis* and *Acropora hyacinthus* prepared by the LSP assay (Venn et al., 2019). Imaging of the ECM has also been made in *Acropora digitifera* juveniles (Ohno et al., 2017a, b). These latter studies found that the calicoblastic ectoderm was lifted in many areas creating pockets of ECM, similar to pockets observed in *S. pistillata* prepared by the LSP assay.

7.4 INSIGHT FROM THE GROWING EDGE (GE) INTO THE MECHANISM OF CALCIFICATION

There is a growing literature on the mechanisms of coral calcification and several reviews are available (Allemand et al., 2011; Drake et al., 2020; Falini et al., 2015; Tambutté et al., 2011; Tresguerres et al., 2017). These reviews bring together data from the literature on different coral species, both on micro-colonies and on entire coral colonies. The aim of the following paragraphs is to focus on the information available for the mechanisms of coral calcification at the GE of *Stylophora pistillata* and discuss them in a wider context.

7.4.1 ECM CARBONATE CHEMISTRY AT THE GROWING EDGE (GE) OF *STYLOPHORA PISTILLATA*

Analysis of the chemical conditions of the ECM where the skeleton develops has the potential to be highly informative about the mechanisms driving calcification. The problem is that in corals and most other calcifying organisms, difficulties in accessing the calcifying medium make non-disruptive, *in-vivo* access to the ECM a major challenge. As such, much of the earlier data on the chemical properties of the ECM in corals were derived from the analysis of dead coral skeletons, using geochemical isotope signatures as indicators of ECM chemistry (e.g., with boron isotopes for pH (Krief et al., 2010; Reynaud et al., 2004; Trotter et al., 2011). While this indirect approach to investigating the ECM is extremely valuable, it relied on assumptions about the chemistry involved in how isotopes were incorporated into the skeleton (DeCarlo et al., 2018a). As such, direct *in vivo* measurements of ECM were much sought after in order to validate indirect isotope methods. Some of the first attempts to obtain *in vivo* data from the ECM involved inserting microelectrodes into the coral tissues by a variety of approaches. These include puncturing the coral cœnosarc tissue to measure pH in the ECM (pH_{ECM}) (Ries, 2011), making incisions with a scalpel (Al-Horani et al., 2003b), and inserting the sensor through the polyp mouth (Cai et al., 2016). The difficulty with these microelectrode measurements was that they were obtained without optical verification that the microelectrode was correctly placed in the ECM throughout the duration of the study because visualization of the ECM is extremely difficult within the complex architecture of most corals. As we saw in the previous section, analysis of the GE overcomes this problem, permitting visual access to the ECM in *S. pistillata* and other corals, greatly facilitating its analysis *in vivo*. Taking advantage of this (Venn et al., 2011) incubated *S. pistillata* grown on glass coverslips with the fluorescent pH-sensitive probe SNARF-1 and measured pH_{ECM} with a confocal microscope (Figure 7.9). This work was able to simultaneously determine pH in the ECM and the diffusive boundary layer of seawater surrounding the coral, showing that corals elevate pH in the ECM above seawater values. Later, Ohno et al. (2017b) also worked with pH-sensitive dyes and confocal microscopy and determined that pH_{ECM} was elevated in juvenile *A. digitifera*. Following that, the confocal-SNARF-1 technique was used in tandem with microelectrodes with 2–3 µm diameter tips inserted into pockets of ECM at the GE of *S. pistillata* (Sevilgen et al., 2019). This approach allowed optical verification of the correct insertion of the microelectrode into the ECM and

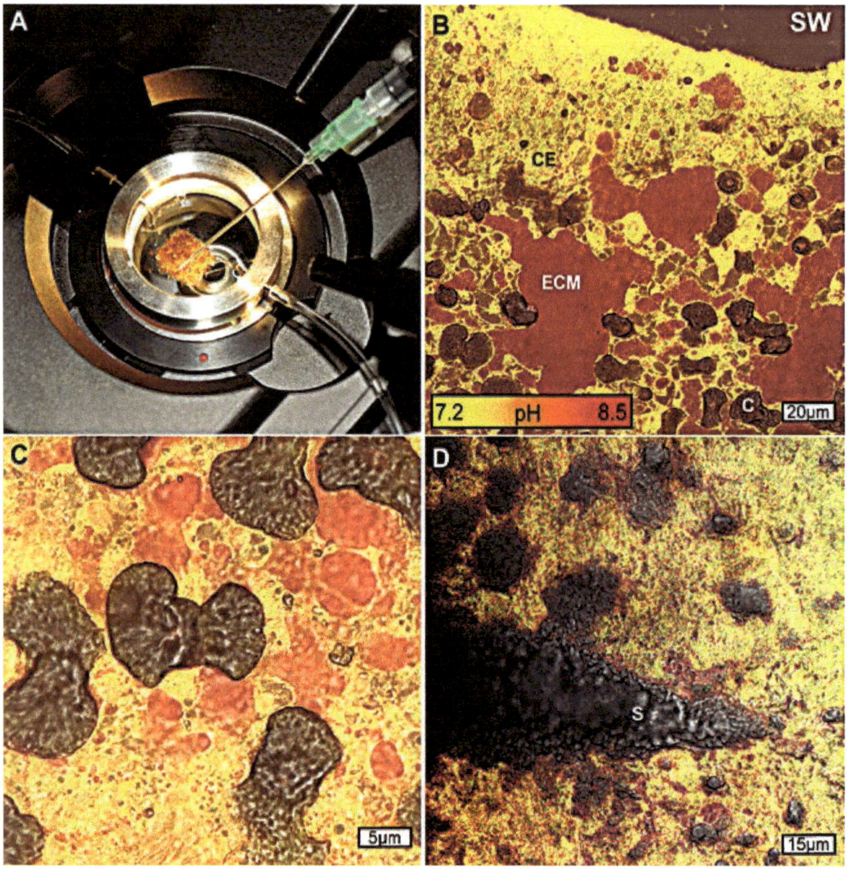

FIGURE 7.9 pH in the calicoblastic epithelium of the growing edge of *Stylophora pistillata*. (A) Microcolony grown on a glass coverslip in the perfusion chamber and mounted on the confocal microscope. (B)–(D) confocal images of SNARF-1 fluorescence merged with transmitted light images to show pH in the extracellular calcifying medium (ECM) (red), the calicoblastic epithelium (CE) (yellow), and isolated calcium carbonate crystals (C) (black/grey). (D) shows developing coenosteal spine (S) surrounded by ECM and CE.

TABLE 7.1 Carbonate Chemistry Parameters Obtained in the Extracellular Calcifying Medium (ECM) at the Growing Edge (GE) of *Stylophora pistillata* in the Light*

	pH (NBS)	CO_3^{2-} (μM kg^{-1} SW)	Ca^{2+} (mM)	DIC (μM kg^{-1} SW)	Ω_{ar}
ECM	8.54 ± 0.10	679 ± 183	13 ± 2	3,133 ± 1,046	12.1 ± 3.6
Seawater	8.09 ± 0.01	196 ± 15	11 ± 0	2,145 ± 279	2.9 ± 0.2

*In the ECM the parameters pH, [Ca^{2+}], [CO_3^{2-}], [DIC], and aragonite saturation state (Ω_{ar}) are elevated relative to seawater (means ± SD).

Source: Values from Sevilgen et al. (2019).

demonstrated that pH_{ECM} measurements by SNARF-1 and by microelectrode agree. In the same study, the use of microscope-guided microelectrodes at the GE enabled calcium and carbonate measurements to be made in the ECM, establishing that [Ca^{2+}] and [CO_3^{2-}] were also elevated relative to seawater (Sevilgen et al., 2019). Together, these results provided the full characterization of ECM carbonate chemistry including its dissolved inorganic carbon (DIC) concentration and aragonite saturation state (Ω_{ar}) (Table 7.1).

The data obtained in these *in vivo* studies are in broad agreement with values of pH, [DIC] and [Ca^{2+}] obtained from the analysis of skeletal isotopes in *S. pistillata* and other coral species (DeCarlo et al., 2018b; Holcomb et al., 2014; McCulloch et al., 2017). However, no study has directly compared geochemical techniques on the same area of the skeleton in the same samples. One study, by Holcomb et al. (2014) combined measurements of boron isotopes and SNARF-1 measurements on *S. pistillata* growing on glass coverslips (LSP assay), but in different zones of the skeleton. SNARF-1 was used to measure pH_{ECM} at the GE, and boron isotope measurements were made on skeletal calcium carbonate samples from two different regions: the lateral region including the GE and fully-formed flat mature skeleton; and the apex (growing tip) that give rise to branches. The study showed that pH_{ECM} values obtained with SNARF-1 at the GE are lower (0.2 Units) than pH_{ECM} in other regions of the microcolony. It is not possible from these data to determine whether this difference is due to the different technologies used (SNARF-1 vs boron isotopes) or to real differences in pH in the different locations where measurements were performed (i.e., pH_{ECM} could be slightly lower at the GE than the rest of the colony). Future research is required to resolve this issue. In the meantime, even if there is uncertainty about how absolute values of pH_{ECM} and other parameters vary between different locations of the skeleton, work with the GE has determined some key principle characteristics of ECM carbonate chemistry: that corals elevate pH, [CO_3^{2-}] and [Ca^{2+}] in the ECM above seawater, providing an Ω_{ar} in the ECM that is itself several times higher above seawater (Table 7.1), thus providing a favorable environment for $CaCO_3$ precipitation.

7.4.2 CALICOBLASTIC CELLS AND CONTROL OF ECM COMPOSITION

As described in previous sections, the ECM is spatially separated from the surrounding seawater by several cell layers and is in direct contact with the calicoblastic ectoderm. The concentration of protons, calcium, and carbonate in the ECM is different from seawater and strongly suggests that the calicoblasts exert biological control on ECM chemistry by active transport, i.e., via transporters on cell membranes. This view is consistent with the results of pharmacological and stable isotope pulse-chase experiments that have indicated that both calcium and DIC are transported to the skeleton by active, transcellular processes (Furla et al., 2000; Tambutté et al., 1996). A number of candidate ion transporters have been characterized at the molecular level in corals, that are believed to be involved in ion transport linked to calcification. Certain of these have been localized to the calicoblastic ectoderm by means of immunolabelling or *in situ* hybridization (ISH) in tissue sections taken from areas of colonies overlying mature skeleton in *Stylophora pistillata* (e.g., Zoccola et al., 2004, 2015). The identity, putative role, and localization of these transporters form an important part of current models of ion transport linked to the coral calcification mechanism. It is important to note that in polarized cells organized in epithelia, specific transporters are generally localized to either apical or basal membranes (with some transporters being latero-basal) (Okamoto, 2017) and paracellular ion diffusion is controlled by junctions. However, in corals, we do not yet know the apical and basal positions of the ion transporters that have been characterized to date, and their proposed localization in Figure 7.10 is based on what is known in other organisms and/or their suggested role in the calcification process. The schematic representation (Figure 7.10) of ion transport mechanisms proposes that calcium enters the calicoblasts via a calcium channel at the basal side (Zoccola et al., 1999) and is released in exchange for protons to the ECM through a Ca^{2+} ATPase (plasma membrane calcium ATPase (PMCA) (Zoccola et al., 2004)) at the apical side. It has been suggested that this calcium/proton exchanger increases the concentration of calcium and decreases the concentration of protons in the ECM, promoting an environment favorable for calcification (Tambutté et al., 2011). DIC used for calcification could come from external (e.g., from seawater) or internal sources (e.g., metabolism) (Furla et al., 2000). CO_2 could diffuse out of the calicoblastic ectoderm (Allison et al., 2014) or/and HCO_3^- could be transported via bicarbonate transporters belonging to SLC26 and SLC4 families. Notably, the HCO_3^-/Cl^- exchanger, SLC4ɣ, is specific to the calicoblastic epithelium suggesting its role involves the supply of DIC to the ECM (Figure 7.10; Zoccola et al., 2015) the kinetics of the reaction of DIC interconversion is favored by intra- and extra-cellular carbonic anhydrases (Bertucci et al., 2013). Recently, an ammonium transporter (AMT1d), has been located on the apical side of the calicoblastic cells and is suggested to supply or remove ammonium (NH_4^+) from the ECM (Capasso et al., 2022).

Although its transport direction has yet to be determined, two roles related to calcification have been proposed for AMT1d: (i) regulation of the conformational stability of organic matrix proteins, by supplying NH_4^+ to the ECM; and (ii) increase in pH_{ECM}, by removing NH_4^+ from the ECM (Capasso et al., 2021, submitted). Additionally several H^+ transporters (including Na^+/H^+ exchanger-SLC9, vacuolar-type electrogenic H^+-ATP hydrolase-V-ATPase and voltage-gated proton channels: HvCN), with a potential role in pH regulation of the calicoblasts, have been identified and molecularly characterized (Capasso et al., 2021), but their specific localization in the coral calicoblastic cells is still undetermined. The calicoblasts also possess an acid-base cellular sensor (e.g., the acid-base sensing enzyme soluble adenylyl cyclase (sAC) (Barott et al., 2020)). A number of other transporters involved in ion trafficking have also been molecularly characterized but their role in the calcification mechanism is not clear (Bhattacharya et al., 2016).

Localization of all the aforementioned ion transporters in the tissues of the GE has not been published, but the *in vivo* work with the GE reported in the literature permits some physiological insight into ion transport linked to calcification, supporting the model proposed in Figure 7.10. First, all carbonate chemistry parameters measured in the ECM are more elevated than in seawater implying that active transport is required, likely involving some of the candidate transporters listed above. There are experiments that support this assertion. For example, Comeau et al. (2017) investigated pH regulation at the GE under different seawater DIC concentrations. Their findings showed that under ambient seawater conditions, the calicoblasts maintain a pH of 7.4, as observed in (Venn et al., 2011). However, the removal of DIC from seawater led to a depression in both pH of the calicoblasts and pH_{ECM} independent of changes in respiration rate. These results indicate that seawater DIC is somehow directly involved in the pH regulation of the calicoblastic ectoderm, either through its capacity to buffer protons exported from this epithelium or in its import into the calicoblasts via a DIC transporter (Figure 7.10). Other work with the GE has revealed that the acid-base sensing enzyme soluble adenylyl cyclase (sAC), which is expressed in the calicoblastic ectoderm (Barott et al., 2020), is involved in pH regulation of the ECM. Inhibition of this enzyme leads to acidification of the ECM as measured by SNARF-1 and decreases in calcification rate measured by calcein incorporation. Together, these findings underscore the importance of the biological control of the calicoblastic ectoderm on ECM chemistry for calcification. As more work is carried out on the GE in the coming years, the model of ion transport in Figure 7.10 is certainly likely to evolve with the integration of discoveries.

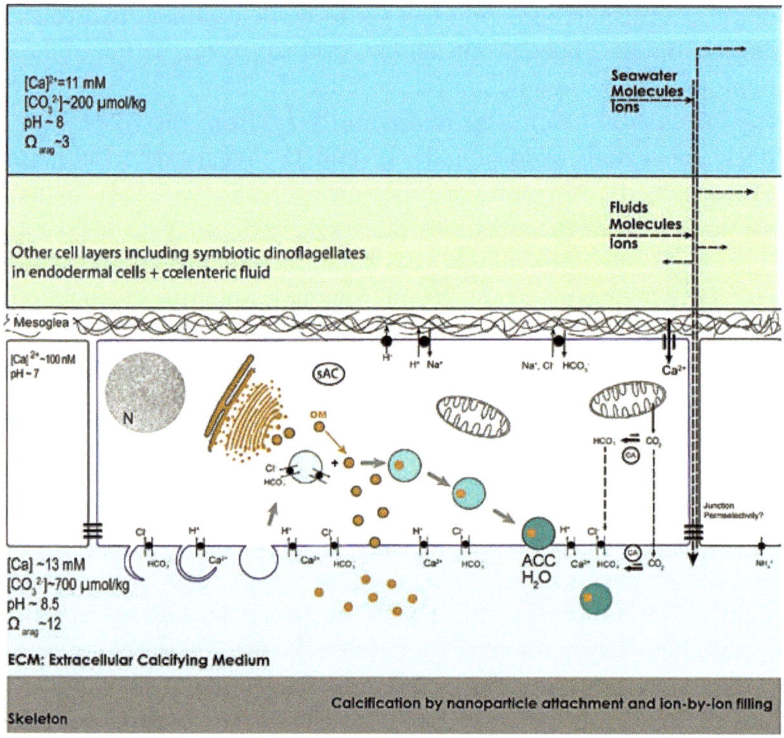

FIGURE 7.10 Schematic representation of proposed cellular mechanisms of coral calcification in *Stylophora pistillata* (adapted from Sun et al., 2020). The values for calcium and carbonate concentrations, pH, and supersaturation with respect to aragonite (Ωarag) in the extracellular calcifying medium (ECM) and seawater were taken from Sevilgen et al. (2019). The source of dissolved inorganic carbon for calcification is partial cell respiration and an external source of bicarbonate (Furla et al., 2000). Basal membrane transporters (located on the cell side facing the mesoglea) are proposed to include a bicarbonate transporter (Zoccola et al., 2015), a sodium proton exchanger, a proton ATPase (Capasso et al., 2021) and an L-type calcium channel (Zoccola et al., 1999). Dissolved inorganic carbon is supplied to the ECM through CO_2 diffusion (Allison et al., 2014) and bicarbonate transport (Zoccola et al., 2015) at the apical side of calicoblastic cells (side of the cells facing the ECM). An ammonium transporter (AMT1) located on the apical calicoblastic membrane either adds or removes NH_4^+ from the ECM. A calcium ATPase supplies calcium to the ECM and removes protons (Zoccola et al., 2004). A cytosolic soluble adenyl cyclase is involved in Bicarbonate/pH sensing (Barrott et al., 2020). Carbonic anhydrases are involved in CO_2/bicarbonate interconversion and catalyzing DIC supply to the ECM (Bertucci et al., 2013). The ECM is endocytosed into a calicoblastic cell from its apical side by macropinocytosis (Ganot et al., 2019), enriched in Ca, HCO_3^-, other ions, organic matrix, and protons removed, in order to promote the formation of hydrated amorphous calcium carbonate nanoparticles. These nanoparticles are then exocytosed into the ECM and attached to the growing skeleton. Gradually the solid aggregate of nanoparticles and ions dehydrates, crystallizes, and ion attachment fills interstitial spaces (Sun et al., 2020).

Crucial to understanding the control of ECM composition, is an appreciation of not just transcellular ion transport but also the passive exchange of ions via the paracellular pathway, which is depicted in Figure 7.10. Indeed, understanding the extent to which the ECM is a closed, semi-open, or open compartment with regard to its exchange with the exterior seawater has major implications for our understanding of how calcification works (Tambutté et al., 2011). Many models of coral calcification invoke the direct transport of seawater to the site of calcification implying the existence of a leaky, open paracellular pathway between the ECM and seawater (Adkins et al., 2003; Cohen & McConnaughey, 2003; Saenger & Erez, 2016). This view is supported by studies measuring the uptake of metal/calcium isotopes from seawater into the coral skeleton and by the passage of fluorescent dyes via the paracellular pathway (Erez & Braun, 2007; Gagnon et al., 2012; Tambutté et al., 2012). The paracellular pathway involves the movement of ions via the intercellular septate junctions and can be visualized at the GE by incubating samples with the fluorescent cell impermeant dye calcein (ionic radius of 6.5 Å (Zarnitsyn et al., 2008)). At the GE, using an inverted confocal microscope, calcein added to seawater can be observed in the ECM and incorporated in the crystals, as well as passing between the calicoblastic cells (Figure 7.11(A)–(D) (Tambutté et al., 2012; Venn et al., 2020)). Calcein can also be observed very clearly in the intercellular spaces between the overlying endoderm cells (Figure 7.11(E) and (F)). Recently, time-lapse imaging to measure the rate at which calcein enters the ECM from the surrounding seawater (Figure 7.11(A)–(D)) gives half-time of calcein influx of 6.37±0.73 min, which can be taken as an indicator of paracellular permeability of the tissue at the GE (Venn et al., 2020). Previous electrophysiological studies performed on microcolonies mounted in Ussing chambers showed that the tissues have intermediate-to-tight permeability characteristics (Tambutté et al., 2012) and that the calicoblastic ectoderm controls this permeability (Tambutté et al., 2012; Taubner et al., 2017). It is important to note that calcein influx is not necessarily representative of the ionic and molecular components of seawater. Indeed, paracellular transport through septate junctions is perm-selective meaning that it is influenced by ion charge and size (Bénazet-Tambutté et al., 1996; Taubner et al., 2017). Moreover, tricellular junctions (at the junction of 3 cells) could regulate the paracellular passage of macromolecules, such as calcein, without affecting ion permeability (Krug et al., 2009). Tricellular junctions have never been characterized in corals and more general knowledge of coral intercellular junctions and the paracellular pathway is in its infancy in this group of animals. This topic awaits further research and the ability to image the paracellular pathway at the GE (as shown in (Tambutté et al., 2011; Venn et al., 2020)) is an advantage that should be exploited in future studies.

Finally, in addition to transcellular and paracellular modes of ion/molecule transport, vesicular transport must also be considered. Based on observations in adult corals of vesicles in calicoblastic cells (Clode & Marshall, 2002) that have the same size as blue calcein-stained particles and ACC mineral particles in tissues, it was deduced that the primary steps of crystal formation occur under the phase of ACC in vesicles within the calicoblastic cells (Mass et al., 2017b). Vesicles of the same size were also observed to form from macropinocytosis both in nubbins and at the GE (Ganot et al., 2020), a process involving non-specific endocytosis of surrounding fluid and nanoparticles in all cell layers. In the calicoblastic cells, macropinocytosis occurs by plasma membrane ruffling and closing at the apical side of the cells that endocytose the ECM (Ganot et al., 2020). Although no data are currently available on the ion composition of macropinosome content or the fate of macropinosomes after internalization, it is noteworthy that these macropinosomes represent a confined compartment. In this process, the membrane of these large vesicles is made from the cellular phospholipid bilayer cellular membrane, and as such is likely to include the transporters present at the apical side of the calicoblastic cells before endocytosis. Hence, it has been proposed that the content of the macropinosomes becomes actively enriched in Ca^{2+}, HCO_3^-, and organics; and that protons are actively removed thus providing favorable conditions for the formation of intracellular hydrated amorphous calcium carbonate (ACC-H_2O) nanoparticles (Sun et al., 2020). It is further proposed that these ACC-H_2O nanoparticles are released into the ECM via exocytosis arising through the fusion of macropinosomes with the apical membrane of calicoblastic cells (Sun et al., 2020) (Figure 7.10). More generally, we note that the role of vesicles and endocytosis/exocytosis in coral calcification is an evolving area of research and requires further investigation.

7.4.3 MODEL OF CALCIFICATION: ACC NANOPARTICLES TO CRYSTAL GROWTH

For some time there has been debate on the mechanisms of crystal deposition and growth in corals with models that either considered ion-by-ion precipitation from solution (Constantz, 1986; Gagnon et al., 2007; Holcomb et al., 2014) or crystallization by particle attachment (CPA) (De Yoreo et al., 2015; Gal et al., 2014; Gilbert et al., 2019; Mass et al., 2017b). A recent study

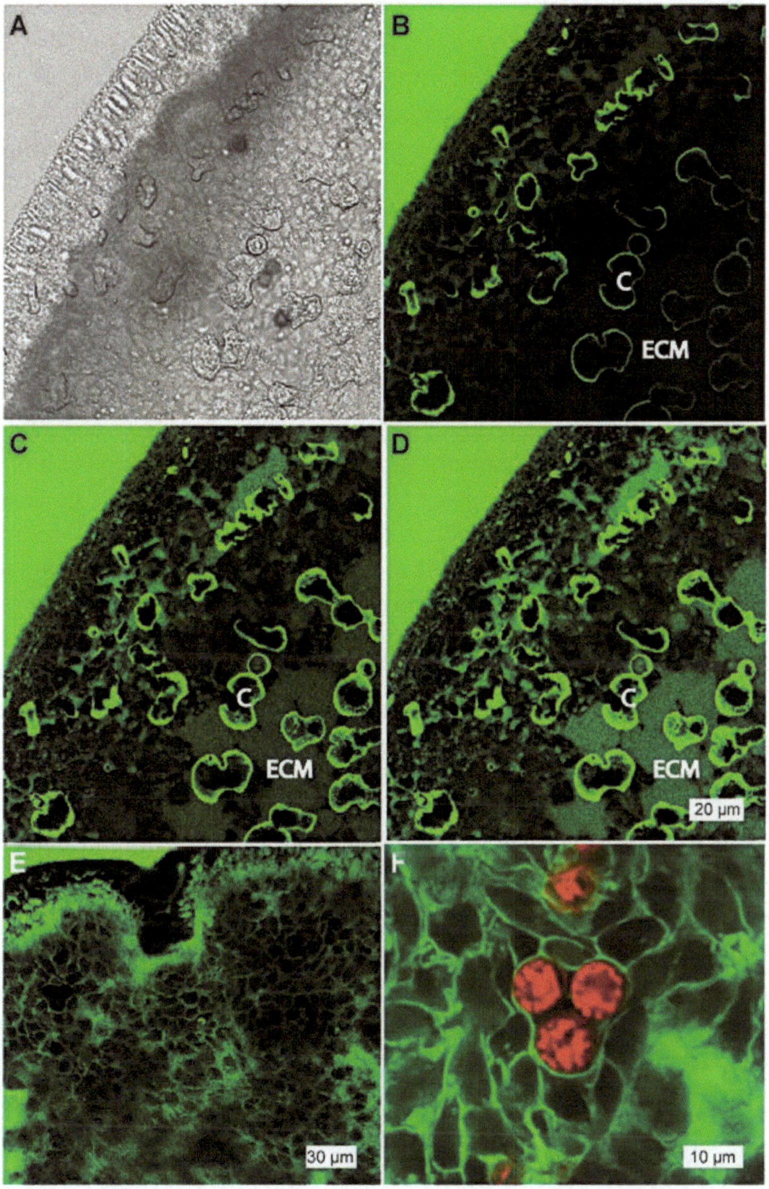

FIGURE 7.11 The paracellular pathway at the growing edge of *Stylophora pistillata.* (A) the transmitted light image of the growing edge. (B)–(D) Time-lapse confocal images of the influx of calcein dye (green) into the extracellular calcifying medium (ECM) from the surrounding seawater. (E) Confocal image acquired in the endoderm cell layer showing calcein dye between the cells in the paracellular pathway. (F) symbiotic algae (red) visible in endoderm cells and calcein dye visible in the paracellular pathway.

argues that these two processes occur concomitantly (Sun et al., 2020). This model proposes that once ACC-H_2O nanoparticles are exocytosed into the ECM (see paragraph above and Figure 7.10), they attach to the growing skeleton and gradually the solid aggregate of nanoparticles and ions dehydrates, crystallizes, and ion attachment fills interstitial spaces (Sun et al., 2020). This model has been derived from observations in different areas of the coral skeleton. Observations at the GE have been critical in showing that ECM carbonate chemistry parameters favor supersaturated conditions that support a role for ions by ion space-filling (Sevilgen et al., 2019; Sun et al., 2020). Other observations at the GE are also consistent with the model proposed by Sun et al. (2020). As described above, macropinocytotic vesicles have been observed at the GE inside the calicoblastic cells that could contain ACC-H_2O nanoparticles (Ganot et al., 2020; Mass et al., 2017b; Sun et al., 2020).

In regards to the mechanisms of crystal deposition and skeleton formation in corals, there is also another parameter that needs to be taken into account: the skeletal organic matrix. It is well-established for all biominerals that the organic matrix can influence many steps of crystal formation and growth. For example, *in vitro* experiments have shown that organic matrix proteins can totally change the thermodynamic equilibrium necessary for mineral precipitation (Njegić Džakula et al., 2019) and can also determine the selection of the $CaCO_3$ polymorph (e.g., calcite or aragonite) (Falini et al., 2013; Rahman & Oomori, 2009)). Additionally, coral acid-rich organic matrix proteins are able to catalyze $CaCO_3$ precipitation *in vitro* probably through the electrostatic interaction of Ca^{2+} ions with the carboxylate groups on the proteins (Mass et al., 2013). It has also been proposed that the organic matrix can stabilize ACC in the intracellular vesicles (Mass et al., 2017b). Despite its multi-faceted role in crystal and skeleton growth, little research has been conducted on the organic matrix at the GE. Part of the reason is that extraction and sequencing procedures for the characterization of organic matrix proteins need more skeletal material than is found in the isolated crystals at the GE and as such, this type of analysis has been performed on mature skeletons.

7.4.4 APPLYING MOLECULAR TOOLS TO STUDY CALCIFICATION AT THE GROWING EDGE (GE)

The use of molecular and genetic approaches is quite recent in Cnidaria, especially in investigating the mechanisms of coral calcification. Although the number of published studies is increasing, the first characterization of ion transporters (Zoccola et al., 1999) and organic matrix molecules (Fukuda et al., 2003) in corals were published only about 20 years ago. It was not until 2007 that the first non-bilaterian animal genome (the sea anemone *Nematostella vectensis*) was sequenced (Putnam et al., 2007) and since then, collections of genomic and transcriptomic data of diverse scleractinian species have been generated. To our knowledge, at present, transcriptomes of 34 species (Bhattacharya et al., 2016; Closek et al., 2014; Karako-Lampert et al., 2014; Kitchen et al., 2015; Ocampo et al., 2015; Zhang et al., 2019), and genomes of 12 scleractinian species are available including *Stylophora pistillata* (Buitrago-Lopez et al., 2020; Consortium, 2015; Cunning et al., 2018; Dziedzic et al., 2019; ReFuGel, 2020; Shinzato et al., 2011; Voolstra et al., 2017; Ying et al., 2019). Without question, these data are important resources to help us investigate calcification, however, their limitation is that they concern bulk tissues, rather than specific cell types. Single-cell RNA-seq (Li et al., 2020), together with a means to identify and sort different cell types (Luecken & Theis, 2019; Saliba et al., 2014) provides a way by which the gene expression of calicoblastic cells can be targeted in order to acquire insight into the calcification process. This has recently been achieved using fluorescence-activated cell sorting in a study by (Levy et al., 2021), in which the authors report gene expression profiles associated with a range of cell types in *S. pistillata* including calcifying cells. Future investigations that aim to target the calicoblastic cells specifically could potentially benefit from using GE in a number of ways. The GE can relatively easily be separated from the rest of the colony with a scalpel or razor blade under a dissecting microscope without the need to separate the tissues from substantial amounts of the skeleton (Pers. Obsv). The lack of fully developed polyps and low density of symbiotic algae at the GE means it is likely to possess fewer cell types, potentially making cell sorting techniques more straightforward.

Another way to study calcification mechanisms at the molecular level is the inactivation of candidate genes via techniques such as morpholino knock-downs, siRNA gene silencing, or genome-editing technology, in tandem with functional experiments (Ikmi et al., 2014). Such experiments have been carried out in Cnidaria (Miljkovic et al., 2002), mainly on hydra and sea anemones (Jakob & Schierwater, 2007; Miljkovic et al., 2002; Quiroga-Artigas et al., 2020; Sinigaglia et al., 2013) to decipher developmental mechanisms at the molecular level (Lommel et al., 2017). Currently, however, there are only two studies that report successful gene inactivation experiments on corals, one using a morpholino (Yasuoka et al., 2016), and the other one using CRISPR/Cas9 genome-editing technology (Cleves et al., 2018). Neither study looked into the calcification process. These experiments were conducted on eggs that developed into larvae. If larvae were allowed to settle and develop into primary polyps, similar gene inactivation experiments approaches with phenotypic observations at the GE could potentially be used to investigate calcification mechanisms. Alternatively, new methods would need to be developed to deliver reagents inside the adult coral cells, for example at the GE. Inspiration may be found in innovative approaches developed during the COVID-19 crisis to deliver DNA and RNA into cells with nanomaterials (Chen et al., 2016; Riley & Vermerris, 2017). The small size of juvenile corals or LSP samples is well-suited to gene inactivation techniques because the high cost of reagents is often technically limiting making small volumes preferable. Finally, a number of other tools await investigation into their utility in coral research. For example, chemically modified siRNA, such as 2′-O-Me-RNA (Kraynack & Baker, 2006) or tricyclo-DNA (Ittig et al., 2010; Renneberg et al., 2002) that do not require a vector to be delivered into the cells could be tested on tissues at the GE. Such vectors used to deliver DNA or RNA hold much potential, although it's not yet known how they behave in seawater.

As described above, the GE model shows characteristics, including the presence in juveniles soon after settlement and small size requiring small volumes of reagents in incubations, that are suitable for the use of advanced molecular techniques that could open new avenues of research on the mechanisms of calcification in corals.

7.5 INSIGHT FROM THE GROWING EDGE (GE) INTO ENVIRONMENTAL EFFECTS ON CALCIFICATION

Coral calcification rates are in decline globally due to many elements of global environmental change, threatening the sustainability of coral reefs (Hoegh-Guldberg et al., 2017). There is an increasing number of research initiatives that seek to define the mechanistic basis underlying the sensitivity of corals to environmental factors in order to understand the scope of corals for physiological acclimatization and genetic adaptation (Anthony et al., 2017; van Oppen et al., 2015). Historically, most investigations concerning the effect of environmental parameters on coral calcification have been carried out at the organism/ecosystem scale, and comparatively limited information is available about how coral calcification responds at the cellular scale (Davy et al., 2012; Weis & Allemand, 2009). In the previous sections, we discussed how the GE model has helped provide cellular-level insight into the mechanism of calcification. In the next section, we will look at what the GE tells us about how environmental parameters affect calcification, including light, seawater carbonate chemistry, temperature, and substrate type.

7.5.1 EFFECT OF LIGHT

It is widely known that exposure to light increases calcification rates in symbiotic corals in a process termed light-enhanced calcification (Gattuso et al., 1999). A number of mechanisms have been proposed to lie behind this phenomenon, many of them linked to photosynthetic algae. These include: that symbiotic algae provide precursors to the organic matrix used in calcification (Allemand et al., 1998; Muscatine et al., 2005; Wainwright, 1963); that photosynthates produced by the algae provide energy for calcification (Chalker & Taylor, 1975; Colombo-Pallotta et al., 2010; Goreau & Goreau, 1959); and that algal photosynthesis elevates pH in the cœlenteron, which sets-up a more favorable gradient for the transport of protons out of the ECM (Al-Horani et al., 2003a; Comeau et al., 2013a, b; Furla et al., 2000). The latter two theories are expected to involve an increased pH_{ECM}, which is anticipated to increase ECM aragonite saturation state thus promoting calcification in light *versus* dark conditions. In work with the GE of *S. pistillata*, Venn et al. (2011) showed that exposure of microcolonies to illumination can elevate pH in the ECM, with respect to dark conditions and therefore supports a role for light-induced pH_{ECM} increases in the mechanism of LEC. However, the light and dark differences in pH_{ECM} cannot always be reproduced between different studies on the GE of *S. pistillata*. While there is often a visible trend of higher light pH_{ECM} than dark pH_{ECM}, the mean values for each condition are sometimes not statically different, for example in the studies of (Sevilgen et al., 2019; Tambutté et al., 2015). One possible reason for the inconsistency is the variation in the density of symbiotic algae at the GE of samples used in different studies. Like certain other parts of the coral colony, such as the branch apex of mature colonies or the cœnosteal spines (Tambutté et al., 2007a), the GE contains relatively few symbionts. These zones are fast-growing (both at the tissue and skeleton level), which demonstrates that the presence of symbiotic algae in the calcifying zone is not directly necessary for calcification to occur. However, sufficient algal density may be required to observe light and dark differences in calcification rate (i.e., light-enhanced calcification). This topic requires further investigation and future work could involve concomitant measurements of symbiotic algal density, pH_{ECM}, and calcification rates at the GE to establish the relationship of these parameters.

Despite this uncertainty, there are advantages to working in a zone (i.e., the GE) where symbiotic algae are in very low density. Low numbers of symbiotic algae make the tissues at the GE relatively transparent and less autofluorescent. This significantly facilitates the visualization of the placement of microelectrodes into pockets of ECM using light microscopy for the measurement of ECM carbonate chemistry parameters (Sevilgen et al., 2019) (Section 7.4.1). In addition, the low density of symbiotic algae can help in experiments designed to disentangle physiological mechanisms related to photosynthesis and calcification. For example, the GE can be dissected from the rest of the colony and analyzed for gene expression specifically related to calcification, in a zone where there are few symbiotic algae and as such little or no photosynthesis.

7.5.2 EFFECT OF OCEAN ACIDIFICATION (OA)

Ocean acidification (OA), which consists of the decrease of seawater pH and associated changes in carbonate chemistry caused by seawater uptake of anthropogenic carbon dioxide, impairs coral calcification rates and can lead to changes in skeleton morphology on micro- and macro-scales (Chan & Connolly, 2013; Fantazzini et al., 2015; Tambutté et al., 2015). The impact of OA can vary greatly among corals of different species and from different locations (Comeau et al., 2014; Pandolfi et al., 2011; Schoepf et al., 2017) and as such there has been much interest in the research community about the mechanistic basis underlying coral vulnerability to OA (Putnam et al., 2017). Research on the GE of *S. pistillata* has provided important *in vivo* physiological insight into this topic. Venn et al. (2013) exposed LSP assay samples to different seawater pH levels: 8.2, 7.8, 7.4, and 7.2, and analyzed the effect on pH_{ECM} and crystal growth at the GE. The small size of LSP assay microcolonies allowed them to be analyzed in cell culture perfusion chambers mounted on the confocal microscope and supplied with flowing seawater in which environmental parameters could be controlled, including seawater carbonate chemistry (including pH), light, and temperature (Figure 7.9(A)). Results showed the pH_{ECM} undergoes a gradual decline across the treatments with respect to the surrounding seawater, leading to an increasing offset in pH between the ECM and seawater (Venn et al., 2013). Crystal growth rates also decreased under acidified conditions but corals were able to continue calcifying even in the lowest pH treatment where seawater aragonite saturation state would predict skeletal dissolution. The same response was observed in lateral growth rates of the entire microcolonies indicating that the crystals at the GE reflect what occurs at the coral microcolony scale. Taken together, these results showed that *S. pistillata* is able to mitigate the effects of OA by maintaining elevated pH_{ECM} despite an increasingly unfavorable pH gradient with the surrounding seawater. Later experiments performed by (Holcomb et al., 2014) measured pH at the GE and used boron isotopes to determine pH_{ECM} in areas of fully formed skeletons. This research confirmed that pH_{ECM} at the GE and other areas of the colony responded in a similar way, although absolute values of pH_{ECM} were lower at the GE (Holcomb et al., 2014) (see discussion in Section 7.4.1). Similar trends of the offset of pH_{ECM} with declining seawater pH were reported by studies that derived pH_{ECM} from skeleton isotopes in the mature coral skeleton of other species (McCulloch et al., 2012a; Trotter et al., 2011), leading to a general consensus that corals possess a degree of physiological resilience against OA through their ability to regulate pH_{ECM} (McCulloch et al., 2012a; Venn et al., 2015; Wall et al., 2016). This physiological resilience against OA was compared in later studies in three coral species *S. pistillata*, *P. damicornis*, and *A. hyacinthus* all grown on glass coverslips by the LSP assay (Venn et al., 2020). Results show that at the GE, *S. pistillata* has a greater capacity than the other two species to maintain pH_{ECM} elevated against seawater acidification in both light and darkness. These observations are broadly in agreement with observations in the literature, which generally describes *S. pistillata* as "environmentally tolerant," whereas acroporids have been found to be susceptible to acidification (Comeau et al., 2013b, 2014; Schoepf et al., 2013).

Analyses performed directly on the calcifying cells are rare in OA investigations and have only been achieved with the use of the GE because this is the only approach where the calicoblastic ectoderm can be investigated without physical disruption to the colony. Both (Venn et al., 2013) and (Venn et al., 2019) used confocal imaging of the calicoblasts to record the impact of OA on intracellular pH (pH_I), in parallel to measurements on pH_{ECM}. Both studies found that significant declines in calicoblast pH_I only occur in the lowest pH treatments, which is not surprising given that intracellular environments are well-buffered and generally more tightly regulated than extracellular media. Disruption to calicoblast pH_I would be predicted to have serious consequences for coral calcification, as cellular processes work within tightly defined pH limits, beyond which the activity of enzymes becomes impaired. Interestingly (Venn et al., 2013) observed significant declines in pH_I in the lowest seawater pH of 7.2 in parallel with a sudden drop in crystal growth, suggesting that depressions in pH_I caused by acidification could contribute to declines in calcification rate (Venn et al., 2013).

Other ways in which the GE has provided insight into the mechanism of how OA impacts coral calcification include work on paracellular permeability using the fluorescent dye calcein (discussed in Section 7.4.2). Following one-week exposure to the same pH 7.2 seawater treatment used in (Venn et al., 2013, 2019), increases in paracellular permeability (leakiness) were observed in *S. pistillata*, as determined by significant decreases in the half-time of calcein influx relative to control treatments (Venn et al., 2020). The causes of this increase in permeability were speculated to involve disruption to the conformation of intercellular junctions between the calicoblasts, perhaps linked to depressions of calicoblast pH_I. If these changes in paracellular permeability extend to the paracellular flux of protons and other ions, then the challenge to acid-base regulation of the ECM under OA would be anticipated to be even greater than previously thought, because increased paracellular leakiness could enhance H^+ entry to the ECM.

In future experiments, the GE might prove useful in understanding mechanisms of acclimatization to OA. Previously, it has been shown that OA can lead to phenotypic acclimatization of coral colonies by epigenetic mechanisms (Liew et al., 2018). Molecular tools could be used on tissues at the GE to isolate nucleic acid from calicoblastic cells and use them in whole-genome bisulfite sequencing (DNA methylation readout) in addition to the single-cell RNA-seq (mentioned in Section 7.4.4). This might help provide insight into the epigenetic mechanisms involved in regulating genes linked to calcification in conditions of OA.

Overall, GE has proven a useful experimental system to examine the effect of OA on the cellular physiology of calcification, but of course, it has its limitations. For example, the incomplete skeleton at the GE makes it impossible to investigate changes in skeletal architecture or porosity. Other skeletal parameters such as qualitative and quantitative measurements of organic matrix content in this zone are challenging due to the low amounts of skeletal material. Measuring photosynthetic rates (by O_2 production), photosynthetic efficiency (by pulse amplitude fluorimetry), and $CaCO_3$ production (by alkalinity anomaly) are also difficult to achieve at the scale of the GE. Investigations using these techniques have deepened our understanding of the effect of OA on physiology and calcification in corals generally, but they cannot easily be applied to GE.

7.5.3 EFFECT OF TEMPERATURE

Studies on coral skeletons from several geographical locations worldwide indicate that increases in sea temperatures associated with climate change are having direct impacts on coral calcification in many reef systems (Cantin et al., 2010; Carricart-Ganivet et al., 2012; De'ath et al., 2009). However, although there has been a substantial amount of research into the cellular mechanisms underlying the thermal sensitivity of the symbiosis (coral bleaching), much less mechanistic work has been directed at the thermal sensitivity of coral calcification. Most existing data on this subject comes from molecular studies (Bernardet et al., 2019; Rocker et al., 2015) and little research has been conducted on thermal effects on the calcifying cells themselves. A previous study on the GE is the exception, in which the same study that investigated how OA affects paracellular permeability at the GE (mentioned above in the previous section) also looked at the effects of temperature (Venn et al., 2020). Working with *S. pistillata* prepared by the LSP assay, the study involved exposure of corals to elevated (31°C) and decreased (20°C) temperatures that were known to cause decreases in calcification rate relative to a controlled temperature of 25°C. A high degree of variability in paracellular permeability was observed among coral colonies at high temperatures whereas at lower temperatures a decrease in (more restrictive) paracellular permeability was observed. In the same study, the measurement of pH in the calcifying cells and the ECM indicated that pH in both of these compartments was unaffected at these temperatures. From these findings, it is difficult to establish any link between the thermal sensitivity of coral calcification and thermal effects on pH regulation and paracellular permeability. However, the observation that temperature variation has little effect on pH_{ECM} at the GE is consistent with data on pH_{ECM} taken from mature coral skeletons (McCulloch et al., 2017). In this work, sampling of skeletal isotope chemistry from corals in the field also indicated that pH_{ECM} remained stable despite seasonal temperature changes, although [DIC] was variable. Recent investigations of *S. pistillata* with microelectrodes came to different conclusions however, observing that corals were not able to maintain pH_{ECM} elevated above seawater at 31°C (Guillermic et al., 2021). Overall, as little work has been conducted on the cellular mechanisms of thermal sensitivity of coral calcification, it is difficult to interpret differences between studies and come to firm conclusions. Future work on temperature impacts is important to carry out. In one possible avenue for research, it will be interesting to build on previous reports that suggest that calcification may be more temperature-sensitive during lateral growth across a substrate than in whole branches of *S. pistillata*. Decreases in the linear extension of LSP samples of *S. pistillata* were observed to occur between 27 and 29°C (Reynaud-Vaganay et al., 1999), but microcolonies made from branches suspended from filaments show no decrease in calcification rate at these temperatures (Bernardet et al., 2019). Additionally, Bernardet (2019) (PhD thesis) reported that calcification ceased completely at the GE of *S. pistillata* LSP samples above and below 31°C and 20°C, respectively, in contrast to microcolonies of branches suspended from filaments that continued to calcify at these temperatures, albeit at much-reduced rates.

7.5.4 EFFECT OF SUBSTRATE

Almost all data pertaining to calcification mechanisms at the GE comes from corals on artificial substrates such as glass supports, including slides (Raz-Bahat et al., 2006), coverslips (Andersson et al., 2019; Venn et al., 2011, 2013, 2020) and Petri dishes (Neder et al., 2019; Ohno et al., 2017a) and glass has different material properties to the natural calcareous substrates that corals usually grow on in their environment. While a body of research has been conducted on the substrate properties and chemical cues that promote planulae settlement and metamorphosis (Heyward & Negri, 1999; Tebben et al., 2015; Whalan et al., 2015), less is known about how different materials promote calcification at the GE. This area constitutes an important area for further work, specifically to test whether the type of substrate influences calcification at the GE. For example, it isn't known how calicoblastic cells interact with glass or other materials and whether these materials promote or restrict the advancement of the tissue and deposition of isolated crystals. Preliminary data suggest that isolated crystals at the GE in samples grown on glass are not peculiar to this substrate, as SEM imaging of *S. pistillata* LSP samples on glass, PVC, and terracotta reveals similar crystal morphologies described in Section 7.3.3 (Pers. Obsv.). Nevertheless, future research is required, particularly on the cellular mechanisms underlying how corals sense the underlying substratum and how this may affect growth rates. Recent work has identified a primary cilium on each of the calicoblast cells that form crystals at the GE (Section 7.3.2) (Tambutté et al., 2021). It has been proposed that these primary cilia could sense the properties of the underlying substrate and relay signals to the calicoblasts which could be a way to modulate cellular calcification activity (Tambutté et al., 2021). These observations are launching pads toward a better cellular-scale understanding of how the GE and the lateral growth response to different substrates, and whether samples grown on glass can be used as a model that represents the GE of corals in their natural environment. Future research on this topic would be valuable for reef restoration initiatives. In particular, this knowledge could be beneficial to initiatives looking to maximize the rates of lateral growth in

TABLE 7.2 Suitability of the Coral Growing Edge (GE) of *Stylophora pistillata* for Investigation into Coral Calcification at the Cellular Scale

Characteristics	Advantages	Limitations
Small size	• The GE can be observed in LSP samples and juveniles in cell culture perfusion chambers (Figure 7.9(A)) or microfluidic set-ups suitable for microscopy (e.g., Shapiro et al., 2016; Venn et al., 2011).	• Physiological responses of small experimental colonies are sometimes not representative of larger colonies (Edmunds & Burgess, 2016).
Availability and easy propagation	• The GE can be analyzed in LSP samples grown on coverslips which are readily propagated from mature coral fragments (Raz-Bahat et al., 2006). • The GE can be analyzed in juvenile corals settled on glass coverslips or Petri dishes (Neder et al., 2019).	• Requirement of seawater aquarium culture facilities to rear corals, e.g., those at www.centrescientifique.mc. • Experiments on GE of juveniles require access to spawning corals in the field and/or aquarium (Neder et al., 2019).
Choice of genetic variation.	• The GE can be analyzed in LSP samples that are prepared as 'genets' from single mother colonies that display little or no intracolony genetic variation. Alternatively, genetically different samples can be prepared from different parent colonies. • The GE can be analyzed in juvenile corals obtained from single or genetically different parent colonies.	• Experiments carried out on single genotypes have reduced ecological relevance relative to experiments on multiple genotypes.
Simplified histology	• Lateral growth at the GE arranges parallel orientation of four tissue layers and skeleton, which makes histology highly amenable for microscopy of fixed (e.g., TEM and SEM; (Tambutté et al., 2007a; Tambutté et al., 2021) and *in vivo* material (e.g., confocal; (Ohno et al., 2017a; Venn et al., 2011)). • Minimal skeleton means that the GE can be cleanly dissected from the colony for transcriptomic or proteomic studies.	• Thickness and arrangement of overlying tissue layers can prevent the direct application of certain drugs (e.g., inhibitors, ionophores, etc.) to the calcifying cells. This limitation applies to all types of whole coral samples and could be overcome by the development of protocols to isolate and/or culture calcifying cells.

TABLE 7.2 *(Continued)*

Characteristics	Advantages	Limitations
Transparency	• Incomplete/thin skeleton allows direct, non-invasive, *in vivo* imaging of calcifying cells, ECM, and $CaCO_3$ by white light and fluorescence-based microscopy techniques, e.g., confocal, with dyes for physiology and cell structure (Neder et al., 2019; Ohno et al., 2017a, b; Venn et al., 2013, 2019). • Incomplete/thin skeleton + low density of symbiotic algae helps microscope-guided placement of microsensors within the ECM, avoiding ambiguity of 'blind' measurements (Sevilgen et al., 2019).	• The extent to which absolute values of physiological parameters obtained in the calicoblastic epithelium and the ECM at the GE are representative of elsewhere in the colony is unknown, although the limited available data suggest that the GE exhibits similar physiological trends to other parts of the colony ((Holcomb et al., 2014).
Rapid calcification	• Isolated $CaCO_3$ crystals grow on hour timescales allowing analysis in real-time or short experimental time frames (Venn et al., 2013). • Simplified 2-dimensional arrangement of isolated $CaCO_3$ crystals facilitates analysis of the early steps of calcification, e.g., crystal morphology and mineralogy (Raz-Bahat et al., 2006).	• Established methods to quantify calcification (e.g., total alkalinity anomaly) are difficult to apply to the GE and microcolonies on glass cover-slips due to low amounts of skeletal material.

coral nurseries and increase the success of substrate colonization by coral fragments that have been out-planted as part of restoration projects (Guest et al., 2011; Lirman & Schopmeyer, 2016).

7.6 CHARACTERISTICS OF THE GROWING EDGE (GE) SUITED TO CELLULAR-SCALE RESEARCH ON CALCIFICATION

The utility of GE as a model to study coral calcification relies on a number of characteristics that have been highlighted in the previous sections. These characteristics are summarized in Table 7.2 and are typical attributes of many models used for cell biology. Size is one characteristic that has historically been a problem for corals which are viewed as colonies of multiple polyps with a massive exoskeleton. Indeed, Weis et al. (2008) highlighted this problem in their review of the suitability of models to investigate coral cell biology. However, we have seen here that the GE can be observed in microcolonies of 1 cm^2 or less, that are fixed on coverslips and maintained in perfusion chambers normally used as cell-culture cultivation systems for mammalian cells (e.g., POC-R2, Pecon, used in (Venn et al., 2011)). Other authors have grown and imaged coral micro-propagates in microfluidic set-ups, referring to these samples as 'coral on a chip'(Shapiro et al., 2016). The rates of physiological processes such as calcification in small coral size samples cannot necessarily be scaled up to represent larger colonies (Edmunds & Burgess, 2016) but this limitation potentially applies to all types of coral samples and is not specific to the GE.

Investigations on corals often involve samples taken from different colonies in order to produce results that may be considered more widely representative of the genetic variation and resulting physiological variation found within a particular species. An alternative approach is to try to minimize genetic diversity by working with coral samples from a single-parent colony in order to reduce as many sources of biological variation as possible. This latter approach is frequently used in cell biology investigations on other model organisms that place a priority on deciphering the fundamentals of a mechanism before establishing its wider relevance to a population. When working with the GE, both options are available. Genetic variation can be maximized by working with newly-settled juveniles or LSP samples prepared from different parent colonies. Alternatively, genetic variation among samples can be minimized by working with LSP microcolonies prepared from a single-parent colony. It should be noted however, that preparing samples from a single-parent colony does not necessarily ensure genetic uniformity, as intracolonial genetic variation is surprisingly common in scleractinians, arising through both mosaicism (somatic mutations) and chimerism (fusion of two individuals) (Schweinsberg et al., 2015).

The most important characteristics of the GE for studying the cell biology of calcification are probably its transparency and simplified arrangement of tissues. These characteristics facilitate various microscopy approaches, the use of fluorescent dyes, and microelectrodes that allow direct *in vivo* investigations of cell biology linked to calcification (e.g., Sevilgen et al., 2019; Venn et al., 2011). Transparency and simplified tissue arrangement at the GE occurs due to the incomplete nature of the skeleton and the low density of symbiotic algae in this zone. This contrasts with other parts of the colony with a fully-formed skeleton and denser population of symbiotic algae. The degree to which physiological parameters vary due to these differences between the GE and other areas of the colony is not clear (discussed in Section 7.4.1). However, if the aim of working with the GE is to decipher fundamental cellular-scale mechanisms of coral calcification, then the results obtained at the GE are envisaged to have relevance to the entire coral colony. This viewpoint has parallels with successful invertebrate models used for investigations into biomineralization, such as the sea urchin larva (Matranga et al., 2013; Wilt, 2002). Whereas the large $CaCO_3$ test of adult sea urchins makes them unsuitable for certain cell biology approaches, investigations have long been carried out on sea urchin larvae (Decker & Lennarz, 1988). Urchin larvae are transparent and morphologically simple allowing access to the primary mesenchyme cells (PMCs) responsible for skeleton formation and to the developing spicule (Decker & Lennarz, 1988). Although the spicule is lost during metamorphosis and never forms part of the adult urchin skeleton, investigations on sea urchin larvae have greatly advanced our understanding of the fundamentals of calcification in invertebrates. Similarly, the cellular scale biology underlying the primary steps of skeleton growth at GE has the potential to teach us much about the fundamentals of coral calcification.

Finally, we highlight the relatively rapid rate of calcification at the GE in Table 7.2 as an important characteristic of studying coral calcification. The ability to observe the growth of skeletal crystals within workable experimental time frames (Figure 7.8) and tie these observations to physiological measurements in the same tissue (e.g., declines in crystal growth and disruption of intracellular pH (Venn et al., 2013; Barott et al., 2020)) is an important advantage. This advantage of relating changes in skeletal growth rates to cellular events has already been exploited for decades in experiments in sea urchin larvae, in which skeletal growth can also be imaged on hour-to-hour time scales (Guss & Ettensohn, 1997; Sun & Ettensohn, 2016). Additionally, growth rates of crystals at the GE can be compared to wider colony-scale measurements of tissue and skeleton lateral extension if desired. For example, Venn et al. (2013) compared the effect of OA on crystal growth at the GE with changes in the colony surface area of microcolonies on glass coverslips grown in the same conditions.

7.7 CONCLUSIONS

In the sections above we have discussed how the GE has opened a window of insight into several key aspects of coral calcification at the cellular scale. While there is still much to learn, the use of the growing edge (GE) in research on coral calcification in recent years has helped advance understanding of the physiology of the calcifying cells (e.g., Venn et al., 2020), the extracellular calcifying medium (ECM) at the interface between the calicoblastic epithelium and newly forming skeleton (e.g., Ohno et al., 2017b), and the formation of the mineral itself, particularly in its early phases (e.g., Neder et al., 2019). In addition to advances made on fundamental aspects of calcification, the GE has facilitated research into the effects of environmental threats to calcification, including OA (Tambutté et al., 2015).

A large majority of the research we have reviewed concerns *Stylophora pistillata*, with the GE observed in both juveniles of this species (Neder et al., 2019) and fragments of adult colonies [LSP assay, (Raz-Bahat et al., 2006)]. However, as we have seen, cell biology investigations and studies on the newly forming crystals at the GE have also been achieved in other pocilloporid species (Gilis et al., 2014; Venn et al., 2019) and acroporids (Ohno et al., 2017a, b; Venn et al., 2019). This underscores the potential utility of the GE to investigate calcification in corals of divergent evolutionary history, including representatives of both robust and complex corals that may have evolved calcification independently having diverged at least 245 million years ago (Park et al., 2012; Romano & Palumbi, 1996; Simpson et al., 2011).

With the characteristics of the GE given in Table 7.2 lending themselves to a cellular-level approach to investigate coral calcification, we are poised to make significant advances in this domain. It remains now to define the paths to take forward. Recent discoveries act as potential signposts. For example, the identification of a primary cilium in the calicoblastic epithelium (Section 7.3.2; (Tambutté et al, 2021)) opens up a spectrum of possible enquires. This organelle is a potentially key piece of the puzzle of how calicoblastic cells interact with the skeleton and the ECM, and perhaps how calicoblastic cells sense and respond to underlying substrates during lateral growth at the GE. Additionally, the recent evidence of ACC formation and its putative deployment from cells to the developing skeleton (Mass et al., 2017b) also requires substantial future investigation, and the GE is ideally suited to this challenge. Indeed, microscopy of the GE is currently the only way in which vesicles and other subcellular structures can be observed in the calicoblastic cells in a living, actively calcifying coral (Neder et al., 2019). The challenges that these lines of inquiry bring would benefit enormously from the application of molecular approaches outlined in Section 7.4.4. Indeed, gene editing and gene expression knockdown protocols have been long awaited to fulfill their potential in corals. The development of this technology in corals and its application to GE should be one of the priority areas for future research.

In conclusion, GE helps us study calcification at the cellular scale. At a time when coral reefs are experiencing rapid environmental change and rapid degradation due to physiological stress imposed on corals themselves, a better mechanistic understanding at the cellular scale is critical to determining what is detrimental to the calcification process and what promotes it. Applied environmental research on corals is now considering innovative but challenging strategies to increase coral reef resilience such as assisted evolution and synthetic biology (including gene editing) of corals (Anthony et al., 2017; van Oppen et al., 2015). Ultimately, advancing understanding of the cell biology of key physiological processes like calcification can only serve to aid these initiatives that hope to protect and restore populations of corals on which reef ecosystems depend.

KEYWORDS

- **calcification**
- **cellular scale**
- **coral reef resilience**
- **coral reefs**
- **corals**
- **gene editing**

REFERENCES

Adkins, J. F., Boyle, E. A., Curry, W. B., & Lutringer, A., (2003) Stable isotopes in deep-sea corals and a new mechanism for "vital effects". *Geochimica et Cosmochimica Acta, 67*(6), 1129–1143.

Akiva, A., Neder, M., Kahil, K., Gavriel, R., Pinkas, I., Goobes, G., & Mass, T., (2018). Minerals in the pre-settled coral *Stylophora pistillata* crystallize via protein and ion changes. *Nature Communications*, *9*, 1880.

Al-Horani, F. A., Al-Moghrabi, S. M., & de Beer, D., (2003a). The mechanism of calcification and its relation to photosynthesis and respiration in the scleractinian coral *Galaxea fascicularis*. *Marine Biology*, *142*(3), 419–426.

Al-Horani, F. A., Al-Moghrabi, S. M., & De Beer, D., (2003b). Microsensor study of photosynthesis and calcification in the scleractinian coral, *Galaxea fascicularis*: active internal carbon cycle. *Journal of Experimental Marine Biology and Ecology*, *288*(1), 1–15.

Allemand, D., Tambutté, É., Girard, J.-P., & Jaubert, J., (1998). Organic matrix synthesis in the scleractinian coral *Stylophora pistillata*: role in biomineralization and potential target of the organotin tributyltin. *Journal of Experimental Biology*, *201*, 2001–2009.

Allemand, D., Tambutté, É., Zoccola, D., & Tambutté, S., (2011). Coral Calcification, Cells to Reefs. In Z. Dubinsky & N. Stambler (Éds.), *Coral Reefs: An Ecosystem in Transition* (p. 119–150). Springer Netherlands.

Allison, N., Cohen, I., Finch, A. A., Erez, J., & Tudhope, A. W., (2014). Corals concentrate dissolved inorganic carbon to facilitate calcification. *Nature Communications*, *5*(1), 5741.

Andersson, A. J., Venn, A. A., Pendleton, L., Brathwaite, A., Camp, E. F., Cooley, S., Gledhill, D., Koch, M., Maliki, S., & Manfrino, C., (2019). Ecological and socioeconomic strategies to sustain Caribbean coral reefs in a high-CO_2 world. *Regional Studies in Marine Science*, *29*, 100677.

Anthony, K., Bay, L. K., Costanza, R., Firn, J., Gunn, J., Harrison, P., Heyward, A., Lundgren, P., Mead, D., Moore, T., Mumby, P. J., Van Oppen, M. J. H., Robertson, J., Runge, M. C., Suggett, D. J., Schaffelke, B., Wachenfeld, D., & Walshe, T., (2017). New interventions are needed to save coral reefs. *Nature Ecology & Evolution*, *1*(10), 1420–1422.

Barnes, D. J., (1970). Coral Skeletons: An explanation of their growth and structure. *Science*, *170*(3964), 1305–1308.

Barnes, D. J., (1972). The structure and formation of growth-ridges in scleractinian coral skeletons. *Proceedings of the Royal Society of London. Series B. Biological Sciences*, *182*(1068), 331–350.

Barott, K. L., Venn, A. A., Thies, A. B., Tambutté, S., & Tresguerres, M., (2020). Regulation of coral calcification by the acid-base sensing enzyme soluble adenylyl cyclase. *Biochemical and Biophysical Research Communications*, *525*(3), 576–580.

Bénazet-Tambutté, S., Allemand, D., & Jaubert, J., (1996). Permeability of the oral epithelial layers in cnidarians. *Marine Biology*, *126*(1), 43–53.

Bernardet, C., (2019). Physiologie des transports ioniques et moléculaires chez les coraux, implications environnementales. Thèse. Sorbonne Université - Ecole doctorale Complexité du Vivant.

Bernardet, C., Tambutté, E., Techer, N., Tambutté, S., & Venn, A. A., (2019). Ion transporter gene expression is linked to the thermal sensitivity of calcification in the reef coral *Stylophora pistillata*. *Scientific Reports*, *9*(1), 18676.

Bertucci, A., Moya, A., Tambutté, S., Allemand, D., Supuran, C. T., & Zoccola, D., (2013). Carbonic anhydrases in anthozoan corals—A review. *Bioorganic & Medicinal Chemistry*, *21*(6), 1437–1450.

Bhattacharya, D., Agrawal, S., Aranda, M., Baumgarten, S., Belcaid, M., Drake, J. L., Erwin, D., Foret, S., Gates, R. D., Gruber, D. F., Kamel, B., Lesser, M. P., Levy, O., Liew, Y. J., MacManes, M., Mass, T., Medina, M., Mehr, S., Meyer, E., … Falkowski, P. G., (2016). Comparative genomics explains the evolutionary success of reef-forming corals. *ELife*, *5*.

Bockel, T. & Rinkevich B., (2019). Rapid Recruitment of Symbiotic Algae into Developing Scleractinian Coral Tissues. *Journal of Marine Science and Engineering*, *7*(9), 306.

Buitrago-López, C., Mariappan, K. G., Cárdenas, A., Gegner, H. M., & Voolstra, C. R., (2020). The genome of the cauliflower coral *Pocillopora verrucosa. Genome Biology and Evolution*, *12*(10), 1911–1917.

Byler, K. A., Carmi-Veal, M., Fine, M., & Goulet, T. L., (2013). Multiple symbiont acquisition strategies as an adaptive mechanism in the coral *Stylophora pistillata. PLoS ONE*, *8*(3), e59596.

Cai, W.-J., Ma, Y., Hopkinson, B. M., Grottoli, A. G., Warner, M. E., Ding, Q., Hu, X., Yuan, X., Schoepf, V., Xu, H., Han, C., Melman, T. F., Hoadley, K. D., Pettay, D. T., Matsui, Y., Baumann, J. H., Levas, S., Ying, Y., & Wang, Y., (2016). Microelectrode characterization of coral daytime interior pH and carbonate chemistry. *Nature Communications*, *7*(1), 11144.

Cantin, N. E., Cohen, A. L., Karnauskas, K. B., Tarrant, A. M., & McCorkle, D. C., (2010). Ocean warming slows coral growth in the central Red Sea. *Science*, *329*(5989), 322–325.

Capasso, L., Zoccola, D., Ganot, P., Aranda, M., Tambutté, S. (2022). SpiAMT1d: molecular characterization, localization, and potential role in coral calcification of an ammonium transporter in *Stylophora pistillata. Coral Reefs*, 41, 1187–1198.

Capasso, L., Ganot, P., Planas-Bielsa, V., Tambutté, S., & Zoccola, D., (2021). Intracellular pH regulation: characterization and functional investigation of H^+ transporters in *Stylophora pistillata. BMC Molecular and Cell Biology*, *22*(1), 18.

Carricart-Ganivet, J. P., Cabanillas-Terán, N., Cruz-Ortega, I., & Blanchon, P., (2012). Sensitivity of Calcification to Thermal Stress Varies among Genera of Massive Reef-Building Corals. *PLoS ONE*, *7*(3), e32859.

Chalker, B. E., & Taylor, D. L., (1975). Light-enhanced calcification, and the role of oxidative phosphorylation in calcification of the coral *Acropora cervicornis*. *Proceedings of the Royal Society of London. Series B. Biological Sciences*, *190*(1100), 323–331.

Chan, N. C. S., & Connolly, S. R., (2013). Sensitivity of coral calcification to ocean acidification: a meta-analysis. *Global Change Biology*, *19*(1), 282–290.

Chen, J., Guo, Z., Tian, H., & Chen, X., (2016). Production and clinical development of nanoparticles for gene delivery. *Molecular Therapy - Methods & Clinical Development*, *3*, 16023.

Cleves, P. A., Strader, M. E., Bay, L. K., Pringle, J. R., & Matz, M. V., (2018). CRISPR/Cas9-mediated genome editing in a reef-building coral. *Proceedings of the National Academy of Sciences*, *115*(20), 5235–5240.

Clode, P. L., & Marshall, A. T., (2002). Low temperature FESEM of the calcifying interface of a scleractinian coral. *Tissue and Cell*, *34*(3), 187–198.

Closek, C. J., Sunagawa, S., DeSalvo, M. K., Piceno, Y. M., DeSantis, T. Z., Brodie, E. L., Weber, M. X., Voolstra, C. R., Andersen, G. L., & Medina, M., (2014). Coral transcriptome and bacterial community profiles reveal distinct Yellow Band Disease states in *Orbicella faveolata. The ISME Journal*, *8*(12), 2411–2422.

Cohen, A. L., (2003). Geochemical perspectives on coral mineralization. *Reviews in Mineralogy and Geochemistry*, *54*(1), 151–187.

Cohen, A. L., McCorkle, D. C., De Putron, S., Gaetani, G. A., & Rose, K. A., (2009). Morphological and compositional changes in the skeletons of new coral recruits reared in acidified seawater: Insights into the biomineralization response to ocean acidification. *Geochemistry, Geophysics, Geosystems*, *10*(7), 1-12.

Colombo-Pallotta, M. F., Rodríguez-Román, A., & Iglesias-Prieto, R., (2010). Calcification in bleached and unbleached *Montastraea faveolata*: Evaluating the role of oxygen and glycerol. *Coral Reefs*, *29*(4), 899–907.

Comeau, S., Carpenter, R. C., & Edmunds, P. J., (2013a). Coral reef calcifiers buffer their response to ocean acidification using both bicarbonate and carbonate. *Proceedings of the Royal Society B: Biological Sciences*, *280*(1753), 20122374.

Comeau, S., Edmunds, P. J., Spindel, N. B., & Carpenter, R. C., (2013b). The responses of eight coral reef calcifiers to increasing partial pressure of CO_2 do not exhibit a tipping point. *Limnology and Oceanography*, *58*(1), 388–398.

Comeau, S., Edmunds, P. J., Spindel, N. B., & Carpenter, R. C., (2014). Fast coral reef calcifiers are more sensitive to ocean acidification in short-term laboratory incubations. *Limnology and Oceanography*, *59*(3), 1081–1091.

Comeau, S., Tambutté, E., Carpenter, R. C., Edmunds, P. J., Evensen, N. R., Allemand, D., Ferrier-Pagès, C., Tambutté, S., & Venn, A. A., (2017). Coral calcifying fluid pH is modulated by seawater carbonate chemistry not solely seawater pH. *Proceedings of the Royal Society B: Biological Sciences*, *284*(1847), 20161669.

Constantz, B. R., (1986). Coral skeleton construction: A physiochemically dominated process. *Palaios*, *1*(2), 152-157.

Cunning, R., Bay, R. A., Gillette, P., Baker, A. C., & Traylor-Knowles, N., (2018). Comparative analysis of the *Pocillopora damicornis* genome highlights role of immune system in coral evolution. *Scientific Reports*, *8*(1), 16134.

Davy, S. K., Allemand, D., & Weis, V. M., (2012). Cell biology of cnidarian-dinoflagellate symbiosis. *Microbiology and Molecular Biology Reviews*, *76*(2), 229–261.

De Yoreo, J. J., Gilbert, P. U. P. A., Sommerdijk, N. A. J. M., Penn, R. L., Whitelam, S., Joester, D., Zhang, H., Rimer, J. D., Navrotsky, A., Banfield, J. F., Wallace, A. F., Michel, F. M., Meldrum, F. C., Cölfen, H., & Dove, P. M., (2015). Crystallization by particle attachment in synthetic, biogenic, and geologic environments. *Science*, *349*(6247), aaa6760.

De'ath, G., Lough, J. M., & Fabricius, K. E., (2009). Declining coral calcification on the Great Barrier Reef. *Science*, *323*(5910), 116–119.

DeCarlo, T. M., Comeau, S., Cornwall, C. E., & McCulloch, M. T., (2018b). Coral resistance to ocean acidification linked to increased calcium at the site of calcification. *Proceedings of the Royal Society B: Biological Sciences*, *285*(1878), 20180564.

DeCarlo, T. M., Holcomb, M., & McCulloch, M. T., (2018a). Reviews and syntheses: Revisiting the boron systematics of aragonite and their application to coral calcification. *Biogeosciences*, *15*(9), 2819–2834.

Decker, G. L., & Lennarz, W. J., (1988). Skeletogenesis in the sea urchin embryo. *Development*, *103*(2), 231–247.

Drake, J. L., Mass, T., & Falkowski, P. G., (2014). The evolution and future of carbonate precipitation in marine invertebrates: Witnessing extinction or documenting resilience in the Anthropocene? *Elementa: Science of the Anthropocene*, *2*, 000026.

Drake, J. L., Mass, T., Haramaty, L., Zelzion, E., Bhattacharya, D., & Falkowski, P. G., (2013). Proteomic analysis of skeletal organic matrix from the stony coral *Stylophora pistillata*. *Proceedings of the National Academy of Sciences*, *110*(10), 3788–3793.

Drake, J. L., Mass, T., Stolarski, J., Von Euw, S., Schootbrugge, B., & Falkowski, P. G., (2020). How corals made rocks through the ages. *Global Change Biology*, *26*(1), 31–53.

Dziedzic, K., (2019). Heritable variation in bleaching responses and its functional genomic basis in reef-building corals (*Orbicella faveolata*). *Molecular Ecology*, *28*(9), 2238–2253.

Edmunds, P. J., & Burgess, S. C., (2016). Size-dependent physiological responses of the branching coral *Pocillopora verrucosa* to elevated temperature and pCO2. *Journal of Experimental Biology*, 219, 3896-3906.

Erez, J., Braun, A., Ersoy, H., Ersoy, A. F., Da, S., & Berkün, M., (2007). Calcification in hermatypic corals is based on direct seawater supply to the biomineralization site. *Goldschmidt Conference Abstracts*, A260.

Ezzat, L., Fine, M., Maguer, J.-F., Grover, R., & Ferrier-Pagès, C., (2017). Carbon and nitrogen acquisition in shallow and deep holobionts of the scleractinian coral *S. pistillata*. *Frontiers in Marine Science*, *4*, 102.

Falini, G., Fermani, S., & Goffredo, S., (2015). Coral biomineralization: A focus on intra-skeletal organic matrix and calcification. *Seminars in Cell & Developmental Biology*, *46*, 17–26.

Falini, G., Reggi, M., Fermani, S., Sparla, F., Goffredo, S., Dubinsky, Z., Levi, O., Dauphin, Y., & Cuif, J.-P., (2013). Control of aragonite deposition in colonial corals by intra-skeletal macromolecules. *Journal of Structural Biology*, *183*(2), 226–238.

Fantazzini, P., Mengoli, S., Pasquini, L., Bortolotti, V., Brizi, L., Mariani, M., Di Giosia, M., Fermani, S., Capaccioni, B., Caroselli, E., Prada, F., Zaccanti, F., Levy, O., Dubinsky, Z., Kaandorp, J. A., Konglerd, P., Hammel, J. U., Dauphin, Y., Cuif, J.-P., … Goffredo, S., (2015). Gains and losses of coral skeletal porosity changes with ocean acidification acclimation. *Nature Communications*, *6*(1), 7785.

Fautin, D. G., & Mariscal, R. N., (1991). Cnidaria: Anthozoa. In *Microscopic anatomy of Invertebrates* (Harrison, F.W. and Westfall, J.A., Vol. 2, p. 267–358). Wiley-Liss.

Fukuda, I., Ooki, S., Fujita, T., Murayama, E., Nagasawa, H., Isa, Y., & Watanabe, T., (2003). Molecular cloning of a cDNA encoding a soluble protein in the coral exoskeleton. *Biochemical and Biophysical Research Communications*, *304*(1), 11–17.

Furla, P., Galgani, I., Durand, I., & Allemand, D., (2000). Sources and mechanisms of inorganic carbon transport for coral calcification and photosynthesis. *Journal of Experimental Biology*, *203*(22), 3445–3457.

Gagnon, A. C., Adkins, J. F., & Erez, J., (2012). Seawater transport during coral biomineralization. *Earth and Planetary Science Letters*, *329–330*, 150–161.

Gagnon, A. C., Adkins, J. F., Fernandez, D. P., & Robinson, L. F., (2007). Sr/Ca and Mg/Ca vital effects correlated with skeletal architecture in a scleractinian deep-sea coral and the role of Rayleigh fractionation. *Earth and Planetary Science Letters*, *261*(1–2), 280–295.

Gal, A., Kahil, K., Vidavsky, N., DeVol, R. T., Gilbert, P. U. P. A., Fratzl, P., Weiner, S., & Addadi, L., (2014). Particle Accretion Mechanism underlies biological crystal growth from an amorphous precursor Phase. *Advanced Functional Materials*, *24*(34), 5420–5426.

Galloway, S. B., Work, T. M., Bochsler, V. S., Harley, R. A., Kramarsky-Winters, E., Mc Laughlin, S. M., Meteyer, C. U., Morado, J. F., Nicholson, J. H., Parnell, P. G., Peters, E. C., Reynolds, T. L., Rotstein, D. S., Sileo, L., & Woodley, C. M., (2007). *Coral Disease and Health Workshop: Coral Histopathology II*. National Oceanic and Atmospheric Administration, Silver Spring, MD. 84pp.

Ganot, P., Tambutté, E., Caminiti-Segonds, N., Toullec, G., Allemand, D., & Tambutté, S., (2020). Ubiquitous macropinocytosis in anthozoans. *eLife*, *9*, e50022.

Ganot, P., Zoccola, D., Tambutté, E., Voolstra, C. R., Aranda, M., Allemand, D., & Tambutté, S., (2015). Structural molecular components of septate junctions in cnidarians point to the origin of epithelial junctions in eukaryotes. *Molecular Biology and Evolution*, *32*(1), 44–62.

Gattuso, J.-P., Allemand, D., & Frankignoulle, M., (1999). Photosynthesis and calcification at cellular, organismal and community levels in coral reefs: a review on interactions and control by carbonate chemistry. *American Zoologist*, *39*(1), 160–183.

Gilbert, P. U. P. A., Porter, S. M., Sun, C.-Y., Xiao, S., Gibson, B. M., Shenkar, N., & Knoll, A. H., (2019). Biomineralization by particle attachment in early animals. *Proceedings of the National Academy of Sciences*, *116*(36), 17659–17665.

Gilis, M., Meibom, A., Domart-Coulon, I., Grauby, O., Stolarski, J., & Baronnet, A., (2014). Biomineralization in newly settled recruits of the scleractinian coral *Pocillopora damicornis*. *Journal of Morphology*, *275*(12), 1349–1365.

Gleason, D. F., & Hofmann, D. K., (2011). Coral larvae: From gametes to recruits. *Journal of Experimental Marine Biology and Ecology*, *408*(1–2), 42–57.

Goreau, T. F., & Goreau, N. I., (1959). The physiology of skeleton formation in corals. II. Calcium deposition by hermatypic corals under various conditions in the reef. *The Biological Bulletin*, *117*(2), 239–250.

Guest, J. R., Dizon, R. M., Edwards, A. J., Franco, C., & Gomez, E. D., (2011). How quickly do fragments of coral "self-attach" after transplantation? *Restoration Ecology*, *19*(2), 234–242.

Guillermic, M., Cameron, L. P., De Corte, I., Misra, S., Bijma, J., de Beer, D., Reymond, C. E., Westphal, H., Ries, J. B., & Eagle, R. A., (2021). Thermal stress reduces pocilloporid coral resilience to ocean acidification by impairing control over calcifying fluid chemistry. *Science Advances*, *7*(2), eaba9958.

Guss, K. A., & Ettensohn, C. A., (1997). Skeletal morphogenesis in the sea urchin embryo: regulation of primary mesenchyme gene expression and skeletal rod growth by ectoderm-derived cues. *Development*, *124*(10), 1899–1908.

Heyward, A. J., & Negri, A. P., (1999). Natural inducers for coral larval metamorphosis. *Coral Reefs*, *18*(3), 273–279.

Highsmith, R., (1982). Reproduction by fragmentation in corals. *Marine Ecology Progress Series*, *7*, 207–226.

Hoegh-Guldberg, O., Pendleton, L., & Kaup, A., (2019). People and the changing nature of coral reefs. *Regional Studies in Marine Science*, *30*, 100699.

Hoegh-Guldberg, O., Poloczanska, E. S., Skirving, W., & Dove, S., (2017). Coral reef ecosystems under climate change and ocean acidification. *Frontiers in Marine Science*, *4*, 158.

Holcomb, M., Cohen, A. L., & McCorkle, D. C., (2013). An evaluation of staining techniques for marking daily growth in scleractinian corals. *Journal of Experimental Marine Biology and Ecology*, *440*, 126–131.

Holcomb, M., Venn, A. A., Tambutté, E., Tambutté, S., Allemand, D., Trotter, J., & McCulloch, M., (2014). Coral calcifying fluid pH dictates response to ocean acidification. *Scientific Reports*, *4*(1), 5207.

Ikmi, A., McKinney, S. A., Delventhal, K. M., & Gibson, M. C., (2014). TALEN and CRISPR/Cas9-mediated genome editing in the early-branching metazoan *Nematostella vectensis*. *Nature Communications*, *5*(1), 5486.

Isa, Y., (1986). An electron microscope study on the mineralization of the skeleton of the staghorn coral *Acropora hebes*. *Marine Biology*, *93*(1), 91–101.

Ittig, D., Luisier, S., Weiler, J., Schumperli, D., & Leumann, C., (2010). Improving gene silencing of siRNAs via tricyclo-DNA modification. *Artificial DNA: PNA & XNA*, *1*(1), 9–16.

Jakob, W., & Schierwater, B., (2007). Changing hydrozoan bauplans by silencing Hox-like genes. *PLoS One*, *2*(8), e694.

Johnston, I. S., (1980). The ultrastructure of skeletogenesis in hermatypic corals. *International Review of Cytology*, *67*, 171–214.

Karako-Lampert, S., Zoccola, D., Salmon-Divon, M., Katzenellenbogen, M., Tambutté, S., Bertucci, A., Hoegh-Guldberg, O., Deleury, E., Allemand, D., & Levy, O., (2014). Transcriptome analysis of the scleractinian coral *Stylophora pistillata*. *PLoS One*, *9*(2), e88615.

Keshavmurthy, S., Yang, S.-Y., Alamaru, A., Chuang, Y.-Y., Pichon, M., Obura, D., Fontana, S., De Palmas, S., Stefani, F., Benzoni, F., MacDonald, A., Noreen, A. M. E., Chen, C., Wallace, C. C., Pillay, R. M., Denis, V., Amri, A. Y., Reimer, J. D., Mezaki, T., … Chen, C. A., (2013). DNA barcoding reveals the coral "laboratory-rat", *Stylophora pistillata* encompasses multiple identities. *Scientific Reports*, *3*(1), 1520.

Kitchen, S. A., Crowder, C. M., Poole, A. Z., Weis, V. M., & Meyer, E., (2015). *De novo* assembly and characterization of four anthozoan (Phylum Cnidaria) transcriptomes. *G3 (Bethesda)*. *5*(11), 2441–2452.

Kraynack, B. A., & Baker, B. F., (2006). Small interfering RNAs containing full 2′- *O* -methylribonucleotide-modified sense strands display Argonaute2/eIF2C2-dependent activity. *RNA*, *12*(1), 163–176.

Krief, S., Hendy, E. J., Fine, M., Yam, R., Meibom, A., Foster, G. L., & Shemesh, A., (2010). Physiological and isotopic responses of scleractinian corals to ocean acidification. *Geochimica et Cosmochimica Acta*, *74*(17), 4988–5001.

Krueger, T., Horwitz, N., Bodin, J., Giovani, M.-E., Escrig, S., Meibom, A., & Fine, M., (2017). Common reef-building coral in the Northern Red Sea resistant to elevated temperature and acidification. *Royal Society Open Science*, *4*(5), 170038.

Krug, S. M., Amasheh, S., Richter, J. F., Milatz, S., Günzel, D., Westphal, J. K., Huber, O., Schulzke, J. D., & Fromm, M., (2009). Tricellulin forms a barrier to macromolecules in tricellular tight junctions without affecting ion permeability. *Molecular Biology of the Cell*, *20*(16), 3713–3724.

LaJeunesse, T. C., Parkinson, J. E., Gabrielson, P. W., Jeong, H. J., Reimer, J. D., Voolstra, C. R., & Santos, S. R., (2018). Systematic revision of symbiodiniaceae highlights the antiquity and diversity of coral endosymbionts. *Current Biology*, *28*(16), 2570-2580.e6.

Le Tissier, M. D. A., (1990). The ultrastructure of the skeleton and skeletogenic tissues of the temperate coral *Caryophyllia smithii*. *Journal of the Marine Biological Association of the United Kingdom*, *70*(2), 295–310.

Levy, S., Elek, A., Grau-Bové, X., Menéndez-Bravo, S., Iglesias, M., Tanay, A., Mass, T., & Sebé-Pedrós, A., (2021). A stony coral cell atlas illuminates the molecular and cellular basis of coral symbiosis, calcification, and immunity. *Cell*, *184*(11), 2973-2987.e18.

Li, M., Liu, H., Guo, Y., Chen, F., Zi, X., Fan, R., Li, H., Cai, Y., He, C., Lu, Z., & Zhao, X., (2020). Single symbiotic cell transcriptome sequencing of coral. *Genomics*, *112*(6), 5305–5312.

Liew, Y. J., Zoccola, D., Li, Y., Tambutté, E., Venn, A. A., Michell, C. T., Cui, G., Deutekom, E. S., Kaandorp, J. A., Voolstra, C. R., Forêt, S., Allemand, D., Tambutté, S., & Aranda, M., (2018). Epigenome-associated phenotypic acclimatization to ocean acidification in a reef-building coral. *Science advances (*4), eaar8028.

Lirman, D., & Schopmeyer, S., (2016). Ecological solutions to reef degradation: optimizing coral reef restoration in the Caribbean and Western Atlantic. *PeerJ*, *4*, e2597.

Lommel, M., Tursch, A., Rustarazo-Calvo, L., Trageser, B., & Holstein, T. W., (2017). Genetic knockdown and knockout approaches in *Hydra*. *BioRxiv*.

Loya, Y., (1976). The Red Sea coral *Stylophora pistillata* is an r strategist. *Nature*, *259*(5543), 478–480.

Luecken, M. D., & Theis, F. J., (2019). Current best practices in single-cell RNA-seq analysis: A tutorial. *Molecular Systems Biology*, *15*(6), e8746.

Marchitto, T. M., Bryan, S. P., Doss, W., McCulloch, M. T., & Montagna, P., (2018). A simple biomineralization model to explain Li, Mg, and Sr incorporation into aragonitic foraminifera and corals. *Earth and Planetary Science Letters*, *481*, 20–29.

Marubini, F., Ferrier-Pagès, C., Furla, P., & Allemand, D., (2008). Coral calcification responds to seawater acidification: A working hypothesis towards a physiological mechanism. *Coral Reefs*, *27*(3), 491–499.

Mass, T., Drake, J. L., Haramaty, L., Kim, J. D., Zelzion, E., Bhattacharya, D., & Falkowski, P. G., (2013). Cloning and characterization of four novel coral acid-rich proteins that precipitate carbonates *in vitro*. *Current Biology*, *23*(12), 1126–1131.

Mass, T., Drake, J. L., Haramaty, L., Rosenthal, Y., Schofield, O. M. E., Sherrell, R. M., & Falkowski, P. G., (2012). Aragonite precipitation by "proto-polyps" in coral cell cultures. *PLoS One*, *7*(4), e35049.

Mass, T., Drake, J. L., Heddleston, J. M., & Falkowski, P. G., (2017a). Nanoscale visualization of biomineral formation in coral proto-polyps. *Current Biology*, *27*(20), 3191-3196.e3.

Mass, T., Giuffre, A. J., Sun, C.-Y., Stifler, C. A., Frazier, M. J., Neder, M., Tamura, N., Stan, C. V., Marcus, M. A., & Gilbert, P. U. P. A., (2017b). Amorphous calcium carbonate particles form coral skeletons. *Proceedings of the National Academy of Sciences*, *114*(37).

Matranga, V., Pinsino, A., Bonaventura, R., Costa, C., Karakostis, K., Martino, C., Russo, R., & Zito, F., (2013). Cellular and molecular bases of biomineralization in sea urchin embryos. *Cah. Biol. Mar.*, *54*, 467–478.

McCulloch, M. T., D'Olivo, J. P., Falter, J., Holcomb, M., & Trotter, J. A., (2017). Coral calcification in a changing orld and the interactive dynamics of pH and DIC upregulation. *Nature Communications*, *8*(1), 15686.

McCulloch, M., Falter, J., Trotter, J., & Montagna, P., (2012a). Coral resilience to ocean acidification and global warming through pH up-regulation. *Nature Climate Change*, *2*(8), 623–627.

McCulloch, M., Trotter, J., Montagna, P., Falter, J., Dunbar, R., Freiwald, A., Försterra, G., López Correa, M., Maier, C., Rüggeberg, A., & Taviani, M., (2012b). Resilience of cold-water scleractinian corals to ocean acidification: Boron isotopic systematics of pH and saturation state up-regulation. *Geochimica et Cosmochimica Acta*, *87*, 21–34.

Meibom, A., Yurimoto, H., Cuif, J., Domart-Coulon, I., Houlbreque, F., Constantz, B., Dauphin, Y., Tambutté, E., Tambutté, S., Allemand, D., Wooden, J., & Dunbar, R.,

(2006). Vital effects in coral skeletal composition display strict three-dimensional control. *Geophysical Research Letters*, *33*(11), 2006GL025968.

Mieog, J. C., Van Oppen, M. J. H., Cantin, N. E., Stam, W. T., & Olsen, J. L., (2007). Real-time PCR reveals a high incidence of *Symbiodinium* clade D at low levels in four scleractinian corals across the Great Barrier Reef: implications for symbiont shuffling. *Coral Reefs*, *26*(3), 449–457.

Miljkovic, M., Mazet, F., & Galliot, B., (2002). Cnidarian and bilaterian promoters can direct GFP expression in transfected *Hydra*. *Developmental Biology*, *246*(2), 377–390.

Muscatine, L., Goiran, C., Land, L., Jaubert, J., Cuif, J.-P., & Allemand, D., (2005). Stable isotopes (δ^{13}C and δ^{15}N) of organic matrix from coral skeleton. *Proceedings of the National Academy of Sciences*, *102*(5), 1525–1530.

Muscatine, L., Tambutte, E., & Allemand, D., (1997). Morphology of coral desmocytes, cells that anchor the calicoblastic epithelium to the skeleton. *Coral Reefs*, *16*(4), 205–213.

Neder, M., Laissue, P. P., Akiva, A., Akkaynak, D., Albéric, M., Spaeker, O., Politi, Y., Pinkas, I., & Mass, T., (2019). Mineral formation in the primary polyps of pocilloporoid corals. *Acta biomaterialia*, *96*, 631–645.

Njegić Džakula, B., Fermani, S., Dubinsky, Z., Goffredo, S., Falini, G., & Kralj, D., (2019). *In vitro* coral biomineralization under relevant aragonite supersaturation conditions. *Chemistry – A European Journal*, *25*(45), 10616–10624.

Nowotny, J. D., Connelly, M. T., & Traylor-Knowles, N., (2021). Novel methods to establish whole-body primary cell cultures for the cnidarians *Nematostella vectensis* and *Pocillopora damicornis*. *Scientific Reports*, *11*(1), 4086.

Ocampo, I. D., Zárate-Potes, A., Pizarro, V., Rojas, C. A., Vera, N. E., & Cadavid, L. F., (2015). The immunotranscriptome of the Caribbean reef-building coral *Pseudodiploria strigosa*. *Immunogenetics*, *67*(9), 515–530.

Ohno, Y., Iguchi, A., Shinzato, C., Gushi, M., Inoue, M., Suzuki, A., Sakai, K., & Nakamura, T., (2017a). Calcification process dynamics in coral primary polyps as observed using a calcein incubation method. *Biochemistry and Biophysics Reports*, *9*, 289–294.

Ohno, Y., Iguchi, A., Shinzato, C., Inoue, M., Suzuki, A., Sakai, K., & Nakamura, T., (2017b). An aposymbiotic primary coral polyp counteracts acidification by active pH regulation. *Scientific Reports*, *7*(1), 40324.

Okamoto, C. T., (2017). Regulation of transporters and channels by membrane-trafficking complexes in epithelial cells. *Cold Spring Harbor Perspectives in Biology*, *9*(11), a027839.

Pandolfi, J. M., Connolly, S. R., Marshall, D. J., & Cohen, A. L., (2011). Projecting coral reef futures under global warming and ocean acidification. *Science*, *333*(6041), 418–422.

Park, E., Hwang, D.-S., Lee, J.-S., Song, J.-I., Seo, T.-K., & Won, Y.-J., (2012). Estimation of divergence times in cnidarian evolution based on mitochondrial protein-coding genes and the fossil record. *Molecular Phylogenetics and Evolution*, *62*(1), 329–345.

Peled, Y., Drake, J. L., Malik, A., Almuly, R., Lalzar, M., Morgenstern, D., & Mass, T., (2020). Optimization of skeletal protein preparation for LC–MS/MS sequencing yields additional coral skeletal proteins in *Stylophora pistillata*. *BMC Materials*, *2*(1), 8.

Putnam, H. M., Barott, K. L., Ainsworth, T. D., & Gates, R. D., (2017). The vulnerability and resilience of reef-building corals. *Current Biology*, *27*(11), R528–R540.

Putnam, N. H., Srivastava, M., Hellsten, U., Dirks, B., Chapman, J., Salamov, A., Terry, A., Shapiro, H., Lindquist, E., Kapitonov, V. V., Jurka, J., Genikhovich, G., Grigoriev, I. V., Lucas, S. M., Steele, R. E., Finnerty, J. R., Technau, U., Martindale, M. Q., & Rokhsar, D.

S., (2007). Sea anemone genome reveals ancestral eumetazoan gene repertoire and genomic organization. *Science*, *317*(5834), 86–94.

Quiroga-Artigas, G., Duscher, A., Lundquist, K., Waletich, J., & Schnitzler, C. E., (2020). Gene knockdown via electroporation of short hairpin RNAs in embryos of the marine hydroid *Hydractinia symbiolongicarpus*. *Scientific Reports*, *10*(1), 12806.

Rahman, M. A., & Oomori, T., (2009). *In vitro* regulation of $CaCO_3$ crystal growth by the highly acidic proteins of calcitic sclerites in soft coral, *Sinularia Polydactyla*. *Connective Tissue Research*, *50*(5), 285–293.

Ramos-Silva, P., Kaandorp, J., Herbst, F., Plasseraud, L., Alcaraz, G., Stern, C., Corneillat, M., Guichard, N., Durlet, C., Luquet, G., & Marin, F., (2014). The skeleton of the staghorn coral *Acropora millepora*: molecular and structural characterization. *PLoS One*, *9*(6), e97454.

Raz-Bahat, M., Erez, J., & Rinkevich, B., (2006). *In vivo* light-microscopic documentation for primary calcification processes in the hermatypic coral *Stylophora pistillata*. *Cell and Tissue Research*, *325*(2), 361–368.

Reaka-Kudla, M. L., (2005). Biodiversity of Caribbean coral reefs. In *Caribbean Marine Biodiversity* (Miloslavich, P. and Klein, E., (eds), p. 259–276). Des Tech Publishers, Lancaster, PA.

ReFuGe 2020 Consortium., (2015). The ReFuGe 2020 Consortium—using "omics" approaches to explore the adaptability and resilience of coral holobionts to environmental change. *Frontiers in Marine Science*, *2*.

Renneberg, D., (2002). Antisense properties of tricyclo-DNA. *Nucleic Acids Research*, *30*(13), 2751–2757.

Reynaud-Vaganay, S., Gattuso, J.-P., Jaubert, J., & Juillet-Leclerc, A., (1999). A novel culture technique for scleractinian corals: Application to investigate changes in skeletal $\delta^{18}O$ as a function of temperature. *Marine Ecology Progress Series*, *180*, 121–130.

Reynaud, S., Hemming, N. G., Juillet-Leclerc, A., & Gattuso, J.-P., (2004). Effect of pCO_2 and temperature on the boron isotopic composition of the zooxanthellate coral Acropora sp. *Coral Reef*, *23*, 539-546.

Ries, J. B., (2011). A physicochemical framework for interpreting the biological calcification response to CO_2-induced ocean acidification. *Geochimica et Cosmochimica Acta*, *75*(14), 4053–4064.

Riley, M., & Vermerris, W., (2017). Recent advances in nanomaterials for gene delivery—A review. *Nanomaterials*, *7*(5), 94.

Rinkevich, B., (2011). Cell cultures from marine invertebrates: New insights for capturing endless stemness. *Marine Biotechnology*, *13*(3), 345–354.

Rinkevich, B., & Loya, Y., (1979). The reproduction of the Red Sea coral *Stylophora pistillata*. I. Gonads and Planulae. *Marine Ecology Progress Series*, *1*, 133–144.

Rinkevich, B., & Loya, Y., (1987). Variability in the pattern of sexual reproduction of the coral *Stylophora pistillata* at Eilat, Red Sea: A long term study. *The Biological Bulletin*, *173*(2), 335–344.

Rocker, M. M., Noonan, S., Humphrey, C., Moya, A., Willis, B. L., & Bay, L. K., (2015). Expression of calcification and metabolism-related genes in response to elevated pCO_2 and temperature in the reef-building coral *Acropora millepora*. *Marine Genomics*, *24*, 313–318.

Roff, G., Dove, S. G., & Dunn, S. R., (2009). Mesenterial filaments make a clean sweep of substrates for coral growth. *Coral Reefs*, *28*(1), 79–79.

Romano, S. L., & Palumbi, S. R., (1996). Evolution of scleractinian corals inferred from molecular systematics. *Science*, *271*(5249), 640–642.

Saenger, C., & Erez, J., (2016). A non-traditional stable isotope perspective on coral calcification. In S. Goffredo & Z. Dubinsky (Eds.), *The Cnidaria, Past, Present and Future* (p. 181–205). Springer International Publishing.

Saliba, A.-E., Westermann, A. J., Gorski, S. A., & Vogel, J., (2014). Single-cell RNA-seq: advances and future challenges. *Nucleic Acids Research*, *42*(14), 8845–8860.

Sawall, Y., Teichberg, M. C., Seemann, J., Litaay, M., Jompa, J., & Richter, C., (2011). Nutritional status and metabolism of the coral *Stylophora subseriata* along a eutrophication gradient in Spermonde Archipelago (Indonesia). *Coral Reefs*, *30*(3), 841–853.

Schoepf, V., Grottoli, A. G., Warner, M. E., Cai, W.-J., Melman, T. F., Hoadley, K. D., Pettay, D. T., Hu, X., Li, Q., Xu, H., Wang, Y., Matsui, Y., & Baumann, J. H., (2013). Coral energy reserves and calcification in a high-CO_2 world at two temperatures. *PLoS One*, *8*(10), e75049.

Schoepf, V., Jury, C. P., Toonen, R. J., & McCulloch, M. T., (2017). Coral calcification mechanisms facilitate adaptive responses to ocean acidification. *Proceedings of the Royal Society B: Biological Sciences*, *284*(1868), 20172117.

Schweinsberg, M., Weiss, L. C., Striewski, S., Tollrian, R., & Lampert, K. P., (2015). More than one genotype: How common is intracolonial genetic variability in scleractinian corals? *Molecular Ecology*, *24*(11), 2673–2685.

Scucchia, F., Nativ, H., Neder, M., Goodbody-Gringley, G., & Mass, T., (2020). Physiological characteristics of *stylophora pistillata* larvae across a depth gradient. *Frontiers in Marine Science*, *7*, 13.

Sevilgen, D. S., Venn, A. A., Hu, M. Y., Tambutté, E., de Beer, D., Planas-Bielsa, V., & Tambutté, S., (2019). Full *in vivo* characterization of carbonate chemistry at the site of calcification in corals. *Science Advances*, *5*(1), eaau7447.

Shafir, S., (2001). Nubbing of coral colonies: A novel approach for the development of inland broodstocks. *Aquarium Sciences and Conservation*, *3*. 183-190.

Shapiro, O. H., Kramarsky-Winter, E., Gavish, A. R., Stocker, R., & Vardi, A., (2016). A coral-on-a-chip microfluidic platform enabling live-imaging microscopy of reef-building corals. *Nature Communications*, *7*(1), 10860.

Shinzato, C., Shoguchi, E., Kawashima, T., Hamada, M., Hisata, K., Tanaka, M., Fujie, M., Fujiwara, M., Koyanagi, R., Ikuta, T., Fujiyama, A., Miller, D. J., & Satoh, N., (2011). Using the *Acropora digitifera* genome to understand coral responses to environmental change. *Nature*, *476*(7360), 320–323.

Shlesinger, Y., & Loya, Y., (1991). Larval development and survivorship in the corals *Favia favus* and *Platygyra lamellina*. *Hydrobiologia*, *216–217*(1), 101–108.

Simpson, C., Kiessling, W., Mewis, H., Baron-Szabo, R. C., & Müller, J., (2011). Evolutionary diversification of reef corals: a comparison of the molecular and fossil records. *Evolution*, *65*(11), 3274–3284.

Sinigaglia, C., Busengdal, H., Leclère, L., Technau, U., & Rentzsch, F., (2013). The bilaterian head patterning gene six3/6 controls aboral domain development in a cnidarian. *PLoS Biology*, *11*(2), e1001488.

Spalding, M. D., Ravilious, C., & Green, E. P., (2001). *World Atlas of Coral Reefs* (p 424) (University of California Press).

Stefani, F., Benzoni, F., Yang, S.-Y., Pichon, M., Galli, P., & Chen, C. A., (2011). Comparison of morphological and genetic analyses reveals cryptic divergence and morphological plasticity in *Stylophora* (*Cnidaria, Scleractinia*). *Coral Reefs*, *30*(4), 1033–1049.

Sun, C.-Y., Marcus, M. A., Frazier, M. J., Giuffre, A. J., Mass, T., & Gilbert, P. U. P. A., (2017). Spherulitic growth of coral skeletons and synthetic aragonite: nature's three-dimensional printing. *ACS Nano*, *11*(7), 6612–6622.

Sun, C.-Y., Stifler, C. A., Chopdekar, R. V., Schmidt, C. A., Parida, G., Schoeppler, V., Fordyce, B. I., Brau, J. H., Mass, T., Tambutté, S., & Gilbert, P. U. P. A., (2020). From particle attachment to space-filling coral skeletons. *Proceedings of the National Academy of Sciences*, *117*(48), 30159–30170.

Sun, Z., & Ettensohn, C. A., (2017). TGF-β *sensu stricto* signaling regulates skeletal morphogenesis in the sea urchin embryo. *Developmental Biology*, *421*(2), 149–160.

Tambutté, E., (1996). Processus de calcification d'un scleractinaire hermatypique *Stylophora pistillata (Esper, 1797)*. Etude de la croissance *in situ* sur l'atoll de Mururoa. Thèse. Université de Nice Sophia-Antipolis – UFR faculté des sciences

Tambutté, E., Allemand, D., Bourge, I., Gattuso, J.-P., & Jaubert, J., (1995). An improved ^{45}Ca protocol for investigating physiological mechanisms in coral calcification. *Marine Biology*, *122*(3), 453–459.

Tambutté, É., Allemand, D., Mueller, E., & Jaubert, J., (1996). A compartmental approach to the mechanism of calcification in hermatypic corals. *Journal of Experimental Biology*, *199*(5), 1029–1041.

Tambutté, E., Allemand, D., Zoccola, D., Meibom, A., Lotto, S., Caminiti, N., & Tambutté, S., (2007a). Observations of the tissue-skeleton interface in the scleractinian coral *Stylophora pistillata*. *Coral Reefs*, *26*(3), 517–529.

Tambutté, E., Ganot, P., Venn, A. A., & Tambutté, S., (2021). A role for primary cilia in coral calcification? *Cell and Tissue Research*, *383*(3), 1093–1102.

Tambutté, É., Tambutte, S., Segonds, N., Zoccola, D., Erez, J., & Allemand, D., (2012). Calcein labelling and electrophysiology: insights on coral tissue permeability and calcification. *Proceedings of the Royal Society B: Biological Sciences*, *279*, 19–27.

Tambutté, E., Venn, A. A., Holcomb, M., Segonds, N., Techer, N., Zoccola, D., Allemand, D., & Tambutté, S., (2015). Morphological plasticity of the coral skeleton under CO_2-driven seawater acidification. *Nature Communications*, *6*(1), 7368.

Tambutté, S., Holcomb, M., Ferrier-Pagès, C., Reynaud, S., Tambutté, É., Zoccola, D., & Allemand, D., (2011). Coral biomineralization: from the gene to the environment. *Journal of Experimental Marine Biology and Ecology*, *408*(1–2), 58–78.

Tambutté, S., Tambutté, E., Zoccola, D., & Allemand, D., (2007c). Organic matrix and biomineralization of scleractinian corals. In E. Buerlein (Éd.), *Handbook of Biomineralization* (p. 243–259). Wiley-VCH Verlag GmbH.

Tambutté, S., Tambutté, E., Zoccola, D., Caminiti, N., Lotto, S., Moya, A., Allemand, D., & Adkins, J., (2007b). Characterization and role of carbonic anhydrase in the calcification process of the azooxanthellate coral *Tubastrea aurea*. *Marine Biology*, *151*(1), 71–83.

Taubner, I., Böhm, F., Eisenhauer, A., Tambutté, E., Tambutté, S., Moldzio, S., & Bleich, M., (2017). An improved approach investigating epithelial ion transport in scleractinian corals: Coral epithelial transport measurement. *Limnology and Oceanography: Methods*, *15*(9), 753–765.

Tebben, J., Motti, C. A., Siboni, N., Tapiolas, D. M., Negri, A. P., Schupp, P. J., Kitamura, M., Hatta, M., Steinberg, P. D., & Harder, T., (2015). Chemical mediation of coral larval settlement by crustose coralline algae. *Scientific Reports*, *5*(1), 10803.

Tresguerres, M., Barott, K. L., Barron, M. E., Deheyn, D. D., Kline, D. I., & Linsmayer, L. B., (2017). Cell biology of reef-building corals: ion transport, acid/base regulation, and energy

metabolism. In D. Weihrauch & M. O'Donnell (Éds.), *Acid-Base Balance and Nitrogen Excretion in Invertebrates* (p. 193–218). Springer International Publishing.
Trotter, J., Montagna, P., McCulloch, M., Silenzi, S., Reynaud, S., Mortimer, G., Martin, S., Ferrier-Pagès, C., Gattuso, J.-P., & Rodolfo-Metalpa, R., (2011). Quantifying the pH 'vital effect' in the temperate zooxanthellate coral *Cladocora caespitosa*: Validation of the boron seawater pH proxy. *Earth and Planetary Science Letters*, *303*(3–4), 163–173.
van Oppen, M. J. H., Oliver, J. K., Putnam, H. M., & Gates, R. D., (2015). Building coral reef resilience through assisted evolution. *Proceedings of the National Academy of Sciences*, *112*(8), 2307–2313.
Venn, A. A., Bernardet, C., Chabenat, A., Tambutté, E., & Tambutté, S., (2020). Paracellular transport to the coral calcifying medium: effects of environmental parameters. *Journal of Experimental Biology*, jeb.227074.
Venn, A. A., Tambutté, E., & Tambutté, S., (2015). Plasticity of coral physiology under ocean acidification. *Oncotarget*, *6*(21), 18248–18249.
Venn, A. A., Tambutté, E., Caminiti-Segonds, N., Techer, N., Allemand, D., & Tambutté, S., (2019). Effects of light and darkness on pH regulation in three coral species exposed to seawater acidification. *Scientific Reports*, *9*(1), 2201.
Venn, A. A., Tambutté, E., Holcomb, M., Laurent, J., Allemand, D., & Tambutté, S., (2013). Impact of seawater acidification on pH at the tissue–skeleton interface and calcification in reef corals. *Proceedings of the National Academy of Sciences*, *110*(5), 1634–1639.
Venn, A., Tambutté, E., Holcomb, M., Allemand, D., & Tambutté, S., (2011). Live tissue imaging shows reef corals elevate ph under their calcifying tissue relative to seawater. *PLoS One*, *6*(5), e20013.
Veron, J. E. N., (1993). *Corals of Australia and the Indo-Pacific*. University of Hawaii Press, Honolulu.
Voolstra, C. R., Li, Y., Liew, Y. J., Baumgarten, S., Zoccola, D., Flot, J.-F., Tambutté, S., Allemand, D., & Aranda, M., (2017). Comparative analysis of the genomes of *Stylophora pistillata* and *Acropora digitifera* provides evidence for extensive differences between species of corals. *Scientific Reports*, *7*(1), 17583.
Wainwright, S. A., (1963). Skeletal organization in the coral, *Pocillopora damicornis*. *Quarterly Journal Of Microscopical Science*, *104*, 169–183.
Wall, M., Fietzke, J., Schmidt, G. M., Fink, A., Hofmann, L. C., De Beer, D., & Fabricius, K. E., (2016). Internal pH regulation facilitates *in situ* long-term acclimation of massive corals to end-of-century carbon dioxide conditions. *Scientific Reports*, *6*(1), 30688.
Weis, V. M., & Allemand, D., (2009). What Determines Coral Health? *Science*, *324*(5931), 1153–1155.
Weis, V. M., Davy, S. K., Hoegh-Guldberg, O., Rodriguez-Lanetty, M., & Pringle, J. R., (2008). Cell biology in model systems as the key to understanding corals. *Trends in Ecology & Evolution*, *23*(7), 369–376.
Whalan, S., Abdul Wahab, M. A., Sprungala, S., Poole, A. J., & de Nys, R., (2015). Larval settlement: The role of surface topography for sessile coral reef invertebrates. *PLoS One*, *10*(2), e0117675.
Wilkinson, C., (2008). *Status of Coral Reefs of the World: 2008* (p.296) Global Coral Reef Monitoring Network and Reef and Rainforest Research Centre, Townsville, Australia.
Wilt, F. H., (2002). Biomineralization of the spicules of sea urchin embryos. *Zoological Science*, *19*(3), 253–261.

Yasuoka, Y., Shinzato, C., & Satoh, N., (2016). The mesoderm-forming gene brachyury regulates ectoderm-endoderm demarcation in the coral *Acropora digitifera*. *Current Biology*, *26*(21), 2885–2892.

Ying, H., Hayward, D. C., Cooke, I., Wang, W., Moya, A., Siemering, K. R., Sprungala, S., Ball, E. E., Forêt, S., & Miller, D. J., (2019). The whole-genome sequence of the coral *Acropora millepora*. *Genome Biology and Evolution*, *11*(5), 1374–1379.

Young, S. D., (1973). Collagen and other mesoglea protein from the coral *Lobophyllia corymbosa* (Anthozoa, Scleractinia). *International Journal of Biochemistry*. *4* (22). 339-344.

Zarnitsyn, V., Rostad, C. A., & Prausnitz, M. R., (2008). Modeling transmembrane transport through cell membrane wounds created by acoustic cavitation. *Biophysical Journal*, *95*(9), 4124–4138.

Zhang, Y., Chen, Q., Xie, J. Y., Yeung, Y. H., Xiao, B., Liao, B., Xu, J., & Qiu, J.-W., (2019). Development of a transcriptomic database for 14 species of scleractinian corals. *BMC Genomics*, *20*(1), 387.

Zoccola, D., Ganot, P., Bertucci, A., Caminiti-Segonds, N., Techer, N., Voolstra, C. R., Aranda, M., Tambutté, E., Allemand, D., Casey, J. R., & Tambutté, S., (2015). Bicarbonate transporters in corals point towards a key step in the evolution of cnidarian calcification. *Scientific Reports*, *5*(1), 9983.

Zoccola, D., Tambutté, E., Kulhanek, E., Puverel, S., Scimeca, J.-C., Allemand, D., & Tambutté, S., (2004). Molecular cloning and localization of a PMCA P-type calcium ATPase from the coral *Stylophora pistillata*. *Biochimica et Biophysica Acta (BBA) - Biomembranes*, *1663*(1–2), 117–126.

Zoccola, D., Tambutté, E., Sénégas-Balas, F., Michiels, J.-F., Failla, J.-P., Jaubert, J., & Allemand, D., (1999). Cloning of a calcium channel alpha1 subunit from the reef-building coral. *Gene*, *227*, 157–167.

CHAPTER 8

The Placozoa: General Biology, Genomics, Cell Signaling, and Behavior

WASSIM ELKHATIB, JULIA GAUBERG, ANHADVIR SINGH, and ADRIANO SENATORE

Department of Biology, University of Toronto Mississauga, Mississauga, Canada

8.1 GENERAL FEATURES OF THE PLACOZOA

8.1.1 INTRODUCTION

Placozoa is an early-diverging phylum of motile seawater invertebrates with arguably the simplest animal body plan. They are flat, ameboid-like animals that lack body symmetry and are composed of a limited number of cell types that are arranged along ciliated dorsal and ventral epithelia, sandwiching a network of mesenchyme-like cells (Grell & Benwitz, 1971; Grell, 1972; Schulze, 1883, 1892). Placozoans lack a dorsal/ventral body axis, so the commonly used terms dorsal and ventral epithelia should be considered purely descriptive, and alternatively referred to as upper and lower epithelia, respectively.

Placozoans are small with a height of only ~20 μm (i.e., along the dorsal-ventral axis), and an irregularly shaped body with a diameter that ranges between 2 and 10 mm (Osigus et al., 2019) (Figure 8.1(A)). All placozoans described to date, including the most well-studied species *Trichoplax adhaerens*, lack bona fide nerve and muscle cells, which distinguishes them from animals with bilateral, radial, and bi-radial body symmetry that all possess neuromuscular systems (Grell et al., 1991; Grell & Benwitz, 1981). Specimens have been observed to occupy diverse aquatic habitats such as tropical and subtropical marine environments, coral reefs, and mangroves, with salinities that range between 20 and 55 ppt (Eitel et al., 2013; Eitel & Schierwater,

Frontiers in Invertebrate Physiology: A Collection of Reviews, Volume 1: Non-Bilaterian Phyla.
Saber Saleuddin, Sally P. Leys, Robert D. Roer, and Iain C. Wilkie, (Eds.)

2010; Grell, 1980; Pearse & Voigt, 2007). Whether placozoans also inhabit deep-sea benthic zones or freshwater environments remains unknown.

Placozoans are easy to maintain in the laboratory, where they are most often fed microalgae that they consume via secretion of digestive enzymes along the ventral/lower epithelium followed by pinocytosis (Grell et al., 1991; Novotný et al., 2017; Pearse & Voigt, 2007; Smith et al., 2015). For this reason, their lower/ventral epithelium is sometimes referred to as the "nutritive" epithelium, while the upper/dorsal epithelium is referred to as "protective," bearing "shiny spheres" that are thought to contain noxious substances that deter predation (Eitel et al., 2013; Grell & Benwitz, 1971; Jackson & Buss, 2009; Ruthmann et al., 1986). Placozoans are most often observed to reproduce asexually via binary fission, or occasionally through a poorly understood budding process that produces small flagellated "swarmers," presumably for the purpose of dispersal (Eitel et al., 2011; Kamm et al., 2018; Pearse, 1989; Schwartz, 1984; Signorovitch et al., 2005; Thiemann & Ruthmann, 1988, 1990). Nevertheless, some reports have confirmed that sexual reproduction can also take place and embryos developing up to a 128-cell stage have been reported (Eitel et al., 2011; Grell, 1971; Grell & Benwitz, 1974; Pearse, 1989; Ruthmann et al., 1981). In addition, allelic variation has been observed among wild-caught *Trichoplax* specimens consistent with the occurrence of meiotic recombination and sexual reproduction (Eitel et al., 2018; Kamm et al., 2018; Signorovitch et al., 2005).

A particularly interesting feature of placozoans is that despite their lacking a nervous system and synapses, they can conduct motile-directed behavior including gravitaxis, phototaxis, chemotaxis, and feeding (Heyland et al., 2014; Mayorova et al., 2018; Pearse, 1989; Romanova et al., 2020a; Srivastava et al., 2008). Feeding behavior is particularly complex in that it involves stereotyped cellular processes starting with directed ciliary locomotion towards food, pausing of ventral ciliary beating once the food is detected, secretion of hydrolytic enzymes from digestive cells in the vicinity of the food particles, followed by a churning motion near the site of digestion, possibly for enhancing nutrient uptake (Fortunato & Aktipis, 2019; Ruthmann et al., 1986; Smith et al., 2014, 2015; Ueda et al., 1999). Directed locomotion is achieved through asynchronously beating cilia that project from ventral mono-ciliated epithelial cells towards the substrate (Pearse, 1989; Rassat & Ruthmann, 1979; Syed & Schierwater, 2002a). With every stroke, the cilia collectively propel the animal along winding trajectories that become more directional when the animal senses environmental signals (Romanova et al., 2020a; Smith et al., 2019a; Ueda et al., 1999). Given the absence of electrical and chemical synapses, questions emerge as to how placozoan cell types can coordinate with each other for physiology and behavior, including the mechanisms for achieving directed ciliary locomotion (Bull et al., 2021; Mayorova et al., 2018; Smith et al., 2019a).

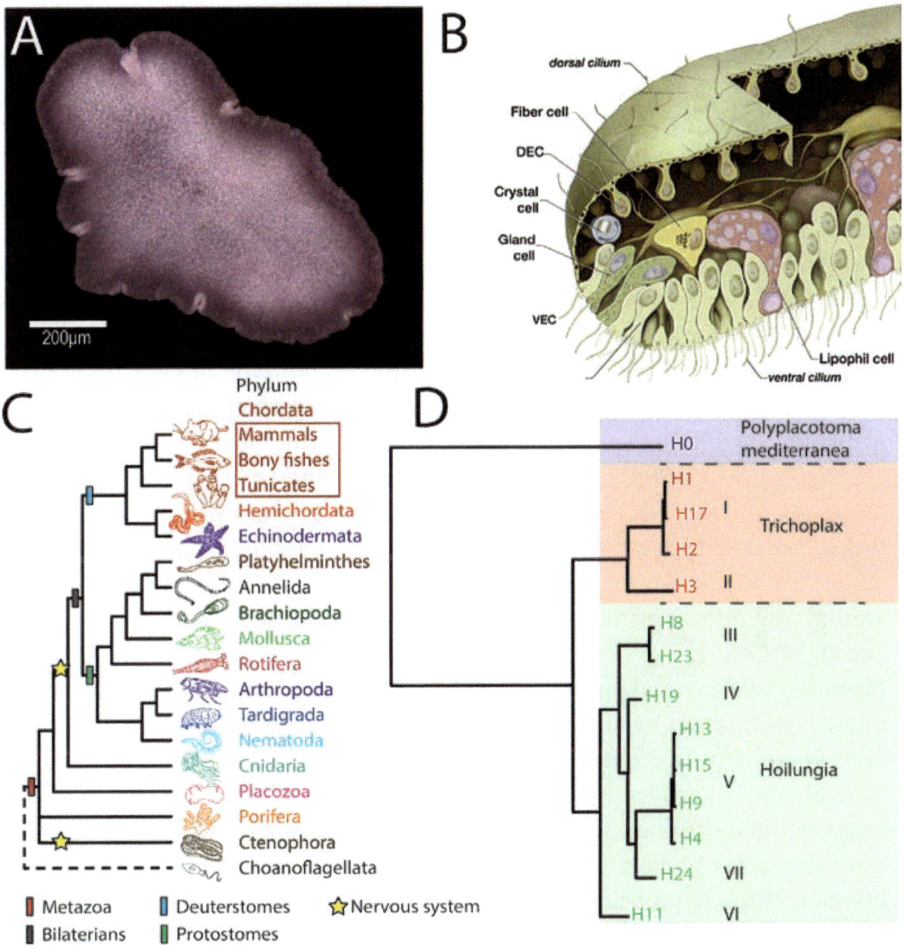

FIGURE 8.1 (A) Top-down micrograph of *Trichoplax adhaerens* fed on the red marine algae *Rhodamona salina.* Scale, 200 μm. (B) Cross-sectional illustration of the six identified cell types for *Trichoplax adhaerens*. This illustration was reprinted with permission from a previous publication (Smith et al., 2014). (C) Cladogram depicting a leading hypothesis about phylogenetic relationships of select metazoan phyla, rooted with choanoflagellates (dashed line). (D) Phylogenetic relationships of the 14 known placozoan haplotypes. Lacozoaan genera are indicated: *Polyplacotoma* (blue background); *Trichoplax* (red background); *Hoilungia* (green background).

Source: B) Reprinted with permission from Smith et al., 2014. © 2014 Elsevier. D) Reprinted from Miyazawa et al., 2020. https://creativecommons.org/licenses/by/4.0/

8.1.2 HISTORY, DIVERSITY, AND TAXONOMY

8.1.2.1 HISTORY

The history of when and how placozoans were discovered, and how the field has evolved over the years, has been eloquently reviewed by others (Schierwater, 2005; Syed & Schierwater, 2002b). Here, we briefly summarize these discussions. In 1883, the German zoologist Franz Eilhard Schulze discovered an amoeba-like animal crawling across the glass of his seawater aquarium (Schulze, 1883). He described it as a sticky, hairy plate, hence the name *Trichoplax adhaerens*, translated from Greek: tricho for "hairiness" and plax for "plate," and Latin: adhaerens for "sticking." In 1893, a second report by Francesco S. Monticelli documented another placozoan specimen isolated from an aquarium in Naples, which he named *Treptoplax reptans* and suggested was a distinct species from *Trichoplax* lacking dorsal cilia (Monticelli, 1893). However, whether this specimen indeed represents a distinct species from *Trichoplax adhaerens* has never been confirmed. Schulze's original work on *Trichoplax* highlighted key histological features including the three-layered organization of cells, the presence of four cell types, and the lack of an extracellular matrix (Schulze, 1883, 1892). The simplicity of *Trichoplax adhaerens* prompted Schulze to propose that the animal resembles a primordial metazoan state (Schulze, 1883). Later, German zoologist Otto Bütschli put forward the hypothesis of the "urmetazoan," a proposed two-layered animal that arose first through the aggregation of flagellated unicellular protozoans into a single-cell layer (akin to the *Trichoplax* ventral epithelium), and subsequently, the emergence of a second cell layer (dorsal epithelium) to form a "placula" (Bütschli, 1884). Soon after, interest in this animal declined, largely due to the incorrect assertion that placozoans represent deformed planula larvae of medusozoan cnidarians (Ender & Schierwater, 2003; Krumbach, 1907), and research ceased until the 1960s–1970s, when *Trichoplax* was rediscovered by Willy Kuhl from the Mediterranean Sea (1963) and Karl Gottlieb Grell from the Red Sea (1969) (Grell, 1971; Grell & Benwitz, 1971; Kuhl & Kuhl, 1963, 1966). Grell's work on placozoa kick-started a new wave of placozoan research. The strain he captured, often referred to as the *Trichoplax adhaerens* Grell strain, is perpetuated in labs across the world and has been the subject of most recent studies. His work encompassed various aspects of *Trichoplax* biology including reproduction and ultrastructure, providing enough support to place it in a phylum of its own, which he named placozoa (derived from Bütschli's placula hypothesis) (Syed & Schierwater, 2002b). In this regard, the establishment of Placozoa as a phylum is atypical where at its inception, it included only a single species known at the time, *Trichoplax adhaerens*.

8.1.2.2 DIVERSITY AND TAXONOMY

Strains of placozoans isolated from the wild generally exhibit considerable overlap in their gross morphological features, which complicates efforts of species separation and identification (Eitel et al., 2018). In addition, the application of the biological species concept, which determines species taxa based on successful interbreeding and offspring viability (Mayr, 1942), is impeded by our limited understanding of placozoan sexual reproduction. As noted, gametocyte production, formation, and fertilization have been documented for *Trichoplax adhaerens* with zygote cleavage going as far as the 128-cell stage, but the embryos perish before reaching maturity (Eitel et al., 2011; Grell, 1971; Ruthmann et al., 1981). As a result, numerous different approaches have been used to distinguish the various strains and assign them to taxonomic rankings. Over the past two decades, a series of genotyping studies using molecular markers, most prominently the mitochondrial 16S rRNA, have uncovered significant diversity among placozoan lineages describing 20 haplotypes (H0 to H19), each of which potentially represents a distinct species (Eitel et al., 2013; Eitel & Schierwater, 2010; Miyazawa & Nakano, 2018; Nakano, 2014; Osigus et al., 2019). A subsequent phylogenetic study of 14 sequenced mitochondrial genomes identified 2 additional haplotypes (H23 and H24) and defined 8 placozoan clades that fall within three proposed genera: (i) a *Trichoplax* genus comprised of 2 clades (i.e., clades I and II) and four haplotypes (H1, H17, H2, and H3); (ii) a *Hoilungia* genus comprised of 5 clades (III, IV, V, VI, and VII) and 9 haplotypes (H4, H8, H9, H11, H13, H15, H19, H23, and H24); and (iii) the most phylogenetically distant *Polyplacotoma* genus, forming an 8th separate clade comprised of a single haplotype (H0) (Miyazawa et al., 2020) (Figures 8.1(D) and 8.3(A) from Miyazawa et al., 2020 paper). To date, only 3 of these haplotypes have been formally accepted as distinct species, *Trichoplax adhaerens* (H1) and *Hoilungia hongkongensis* (H13), which have had their mitochondrial and nuclear genomes sequenced (Dellaporta et al., 2006; Eitel et al., 2018; Miyazawa et al., 2020; Srivastava et al., 2008), and *Polyplacotoma mediterranea* (H0), defined by its unique branched morphology and highly divergent mitochondrial genome (Miyazawa et al., 2020; Osigus et al., 2019). Corroborating these phylogenetic studies, several light and electron microscopy studies have collectively reported differences in the ultrastructural morphology of various placozoan haplotypes (Guidi et al., 2011; Romanova et al., 2021; Smith et al., 2014). Thus, more species are likely to emerge, particularly as more genomes become sequenced.

The relationship of placozoans to other animals has been part of a highly debated topic surrounding the phylogenetic placement of the earliest diverging animal phyla, which includes Ctenophora (comb-jellies), Porifera (sponges), and Cnidaria (jellyfish, hydra, sea anemones, and corals). Central to this debate is which phylum was the first to diverge, with ctenophores and sponges emerging as the two major contenders for this phylogenetic position. Although still under debate, the phylogenetic position of placozoans is less uncertain, with most recent phylogenomic studies placing them in a sister relationship with cnidarians and bilaterians, diverging from these groups after poriferans and ctenophores (Figure 8.1(C)) (Eitel et al., 2018; Hejnol et al., 2009; Kapli & Telford, 2020; Moroz et al., 2014b; Pisani et al., 2015; Ryan et al., 2013; Simion et al., 2017; Whelan et al., 2017). Accordingly, placozoans more resemble cnidarians and bilaterians in their genomic complement of homeodomain-containing proteins associated with development, compared to ctenophores and sponges that lack several key homeodomain classes (Ryan et al., 2010). Clearly, placozoans are in a strategic phylogenetic position for exploring the evolution of complex animal traits such as development, neural signaling, and cell type evolution.

8.1.3 CELL TYPES AND CELLULAR DIVERSITY

Early morphological work by Schulze and subsequent electron microscopy imaging by Grell and colleagues provided major insights into the simple placozoan body plan, initially thought to consist of only four somatic cell types: dorsal and ventral ciliated epithelial cells, ventral gland cells, and an interior network of fiber cells (Figure 8.1(B)) (Grell & Benwitz, 1971, 1974; Grell & Ruthmann, 1991; Rassat & Ruthmann, 1979). Various subsequent observations pointed to the existence of additional cell types, where for example, it was postulated that the difference in vesicle sizes observed within gland cells could reflect multiple cell types (Grell et al., 1991). In 2014, a novel freezing-based fixation technique was used to reinvestigate the ultrastructure of *Trichoplax*, identifying two novel cell types, lipophil cells and crystal cells (Smith et al., 2014) (Figure 8.1(B)). Crystal cells are rare but easily identifiable under the light microscope due to their large cytoplasmic birefringent crystals, and ultrastructurally by their distinct cup-shaped nuclei with "openings" that project towards the outside edges of the animal (Mayorova et al., 2018; Smith et al., 2014). Crystal cells contact fiber cells and dorsal epithelial cells near the edge of the animal, and their cytoplasmic crystal, which consists of calcium carbonate (aragonite), shifts position in the cytoplasm relative to the cup-shaped nucleus with changes in a vertical orientation (Mayorova et al., 2018). Placozoans rely on these cells for gravity sensing, made evident by specimens lacking this cell type that does not react to changes in a vertical orientation (Mayorova et al., 2018). Lipophil cells are non-ciliated and interspersed along the ventral epithelium, extending dorsoventrally with their cell bodies positioned close to fiber cells (Smith et al., 2014). These are proposed to serve the function of enzyme secretion for external digestion, due to their abundant expression of large acidic granules, and their position within a central region of the ventral epithelium which overlaps with the mRNA expression of the digestive enzymes trypsin, chymotrypsin, and phospholipase A2 as determined by *in situ* hybridization (ISH) (Mayorova et al., 2019; Smith et al., 2014). The exact distribution of lipophil cells begins ~100 μm from the outer edge of the animal referred to as the digestive or lipophil zone, surrounded by a concentric zone referred to as the marginal zone (Mayorova et al., 2019) (Figure 8.1(B)).

Fiber cells are the only cell type interspersed between the dorsal and ventral *Trichoplax* epithelia, with extensions that contact all other cell types forming a loosely packed organization described as a mesenchyme-like layer (Buchholz & Ruthmann, 1995). These cells also exhibit syncytial (cytoplasmic) junctions between them, evident as dense septa in electron micrographs (Grell & Benwitz, 1974; Grell & Ruthmann, 1991; Smith et al., 2014). The interconnectedness between fiber cells, coupled with their syncytial junction architecture, is said to resemble the syncytial junctions of glass sponges (Leys, 2003; Mackie & Singla, 1983; Smith & Mayorova, 2019). In their three-dimensional organization and expression of actin, microtubules, and microfilaments, fiber cells are also suggested to play roles in contraction, movement coordination, and integration of complex behaviors (Behrendt & Ruthmann, 1986; Grell & Benwitz, 1971; Thiemann & Ruthmann, 1989). Fiber cells are reported to carry out transepithelial cytophagy, by which algae and other particles get endocytosed, presumably along the dorsal epithelium to accumulate inside vacuoles (Smith et al., 2014; Wenderoth, 1986). Recently, Fiber cells were reported to mediate innate immunity like macrophage cells, and hence, these inclusions likely represent phagocytotic management of foreign cells that have invaded the inner layer (Mayorova et al., 2021). Two classes of intact (living) bacteria, of the order Rickettsiales and the phylum *Margulisbacteria* are reported to form intracellular symbiotic associations within placozoan cells (Grell & Benwitz, 1971; Gruber-Vodicka et al., 2019; Guidi et al., 2011; Kamm et al., 2019; Smith et al., 2014; Wenderoth, 1986). Consistent with symbiosis, the *Rickettsiales* bacteria that reside within the endoplasmic reticulum (ER) of fiber cells do not appear to express genes associated with energy parasitism (Gruber-Vodicka et al., 2019).

Placozoans lack gap junctions and tight junctions, and their epithelia are held together by adherens junctions that provide a leaky barrier between cells (Grell et al., 1991; Smith & Reese, 2016). The dorsal epithelium is formed by a layer of thin and dynamically contractile T-shaped epithelial cells bearing a single cilium as well as microvilli (Armon et al., 2018; Grell & Benwitz, 1971; Smith et al., 2014). Contraction by these cells involves transient rises in cytoplasmic Ca^{2+}, and can occur sparsely in individual cells, or as propagating waves (Armon et al., 2018). Although the function of these contractions is not known, they are proposed to help maintain the integrity of the dorsal epithelium in response to mechanical stress and perhaps contribute to contractile behaviors such as churning, crinkling, and whole-body contractions (Romanova et al., 2020a; Smith et al., 2015; Varoqueaux et al., 2018). A notable feature of the *Trichoplax adhaerens* dorsal epithelium is the presence of shiny spheres (Grell & Benwitz, 1971; Grell & Ruthmann, 1991; Guidi et al., 2011), which are electron-dense extracellular inclusions suggested to play a role in defense from predation, perhaps through chemical agents that paralyze predators (Jackson & Buss, 2009). Interestingly, shiny spheres were not present in the recent study of *Trichoplax* ultrastructure that identified crystal and lipophil cells (Smith et al., 2014), and it is unclear whether strains can lose these structures when cultured in the lab, or perhaps, if they are derived from specific food sources only present in the wild.

The significance of the ventral epithelium is marked by its expanded cell type diversity, harboring nearly 75% of the total cells found in *Trichoplax* (Smith et al., 2014). It is a crucial surface for major processes including feeding and locomotion. Prominent in the ventral epithelium are dorsoventrally extended epithelial cells, each expressing multiple microvilli and a single beating cilium that contributes to locomotion (Grell & Benwitz, 1971; Ruthmann et al., 1986; Smith & Reese, 2016; Smith et al., 2014). Gland cells, also present in the dorsal epithelium but enriched ventrally, contain vesicles of varying sizes, and stain for proteins associated with regulated exocytosis such as syntaxin, SNAP-25, synapsin, complexin, and voltage-gated calcium channels (Gauberg et al., 2020; Mayorova et al., 2019; Smith et al., 2014, 2017). Various studies have also documented the staining of gland cells by antibodies for the amidated regulatory peptides RFamide and FMRFamide (Schuchert, 1993; Smith et al., 2014), altogether suggesting that these cells are neuroendocrine-like. However, bioinformatic prediction of small secreted regulatory peptides within the *Trichoplax adhaerens* genome and transcriptome failed to identify an FMRFamide precursor gene, and instead, predicted numerous other amidated and non-amidated regulatory peptides (Nikitin, 2015; Wong et al., 2019). In accordance, studies have now localized some of these and other endogenous peptides to these same FMRFamide-positive cells, with the non-specific labeling attributed to the cross-reactivity of the FMRFamide antibodies with endogenous *Trichoplax* peptides (Mayorova et al., 2019; Senatore et al., 2017; Varoqueaux et al., 2018). As we discuss in Section 8.2.6, several studies have now established that secreted peptide signaling plays an important role in the regulation of *Trichoplax* motile behavior.

More recently, a single-cell transcriptome study identified a cluster of six cells that uniquely express these identified regulatory peptides (Sebé-Pedrós et al., 2018). These cells were found to be transcriptionally similar to ciliated epithelial cells but lacked mRNA expression of genes required for motile cilia. Whether these peptidergic cells represent the microscopically defined gland cells is uncertain, as the authors of this study used the term gland cells to describe a separate cluster of cells, distinct from ciliated epithelial cells and the 6 peptidergic cell types. Specifically, they reverted to the classic definition that gland cells are digestive in nature (Grell & Ruthmann, 1991), noting mRNA expression of digestive enzymes such as trypsin, chymotrypsin, and phospholipase A2 (Sebé-Pedrós et al., 2018). However, these same enzymes were shown to be restricted to the lipophil zone of *Trichoplax* via mRNA *in situ* localization, consistent with the location of lipophil cells but not gland cells, which are distributed along both the marginal and lipophil zones (Mayorova et al., 2019). Clearly, an important step forward will be to clarify the link between transcriptional profiles of the 11 cell types defined in the single-cell transcriptomic study (Sebé-Pedrós et al., 2018), and cell types defined microscopically in terms of morphology, ultrastructure, location in the body, and expression of marker genes *in situ*.

The various studies described above collectively hint at the existence of yet additional cell types. Indeed, a recent study employing scanning electron microscopy (SEM) and 3D image reconstruction delineated three distinct gland cells types in *Trichoplax* (Mayorova et al., 2019), confirming earlier observations of ultrastructural diversity within this cell type (Grell et al., 1991). Type 1 gland cells comprise roughly 20% of the total gland cell population, while type 2 and type 3 comprise 65% and 15%, respectively. Type 1 gland cells are in the marginal zone (Figure 8.2(B)), are mono-ciliated, bear microvilli, and can be uniquely recognized by their large secretory granules and electron-dense ER. Type 2 cells span the marginal zone and are found sparsely within the outer regions of the lipophil zone. They are distinguished by their small electron-lucent granules, apical microvilli, and the absence of a cilium, the latter reminiscent of peptidergic cells in the single-cell transcriptome study. Type 3 gland cells are also found within the edges of the lipophil zone, and uniquely, within the dorsal epithelium. These cells have both cilium and microvilli-like type 1 gland cells but contain small secretory granules. Type 2 gland cells label for mucin-related carbohydrates and are thus proposed to constitutively secrete mucous to facilitate ciliary locomotion (Mayorova et al., 2019). For this reason, these cells are also referred to as mucocytes and are thought to be functionally analogous to mucus-secreting goblet cells of bilaterians (Mayorova et al., 2019; Takashima et al., 2013). Interestingly, a subset of granules in type 2 gland cells also labels for the regulatory peptide named *Trichoplax adhaerens* endomorphin-like peptide (*Ta*ELP), which is involved in regulating ventral ciliary beating and locomotion (Senatore et al., 2017). This observation provides a possible link between microscopically defined gland cells and the noted peptidergic cells from the single-cell transcriptome study, where in the latter study, the peptide was found to be expressed in one of the 6 peptidergic cell clusters (Mayorova et al., 2019; Sebé-Pedrós et al., 2018). Similarly, the immunolocalization of various other *Trichoplax* peptides, such as SIFGamide, FFNPamide, and RWamide, coincides with the locations of the microscopically-defined gland cells types (Varoqueaux et al., 2018), also detected in peptidergic single-cell transcriptomes (Sebé-Pedrós et al., 2018). Thus, the various observations would suggest that the six peptidergic cells defined in the single-cell transcriptome study are most likely gland cells and that the cells termed gland cells in this study are instead lipophil cells.

An interesting note about gland cells relates to the co-labeling of some type II cells for mucus and *Ta*ELP, resembling a proposed ancestral cell type that could secrete both mucus and peptides, which subsequently evolved in vertebrates to give rise to mucocytes, that are specialized to secrete mucus, enteroendocrine, that secrete peptide hormones, and paneth cells, that secrete antimicrobial peptides (AMPs) to control intestinal flora (Mayorova et al., 2019). Consistent with a common evolutionary origin, the differentiation of vertebrate mucocytes, enteroendocrine cells, and paneth cells occurs through an overlapping developmental program (Yang et al., 2001). Indeed, although paneth-like cells have not been defined in placozoans, whole animal transcriptome sequencing identified a set of highly expressed secreted protein genes bearing lysine and arginine repeats (Wong et al., 2019), a molecular feature of AMPs (Jin et al., 2016). Furthermore, the single-cell transcriptome study identified a set of ciliated epithelial cells expressing one of these lysine/arginine-rich proteins, homologous to the antimicrobial peptide defensin, which was transcriptionally similar to the 6 peptidergic cell types that we suggest are gland cells (Sebé-Pedrós et al., 2018).

Thus far, most of the mentioned morphological ultrastructure is based on *Trichoplax adhaerens* isolated from the Red Sea (Grell strain), and with the growing understanding of placozoan genetic diversity, questions emerge about whether all placozoan lineages share the same cellular features. As noted, a study of various placozoan haplotypes reported several differences in cellular ultrastructure, hinting at the existence of hidden cell diversity among different lineages (Guidi et al., 2011). This was recently confirmed for the species *Hoilungia hongkongensis*, for which several key differences and putative novel cell types were identified (Romanova et al., 2021). The ventral epithelium of *Hoilungia* shares with *Trichoplax* all previously mentioned cell types (Romanova et al., 2021). However, there appear to be several additional cell types in the ventral epithelium, one bearing a unique electron-dense cytoplasm, and another enriched in granular material, with neither cell type contacting the outside environment (i.e., exclusively internal). The middle mesenchyme-like layer also resembles that of *Trichoplax* in bearing fiber cells with similar elaborate extensions. However, an additional star-shaped cell was observed bearing long processes, potentially representing a different cell type, or rather, a fiber cell at a different stage of maturation. Due to the rare appearance of this star-shaped cell, a morphological and molecular characterization is yet lacking and further investigation to support this cell type is needed. Like *Trichoplax*, the *Hoilungia* dorsal epithelium also contains shiny spheres, emerging from a proposed novel cell type named sphere cells (Romanova et al., 2021). However, shiny spheres were previously suggested to originate in fiber cells, within projections that extend towards the surface of the dorsal epithelium (Guidi et al., 2011), and it is unclear if sphere cells as reported in *Hoilungia* are fiber cells, or rather, if fiber cells reported to produce shiny spheres are instead sphere cells. Likely, continued efforts in microscopy and single-celled transcriptomics will elucidate the cellular diversity among placozoans, and by extension, provide inroads for assessing the homology of placozoan cell types with those of other animals.

During development, cell types emerge from one of three germ layers (i.e., ectoderm, endoderm, and mesoderm), with diploblastic animals like cnidarians and arguably ctenophores having endoderm and ectoderm, and triploblastic bilaterian animals having all three germ layers (Martindale et al., 2004). For Placozoa, the dorsal and ventral epithelia are proposed to be homologous to the ectoderm and endoderm cell layers of other animals, respectively (Grell & Benwitz, 1971; Schierwater et al., 2009; Schulze, 1883; Syed & Schierwater, 2002b). However, whether fiber cells constitute a mesoderm, or originate from one of the epithelial germ layers, is still unclear. Considering the noted difficulties in observing placozoan sexual reproduction and development in the lab, researchers have instead attempted to infer developmental processes by exploring mRNA expression patterns and phylogenetic properties of developmental gene homologs in control, regenerating, and dividing presumptive adult specimens. For example, siRNA downregulation of the *Trichoplax* Hox/ParaHox gene homolog (*Trox-2*), a homeodomain transcription factor involved in axial patterning in bilaterians and cnidarians (Ryan et al., 2010), interrupted growth and binary fission, and was found to be expressed along the periphery of the animal marking the boundary between the dorsal and ventral epithelia, consistent with the proposed location of pluripotent stem cells (Jakob et al., 2004; Martinelli & Spring, 2003). The *Trichoplax* PaxA/B/C homolog, another homeodomain gene from a family of genes involved in eye and sensory organ development in cnidarians and bilaterians (Kozmik, 2008), was also localized to the peripheral zone. Similarly, and arguing against the notion that fiber cells constitute a mesoderm, two T-box genes involved in bilaterian mesoderm development, Brachyury, and Tbx2/3, were localized to the periphery, rather than more central regions where fiber cells are located (Martinelli & Spring, 2003). In cnidarians, Brachyury is suggested to be involved in specifying the blastopore (Scholz & Technau, 2003), where the endodermal and ectodermal epithelia meet, which is perhaps similar to the peripheral zone of *Trichoplax* where the dorsal and ventral epithelia intersect. From a morphological perspective, the presence of stem cells along the edge of the placozoan animal makes sense, if we envision that growth occurs outward along the concentric edges. However, in *Hoilungia hongkongensis*, tandem sets of cells with matching ultrastructure have been reported in epithelial regions outside of the proposed marginal proliferation zone, most likely representing cells that recently underwent mitosis (Grell & Benwitz, 1981; Romanova et al., 2021). Such an observation might suggest that placozoan cell types can proliferate via duplication of specific cell types throughout the animal body. Furthermore, the localization of mRNAs along the placozoan marginal zone should perhaps be re-examined in some cases, where possibly the high concentration of mucous carbohydrates in this region, present within type II gland cells, could cause ISH probes to accumulate, leading to non-specific labeling. For example, a recent study utilizing a modified mRNA ISH protocol found that the Trox-2 mRNA is expressed within the central lipophil zone of *Trichoplax adhaerens* (DuBuc et al., 2019), instead of the periphery as documented previously (Jakob et al., 2004), a fact that might alternatively be explained by the developmental status of the animal after a previous binary fission.

In summary, the work done to date highlights a variety of placozoan cell types and has provided insights into their potential functions, but much more work needs to be done to understand the diversity and molecular features of placozoan cell types. Defining cell type homology with other animal cells also remains challenging, particularly with the ambiguity regarding placozoan germ layers and developmental processes. Altogether, this will be a necessary step for understanding the relatedness of *Trichoplax* cell types with those of other animals, the evolution of cell types in general, and by extension, the evolution of animal multicellularity, which requires the coordinated establishment of different cell types at discrete morphological locations. One proposed solution is to study cell type evolution and homology by classifying homology upon the repertoire of expressed transcription factors and regulatory elements which together act as regulatory signatures or expression programs unique to every cell type (Arendt et al., 2016). However, phenotypic convergence or molecular parallelism should also be considered (Rosenblum et al., 2014), where shared selective pressures could in theory lead to parallel solutions at the transcriptional level for the recruitment of genes into parallel functional modules, which could impose cellular features that appear homologous. Perhaps, a deeper understanding of cell type evolution will emerge if in conjunction with exploring transcriptional homology, we also explore homology at the molecular level, that is, the degree to which different sets of proteins form homologous or convergent functional units that are transcriptionally deployable across different cell types.

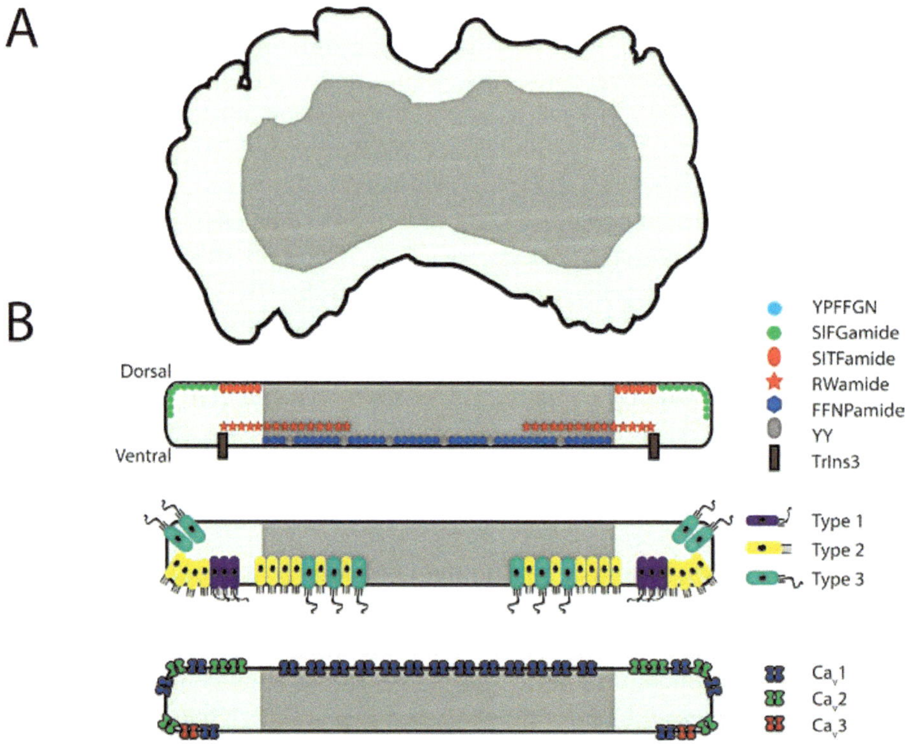

FIGURE 8.2 (A) Schematic representation of *Trichoplax adhaerens* (dorsal-ventral orientation) illustrating the marginal zone (green) and the ipophile zone (grey). (B) Top panel: cross-sectional illustration depicting the locations of various peptides as determined by immunolocalization, superimposed over the marginal zone (green background) and the lipophil zone (grey background). Middle panel: a similar illustration highlighting the locations of type 1 gland cells (purple), type 2 gland cells (yellow), and type 3 gland cells (cyan). Bottom panel: illustration of the locations of Ca$_V$1 (blue), Ca$_V$2 (green), and Ca$_V$3 (orange) voltage-gated calcium channels, as determined by immunolocalization.

8.1.4 GENOMICS AND PROTEOMICS

The *Trichoplax* karyotype indicates a diploid chromosomal arrangement with only 6 chromosome pairs (Birstein, 1989). Shotgun sequencing of the *Trichoplax adhaerens* nuclear genome (Grell strain, H1 haplotype) estimated 11,514 protein-coding genes, 87% of which were found to be homologous with genes from other animals (Srivastava et al., 2008). Subsequent reannotation of the genome increased the total gene count to 12,515 (Kamm et al., 2018), surpassing the number of 12,200 genes predicted in the genome of *Trichoplax* H2, a closely related haplotype (Kamm et al., 2018). In accordance with the phylogenetic proximity of H1 and H2, a comparison of the two genomes revealed highly overlapping features, including 99.1% sequence identity, and almost complete gene synteny (Kamm et al., 2018). Worth noting is that although the H1 and H2 gene counts appear somewhat reduced relative to other animals, a large proportion of orthologous genes are present as single copies (i.e., 9,900 out of 10,030 orthologous gene clusters, or 98.7%) (Kamm et al., 2018), indicative of a general absence of gene paralogues with redundant/overlapping functions. Nevertheless, genome comparison has revealed that a considerable number of novel genes emerged after placozoans diverged from cnidarians and bilaterians, and again after cnidarians diverged, strictly in bilaterians (Paps & Holland, 2018). Interestingly, the intron-exon structures of *Trichoplax* H1 and H2 haplotype genes generally resemble those of human genes, with 82% of human introns having orthologous counterparts in the placozoan genes (Kamm et al., 2018; Srivastava et al., 2008). This is in contrast to other invertebrates with small genomes such as fruit flies, nematode worms, and sea anemones, which appear to have lost numerous introns as their genomes became compacted during the course of evolution (Putnam et al., 2007). Also in contrast to fruit flies and nematodes, the two *Trichoplax* haplotypes exhibit a notable degree of conservation in gene order and synteny with vertebrate and cnidarian genomes (Srivastava et al., 2008).

To date, the only other placozoan species to have had its nuclear genome sequenced is *Hoilungia hongkongensis* (H13 haplotype), which as noted belongs to a different genus than the *Trichoplax* haplotypes. The divergence of these two placozoan lineages is evident in their nuclear genomes, where 2,101 of the 12,010 genes identified in the *Hoilungia* genome were found to be inverted in their chromosomal orientation relative to their *Trichoplax* orthologue (Eitel et al., 2018), resembling the rate of rearrangement reported between mouse and human (Drillon et al., 2014). Furthermore, on average, orthologous gene sequences exhibit genetic distances of roughly 28% at both the protein and DNA levels, resembling genus-level differences observed among ctenophores, and family-level differences among cnidarians (Eitel et al., 2018). Nevertheless, roughly 97% of orthologous genes between *Hoilungia hongkongensis* and *Trichoplax adhaerens* are subject to strong purifying selection, indicative of genetic constraints that may in part account for the phenotypic similarities observed for different placozoan haplotypes and species.

The genome-predicted genes that are actively expressed have been determined through transcriptome sequencing and proteomics (Eitel et al., 2018; Gruber-Vodicka et al., 2019; Kamm et al., 2018; Ringrose et al., 2013; Wong et al., 2019). This includes an unpublished whole-animal transcriptome of *Trichoplax adhaerens*, described in a preprint (Wong et al., 2019), which identified 11,981 genes expressed at the mRNA level, confirming the expression of 9,260 genome-predicted genes. This assembled gene set also identified several 100 novel genes, and to date, appears to be the most complete set of gene sequences available for a placozoan, producing a BUSCO completeness score of 95.1%, and only 0.7% and 4.2% of gene sequences either fragmented or missing, respectively (Table 8.1) (Simão et al., 2015). By comparison, the *Trichoplax adhaerens* gene sequences currently available on the UniProt database, and the *Hoilungia hongkongensis* gene set, both derived from combined genome and transcriptome sequencing and annotation, have lower completeness scores of 89.3% and 92%, respectively (Table 8.1). Using high-resolution mass spectrometry, just over half of the predicted *Trichoplax* genes were also confirmed to be expressed at the protein level (6,516), among which proteins associated with the extracellular matrix and spermatogenesis were observed (Ringrose et al., 2013), despite neither of these cellular structures being apparent under the electron microscope. Altogether, the various placozoan genome and transcriptome studies, as well as the proteomic study have revealed a striking complexity in expressed nuclear gene content, including numerous genes associated with complex animal traits that placozoan lack, such as body patterning, a neuromuscular system, and an extracellular matrix and associated adhesion proteins.

In addition to nuclear genomics, significant efforts have been focused on mitochondrial genomics, with 14 placozoan haplotypes having had their mitochondrial genomes fully sequenced (H0–H4, H8, H9, H11, H15, H17, H19, H23, and H24; Figure 8.1(D)) (Dellaporta et al., 2006; Eitel et al., 2018; Miyazawa et al., 2012, 2020; Osigus et al., 2019; Signorovitch et al., 2007). The first was *Trichoplax adhaerens* (H1), revealing several atypical features including a large size of ~43 kilobase pairs, roughly double the size of most other metazoan mitogenomes, and sequence characteristics of non-metazoan mitochondria (i.e., choanoflagellates and fungi), including large intergenic spacers and additional introns (Dellaporta et al., 2006). However, the *Trichoplax* mitogenome also bears features that are unique to animals, including the absence of ribosomal protein-coding genes, and several introns shared only with cnidarians. In light of these observations, the *Trichoplax* H1 mitogenome is suggested to resemble an intermediate state between those of non-metazoans and metazoans (Signorovitch et al., 2007). Subsequent sequencing of the H3, H4, and H8 haplotype mitogenomes revealed similar attributes including large sizes ranging from ~32 to 37 kilobase pairs,

TABLE 8.1 Comparison of Multiple Non-Bilaterian Genome and Transcriptome Assemblies Using BUSCOv5 Assessment on Metazoan Genes Database

Phylum	Species	Complete	Fragmented	Missing	References
Cnidaria	*Acroporamillepora*	95.90%	2.30%	1.80%	QTZP00000000
Cnidaria	*Alatina Alata*	92.90%	3.80%	3.30%	(Zapata et al., 2015)
Cnidaria	*Clytia hemisphaerica*	87.70%	6.20%	6.10%	GCA_902728285
Cnidaria	*Coralliumrubrum*	88.40%	4.10%	7.50%	(Zapata et al., 2015)
Cnidaria	*Nematostella vectensis*	97.70%	1.20%	1.10%	GCA_000209225.1
Cnidaria	*Rhopilemaesculentum*	97.00%	1.70%	1.30%	(Li et al., 2020)
Placozoa	*Hoilungia hongkongensis*	92.00%	1.00%	7.00%	(Eitel et al., 2018)
Placozoa	*Trichoplax adhaerens*	89.30%	4.60%	6.10%	(Srivastava et al., 2008)
Placozoa	*Trichoplax adhaerens*	95.10%	0.70%	4.20%	(Wong et al., 2019)
Ctenophora	*Pleurobrachiabachei*	70.70%	12.20%	17.10%	(Moroz et al., 2014b)
Ctenophora	*Mnemiopsis leidyi*	80.40%	5.10%	14.50%	(Ryan et al., 2013)
Ctenophora	*Beroeabyssicola*	67.70%	7.70%	24.60%	NEUROBASE
Porifera	*Oscarella carmela*	90.20%	3.20%	6.60%	Compagen
Porifera	*Ephydatia muelleri*	89.80%	1.90%	8.30%	(Kenny et al., 2020)
Porifera	*Amphimedon queenslandica*	89.80%	4.20%	6.00%	(Srivastava et al., 2010)

identical sets of structural RNA and electron transport chain genes, and characteristic introns and intergenic spacers (Signorovitch et al., 2007). A maximum likelihood species phylogeny using 12 concatenated mitochondrial protein-coding sequences suggested an early-branching metazoan position for the four placozoan haplotypes. However, in this study where choanoflagellates and fungi were set as the outgroup, bilaterians were the first to diverge, followed by placozoans forming a sister clade with poriferans and cnidarians. This is altogether inconsistent with more current phylogenetic observations generated using large sets of nuclear-encoded genes. Accordingly, subsequent reanalysis of placozoan mitochondrial phylogenomics with increased taxon sampling indicated that the mitochondrial sequences alone could not resolve the phylogenetic relationship of placozoans with other animals (Burger et al., 2009). Nonetheless, mitochondrial phylogenomics has proved useful for establishing taxonomic relationships among placozoan haplotypes (Miyazawa et al., 2020) (Figure 8.1(D)). Furthermore, in the case of *Polyplacotoma mediterranea* (H0 haplotype), mitochondrial sequencing was the basis for establishing it as a new species with a highly divergent and more compact mitogenome of only ~23 kilobase pairs attributable to smaller intergenic spacer regions, and sequences lacking features known for other placozoans (Osigus et al., 2019).

8.1.5 PROSPECTS FOR GENETIC MANIPULATION STUDIES

Placozoans can be easily cultured and maintained in the lab, a feature that previously drove other invertebrate animal subjects into the mainstream (Neff, 2018). However, a major limitation is our inability to effectively perturb and manipulate their genomes, a requirement for assessing genotype-to-phenotype relationships. The development of technologies for manipulating the placozoan genome will be critical if we are to harness the opportunities that these animals provide as research subjects. Indeed, the advent of broadly applicable genome editing tools such as CRISPR/Cas9, coupled with ever-improving high-throughput DNA sequencing technologies, has greatly contributed to genetic research in novel organismal lineages (Reardon, 2019). However, a common approach in this regard is to genetically modify a single progenitor cell (embryo or stem cell), that will subsequently develop into a complete organism (or organoid), with all derived cells being clonal in their genetic modification. Although mosaic transgenesis has some advantages in some contexts (Fei et al., 2019), clonal transgenesis is advantageous as it removes potential confounding effects of cell genotypic heterogeneity. Indeed, our current inability to observe and control *Trichoplax* sexual reproduction, with a single-celled embryo developing into an "adult" form, precludes the establishment of conventional protocols for generating transgenic specimens with clonal modifications. Any efforts at transgenesis would therefore require the generation of mosaic individuals, where conceivably, genetically modified cells could be enriched by using positive selection markers, or perhaps, though dissociation of cells, followed by cell sorting for transgenic fluorescent markers flowed by reaggregation of selected cells, the latter shown to be a viable manipulation for *Trichoplax adhaerens* (Ruthmann & Terwelp, 1979).

Notably, although CRISPR/Cas9 provides avenues for manipulating genes and their regulatory elements encoded within the genome, it is often advantageous to instead insert an exogenous/recombinant gene, with high-level *in vivo* expression driven by a flanking promoter sequence. For other transgenic animals, expression of exogenous genes is often achieved with strong constitutive promoters isolated from viruses, but no such elements have been discovered for placozoans. Perhaps, the use of endogenous promoters expected to drive high-level transcription in multiple cell types, or ones that drive transcription in specific cell types could be adopted, but these would first have to be functionally characterized using reporter gene assays. Another major hurdle is the transfer of genetic materials into placozoan cells. To date, only two studies report altered gene expression *in vivo* using morpholinos and small interfering RNAs that were transfected into cells (Jakob et al., 2004; Loenarz et al., 2011). Alternative strategies could be the implementation of recombinant baculoviruses, that appear to have broad infectivity (Airenne et al., 2013), nano-biolistics (O'Brien & Lummis, 2011), or cell-penetrating peptides (Milletti, 2012). Although challenging, the prospects of developing genetic tools for *Trichoplax* and Placozoa are tantalizing, where for example, ectopic expression of channelrhodopsins would permit light-induced excitation or inhibition of cells, to examine the role of electrical signaling in their biology, or generation of gene knockout animals, to link genotype to phenotype.

8.2 CELLULAR SIGNALING AND BEHAVIOR

8.2.1 GENERAL CONSIDERATIONS ABOUT PLACOZOAN CELL SIGNALING AND BEHAVIOR

Arguably, the evolution of complex animals with different cell types and organ/tissue systems was made possible through the adaptation of molecular systems that could harness the ionic composition of the cellular environment to generate fast electrochemical signals that could be propagated quickly between cells and over long distances. Although electrical signaling is observed among all cellular lifeforms (Brenner et al., 2006; Eckert & Brehm, 1979; Levina et al., 1995; Prindle et al., 2015), metazoans are perhaps the most complex, evident by the significant expansion of ion channel gene families in animals compared to other eukaryotic lineages (Liebeskind et al., 2015). Accordingly, electrical signaling is ubiquitous in animals, including the early diverging lineages of sponges, ctenophores, and cnidarians where numerous reports have documented electrical signaling in an array of cell types (Anderson, 2004; Horridge, 1965; Leys & Meech, 2006; Mackie, 1990; Moroz et al., 2014a; Moss & Tamm, 1986; Senatore et al., 2016). In the nervous system, fast electrochemical signaling can be broken down into three fundamental cellular processes: (i) generation of resting membrane potential; (ii) generation of fast electrical signals such as action and graded potentials; and (iii) transmission of electrical signals between cells, at sites called synapses.

Of these three fundamental processes, the establishment of resting membrane potential and the generation of electrical impulses is observed broadly in different cells of multicellular organisms and unicellular organisms, whereas synapses are a unique feature of the animal nervous system that permits the discrete transmission of information between cells. Under this definition, placozoans would therefore lack nervous systems because they lack the ultrastructural signatures of the chemical synapse, that is, pre-synaptic active zones with clusters of neurotransmitter-filled vesicles, and corresponding post-synaptic densities where appropriate neurotransmitter receptors and cytoplasmic signaling molecules are localized to receive incoming signals (Smith et al., 2014). Placozoans also lack genes for gap junctions and electrical synapses, which were apparently lost after they diverged from other animals (Eitel et al., 2018; Kamm et al., 2018; Srivastava et al., 2008; Wong et al., 2019). Still, extensive cellular communication in the nervous system occurs outside of the synapse, in the form of volume transmission and paracrine signaling, which nevertheless utilizes overlapping sets of genes involved in fast electrochemical and synaptic transmission (Marder, 2012). Specifically, genes for excitation-induced secretion of neurotransmitters and neuromodulators as observed in neuroendocrine cells and neural varicosities, and those for the detection of secreted signaling molecules (and molecules in the environment) as observed at synaptic and non-synaptic sites (Fuxe et al., 2013). Often, volume transmission and neuromodulation serve to regulate the activity of fast-propagating electrical signals and synapses, causing changes in neural circuit function and behavior (Marder, 2012). Like neural circuits, volume/paracrine signaling in the nervous system permits specific communication between cells, via the paired expression of secreted signaling molecules in one set of cells, coupled with the expression of appropriate receptors for these ligands on target cells (Jékely, 2021). Indeed, although we have come to understand much about hard-wired neural connectomes composed of neurons and synapses, we understand much less about how paracrine communication systems functionally integrate with neural circuits to elicit changes in neural function and behavior (Marder, 2012). Considering suggestions that paracrine communication networks emerged before synapses (Mackie, 1990; Moroz, 2009), placozoans might resemble an ancient metazoan arrangement, over which a faster and more compartmentalized communication system emerged with the advent of synapses (Jékely, 2021; Wong et al., 2019). Accordingly, placozoans express a rich complement of gene homologs for volume transmission, including those for the production and controlled secretion of signaling molecules, and an array of ionotropic and metabotropic receptors (Moroz & Kohn, 2015; Moroz et al., 2021; Wong et al., 2019). Investigations into the molecular processes that underlie cellular communication in placozoans may therefore shed light on evolutionary processes by which fast neural electrogenic/synaptic signaling evolved and became functionally intertwined with the more ancestral, slower, and longer-lasting paracrine communication networks (Katz & Lillvis, 2014).

Compared to other early-diverging animals, we know very little about the prevalence, importance, and mechanisms for cell signaling in placozoans. Based purely on gene content and expression, placozoans likely utilize complex electrochemical and paracrine signaling for coordinating cellular activity and behavior. Going forward, some key questions that need to be answered are: (i) What are the resting membrane properties of the different cell types? (ii) What is the nature and diversity of electrical signaling in the different cell types? (iii) What are the mechanisms by which cells use electrical and paracrine signaling to communicate with each other, and to elicit intracellular changes for the control of physiology and behavior?

8.2.2 FAST ELECTRICAL SIGNALING

Like all other animals, *Trichoplax* expresses genes necessary for establishing ionic gradients and generating resting membrane potential including the Na^+/K^+ ATPase, Na^+ and K^+ leak channels, plasma membrane, and ER Ca^{2+} ATPase, the Na^+/Ca^{2+} exchanger, various other ion pumps and exchangers (Srivastava et al., 2008; Wong et al., 2019). Also, like other animals, perfusion of external K^+ ions, which would diminish the K^+ gradient across the cell membrane and in doing so cause cells to depolarize, causes reversible contractions in *Trichoplax adhaerens* consistent with excitation-contraction coupling (Figure 8.3). For some time, it has been known that placozoans also express voltage-gated potassium (K_V) and sodium (Na_V) channels that would allow them to fire action potentials (Eitel et al., 2018; Srivastava et al., 2008; Wong et al., 2019). However, it was only recently that action potentials were experimentally confirmed, in a study that employed extracellular electrodes to record action potentials from immobilized animals (Romanova et al., 2020b). Four placozoan haplotypes, two from the *Trichoplax* genus (H1 and H2), and two from the *Hoilungia* genus (H13 and H4; Figure 8.1(D)) were found to generate evoked trains of sodium-dependent action potentials when stimulated by a recording electrode. The authors of this study also reported bursts of action potentials from single dissociated crystal cells, noted earlier to be involved in gravitactic behavior (Mayorova et al., 2018).

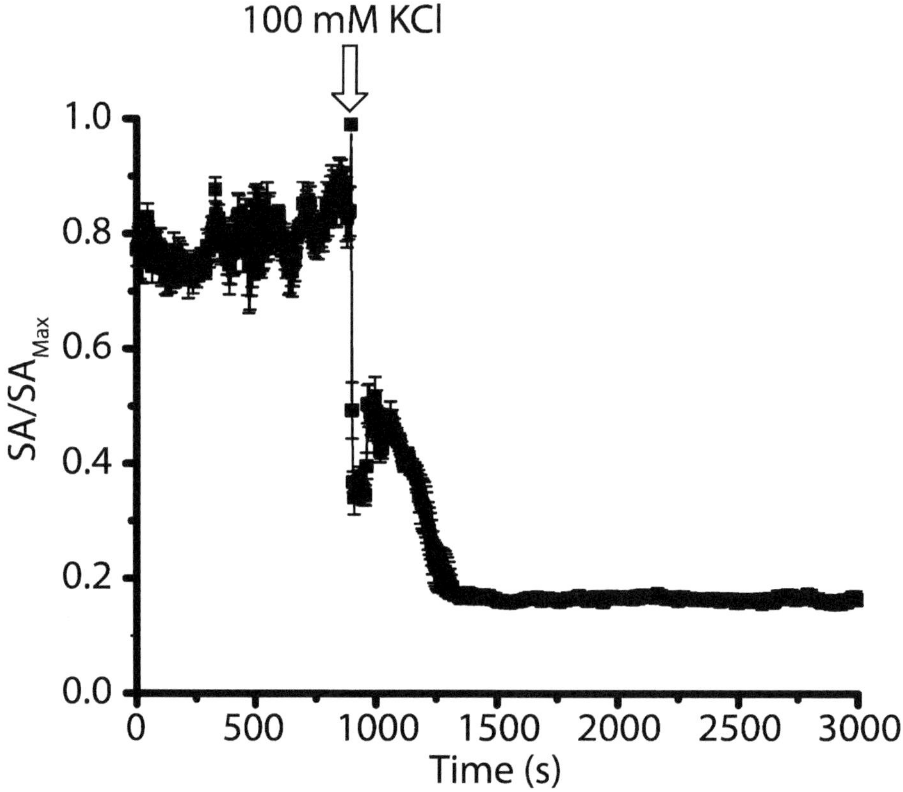

FIGURE 8.3 The plot of the average surface area of *Trichoplax adhaerens* specimens over time, normalized against the maximal surface area, before and after perfusion of a seawater solution containing 100 mM K^+ (i.e., KCl), revealing strong and prolonged contraction after its application.

In contrast to sponges, which appear to have lost Na_V and K_V channels (Moran et al., 2015), placozoans underwent an expansion in Na_V channel genes relative to other early-diverging animal lineages, with 5 paralogues identified in the *Trichoplax* H1 and H2 haplotypes, and 7 in the *Hoilungia* H13 and H4 haplotypes (Romanova et al., 2020b). This unique expansion, coupled with the shared expansion and diversification of shaker family K_V channels with cnidarians and bilaterians (Li et al., 2015), would suggest that placozoans differentially express distinct sets of Na_V and K_V channels in different cell types to produce variations in electrical signaling. Thus, efforts to understand the nuances and significance of electrical signaling in placozoan cells will require an exploration of the varying functional/biophysical properties of their Na_V and K_V channels, as well as other relevant electrogenic genes, coupled with explorations of how these genes are differentially expressed in different cell types.

A notable observation that emerged from the recent identification of placozoan action potentials was the apparent requirement for external Na^+ ions for the occurrence of action potentials, which suggests that the *Trichoplax* Na_V channels are selective for Na^+ and unable to conduct other extracellular cations such as Ca^{2+}. This observation does not fit with the proposed evolution of sodium selective channels in animals. Specifically, ancestral Na_V channels, dubbed Na_V2 channels, possess four extracellular loops (Pore loops or P-loops) each projecting a critical amino acid into the central region of the ion permeation pathway, to give rise to an ion selectivity motif of aspartate (D), glutamate (E), glutamate (E), and alanine (A) (i.e., a DEEA motif; Figure 8.4). The presence of three oxygen-bearing amino acid side chains (i.e., carboxyl groups of the DEE residues), resembles the selectivity filter of voltage-gated calcium channels of EEEE or EEDD, that together produce a high-affinity binding site for external Ca^{2+} ions resulting in exquisite selectivity for Ca^{2+} over the more abundant Na^+ (Sather & McCleskey, 2003; Talavera & Nilius, 2006). Accordingly, Na_V2 channels bearing DEEA selectivity filter motifs have been shown to be quite Ca^{2+} permeable, and in some cases, highly Ca^{2+}-selective (Gosselin-Badaroudine et al., 2016; Zhang et al., 2011; Zhou et al., 2004). Bona fide Na^+-selective Na_V channels are thought to have evolved independently and exclusively in bilaterians and cnidarians via alterations in the selectivity filter motif. In bilaterians, gene duplication gave rise to a clade of Na_V1 channels (Liebeskind et al., 2011), which took up a lysine (K) in the third position of the selectivity filter producing a motif of DEKA. The positive charge of the lysine side chain is thought to disrupt Ca^{2+} binding and promote Na^+ selectivity. Independently in cnidarians, some Na_V2 channel genes adopted a lysine in the second position of the selectivity filter (DKEA) (Liebeskind et al., 2011), and these were shown to be Na^+ selective when expressed *in vitro* (Barzilai et al., 2012) (Figure 8.4). For placozoans, which only possess Na_V2 channels, greater diversity has been reported with selectivity filter motifs of DEEA, DEDA, DEET, DEDT, and DEDS (Romanova et al., 2020b). The ion conductance of these different Na_V2 channels has not been examined but given that these animals fire Na^+-dependent action potentials, it seems likely that at least one of these Na_V2 channel genes may be sodium selective.

In addition to genes for fast action potentials, placozoans express an array of genes capable of generating graded potentials, many of which are utilized for receiving sensory information or detecting secreted signals from other cells. This includes a large set of ionotropic receptors (IRs) that become activated by extracellular ligands, such as ionotropic glutamate receptors (iGluRs) that respond to glutamate and/or glycine, transient receptor potential (TRP) and Degenerin/epithelial (Deg/ENaC) sodium channels that are gated by various ligands, purinergic (P2X) receptors that are gated by adenosine triphosphate (ATP), and mechanically gated channels such as Piezo. Perhaps the most significant absence is the Cys-loop channels, precluding conventional ionotropic responses to the transmitter ligands acetylcholine, GABA, and serotonin.

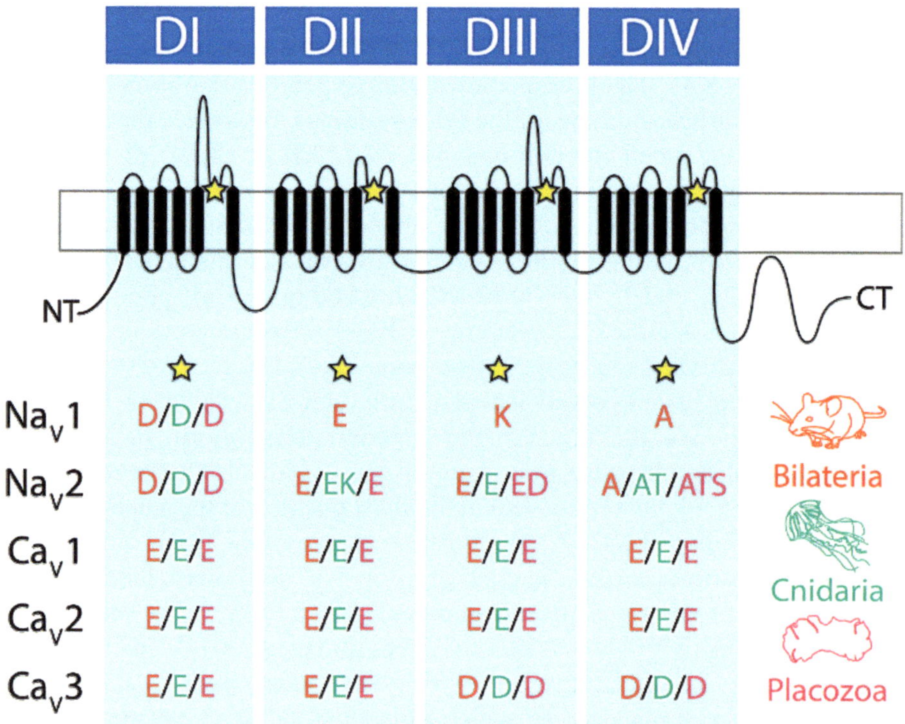

FIGURE 8.4 Top: illustration of the general transmembrane primary structure of voltage-gated sodium (Na_V) and calcium (Ca_V) channels comprised of four repeat domains (DI-DIV) each containing six transmembrane alpha helices. The yellow stars indicate the locations of residues that form the selectivity filter in the pore of the folded channel protein. Bottom: amino acid signatures of Na_V and Ca_V channel selectivity filters, color-coded for bilaterians (orange), cnidarians (teal), and placozoans (pink), showing their variation across the three phyla.

8.2.3 CA^{2+} SIGNALING

Generally, the ionic basis for resting membrane potential, fast action, and graded potentials, involves the monovalent cations sodium (Na^+) and (K^+). That is, pumps and exchangers establish a gradient of these two ions across the cell membrane, hence a voltage, and subsequently, Na^+ and K^+ leak channels create asymmetric ionic charge concentration across the membrane producing a voltage. This stored energy provides the driving force for ions to flow across the membrane through transiently activatable ion channels and neurotransmitter receptors that give rise to fast electrical impulses. The Ca^{2+} ion, which is generally maintained at very low levels within the cell cytoplasm, serves a dual role in electrical signaling in that it directly contributes to action and graded potentials when conducted across the cell membrane, and once in the cytoplasm, to Ca^{2+} signaling by binding oxygen moieties on target proteins to trigger conformation changes (Clapham, 2007). That extensive cytoplasmic Ca^{2+} signaling occurs in placozoans is suggested by the *Trichoplax* transcriptome, where numerous types of Ca^{2+} permeable channels and cytoplasmic Ca^{2+} binding proteins are expressed, including the important Ca^{2+} sensor protein calmodulin (CaM), which is amongst the most highly-expressed mRNAs at the whole animal level, and intracellular ryanodine receptors and IP_3 receptors that serve to release Ca^{2+} from the ER in a highly regulated manner (Wong et al., 2019).

At the cell membrane, the most Ca^{2+}-selective ion channels are the voltage-gated Ca^{2+} (Ca_V) channels, which serve critical functions in converting graded and action potentials into cytoplasmic Ca^{2+} signals. For example, Ca_V channels couple membrane excitation with the control of ciliary beating, regulated exocytosis of neurotransmitters, gene transcription, control of cell cycle, and muscle contraction (Capiod, 2011; Catterall, 2011; Senatore et al., 2016). Ca_V channels are likely derived from the same stem eukaryotic channel as Na_V channels due to their shared transmembrane structure (Figure 8.4) (Pozdnyakov et al., 2018). Humans and most animals possess three types of voltage-gated calcium channels, called Ca_V1, Ca_V2, and Ca_V3. Genome duplications in vertebrates gave rise to several paralogues, producing four Ca_V1 channels (i.e., $Ca_V1.1$–$Ca_V1.4$), three Ca_V2 channels ($Ca_V2.1$–$Ca_V2.3$), and three Ca_V3 channels ($Ca_V3.1$–$Ca_V3.3$) (Catterall, 2011; Ertel et al., 2000). Instead, invertebrates generally possess just single copies of each of the three Ca_V channel types, and the most early-diverging animals to possess all three are the placozoans (Moran & Zakon, 2014; Piekut et al., 2020; Senatore et al., 2012). Since Ca_V1–Ca_V3 channels serve specialized and conserved roles in various metazoan cell types (Senatore et al., 2016), placozoans and their limited cell repertoire provide unique opportunities for exploring the evolution of calcium channel function in animals.

Notably, *Trichoplax* is one of only three invertebrate species for which all three Ca_V channel types have been cloned and expressed *in vitro* for electrophysiological characterization, contributing insights into the evolution of Ca_V channel structure-function properties. Previous work in the other two species, the pond snail *Lymnaea stagnalis*, and the honeybee *Apis melifera*, collectively suggested that the core functional features that distinguish the different Ca_V channel types from each other are deeply conserved, at least present in the common bilaterian ancestor. For example, *Apis* and *Lymnaea* Ca_V channels share with vertebrate homologs the core functional distinction on voltage sensitivity, namely that Ca_V1 and Ca_V2 channels require strong depolarization to become activated, while Ca_V3 channels begin activating at low membrane voltages in response to mild depolarization (Cens et al., 2015; Senatore & Spafford, 2010; Spafford et al., 2003a, 2006). Thus, whereas Ca_V1 and Ca_V2 channels respond to action potentials to exert downstream effects, Ca_V3 channels operate below the action potential threshold, help promote the generation of action potentials, and are often involved in rhythmic excitability such as regulating heartbeat or rhythmically contracting smooth muscle cells (Perez-Reyes, 2003). Furthermore, key modulatory differences between Ca_V1 and Ca_V2 channels are conserved, namely, the selective regulation of Ca_V2 channels by cytoplasmic G proteins, which become activated by neuromodulatory G protein-coupled receptors (de Hoog et al., 2019; Dong et al., 2018; Huang et al., 2010; Tedford & Zamponi, 2006), and Ca^{2+}/calmodulin feedback regulation of Ca_V1 channels (inhibition), elicited by local rises in cytoplasmic Ca^{2+} emanating from the channel pore (i.e., transient local Ca^{2+} signals, as opposed to global longer lasting Ca^{2+} signals that occur for both Ca_V1 and Ca_V2 channels) (Ben-Johny & Yue, 2014; Spafford et al., 2003b; Taiakina et al., 2013). These differences in modulation have important implications for the unique contributions of Ca_V1 and Ca_V2 channels in animal cells. Bilaterian Ca_V2 channels, for example, are the primary drivers for the exocytosis of neurotransmitters at the pre-synaptic terminal (Senatore et al., 2016). Thus, regulation by G proteins provides avenues for controlling synaptic transmission through various paracrine cell signaling systems, important for the neuromodulation of neural circuits and behavior. Instead, Ca_V1 channels are the central drivers for muscle contraction in bilaterians (Senatore et al., 2016), where their feedback sensitivity to cytoplasmic Ca^{2+} perhaps makes them better suited to dynamically regulate Ca^{2+} influx for the activation of contractile filaments.

The first *Trichoplax* Ca_V channel to be functionally characterized was the Ca_V3 channel, prompted by its structural resemblance to a hypothesized ancestral Ca_V3 channel proposed to be highly permeable to Na^+. That is, work on the *Lymnaea* Ca_V3 channel identified alternative splicing in a region of the pore that rendered this particular channel highly permeable to Na^+, despite retaining a canonical Ca_V3 channel selectivity filter motif of EEDD (Senatore et al., 2014) (Figure 8.4). These findings went against the convention that four key amino acids in the selectivity filter of Ca_V channels are the central determinants for Ca^{2+} selectivity because the alternative splicing that altered ion selectivity did not alter the selectivity filter amino acids, but rather, a position adjacent to the second glutamate of the EEDD motif (Senatore et al., 2014). In this study, structural observations were made suggesting that ancestral Ca_V3 calcium channels were originally sodium channels and that Ca^{2+} selectivity evolved secondarily in bilaterians, after the placozoan divergence. Cloning and electrophysiological characterization of the *Trichoplax adhaerens* Ca_V3 channel *in vitro* revealed that despite considerable divergence from other Ca_V3 channels, both phylogenetically and in terms of protein sequence, the placozoan orthologue produced voltage-gated Ca^{2+} currents very similar to those of other Ca_V3 channels, most particularly in their low-voltage activation (Smith et al., 2017). Thus, like other characterized Ca_V3 channels, the *Trichoplax* Ca_V3 channel may contribute to sub-action potential events, such as controlling membrane excitability and driving low threshold exocytosis (Perez-Reyes, 2003). Importantly, the *Trichoplax* channel was selective for Ca^{2+}, going against the hypothesis of high Na^+ permeability. However, the pore itself was found to be less selective for Ca^{2+} than vertebrate Ca_V3 channels, suggesting that Ca^{2+} selectivity was retained via the block of Na^+ permeation by extracellular Ca^{2+} ions binding to the extracellular surface of the channel pore with high affinity (Smith et al., 2017).

Subsequent cloning and functional characterization of the *Trichoplax* Ca_V1 and Ca_V2 channels have provided additional interesting insights. As expected, both channels are high voltage-activated, consistent with roles in coupling action potentials with cytoplasmic Ca^{2+} signaling (Gauberg et al., 2022). However, compared to the cloned channels from *Lymnaea* which exhibited highly overlapping biophysical properties with their vertebrate orthologues, the two *Trichoplax* channels were more functionally divergent, suggesting that the distinguishing biophysical features of Ca_V1 and Ca_V2

channels as observed in vertebrates and bilaterian invertebrates emerged after the placozoan divergence. Interestingly however, like bilaterian Ca_V channels, the *Trichoplax* Ca_V1 channel undergoes both global and local Ca^{2+}/calmodulin feedback regulation, while the Ca_V2 channel only exhibits global regulation (Gauberg et al., 2022). Given that Ca_V1 and Ca_V2 channels are proposed to have evolved by gene duplication from an ancestral $Ca_V1/2$ type channel somewhere along the stem metazoan lineage (Moran et al., 2015), it would appear that this particular distinguishing feature evolved soon after these channel types were established. Conversely, both *Trichoplax* Ca_V1 and Ca_V2 channels were found to lack direct G protein modulation, precluding canonical regulation by neuromodulatory signaling pathways (Gauberg et al., 2022). Altogether, these observations suggest that the divergent modulation of Ca_V1 and Ca_V2 channels by Ca^{2+}/calmodulin evolved in the common ancestor of placozoans, cnidarians, and bilaterians and that Ca_V2 channel regulation by neuromodulatory G proteins emerged after the placozoan divergence.

Future studies aimed at understanding the physiological functions of Ca_V channels in placozoans will be important, as they might shed light on the distinct adaptations of Ca_V1 to Ca_V3 channels for Ca^{2+} signaling in different cell types. This includes studies on the protein-protein interactions that serve to regulate Ca_V channel function, and importantly, localize the channels to discrete cellular compartments to permit discrete, localized Ca^{2+} signaling.

8.2.4 G PROTEIN COUPLED RECEPTOR SIGNALING

G-protein coupled receptors (GPCRs) are unique to eukaryotes, used to detect various extracellular signals including peptides/hormones, neurotransmitters, light, protons/pH, nucleotides, Ca^{2+}, extracellular proteolysis, and contact with extracellular proteins. In these various capacities, GPCRs evolved to play central roles in metazoan multicellularity and cell signaling. GPCRs share a common primary structure of seven transmembrane α-helices, which form a tertiary receptor structure bearing an intracellular domain where the G-protein subunit Gα, Gβ, and Gγ interact. For ligand-activated receptors, extracellular ligands bind either the N-terminus of the GPCR or a pocket within the transmembrane helices. Upon ligand binding, conformational changes in the receptor trigger GTP/GDP nucleotide exchange on the Gα subunit, activating the trimeric G proteins via dissociation of Gα from the Gβγ heterodimer. Because of their anchoring to the intracellular surface of the cell membrane, activated Gα and Gβγ subunits diffuse away from the GPCR strictly along the membrane, binding to nearby target proteins to exert their effects.

GPCRs are typically classified into five families: Rhodopsin (class A), Adhesion (class B), Secretin (class B), Glutamate (class C), and Frizzled (class F) (Schiöth & Fredriksson, 2005), with corresponding distinctions in several functional motifs and their N-terminal architectures (Krishnan et al., 2012). In fact, bioinformatic analyses indicate that the GPCR superfamily is quite ancient, with the common ancestor of animals and fungi bearing homologs of Rhodopsin, Frizzled, Adhesion, and Glutamate GPCR families (Krishnan et al., 2012). Instead, secretins are thought to have evolved from class B Adhesion GPCRs in the common ancestor of cnidarians and bilaterians (Nordström et al., 2009). In *Trichoplax*, GPCR prediction analyses have identified over 665 putative GPCRs (Kamm et al., 2018; Wong et al., 2019), including 533 Rhodopsin, 64 Glutamate, 63 Adhesion, 2 Frizzled, and 3 that fall outside of the noted classification scheme (i.e., 2 cAMP-like and 1 ocular albinism-like) (Wong et al., 2019). This high number of GPCRs resembles the numbers of most other animals, in contrast to sponges and arthropods which have reduced GPCR gene counts, likely through the loss of Rhodopsin and Adhesion/Secretin receptor gene counts (De Mendoza et al., 2014). Within Placozoa, GPCRs have gone through unique expansions with independent duplications occurring in *Hoilungia hongkongensis* and *Trichoplax adhaerens* (Eitel et al., 2018).

Trichoplax also has extensive gene homologs for the intracellular action of GPCRs (De Mendoza et al., 2014), including various classes of G proteins, arrestins, phosducins, G protein-coupled receptor kinases (GRKs), regulators of G protein signaling (RGS), Ric8, and GoLoco (Nordström et al., 2009). Transcriptome sequencing also confirmed the expression of downstream GPCR signaling elements that serve to integrate neuromodulatory GPCR signaling and electrochemical/synaptic neural signaling (Wong et al., 2019), such as the effector enzyme adenylyl cyclase that increases cytoplasmic cAMP levels upon activation by G proteins, and the cAMP-dependent protein kinase A (PKA), which phosphorylates various proteins involved in electrical signalings, such as Cav1 channels, causing long-lasting changes in cellular excitability (Gao et al., 1997; Sassone-Corsi, 2012). Similarly, the G-protein activated effector enzyme phospholipase C hydrolyses phosphatidylinositol 4,5-bisphosphate (PIP_2) to produce diacylglycerol (DAG) and inositol triphosphate (IP_3), which respectively activate the kinase protein kinase C (PKC) and IP_3 receptors at the ER, the latter releasing Ca^{2+} into the cytoplasm for signaling (Gamper & Shapiro, 2007; Prole & Taylor, 2019; Shearman et al., 1989). PKC phosphorylates various targets including ion channels and synaptic proteins (Shearman et al., 1989), and like PKA, acts to exert long-lasting changes in cellular excitability and synaptic signaling. Like Ca_V2 channels, activated G proteins can directly bind and modulate other ion channels (Wickman & Clapham, 1995) including G protein-activated inward rectifying K^+ (GIRK) channels (Lüscher & Slesinger, 2010), and in mammals, $Ca_V3.2$ calcium channels (Perez-Reyes, 2010). In bearing such a rich complement of electrogenic and neuromodulatory genes, placozoans cells are equipped to carry out cell-cell signaling that is exquisitely complex, both spatially and temporally, posing interesting questions about the extent and homology of the integration between fast electrogenic and slow neuromodulatory signaling machinery.

8.2.5 CONSIDERATIONS ON CELL SIGNALING BY PLACOZOAN CELL TYPES

Collectively, our knowledge about placozoan cellular features and genomics would suggest that different cell types can utilize complex electrochemical, neuromodulatory/paracrine, and Ca^{2+} signaling akin to that observed in neurons, muscle, and other excitable cells of cnidarians and bilaterians. Based on the features of the different placozoan cell types described in Section 8.1.3, several suggestions can be put forward in this regard.

Dissociated crystal cells are the only cell type so far confirmed to be excitable, observed to be quiescent but fire tonic bursts of action potentials when stimulated with a recording electrode (Romanova et al., 2020b). This feature would suggest that *in vivo*, crystal cell excitability requires graded depolarization, and given the role of this cell type in gravity sensing (Mayorova et al., 2018), action potentials and electrical signaling might transmit information about body orientation along vertical planes. As noted in Section 8.1.3, crystal cells are sparsely arranged along the periphery of the flat animal, with the rims of their cup-shaped nuclei pointing concentrically outward. This arrangement causes the cytoplasmic calcium carbonate crystals to either fall into the nuclear cup if the cells are located along the top edge of a vertically-oriented animal, or out of the cup and against the cell membrane if the cells are along the bottom edge. Crystal cells bear extensive actin filaments, and form contact associations with epithelial cell extensions also containing a thick network of actin, perhaps for the purpose of anchoring crystal cells such that their nuclei are appropriately positioned (Mayorova et al., 2018). These same cytoskeletal networks might provide the necessary structural elements for mechanotransduction, where for example, mechanosensitive channels expressed at the cell membrane could become activated when the aragonite crystal contacts the cell membrane in cells positioned at the bottom of a vertically oriented individual. If so, ion channels that are known to be mechanosensitive in other animals, such as piezo, TRP, and Deg/ENaC, could be providing the necessary depolarization (graded potentials) for generating action potentials. Through their contacts with both fiber cells and epithelial cells (Mayorova et al., 2018), crystal cell action potentials might transmit local signals that cause, for example, ventral ciliated cells in the vicinity to increase beating frequency, providing a net upward movement relative to more slowly beating cilia along the top edge. Still, given the absence of synapses, it is unclear how crystal cell action potentials, or any form of intracellular signaling, would be transmitted to other cells. Perhaps, secretion or release of diffusible signaling molecules, or extracellular field potentials could be involved. Notably, the presence of mitochondria that surround the aragonite crystals could reflect signal transduction that is not dependent on action potentials and mechanosensitivity per se, but rather, through the cellular properties of mitochondria. For example, it is established that mitochondria and the ER dynamically shuffle Ca^{2+} between them to control intracellular Ca^{2+} signaling (Scheitlin et al., 2016) and that their proximity to each other is a key factor for this process (Rizzuto et al., 1998). Thus, when the aragonite crystal and its surrounding mitochondria are inside the nuclear cup, the mitochondria could either release or take up Ca^{2+} ions to exert effects on the Ca^{2+}-sensitive ryanodine and/or IP_3 receptors of the ER. Mitochondrial Ca^{2+} release would dynamically regulate IP_3 receptors (Finch et al., 1991; Watras et al., 1991), and activate ryanodine receptors through the process of Ca^{2+}-induced Ca^{2+} release (CICR) (Franzini-Armstrong & Protasi, 1997). If instead signal transduction takes place at the cell membrane, positioning of the mitochondria against the membrane could result in the release of ATP into the extracellular environment, activating auto- and/or paracrine purinergic signaling systems, as occurs in vascular endothelial cells where mitochondria play an important role in shear-force induced cytoplasmic signaling, for the production of the diffusible effector signaling molecule nitric oxide (NO) (Andrews et al., 2014; Scheitlin et al., 2016). Notably, electrical excitability and Ca_V channels are also implicated in NO production (Gilbert et al., 2017), a compound that is able to diffuse across the cell membrane to generate local signals (*see* Section 8.2.10.1), making it a viable candidate for local cell signaling by crystal cells given the absence of synapses and electrical junctions with other cells.

As previously discussed, gland cells express both pale and dense core vesicles (Eitel et al., 2018; Mayorova et al., 2019; Romanova et al., 2021; Smith et al., 2014), the former associated with small molecule transmitters, and the latter with peptide transmitters such as neuropeptides and hormones. Also previously noted, gland cells express neuropeptides (Mayorova et al., 2019; Senatore et al., 2017), proteins associated with regulated exocytosis (e.g., complexin, syntaxin, and SNAP-25) (Smith et al., 2014), and Ca_V channels (Gauberg et al., 2020; Gauberg et al., 2022; Smith et al., 2017), altogether consistent with neuroendocrine-like properties and the capacity for regulated secretion of signaling molecules. Although action or graded potentials have yet to be recorded from gland cells, their expression of Ca_V channels, shown *in vitro* to require fast depolarization for activation (Gauberg et al., 2020; Gauberg et al. 2022; Smith et al., 2017), would certainly suggest that these cells utilize fast electrical and transient Ca^{2+} signaling. In canonical neural and neuroendocrine systems, the control of action potentials is directly linked to secretion, through the recruitment of Ca_V channels positioned close to Ca^{2+} sensitive exocytotic proteins such as synaptotagmin and complexin (Südhof, 2012). Immunolabeling studies have shown that type II gland cells express all three Ca_V channels (Gauberg et al., 2020; Gauberg et al., 2022; Smith et al., 2017). Thus, the expressed low voltage-activated Ca_V3 channel could contribute to low threshold exocytosis, occurring below the action potential threshold as observed in mammalian neuroendocrine cells and some neurons (Carbone et al., 2006; Weiss & Zamponi, 2013), while Ca_V1/Ca_V2 channels could drive more pronounced exocytosis in response to one or a burst of action potentials (Mansvelder & Kits, 2000). The other two types of gland cells have been less investigated, but voltage-gated calcium channels (Gauberg et al., 2020) and other peptides (Varoqueaux et al., 2018) are expressed in similar regions as these gland cells (Mayorova et al., 2019) (Figure 8.2(B)), thus they too may be excitable and undergo Ca^{2+}-dependent exocytosis. If so, like crystal cells, extracellular signals would be required to depolarize the cell membrane to recruit Ca_V channels and Ca^{2+} influx for secretion. In bearing cilia, type I and III gland cells may be sensory (Jékely, 2011). Perhaps on these cilia, membrane receptors for secreted ligands and other extracellular compounds (ionotropic or metabotropic), or mechanosensory receptors, can generate membrane depolarization to drive excitation-secretion coupling.

Recently, dorsal epithelial cells were shown to be dynamically contractile, with contractions dependent on corresponding rises in cytoplasmic Ca^{2+} (Armon et al., 2018). Ca_V1 channels have also been immunolocalized to the dorsal epithelium (Gauberg et al., 2022). This data, combined with the observation that perfusion of external K^+ causes whole-body contraction (Figure 8.3), suggests that dorsal epithelial cells are capable of excitation-contraction coupling. In muscle, depolarization and resulting Ca^{2+} influx through Ca_V1 channels activate ryanodine receptors at the ER, releasing ER Ca^{2+} into the cytoplasm to activate contractile myofibers (i.e., CICR) (Collier et al., 2000). Although dense arrangements of myofibers have not been observed in placozoan cells (Smith et al., 2014), *Trichoplax* possess the core genes required for smooth muscle contraction (Steinmetz et al., 2012), and perhaps, myofibers are arranged sparsely making them difficult to observe under the electron microscope. An interesting feature of dorsal contractile cells is that contraction appears to propagate along the dorsal epithelium in waves (Armon et al., 2018), which given the absence of gap/electrical junctions, suggests that mechanotransduction is occurring, leading to waves of depolarization and contraction along adjacent cells. Furthermore, the observed contraction of placozoans in response to applied endogenous neuropeptides (Varoqueaux et al., 2018), glycine (Romanova et al., 2020a), and UV light (Pearse, 1989) would suggest that contraction can be regulated by external signals, perhaps activating membrane receptors on dorsal epithelial cells or other cells that promote network paracrine signaling.

Placozoans exhibit a range of dynamic locomotory behaviors via the regulation of beating cilia on their ventral epithelium. For example, *Trichoplax* will pause ciliary locomotion when it detects food (Smith et al., 2015), and move faster in light compared to dark conditions (Heyland et al., 2014). In other eukaryotic cells, ciliary beating is regulated by a variety of cytoplasmic second messenger systems and Ca^{2+} signaling. In eukaryotes, activation of the enzyme adenylyl cyclase leads to increases in cytoplasmic cAMP levels, in turn activating PKA. This enzyme phosphorylates various microtubule-associated proteins on cilia, such as dynein, to regulate ciliary beating (Hamasaki et al., 1991; Hochstrasser et al., 1996; Salathe, 2007). This observation would link GPCR signaling to the control of cilia, through interactions between activated G proteins and adenylyl cyclase. Notably, neuropeptides, the phospholipase C pathway, and PKC are also implicated in the control of ciliary beating in mammals (Wong et al., 1998). Lastly, Ca^{2+} signaling also plays an important role in the control of ciliary beating, where in mammals, transient rises in cytoplasmic Ca^{2+} lead to increased beating frequency (Salathe, 2007), while in paramecia, the opposite is observed (Naitoh & Kaneko, 1972). Given that ciliary locomotion by placozoans appears to be regulated by both internal and environmental/external signals, it may be that electrochemical and paracrine signaling systems are being utilized as observed for cilia in other organisms.

A consideration of the roles of cellular signaling in Fiber cells is also interesting. In bearing cytoplasmic bridges between them (i.e., dense septa) (Buchholz & Ruthmann, 1995; Smith et al., 2014), fiber cells form a network that perhaps acts as a syncytium for transmitting electrogenic or biochemical signals. In contacting each other and all other cell types (Smith et al., 2014), fiber cells are thus poised to relay information across the animal body, particularly between the loosely connected dorsal and ventral epithelia. Isolated fiber cells form extensions (lamellipodia) that are reported to be contractile, bearing microtubules and dense arrangements of actin filaments (Thiemann & Ruthmann, 1989). However, organized myofilaments (myosin and actin arrays) have not been observed for any placozoan cell type (Smith et al., 2014). Nonetheless, single-cell analysis has found contractile elements expressed in a cell type proposed to be fiber cells (e.g., myosin, tropomyosin, calponin) (Sebé-Pedrós et al., 2018). Based on these various observations, fiber cells are proposed to mediate body contractions, but it is difficult to envision how the thin and relatively sparse extensions that fiber cells use to contact other cells would withstand the forces to support whole-body contractions. Rather, the observed dynamic motility of fiber cell lamellipodia, including contractions, is perhaps more appropriate for establishing dynamic contacts with other cells for cell signaling purposes, or cell migration as observed for fibroblasts and immune cells. In other cells, lamellipodial contraction has been associated with actin depolymerization at the rear end, rather than through canonical muscle filaments (Wolgemuth, 2005), which would be consistent with the absence of myofilaments and the presence of actin bundles in fiber cell extensions. Interestingly, in their morphology fiber cells resemble immune macrophage cells that use lamellipodia to bind and phagocytose foreign cells such as bacteria. This similarity is further reflected by the capacity of fiber cells to phagocytose foreign cells (algae) (Wenderoth, 1986), and their bearing vacuolar (perhaps lysosomal) inclusions of partially degraded algae (Smith et al., 2014). The role of fiber cells in innate immunity is further supported by their expression of marker genes associated with phagocytosis, and electron and light microscopy images that provide strong evidence that these cells can engulf dead and unhealthy/dying cells and heat-killed bacteria. Interestingly, fiber cells were also found quickly migrate towards the experimentally wounded edges of the animal forming a rapid seal to the outside environment and promoting full healing within just one hour. The wound healing process did not cause a significant change in proliferation which supports the motile property of fiber cells and healing by virtue of existing cell rearrangements. The aforementioned role of fiber cells suggests that electrical and GPCR signaling could play similar roles as observed in bilaterian immune cells for directed migration towards wound sites, regulating cellular proliferation, cytokine signaling, and phagocytosis (Feske et al., 2015; Lattin et al., 2007; Wang et al., 2019).

Lastly, the roles of cellular signaling for lipophil cells are perhaps best inferred from the observation that membrane fusion of their large acidic vesicles, containing hydrolytic enzymes for external hydrolysis and digestion of food, only occurs along the ventral epithelium in close vicinity to the detected food particles (Smith et al., 2015). This would certainly suggest that the fusion of these large acidic vesicles is functionally coupled to the detection of extracellular signals, perhaps chemosensory, but also mechanosensory. In cnidarians, a convergence of these two sensory modalities is used to regulate the exocytotic release of the venom-laced stinging barbs from nematocytes, preventing barb release if either modality is presented on its own. Here, chemosensory input arrives via a synaptic pathway that causes the nematocyte cell membrane to become hyperpolarized, permitting mechanically-induced depolarization to activate a Ca_v channel that drives barb exocytosis (Weir et al., 2020). For placozoans that have limited cell types and lack synapses, receptors that can integrate different extracellular signals on the same cell would permit coincidence detection, and hence, more precise control of cellular activity in response to external signals.

8.2.6 PEPTIDE SIGNALING AND BEHAVIOR

Small regulatory peptides and peptide hormones are a diverse class of signaling molecules used in the animal nervous system to elicit changes in physiology and behavior, most often acting as neuromodulatory ligands that regulate long-range neural circuit function (e.g., neurohormones from the lymph and circulatory systems), as well as short-range (e.g., volume transmission, paracrine, and autocrine signaling) (Marder, 2012). Peptides and hormones are generated from pre-propeptide precursors that are cleaved in the ER by the enzyme pro-hormone convertase at mono/di-basic cleavage sites (i.e., lysine or arginine residues). The cleaved peptides can be further processed by carboxypeptidase B, which removes basic residues from C-terminal ends, and subsequently, peptidyl glycine monooxygenase (PAM) amidates C-terminal glycine residues to produce amidated peptides such as RFamides (Eipper et al., 1992). These mature peptidergic signaling molecules are packaged into dense-core vesicles in neurons and neuroendocrine cells, where their regulated exocytosis occurs in a Ca^{2+}-dependent manner, using similar machinery found in the pre-synaptic terminal, but with more latency between membrane excitation and secretion, and a propensity for incomplete vesicle fusion and kiss-and-run exocytosis (Malacombe et al., 2006; Thorn et al., 2016). Peptidergic signaling is presumed to be one of the oldest signaling modalities of cellular communication in the Metazoa, employed by extant species from all major animal phyla. The evolutionary origin of several peptide/hormone families can be traced to the common ancestor of cnidarians and bilaterians, with a few also conserved in placozoans and ctenophores (de Oliveira et al., 2019; Nikitin, 2015; Wong et al., 2019). Sponges are the only exception since their genome lacks bilaterian homologs of known neuropeptides, but harbor the enzymatic machinery required to produce them (Srivastava et al., 2010). Other subsequent studies have reported that cnidarian peptides can induce settling behavior in sponge larvae, and the *Amphimedon queenslandica* genome and other sponge transcriptomes reported hypothetical regulatory peptides with no recognizable homology to bilaterian neuropeptides (Schoofs et al., 2017; Whalan et al., 2012). Moreover, recent studies have indicated that the neuropeptide precursor genes for phoenixin and nesfatin-1 have homologs choanoflagellates, sponges, ctenophores, cnidarians, and bilaterians (lost in placozoans), suggesting these particular peptides are of premetazoan origin (Luis Alfonso et al., 2021).

The machinery involved in producing peptides is believed to have even deeper ancestral roots, with putative enzyme homologs of prohormone convertase, carboxypetidase B, and PAM, as well as amidated peptides identified in the green algal species *Chlamydomonas reinhardtii* (Kumar et al., 2016; Luxmi et al., 2019). In *Chlamydomonas*, planaria, and mice, PAM plays a conserved role in ciliogenesis (Kumar et al., 2017), and remarkably in *Chlamydomonas*, the enzyme amidates pro-peptides that modulate gamete chemotactic behavior during sexual reproduction (Luxmi et al., 2019). The deep ancestry of peptide signaling points to an advantage for cell communication where because they are genetically encoded, they permit diversification of signals through the emergence of distinct peptide sequences and corresponding receptors (Jékely, 2021; Mirabeau & Joly, 2013; Schmidt et al., 2018). This provides the ability to discriminate signals emanating from the self or conspecifics, as opposed to other signaling systems based on amino acids or other small molecule transmitters that are broadly utilized by different organisms.

The presence of peptide-expressing gland cells in placozoans, and their observed behavioral responses to applied endogenous peptides, indicates the use of regulated secretion of peptides for cellular communication and behavior (Senatore et al., 2017; Smith et al., 2014; Varoqueaux et al., 2018). The *Trichoplax* genome encodes all essential enzymes involved in peptide processing including three homologs of prohormone convertase, seven carboxypeptidases, and PAM (Nikitin, 2015; Srivastava et al., 2008). Using homology-based searches and various peptide prediction methods, 18 regulatory peptides and hormones have been identified, including numerous 6 amidated peptides (RWamide, SITFamide, SIFGamide, FFNPamide, YYamide, and DAYQamide), 4 insulin prohormones, and one granulin (Nikitin, 2015; Varoqueaux et al., 2018; Wong et al., 2019). Several of the *Trichoplax* peptides were proposed to be homologous to peptides from other animals: RWamide, similar to *Nematostella vectensis* antho-RFamide, FFNPamide, similar to mollusk (*Aplysia californica*) PRQFVamide, and LFNE, similar to nematode (*C. elegans*) flp-18 FMRFamide-related peptide (Nikitin, 2015). Notably, none of the predicted placozoan peptides have been validated through proteomics, and most of our understanding of placozoan peptide signaling comes from behavioral studies where predicted peptides have been applied ectopically.

The first study implicating peptide signaling in placozoan behavior reported that the invertebrate peptide FMRFamide, and mammalian mu opioid receptor peptides endomorphin-1 (YPWFamide) and endomorphin-2 (YPFFamide), could elicit cessation of ventral ciliary beating and corresponding pauses in *Trichoplax* locomotion when applied exogenously in the bath (Senatore et al., 2017). A second peptide, SIFGamide predicted from the *Trichoplax* genome (Nikitin, 2015), elicited whole-body contractions (writhing) and detachment from the substrate. Although genes encoding FMRFamide-like or endomorphin-like peptides were not identified in the *Trichoplax* genome, transcriptome sequencing led to the identification and cloning of a pre-pro-protein encoding a short peptide with a C-terminal sequence of YPFFGN/S, resembling that of endomorphin-2 (i.e., YPFFamide). It is unclear whether the *Trichoplax* peptide is amidated by PAM *in vivo*, as it bears an asparagine or serine residue downstream of the glycine. However, both the amidated and non-amidated versions could elicit pausing, albeit requiring a higher concentration for the non-amidated peptide (Senatore et al., 2017). As noted earlier, the identified endomorphin-like peptide, named *Ta*ELP, was found to be expressed in neuroendocrine-like type II gland cells (Senatore et al., 2017; Smith et al., 2015). This same study documented naturally occurring pausing behavior that propagates between adjacent individuals, a process that perhaps contributes to the recently described social feeding behavior (Fortunato & Aktipis, 2019). The ability of animals to pause in concert strongly suggests that regulated secretion of *Ta*ELP, or another signaling molecule, is occurring, and affecting not only the individual that secreted the substance but also others nearby. Furthermore, the propagation of pausing behavior among individuals would suggest a positive feedback mechanism of peptide-induced peptide release, given that the estimated abundance of *Ta*ELP in gland cells of a single individual would not suffice to reach a concentration required to elicit pausing in an adjacent animal (Senatore et al., 2017). *In vivo*, *Ta*ELP secretion mostly likely occurs when the animal detects food along its ventral epithelium, activating gland cell secretion, prompting the animal to stop moving and subsequently release digestive enzymes through their granular lipophil cells. Notably, however, certain behaviors observed during feeding are not elicited by *Ta*ELP, including the stretching and flattening of the periphery before lipophil secretion, or churning and rotational movements of the inner parts of the animal after secretion has taken place (Smith et al., 2015). Therefore, other processes and/or signaling molecules must regulate those behaviors.

Knowledge of peptidergic signaling in *Trichoplax* was expanded in a subsequent study investigating the expression patterns and behavioral responses to other predicted regulatory peptides (Varoqueaux et al., 2018). Exogenous application of 11 regulatory peptides (FFNPamide, RWamide, SIFGamide, SITFamide, ELPE, LF, LFNE, MFPF, PWN, WPPF, and YYamide) elicited varying behavioral responses that were categorized into three general types: (i) crinkling/contracting; (ii) rounding up; and rotating; and (iii) flattening and churning. The peptides PWN and SIFGamide induced crinkling of *Trichoplax* lasting for 10 minutes or longer, with SIFGamide causing the most pronounced response often causing the animals to detach from the substrate, as was first reported in the study that identified *Ta*ELP (Senatore et al., 2017). LF and LFNE peptides triggered rounding up and rotational movements, with LF specifically inducing a pronounced rounded shape. Collectively, the FFNPamide, ELPE, MPFP, and WPPF peptides elicited responses resembling natural feeding behavior, all causing *Trichoplax* to flatten with occasional concomitant pausing. FFNPamide caused the most extreme flattening, while the ELPE, MFPF, and WPFF peptides caused animals to adhere more tightly to the substrate at the perimeter, and all four of these peptides caused churning movements. Two other peptides, SITFamide and YYamide, caused minor speed changes with no other significant effects, while RWamide was found to have no behavioral effect. Using custom antibodies, the localization profiles of five *Trichoplax* peptides were examined (i.e., SIFGamide, SITFamide, YYamide, RWamide, FFNPamide), along with the predicted insulin-like hormone (TrIns3), revealing that each peptide is expressed in a unique set of cells arranged along concentric regions within the ventral epithelium (YYamide, RWamide, FFNPamide, TrIns3) and the dorsal epithelium (SIFGamide, SITFamide) (Figure 8.2(B)). Moreover, the immunofluorescence of several peptides resembled that of different gland cell types, with TrIns3 expression found in regions where type 1 gland cells are located, and FFNPamide and YYamide in regions corresponding to type 3 and central type 2 gland cells (Mayorova et al., 2019). Unique peptidergic cell type diversity is further supported by the single-cell transcriptome study described previously (Sebé-Pedrós et al., 2018), a characteristic seen across many animals, for example in peptidergic neurons of the mouse cortex and spinal cord (Das Gupta et al., 2021; Jékely, 2021; Roller et al., 2008; Smith et al., 2019b; Williams et al., 2017).

In most animal lineages, secreted regulatory peptides bind and activate GPCRs, which as noted, are abundant in placozoans along with corresponding intracellular effectors. Thus, the two most obvious placozoan behaviors that appear to be regulated by peptides, namely ciliary locomotion and body contractions, most likely involve peptide-GPCR signaling. A growing body of research supports the role of peptide-mediated GPCR signaling in ciliated cells with many studies identifying GPCRs and components of GPCR signaling in cilia of bilaterian and non-bilaterian animals (Luxmi et al., 2019; Shah et al., 2009; Sigg et al., 2017). GPCRs also regulate contraction and relaxation of smooth muscle, through pathways that can operate independently of membrane excitation. For example, GPCR signaling through the phospholipase C pathway leads to the activation of IP_3 receptors at the ER, causing the release of Ca^{2+} into the cytoplasm. Ca^{2+} ions then bind to the calcium sensor protein calmodulin, which then interacts with myosin light chain kinase to activate its phosphorylation of actinomyosin filaments to elicit contraction (Kuo & Ehrlich, 2015). GPCR-mediated relaxation on the other hand is accomplished primarily through the cyclic adenosine monophosphate (cAMP) pathway, activating the enzyme PKA that phosphorylates numerous proteins involved in contraction (Kuo & Ehrlich, 2015).

A key step forward for understanding the role of GPCRs and peptide signaling in placozoan physiology and behavior will be to functionally link specific receptors to specific peptide ligands. Recently, a combinatorial cell culture-based screening method was used to screen 126 predicted neuropeptides against 87 cloned GPCRs from the marine annelid *Platynereis dumerilii*, producing 19 peptide-GPCR matches (Bauknecht & Jékely, 2015). Such an approach would seem viable for deorphanizing placozoan GPCRs. Furthermore, the determination of cell type expression of ligands and corresponding receptors will permit detangling the paracrine network used to coordinate cellular activity for the purpose of behavior (Jékely, 2021). This may become possible through advances in single-cell transcriptome sequencing that would permit deeper sequencing depths, such that more exhaustive identification of expressed transcripts can be identified.

In specific animal lineages, IRs belonging to the degenerin/epithelial sodium channel (DEG/ENaC) superfamily have adopted peptides as modulatory or primary activating ligands. DEG/ENaC channels are a family of IRs sharing a hetero/homotrimeric structure, with each subunit consisting of a large extracellular ligand-binding domain and two pore-forming transmembrane alpha helices (Kellenberger & Schild, 2002). They are unique to animals (Liebeskind et al., 2015) and have gone through extensive lineage-specific gene duplication presenting phylogenetic relationships that do not correspond with the species phylogeny (Elkhatib et al., 2019). They have also evolved to respond to an array of ligands, and their sensitivity to these ligands does not appear to be phylogenetically constrained. Such diversity is common to somatosensory receptors such as transient receptor channels (TRP channels) and GPCRs, often resulting in channels capable of detecting various stimuli and serving numerous functions (Peng et al., 2015; van Giesen & Garrity, 2017). For instance, DEG/ENaC homologs in invertebrates such as *C. elegans* are mechanically gated and responsible for touch sensitivity, while arthropods such as *Drosophila melanogaster* have vastly expanded their repertoire of DEG/ENaC homologs (pickpocket, PPKs), which in addition to mechanosensation, were adapted for roles in pheromone sensitivity, salt taste, and water sensing (Cameron et al., 2010; Lin et al., 2005; Liu et al., 2003; O'Hagan et al., 2005). Deuterostome homologs of DEG/ENaCs form acid-sensing ion channels (ASICs), which are activated by low extracellular pH (i.e., protons), and are modulated by the neuropeptides FF and FMRFamide (Askwith et al., 2000; Lynagh et al., 2018; Vick & Askwith, 2015). Interestingly, certain DEG/ENaC channels are directly gated by neuropeptides. This includes the gastropod mollusk (e.g., *Aplysia californica*, *Lymnaea stagnalis*) FMRFamide-gated sodium channels (FaNaCs), the annelid *Platynereis dumerilii* myoinhibitory peptide

(MGIC)-gated channel, and the RFamide-gated sodium channels from the cnidarian *Hydra magnipapillata* (HyNaCs) (Cottrell et al., 1990; Golubovic et al., 2007; Lingueglia et al., 1995). All homologs are activated by RFamide ligands, which are conserved across vertebrates and invertebrates (Jékely, 2013). Based on phylogeny, the peptide-gated DEG/ENaC channels from mollusks and annelids most likely share a common ancestry, but the apparent absence of peptide-gated DEG/ENaC channels in phyletic groups that separate lophotrochozoans and cnidarians would suggest that this gating modality evolved independently in cnidarians. For both animal groups, the receptors appear to act as IRs that drive fast synaptic signaling in snail neurons and neuromuscular junctions in *Hydra* tentacles (Assmann et al., 2014; Davey et al., 2001; Green et al., 1994). Interestingly, the *Platynereis dumerilii* peptide MGIC was found to also activate GPCRs, representing a dual signaling mechanism for this peptide involved in larval settlement initiation and feeding (Schmidt et al., 2018). Thus, apparently, peptidergic networks can utilize both ionotropic and metabotropic receptors contributing to the functional diversity of peptide signaling.

Although we have come to understand much about DEG/ENaC channels in vertebrates and bilaterian invertebrates, we know little about their modalities for activation and physiological functions in early-diverging animals. Given the diversity of ligands for activation and the incomplete lineage sorting of DEG/ENaC channels apparent on phylogenetic trees, it is difficult to envision what the ancestral ligands might have been when these channels emerged in the common metazoan ancestor. The most comprehensively studied non-bilaterian DEG/ENaC channels are the aforementioned HyNaC channels from *Hydra*, that form heterotrimeric peptide-gated channels with no response to extracellular pH (Assmann et al., 2014), the latter notable because of their sister relationship with deuterostome ASIC channels (Gründer & Assmann, 2015). Considering these findings, coupled with the observation that ctenophores significantly expanded their neuropeptide DEG/ENaC channel gene content, it has been proposed that DEG/ENaC channels were involved in synaptic signaling in the early animal nervous system (Assmann et al., 2014; Moroz et al., 2014a). Around 10 of the 11 DEG/ENaC channels identified for *Trichoplax adhaerens* (named TadNaC1-9 and TadNaC11) form a clade with ASIC and HyNaC channels on phylogenetic trees derived from protein alignments, while the single homolog TadNaC10 forms a clade with vertebrate ENaCs, the lophotrochozoan FMRFamide-gated channels, and the *C. elegans* mechanically gated channels (Elkhatib et al., 2019). If we infer putative gating ligands for the various placozoan receptors based on phylogeny, the ones that form a clade with ASIC and HyNaC channels would perhaps be activated by peptides and/or protons, while TadNaC10 could be gated by peptides and/or mechanical stimuli.

Recent work on DEG/ENaC channels from *Trichoplax* has provided first insights (Elkhatib et al., 2019). A single channel belonging to the ASIC/HyNaC clade, named TadNaC6, was cloned and expressed in mammalian cells for electrophysiological characterization. The channel was found to produce robust Na^+ leak currents that could be fully blocked by extracellular protons, thus resembling ASIC channels in their sensitivity to pH but differing in their gating. Whereas ASIC channels remain closed under baseline conditions and become activated by protons to produce transient depolarizing currents, TadNaC6 conducts constitutive leak Na^+ currents, and protons block the channels which would lead to transient membrane hyperpolarization. Interestingly, constitutively open DEG/ENaC channels are found in other animals, including the vertebrate ENaCs, and it appears as though this particular modality for DEG/ENaC channel gating has evolved numerous times independently (Elkhatib et al., 2019). Hopefully, future work will establish whether the remaining *Trichoplax* DEG/ENaC channels are also sensitive to protons, mechanical stimuli, or any of the various regulatory peptides that have been identified.

8.2.7 DIRECTED MOVEMENT TOWARDS GLYCINE AND FOOD MICROALGAE

Placozoan genome and transcriptome studies reveal a complete repertoire of glutamate and GABA synthesis and degradation pathway components. This includes glutamic acid decarboxylase (GAD), GABA alpha-oxoglutarate transaminase, glutaminase, and glutamine synthase. However, placozoans, sponges, and ctenophores all lack GABA vesicular transporters, indicating that GABA might not function as a transmitter in these animals and could have different molecular functions (Moroz et al., 2021). Other biogenic amine and catecholamine transmitters such as serotonin, octopamine, dopamine, histamine, epinephrine, and norepinephrine are missing essential synthesis and degradation components and are most likely not utilized in placozoan signaling systems (Wong et al., 2019). Placozoans also have two homologs of choline acetyltransferase (ChAT), an enzyme involved in acetylcholine synthesis, but are missing the acetylcholine degradation enzyme acetylcholinesterase.

Microchemical analyses of placozoan low molecular weight transmitters corroborate the genomic inferences detecting high levels of glycine (~3 mM), L-glutamate (~1.3 mM), D/L-aspartate (0.5 mM), and GABA (~50 μM) with no trace of serotonin or dopamine (Moroz et al., 2020b; Romanova et al., 2020a). D-glutamate, an enantiomer common in bilaterians and ctenophores, was not detected in *Trichoplax*, sponges, or cnidarians, though they all had high levels of D-aspartate (Moroz et al., 2020b). Although behavioral responses to glutamate, GABA, and aspartate have not been reported, the application of 1 mM glycine on *Trichoplax* (H1 and H2) and *Hoilungia hongkongensis* was found to elicit contractions, with longer incubation causing the animals to break down into individual cells or smaller fragments (Romanova et al., 2020a). Interestingly, glycine was also found to act as a chemoattractant, where a block of agar infused with glycine attracted nearby animals (Heyland et al., 2014; Romanova et al., 2020a). Positive chemotaxis has also been reported towards food, where *Trichoplax* animals positioned at various distances from a clump of agar-embedded microalgae (*Rhodamonas salina*) were found to slowly congregate over the clump, presumably in response to a chemoattractant being emitted by the algae (Smith et al., 2019a). Whether glycine or another compound mediates this behavior remains to be determined. However, the proposed mechanism for directed movement might apply regardless, where based on mathematical modeling, it was proposed that individual ciliated epithelial cells respond to a chemoattractant concentration gradient independently, perhaps by modulating ciliary beating, and that their collective propulsive forces act on the whole body to govern locomotive dynamics (Smith et al., 2019a). If so, regulation of ciliary beating could occur in a concentration-dependent manner, encoding information about the distance of the animals from the chemoattractant source, or perhaps, about the orientation of the animal, where cilia at the edge closest to the source could exhibit different responses than those farther away.

In lacking Cys-loop channels, placozoans are likely unable to generate canonical responses to glycine as observed in vertebrates, where these channels conduct chloride (Cl^-) to elicit transient membrane hyperpolarization and inhibition, or invertebrates, where the channel evolved to be cationic or anionic (Cymes & Grosman, 2016). Furthermore, GPCR receptors that are selectively activated by glycine have not been identified, although the promiscuous mammalian T1R1/T1R3 heterodimeric receptors, involved in taste, and the receptor GPRC6A, involved in androgen signaling, can be activated by glycine (Wellendorph & Bräuner-Osborne, 2009). Thus, it is certainly conceivable that placozoans express a GPCR that can detect glycine in the extracellular environment. *Trichoplax adhaerens* do possess at least 2 glutamatergic GPCR orthologues, which appear to have duplicated strictly in bilaterians to produce classes I-IV (Ramos-Vicente et al., 2018).

Another viable set of receptors for glycine is the ionotropic glutamate receptors (iGluRs), which in both vertebrate and invertebrate nervous systems, can be activated by glycine and/or glutamate. In the mammalian nervous system, three classes of iGluRs, NMDA, AMPA, and Kainate receptors, are responsible for the majority of fast, excitatory signaling across the synapse, classified both phylogenetically and according to their namesake selective agonists (i.e., **N**-**M**ethyl-**D**-**A**spartate, α-**A**mino-3-hydroxy-5-**M**ethyl-4-isoxazole**P**ropionic **A**cid, and **Kainate**, respectively) (Mayer, 2021). Whereas AMPA and Kainate receptors are activated solely by glutamate (Bettler & Mulle, 1995), NMDA receptors require both glutamate and glycine for activation (Iacobucci & Popescu, 2017). Once activated, these receptors open to conduct depolarizing cation currents into the post-synaptic cell, triggering electrical signals, and in some cases cytoplasmic Ca^{2+} signals attributable to the high Ca^{2+} permeability of select receptor subtypes (Paoletti et al., 2013). Like Deg/ENaC channels, iGluRs can be homo- or heteromeric but are made up of four subunits (i.e., tetrameric). Mammalian iGluRs have been subject to extensive scientific study due to their key roles in synaptic transmission, learning and memory, and diseases of the nervous system (Bowie, 2008). Notably, studies on synaptic signaling and learning and memory in the sea hare *Aplysia californica*, pioneered by Eric Kandel, have revealed that function and dynamic regulation of these receptors at the synapse is conserved between vertebrates and bilaterian invertebrates (Glanzman, 2010).

More recently, phylogenomic studies have revealed the evolutionary history of iGluRs in animals (Ramos-Vicente et al., 2018), which in fact, are of premetazoan origin and even present in bacteria (Chen et al., 1999; Price et al., 2012). Based on a leading hypothesis about animal phylogenetic relationships, one key observation is that the ancestral metazoan likely possessed a single type of iGluR called epsilon (Figure 8.5(A)), still present in the genomes of most animals (e.g., present in chordates but lost in vertebrates) (Mayer, 2021; Mayer & Jegla, 2018; Ramos-Vicente et al., 2018). Functional characterization of various epsilon receptors revealed that these are activated by glycine and/or glutamate (Alberstein et al., 2015; Ramos-Vicente et al., 2018), thus resembling NMDA receptors in their bimodal ligand sensitivity. After the divergence of ctenophores, an ancestral AMPA-Kainate-Delta-Phi (AKDF) receptor emerged, which via gene duplication, subsequently gave rise to true AMPA and Kainate receptors. Interestingly, in protostomes significant expansion of iGluRs resulted in a large repertoire of IRs, that form a sister clade with AMPA and Kainate receptors, involved in chemosensation, thermosensation, and hygrosensation (Croset et al., 2010; Ni, 2020). Lastly, a separate receptor type, called lambda, appears to have emerged exclusively in sponges, while NMDA receptors evolved in the common ancestor of cnidarians and bilaterians after placozoans diverged (Figure 8.5(A)).

A cursory survey of iGluR paralogues from *Trichoplax adhaerens* and *Hoilungia hongkongensis* reveals 4 conserved AKDF subunit orthologues, compared to 5 previously identified for the cnidarian *Acropora digitifera*, and 12 for the sponge *Oscarella carmela* (Ramos-Vicente et al., 2018). Additionally, lineage-specific adaptations in epsilon subunit genes resulted in 9 subunits for *Trichoplax* and only 6 for *Hoilungia* (Figure 8.5(B)). These gene numbers are comparable to those of *Oscarella* (7) and *Acropora* (5), in contrast to ctenophores that appear to have significantly expanded iGluRs (Liebeskind et al., 2015; Moroz et al., 2014a), with 17 identified for *Mnemiopsis leidyi* (Ramos-Vicente et al., 2018). iGluRs contain three transmembrane alpha helices that contribute to the cation permeation pathway of the tetrameric receptor, an extracellular amino-terminal domain (ATD), and an adjacent clam-shaped ligand binding domain (LBD). Both the ATD and the LBD form dimers with corresponding ATD and LBD dimers from two separate subunits, giving rise to a "dimer of dimers" quaternary structure (Gan et al., 2015; Tichelaar et al., 2004). Because of the amenability of the iGluRs LDB for X-ray crystallography, we have come to understand considerable details about the atomic interactions that glycine-sensitive receptors use to bind glycine, and for those that are highly selective for glycine, the mechanisms for preventing activation by glutamate (Mayer, 2021). For all iGluRs, an aromatic residue (phenylalanine/F469 for the *Mnemiopsis leidyi* epsilon receptor ML032222a; Figure 8.5(B)) forms crucial van der Waals interactions with the glycine ligand. Furthermore, a salt bridge formed by the side chains of residues glutamate/E423 and arginine/R703 prevents the larger glutamate molecule from entering the ligand-binding cavity (Alberstein et al., 2015; Yu et al., 2016). Two additional positions have been suggested to play a role in glutamate/glycine specificity, a serine/S653, and a nonpolar residue at position 655 (both numbers corresponding to the rat GluA2 subunit sequence). Instead, if these respective positions contain glycine and threonine residues, glutamate will be the preferred ligand (Ramos-Vicente et al., 2021). Based on the corresponding sequences in the placozoan iGluRs (Figure 8.5(B)), it appears as though the combined sets of AKDF and epsilon receptors would be sensitive to both glycine and glutamate, the latter notable because only glycine-mediated behaviors have been observed (Romanova et al., 2020a).

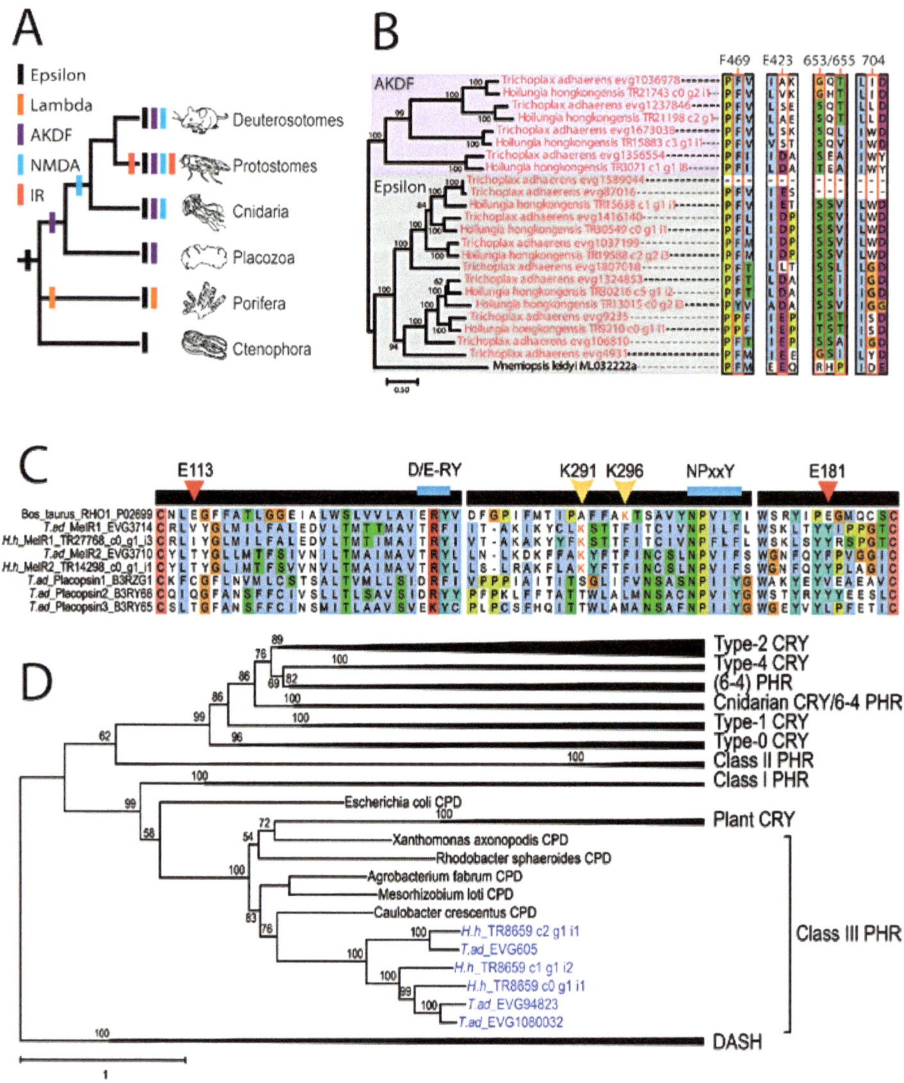

FIGURE 8.5 (A) Illustration of metazoan phylogenetic relationships, with colored bars indicating the proposed emergence of different iGluR sub-types (Mayer & Jegla, 2018; Ramos-Vicente et al., 2021). (B) Maximum likelihood phylogenetic tree of placozoan iGluR homologs, revealing distinct clades of AKDF and epsilon receptors homologs, rooted with the glycine-sensitive epsilon receptor from the ctenophore *Mnemiopsis leidyi* (sequence ID ML032222a). On the right are a series of multiple sequence alignments illustrating key amino acid positions associated with selective glycine/glutamate binding. (C) Multiple sequence alignment of the *Bos taurus* rhodopsin, two melatonin receptors from *Trichoplax adhaerens* (*T.ad*) (Wong et al., 2019) and *Hoilungia hongkongensis* (*H.h*) (Eitel et al., 2018), and three placopsin genes (UniProt IDs are listed). Red arrows indicate locations of noted chromophore

counterion residues (E113, E181), blue bars are motifs that are broadly conserved among class A GPCRs, and yellow arrows indicate the location of the retinal binding lysine (K296, according to the *Bos taurus* rhodopsin numbering), and the noted lysine K291 in the placozoan melatonin receptors. (D) A maximum likelihood phylogenetic tree of Cryptochrome (CRY)/ Photolyases (PHR) proteins inferred from a muscle alignment (Edgar, 2004) trimmed with trimAl (Capella-Gutiérrez et al., 2009). The tree was built using IQ-Tree (Nguyen et al., 2015) with a best-fit model of LG+I+G4. Bootstrap values for 1,000 ultrafast replicates are indicated. The sequence identifiers for identified *Trichoplax adhaerens* and *Hoilungia hongkongensis* PHR homologs are in blue font, and these were extracted from the *Trichoplax* transcriptome (Wong et al., 2019) and the *Hoilungia hongkongensis* (Eitel et al., 2018) genome sequences using BLAST searches.

8.2.8 BEHAVIOR IN RESPONSE TO GRAVITY

Gravity is a ubiquitous force that organisms experience as they orient themselves and navigate their environments. The ability to perceive gravity is observed broadly among plants, animals, fungi, and protists, and various cellular mechanisms have evolved to accomplish this important task (Häder et al., 2017; Hemmersbach et al., 1999). Many aquatic invertebrates use a specialized mechanical receptor organ called a statocyst to sense gravity and spatial orientation changes (Panchin et al., 1995; Singla, 1975; Young, 1960). Statocysts consist of a heavy mineralized mass or statolith enclosed by ciliated mechanosensory cells, where contact between the mass and cilia transduces information about gravity, as well as changes in orientation and momentum caused by movement. Most non-bilaterian animals use statocysts to sense gravity and adjust their movements accordingly (Jékely et al., 2021). Ctenophores, for example, prefer to maintain a vertical position with their mouths facing up as they feed near the water surface. However, when water conditions become undesirable, they swim downwards and rotate so that their mouth points downward (Tamm, 2014). Ctenophores can correct their orientation if forcibly tilted and return to their desired vertical orientation by modulating their comb row beating (Tamm, 2014). Interestingly, gravisensing can also be mediated by single cells such as statocytes in plant root cells (Driss-Ecole et al., 2003), and statocytes or lithocytes in some animals (Eakin & Kuda, 1970; Ehlers, 1991; Rohde et al., 1993). In contrast, placozoans are benthic and spend most of their time grazing on algae biofilms on rocks and other surfaces. When placed in a dish, they sink and stick to the bottom, indicating that they are denser than water and would require compensatory locomotion to remain within their preferred vertical habitat. For yet unknown reasons, placozoans are occasionally found floating on the water surface, although this may not necessarily reflect a gravity-sensing behavior per se, but rather, a response to hypoxic conditions as observed for other aquatic animals (Bell et al., 2003; Riedel et al., 2014). Still, both placozoan ciliary locomotion and migration to the surface would involve some degree of gravity perception.

As discussed previously, placozoans possess non-ciliated crystal cells that are used by the animals to maintain their positioning along vertical surfaces, composed of a cytoplasmic aragonite crystal surrounded by mitochondria, that falls into a cup-shaped nucleus depending on their vertical orientation (*see* Sections 8.1.3 and 8.2.5). Evidence that crystal cells are indeed gravity sensing came from observations that a minority of *Trichoplax adhaerens* specimens have a reduced number of crystal cells (i.e., less than 8), attributed to a temporary developmental process since the animals would eventually produce a normal number of crystal cells (Mayorova et al., 2018). Interestingly, these individuals were unable to maintain their position along a vertical plane, moving directly downward with the force of gravity, compared to specimens with normal crystal cell numbers. Further evidence came from the observation that in vertically oriented animals, the aragonite crystal deposits within the nuclear cup in crystal cells along the top edge of the animal, and out of the cup in cells along the bottom, thus providing a possible mechanism for proprioception respect to orientation and gravity. Interestingly, lithocytes resembling placozoan crystal cells are also found in various other animals including cnidarians, vertebrates, and Platyhelminthes, the latter notable because the cytoplasmic crystal is similarly surrounded by mitochondria (Campbell, 1972; Eakin & Kuda, 1970; Häder et al., 2017; Hemmersbach & Braun, 2006; Mayorova et al., 2018; Rohde et al., 1993; Vinnikov, 1974).

8.2.9 BEHAVIORS IN RESPONSE TO LIGHT

Placozoans can sense light in their environment and most likely do so to modify locomotion and behavior. In light conditions, *Trichoplax* ciliary gliding/displacement is significantly increased compared to dark (Heyland et al., 2014), and it has been noted that positive and negative phototactic responses occur, suggestive of directional movement (Srivastava et al., 2008). Placozoans also react to UV light with robust writhing-like contractions (Pearse, 1989), akin to what is observed after the application of the peptide SIFGamide (Senatore et al., 2017). Still unknown is whether these light-mediated behaviors require specific wavelengths or intensities of light if they intersect with other behaviors such as chemotaxis and feeding, or which cell types might be involved in sensing light and producing the behavioral responses. Given what we know about placozoan cell types, it would seem likely that the observed contractions in response to UV light could be mediated by dorsal epithelial cells, while changes in locomotion in light vs. dark could be through an altered ciliary beating at the ventral epithelium. If so, then a key question emerges: are these cells themselves able to detect light through the expression of light-sensitive proteins, or are other cell types responsible for phototransduction? Notably, placozoan cell types lack ultrastructural features characteristic of photoreceptor cells, such as clustered microvilli of rhabdomeric photoreceptors, or elaborate membrane structures over a basal body of ciliary photoreceptors, both with intracellular inclusions of shading photopigments that provide information about the light direction (Fain et al., 2010; Smith et al., 2014). Furthermore, the absence of anterior-posterior body symmetry would preclude non-visual scanning phototactic behavior as is observed broadly in animals that lack eyes, where rhythmic body movements (scanning) move photoreceptor cells through space to provide information about the direction of incoming light (Jékely, 2009; Nilsson, 2009). The only cell type that could conceivably convey spatial information about incoming light is the crystal cells with their radial, outward-facing nuclei (Mayorova et al., 2018). However, based on their ultrastructure it is difficult to envision how these cells might differentially detect light. Indeed, the proposed directional movement for placozoans in response to light could be achieved without a requirement for orienting photoreceptive cells, where for example, increased ciliary beating in the light could cause animals to accumulate in darker regions where the cilia slow down. Another important consideration applies to the genes/proteins that placozoans use to detect light. In other animals, opsins and cryptochromes (CRYs) are the two photopigments associated with light detection and vision. Thus, below we explore the potential for opsins in mediating placozoan light-sensing behavior, and later, CRYs, which are generally less studied and have more diverse roles outside of metazoan visual/light sensing systems.

Opsins belong to the α-group of the rhodopsin-like GPCRs and are uniquely coupled to a chromophore, commonly retinal, which is crucial for photon activation. This linkage, between the chromophore and the opsin, is dependent on a highly conserved lysine/K296 residue (bovine rhodopsin numbering), considered central for the functioning of opsins as true light receptors (Peirson et al., 2009) (Figure 8.5(C)). The K296 residue forms a covalent Schiff-base linkage with the chromophore, and this bond is present in a protonated state and stabilized by a counterion at position 113 (predominantly glutamate/E113) (Nathans, 1990; Sakmar et al., 1989). In most invertebrate opsins, the residues occupying the counterion position are phenylalanine (F) or tyrosine (Y), while invertebrate Gq and Go-coupled opsins have a displaced glutamate counterion at position 181 (Terakita et al., 2004) (Figure 8.5(C)). The importance of the counterion is evident in mutation experiments, where the replacement of this residue causes significant changes in opsin photoactivation (Kojima et al., 2017). G-proteins interaction and signal transduction occurs at a glutamate/aspartate, arginine, and tyrosine motif (D(E)-RY, positions 134 to 136), also conserved across all animal opsins (Peirson et al., 2009; Shichida & Matsuyama, 2009) and more broadly among class A GPCRs (Rovati et al., 2007).

Despite lacking the retinal binding domain (K296), placozoans possess opsin-like receptors that are phylogenetically related to opsins, hence their name placopsins (Feuda et al., 2012). These GPCRs form a sister clade with all other metazoan opsins, indicating an ancestral relationship (Feuda et al., 2012, 2014). However, this phylogeny does not match the expected species phylogeny, where either sponges or ctenophores are the most early-diverging animals (*see* Section 8.1.2.2). It is, therefore, possible that this anomaly represents incomplete lineage sorting, where the placopsins underwent lineage-specific diversification and lost their retinal-binding lysine in the process (Feuda et al., 2014; Schnitzler et al., 2012). The placopsins also bear conserved NPxxY and X-RY motifs for interacting with G proteins but are missing the counterions required for stabilizing the retinal chromophore (E113 or E181; Figure 8.5(C)) (Audet & Bouvier, 2012; Fritze et al., 2003; Han et al., 1998; Kim et al., 2004). Fundamentally, the lack of K296 in placopsins indicates improbable light-sensing properties and insights into their physiological roles would require functional characterization.

Notably, the most closely-related GPCRs to the opsins are the melatonin receptors, involved in the regulation of circadian rhythm (Reppert, 1997), which form a sister clade with opsins and placopsins. Like placopsins, melatonin receptors lack the K296 residue and hence the ability to detect light (Feuda et al., 2012; Fleming et al., 2020; Ramirez et al., 2016). However, it is interesting that they are nevertheless involved in a cellular regulatory system that is highly entrained by light. Strikingly, a cursory analysis of melatonin receptors in the *Trichoplax adhaerens* and *Hoilungia hongkongensis* gene datasets (Eitel et al., 2018; Wong et al., 2019) uncovered a set of conserved melatonin-like receptors with a lysine residue four positions upstream of the bovine K296 residue (i.e., position 291 in the bovine rhodopsin sequence:

Figure 8.5(C)). This is interesting because bilaterian melatonin receptors lack this lysine residue, and experiments have shown that movement of the K296 residue to other positions in the receptors can produce functional opsins that bind the chromophores, albeit with alterations in their phototransduction properties (Devine et al., 2013). Perhaps, these two melatonin-like genes, conserved between *Trichoplax* and *Hoilungia*, can bind a chromophore and function as photosensitive receptors.

CRYs and photolyases (PHR) represent the two major functionally distinct light-sensitive proteins belonging to a superfamily of flavin adenine dinucleotide (FAD)-based blue-light receptors, found across all forms of life (Cashmore et al., 1999; Lin, 2000; Todo, 1999). Photolyases are DNA repair enzymes that either repair pyrimidine-pyrimidine dimers (i.e., 6-4 PHRs), or cyclobutane pyrimidine dimers (CPD-PHRs), through cofactors such as (FAD) that absorb blue light (Hitomi et al., 2009; Lin & Shalitin, 2003; Sancar, 2003; Suter et al., 1997). CRYs s also utilize FAD cofactors but differ from photolyases in having variable C-terminal extensions. They are classified as plant CRY, animal CRY (types 1, 2, and 4) (Cashmore et al., 1999; Yuan et al., 2007), and sponge CRY (type-0) (Figure 8.5(D)) (Rivera et al., 2012). However, the phylogenetic relationships within this protein superfamily do not follow the mentioned functional classifications (i.e., CRYs vs. PHRs; Figure 8.5(D)). In animals, CRYs are involved in various processes such as resetting of circadian rhythm and magnetoreception in *Drosophila* (Emery et al., 1998; Gegear et al., 2008), light-dependent gene expression in cnidarians (Reitzel et al., 2010), and light-sensing in the primordial eyes of sponge larvae (Rivera et al., 2012). In plants, CRYs also carry out light-dependent functions such as inhibition of stem elongation by blue light (Ahmad & Cashmore, 1993). Since sponges lack opsins and utilize CRYs for light detection, it is reasonable to speculate that placozoans might also utilize CRY homologs for light sensing.

Building on previous phylogenetic work (Ozturk, 2017; Rivera et al., 2012), a protein phylogeny incorporating CRY homologs from *Trichoplax adhaerens* and *Hoilungia hongkongensis* reveals an unexpected grouping of placozoan CRY homologs with Class III CPD-PHRs, including homologs of eubacteria PHR genes, forming a sister clade relationship with plant CRYs. It has been proposed that PHR genes in *Trichoplax* are a result of a lateral gene transfer event from non-rickettsial bacteria (Driscoll et al., 2013), which is consistent with the observed phylogeny (Figure 8.5(D)). Moreover, the type III CPD-PHR from the bacterium *Caulobacter crescentus*, which has a close phylogenetic relationship with the placozoan CRY/PHR homologs, functions as a bona fide CPD repairing photolyase with no evidence for a role in phototactic behavior (Ozturk et al., 2008). The plant CRYs, on the other hand, function as photo-sensors for regulating photoperiodic flowering in *Arabidopsis thaliana* (Mockler et al., 2003), and as noted, modulation of growth via inhibition of hypocotyl elongation by blue light (Ahmad & Cashmore, 1993; Lin et al., 1998). Overall, the *Trichoplax* CRY/PHR genes are more likely to function as DNA-repairing photolyases since they have a monophyletic relationship with the CPD-PHRs, and it is unclear if the phylogenetically related plant CRYs also have this enzymatic capability, which forms the basis for phototactic signals.

Overall, photosensory receptors in placozoans remain enigmatic. Nevertheless, until placopsins, placozoan melatonin receptors, and placozoan CRY/PHR homologs are characterized, they cannot be ruled out. In addition, other proteins such as phototropins or phytochromes can transduce light into signals, but whether they are expressed in placozoans and perform light-sensing functions requires further investigation.

8.2.10 OTHER CHEMICAL SIGNALS

8.2.10.1 NITRIC OXIDE (NO)

Nitric oxide (NO) is a free radical gaseous molecule that can diffuse across the cell membrane for cellular communication in a process akin to volume transmission, but without a need for regulated secretion (Liaudet et al., 2000). In bilaterians, NO is involved in numerous important physiological processes including vasodilation (Donald et al., 2015) and regulation of synaptic transmission and neurotransmitter release (Jacklet, 1997). Because the enzymes that synthesize NO are associated with blood pressure and cardiovascular disease (Farah et al., 2018), they have been most extensively studied in vertebrates, where three NO synthase (NOS) enzymes were identified. Two of these are Ca^{2+}/calmodulin dependent (hence termed cNOS for calcium-dependent) (Nathan & Xie, 1994), and the other is Ca^{2+}-independent, regulated by various proinflammatory signals (hence term iNOS for inducible) (Szabo, 1995). Most invertebrates on the other hand generally possess one or two NOS enzymes, that expanded independently of the three vertebrate NOS enzymes (Moroz et al., 2020a). For vertebrate cNOS enzymes, calcium regulation of enzymatic activity involves rises in cytoplasmic Ca^{2+} ions that bind calmodulin, followed by binding of Ca^{2+}/calmodulin protein to the enzyme. Instead, the iNOS isoform lacks an autoinhibitory N-terminal loop present in cNOS that restricts calmodulin binding, thereby allowing constitutive binding of calmodulin to cNOS irrespective of cytoplasmic Ca^{2+} levels (Daff et al., 1999; Knudsen et al., 2003; Nishida & de Montellano, 1999, 2001; Salerno et al., 1997). Once synthesized, NO is highly reactive and can directly exert its effects at the site of synthesis or diffuse through the cell membrane into neighboring cells acting on various biomolecules and receptors. For example, a highly studied NO signaling pathway involves a primary soluble receptor known as soluble guanylyl cyclase, which upon stimulation by NO, causes an increase in cGMP levels and downstream signaling effects (i.e., smooth muscle relaxation) (Murad et al., 1987, 1978).

NO signaling also serves important signaling functions in non-bilaterians, such as cnidarians where it is implicated in the neural control of locomotory behavior of the hydrozoan *Aglantha digitale* (Moroz et al., 2004), and head regeneration and feeding behavior of *Hydra vulgaris* (Colasanti et al., 1997, 2009). In sponges, NO signaling is involved in the larval settlement, which, as mentioned earlier, involves the light-sensing CRYs and is thought to integrate with the NOS pathway (Say & Degnan, 2020). Although the role of NO signaling in ctenophores is uncertain, recent work has suggested that placozoans utilize NO signaling. First, placozoans resemble vertebrates in having duplicated their NOS genes to produce three copies in *Trichoplax* haplotypes H1 and H2, and two copies in *Hoilungia* haplotypes H4 and H13 (Moroz et al., 2020a). The *Trichoplax* enzymes, dubbed NOS-1, NOS-2, and NOS-3, all have NO synthase, flavodoxin, FAD, and NAD domains conserved among all metazoan NOS enzymes except for the homolog from the ctenophore *Mnemiopsis leidyi* that lacks the FAD domain. Interestingly, the NOS-1 protein from the four placozoan haplotypes bears an N-terminal PDZ domain, a motif found in vertebrate neuronal cNOS and select groups of invertebrates, including sponges and cnidarians. In mammals, the cNOS PDZ domain mediates protein-protein interactions with syntrophin (Hillier et al., 1999), PSD95 (Brenman et al., 1996; Tochio et al., 2000), and other adapter proteins, facilitating subcellular localization and distribution of the enzyme (Costas-Insua et al., 2018; Zhou & Zhu, 2009). In addition, autoinhibitory loops and calmodulin-binding sites are present in all placozoan NOS genes, indicating potential regulation by Ca^{2+}/calmodulin as occurs for the cNOS enzymes.

Employing capillary electrophoresis microchemical assays, endogenous NOS activity was confirmed by monitoring the conversion ratio of L-arginine to L-citrulline in *Trichoplax*, which doubled in the presence of NOS inhibitor. Another assay examining the NO metabolite product nitrite

(NO_2^-) in inhibitor conditions yielded no NO_2, further confirming NOS gene activity *in vivo*. Moreover, *Trichoplax* NOS-1 and NOS-3 were found to have differential cell-specific expression near the dorsal epithelial layer, and occasionally NOS-1 and NOS-3 were found to be co-expressed in certain cells near the edge of the animal (although the identity of these cells remains uncertain). Placozoans also possess target receptors for NO signaling, with seven homologs of soluble guanylate cyclases, five membrane-associated atrial natriuretic peptide-like receptors, and 12 guanylate cyclases containing unique NIT domains associated with nitrite/nitrate sensing in bacteria (Moroz et al., 2020a). Overall, it appears as though NO signaling plays a role in placozoan biology and cellular communication, providing clear advantages, especially given the absence of synapses and gap junctions.

8.2.10.2 OXYGEN SENSING

Sensing oxygen and oxygen homeostasis are fundamental processes for all aerobic organisms. At the cellular level, the heterodimeric transcription factor hypoxia-inducible transcription factor (HIFα) is directly regulated by oxygen levels. Under normoxic conditions, HIFα is constitutively targeted for proteosomal degradation through post-translational modifications by the prolyl-hydroxylase domain (PHD) enzymes and the von Hippel Lindau (VHL) ubiquitin ligase complex. Under hypoxic conditions, PHD activity is diminished allowing HIFα to persist and heterodimerize with HIFβ to drive the expression of various genes involved in low oxygen responses (Bargiela et al., 2018; Kaelin & Ratcliffe, 2008; Tarade et al., 2019). Most animals have conserved this oxygen-sensing system but sponges and ctenophores seem to be missing key components (Mills et al., 2018).

Interestingly, a study found that *Trichoplax adhaerens* can survive hypoxic conditions for ~1.5 days (i.e., 5% atmospheric O_2), and up to 5 hours in even extreme hypoxic conditions (i.e., 1% O_2) (Loenarz et al., 2011). Furthermore, the animal was found to possess functional components of the HIF system. Specifically, hypoxic conditions caused the upregulation of several genes with predicted hypoxia response elements in their promoter sequences and feedback upregulation of the PHD hydroxylase mRNA as observed in mammalian cells under hypoxic conditions (Minamishima et al., 2009). Consistent with PHD promoting HIFα degradation under normoxic conditions (i.e., ~20% atmospheric O_2), gene silencing of the *Trichoplax* PHD homolog led to upregulation of similar genes that were upregulated under hypoxic conditions, when PHD activity is expected to be diminished.

The PHD enzyme was also shown to be catalytically active *in vitro*, and altogether, convincing evidence was provided that placozoans can sense oxygen levels to elicit changes in gene expression. Moving forward, it will be interesting to understand the downstream behavioral and physiological implications of hypoxia signaling in placozoans. Of course, mitochondria should also be considered as they play crucial roles in oxygen sensing and homeostasis, including direct regulation of PHD enzymes and the HIF pathway (Bargiela et al., 2018), but also through the production of reactive oxygen species and release of Ca^{2+} into the cytoplasm (Chandel & Schumacker, 2000; Jonz et al., 2016).

8.2.10.3 NUCLEOTIDES

ATP is more commonly known for its role in energy storage but early work has shown that it forms an ancient sophisticated intercellular signaling network across many metazoan lineages with secondary losses in some animals (Verkhratsky, 2021). Purinergic receptors are divided into IRs (P2X) and metabotropic receptors (P2Y and P1), the former being non-selective cation homo- or heterotrimeric channels gated by ATP and ADP, and the latter are GPCRs activated by ATP/ADP (P2Y receptors) or adenosine (P1 receptors) (Burnstock, 2018). Phylogenetic analysis indicates that ionotropic P2X receptors predate metabotropic P2Y receptors, and are found across many kingdoms of life including P2X-like genes detected in protozoans, algae, fungi, and non-bilaterian metazoans (Verkhratsky & Burnstock, 2014). Accordingly, a P2Y receptor is not apparent in the *Trichoplax* gene datasets, but the animal does possess two P2X-like genes encoding receptors with conserved extracellular and transmembrane domain regions important for receptor function and ATP binding (Hanmer, 2014). However, heterologous expression of these channels in mammalian cells did not produce ATP-sensitive currents (Hanmer, 2014), and more work needs to be done to understand the functional properties of P2X receptors in placozoans.

8.2.10.4 NUCLEAR RECEPTORS (NRS)

Nuclear receptors (NRs) are transcription factors involved in the regulation and maintenance of cellular function across all metazoans (Laudet, 1997; McEwan et al., 2004). Activation of NRs is mainly dependent on the binding of specific ligands such as retinoids, fatty acids, thyroid hormones, steroids, and environmental ligands, though ligands for several NRs remain unknown

(Delfosse et al., 2015; Mazaira et al., 2019; Sever & Glass, 2013). NRs share a common structure consisting of an N-terminal A/B domain, a ligand-binding domain, a DNA-binding domain, a hinge region, and a C-terminal domain. Ligand binding causes NRs to dimerize with their NR partners and bind to specific nuclear DNA response elements, modulating the transcription of target genes (Evans & Mangelsdorf, 2014; Lefebvre et al., 2010). The diversification of ligand-NR pairs during evolution gave rise to six subfamilies (Escriva et al., 1998; Laudet, 1997). The tendency for these receptors to have multiple low-affinity binding partners has led to the proposal that they act as environmental sensors of various compounds, allowing for dynamic regulation of gene expression, metabolism, and development in response to signals emanating from the environment (Bridgham et al., 2010; Novotný et al., 2017).

Studies on NRs from non-bilaterian phyla have suggested that the NR2 subfamily is the most like the ancestral NRs and that diversification of these receptors began at the stem lineage of metazoans. The Ctenophore *Mnemiopsis leidyi*, for example, harbors two NR2A (HNF4) orthologues, though they lack the DNA binding domain (Reitzel et al., 2011), while the sponge *Amphimedon queenslandica* contains a fatty-acid binding NR2 (HNF4) and a more ancestral NR2 ortholog (Bridgham et al., 2010). Cnidarians such as *Nematostella vectensis* contain 17 NRs belonging to the NR2 subfamily (Reitzel & Tarrant, 2009), of which the retinoid X receptor ortholog is capable of binding 9-cis-retinoic acid with high affinity, similar to vertebrate counterparts (Kostrouch et al., 1998). Retinoic acid (RA) is a vitamin A metabolite with multiple isomers, including all-trans RA and 9-cis-RA which have been detected endogenously in the nervous system of multiple vertebrates (Maden et al., 1998; Wagner et al., 1992) and invertebrates (Dmetrichuk et al., 2008). RA induces axonal outgrowth and plays an important role in modulating synapses and electrical properties of neurons in both vertebrate and invertebrate nervous systems (Chiang et al., 1998; Clagett-Dame et al., 2006; Rothwell et al., 2017; Tonini et al., 1999; Vesprini et al., 2014). This compound is a key regulator of cellular differentiation and patterning during development by acting on NRs (Han et al., 1995; Maden & Hind, 2003; Maden et al., 1998; Wuarin et al., 1990).

Recent work on *Trichoplax adhaerens* uncovered four highly conserved NR2-related *Ta*NRs orthologous to NR2A (*Ta*HNF4), NR2B (*Ta*RXR), NR3B (*Ta*ERR), and NR2F (*Ta*COUP-TF) (Novotný et al., 2017). Interestingly, the *Ta*RXR has a highly conserved protein sequence and harbors all domains found in vertebrate and cnidarian RXRs (Novotný et al., 2017).

*Ta*RXR was found to have a high binding affinity for 9-cis-retinoic acid, and specificity over all trans-retinoic acid (Novotný et al., 2017; Reitzel et al., 2018). The downstream functional effects of 9-cis retinoic acid activation of *Ta*RXR activity were verified by the upregulation of the metabolic enzyme L-malate-NADP+ oxidoreductase, which in vertebrates, is an enzyme regulated by thyroid hormone receptors and RXR heterodimers (Dozin et al., 1985; Petty et al., 1990, 1989). *Ta*RXR activation was found to also auto-regulate its transcription, and the transcription of other *Ta*NRs, indicative of an early establishment of regulatory NR networks in a pre-bilaterian ancestor (Novotný et al., 2017).

In other animals, 9-cis-RA and RXR receptors regulate cell growth and differentiation (Maeng et al., 2012; Rhinn & Dollé, 2012; Schubert & Gibert, 2020). Interestingly, this function is also apparent in placozoans. First, certain algae containing red pigments such as *Rhodamona salina* and *Porphyridium cruentum* increase the size and growth rate of the animals, while green algal species such as *Chlorella* sp. show no growth unless combined with other red algae (Novotný et al., 2017), and this has been suggested to occur due to specific ligands emanating from the red algae that act on *Ta*NRs to promote growth. Remarkably, the addition of 3.3 nM of 9-cis-retinoic acid to animals fed *Porphyridium cruentum* resulted in a balloon-like phenotypic change and consequently death, while animals fed on the green *Chlorella* sp. remained alive and showed a small but statistically insignificant increase in size (Novotný et al., 2017). This study thus uncovered a functional intersect between environmental stimuli and food consumption, with the regulation of growth and physiology. Key questions remain including the identity of genes whose transcription is regulated by NRs in placozoans, and the identity of ligands that activate these receptors, perhaps originating from the animal itself, or the environment. With respect to the latter, a first insight has come through *in silico* prediction of *Ta*RXR DNA binding motifs in the *Trichoplax* genome, identifying sets of genes involved in intracellular signal transduction, ubiquitin-dependent protein catabolic process, and protein translation (Reitzel et al., 2018).

8.3 CONCLUSIONS

To date, much has been learned about placozoan ecology, phylogeny, genomics, cell biology, cell signaling, and behavior. However, many questions remain. In their capacity to coordinate motile behaviors and respond to environmental stimuli, while lacking canonical neuromuscular systems generally associated with these capabilities, placozoans represent a unique and perhaps ancestral manifestation of animal life. In that, they provide exciting opportunities for uncovering steps that might have given rise to complex animal traits, including neural signaling, through modifications of an ancestral genetic toolkit, and its varying implementation across different cell types.

ACKNOWLEDGMENTS

We would like to thank Luis Yanez-Guerra for providing the BUSCO analysis on select gene datasets.

AUTHOR CONTRIBUTIONS

All authors contributed to the writing of the manuscript.

KEYWORDS

- **cell types**
- **gene datasets**
- **motile seawater invertebrates**
- **muscle cells**
- **neural signaling**
- **neuromuscular systems**

REFERENCES

Ahmad, M., & Cashmore, R. A., (1993). HY4 gene of *A. thaliana* encodes a protein with characteristics of a blue-light photoreceptor. *Nature, 366*, 162–166.

Airenne, K. J., et al., (2013). Baculovirus: An insect-derived vector for diverse gene transfer applications. *Molecular Therapy, 21*, 739–749.

Alberstein, R., Grey, R., Zimmet, A., Simmons, D. K., & Mayer, L. M., (2015). Glycine activated ion channel subunits encoded by ctenophore glutamate receptor genes. *Proceedings of the National Academy of Sciences, 112*, E6048.

Anderson, P. A., (2004). Cnidarian neurobiology: What does the future hold? *Hydrobiologia, 530*, 107–116.

Andrews, A. M., Jaron, D., Buerk, D. G., & Barbee, A. K., (2014). Shear stress-induced NO production is dependent on ATP autocrine signaling and capacitative calcium entry. *Cellular and Molecular Bioengineering, 7*, 510–520.

Arendt, D., et al., (2016). The origin and evolution of cell types. *Nature Reviews Genetics, 17*, 744–757.

Armon, S., Bull, M. S., Aranda-Diaz, A., & Prakash, M., (2018). Ultrafast epithelial contractions provide insights into contraction speed limits and tissue integrity. *Proceedings of the National Academy of Sciences, 115*, E10333–E10341.

Askwith, C. C., Cheng, C., Ikuma, M., Benson, C., Price, M. P., & Welsh, J. M., (2000). Neuropeptide FF and FMRFamide potentiate acid-evoked currents from sensory neurons and proton-gated DEG/ENaC channels. *Neuron, 26*, 133–141.

Assmann, M., Kuhn, A., Dürrnagel, S., Holstein, T. W., & Gründer, S., (2014). The comprehensive analysis of DEG/ENaC subunits in hydra reveals a large variety of peptide-gated channels, potentially involved in neuromuscular transmission. *BMC Biology, 12*, 84–84.

Audet, M., & Bouvier, M., (2012). Restructuring G-protein- coupled receptor activation. *Cell, 151*, 14–23.

Bargiela, D., Burr, S. P., & Chinnery, F. P., (2018). Mitochondria and hypoxia: Metabolic crosstalk in cell-fate decisions. *Trends in Endocrinology & Metabolism, 29*, 249–259.

Barzilai, M. G., et al., (2012). Convergent evolution of sodium ion selectivity in metazoan neuronal signaling. *Cell Reports, 2*, 242–248.

Bauknecht, P., & Jékely, G., (2015). Large-scale combinatorial deorphanization of *Platynereis* neuropeptide GPCRs. *Cell Reports, 12*, 684–693.

Behrendt, G., & Ruthmann, A., (1986). The cytoskeleton of the fiber cells of *Trichoplax adhaerens* (Placozoa). *Zoomorphology, 106*, 123–130.

Bell, G. W., Eggleston, D. B., & Wolcott, G. T., (2003). Behavioral responses of free-ranging blue crabs to episodic hypoxia. I. movement. *Marine Ecology Progress Series, 259*, 215–225.

Ben-Johny, M., & Yue, T. D., (2014). Calmodulin regulation (calmodulation) of voltage-gated calcium channels. *Journal of General Physiology, 143*, 679–692.

Bettler, B., & Mulle, C., (1995). AMPA and kainate receptors. *Neuropharmacology, 34*, 123–139.

Birstein, V., (1989). On the karyotype of *Trichoplax* sp. (Placozoa). *Biol. Zent. Bl, 108*, 63–67.

Bowie, D., (2008). Ionotropic glutamate receptors & CNS disorders. *CNS & Neurological Disorders-Drug Targets (Formerly Current Drug Targets-CNS & Neurological Disorders), 7*, 129–143.

Brenman, J. E., Christopherson, K. S., Craven, S. E., McGee, A. W., & Bredt, S. D., (1996). Cloning and characterization of postsynaptic density 93, a nitric oxide synthase interacting protein. *Journal of Neuroscience, 16*, 7407–7415.

Brenner, E. D., Stahlberg, R., Mancuso, S., Vivanco, J., Baluška, F., & Van, V. E., (2006). Plant neurobiology: An integrated view of plant signaling. *Trends in Plant Science, 11*, 413–419.

Bridgham, J. T., et al., (2010). Protein evolution by molecular tinkering: Diversification of the nuclear receptor superfamily from a ligand-dependent ancestor. *PLoS Biology, 8*, e1000497.

Buchholz, K., & Ruthmann, A., (1995). The mesenchyme-like layer of the fiber cells of *Trichoplax adhaerens* (Placozoa), a syncytium. *Zeitschrift für Naturforschung C, 50*, 282–285.

Bull, M. S., Kroo, L. A., & Prakash, M., (2021). Excitable Mechanics Embodied in a Walking Cilium. arXiv preprint arXiv:2107.02930.

Burger, G., Yan, Y., Javadi, P., & Lang, F. B., (2009). Group I-intron trans-splicing and mRNA editing in the mitochondria of placozoan animals. *Trends in Genetics, 25*, 381–386.

Burnstock, G., (2018). Purine and purinergic receptors. *Brain and Neuroscience Advances, 2*, 2398212818817494.

Bütschli, G., (1884). XLIII.—Remarks on the gastræa-theory. *Journal of Natural History, 13*, 372–383.

Cameron, P., Hiroi, M., Ngai, J., & Scott, K., (2010). The molecular basis for water taste in drosophila. *Nature, 465,* 91–95.

Campbell, R. D., (1972). Statocyst lacking cilia in the coelenterate *Corymorpha palma. Nature, 238*, 49–51.

Capella-Gutiérrez, S., Silla-Martínez, J. M., & Gabaldón, T., (2009). trimAl: A tool for automated alignment trimming in large-scale phylogenetic analyses. *Bioinformatics, 25*, 1972–1973.

Capiod, T., (2011). Cell proliferation, calcium influx and calcium channels. *Biochimie, 93*, 2075–2079.

Carbone, E., Giancippoli, A., Marcantoni, A., Guido, D., & Carabelli, V., (2006). A new role for T-type channels in fast "low-threshold" exocytosis. *Cell Calcium, 40*, 147–154.

Cashmore, A. R., Jarillo, J. A., Wu, Y. J., & Liu, D., (1999). Cryptochromes: Blue light receptors for plants and animals. *Science, 284*, 760–765.

Catterall, W. A., (2011). Voltage-gated calcium channels. *Cold Spring Harbor Perspectives in Biology, 3,* a003947.

Cens, T., et al., (2015). Molecular characterization and functional expression of the *Apis mellifera* voltage-dependent Ca^{2+} channels. *Insect Biochemistry and Molecular Biology, 58*, 12–27.

Chandel, N. S., & Schumacker, T. P., (2000). Cellular oxygen sensing by mitochondria: Old questions, new insight. *Journal of Applied Physiology, 88*, 1880–1889.

Chen, G. Q., Cui, C., Mayer, M. L., & Gouaux, E., (1999). Functional characterization of a potassium-selective prokaryotic glutamate receptor. *Nature, 402*, 817–821.

Chiang, M. Y., et al., (1998). An essential role for retinoid receptors RARβ and RXRγ in long-term potentiation and depression. *Neuron, 21*, 1353–1361.

Clagett-Dame, M., McNeill, E. M., & Muley, D. P., (2006). Role of all-trans retinoic acid in neurite outgrowth and axonal elongation. *Journal of Neurobiology, 66*, 739–756.

Clapham, D. E., (2007). Calcium signaling. *Cell, 131*, 1047–1058.

Colasanti, M., Mazzone, V., Mancinelli, L., Leone, S., & Venturini, G., (2009). Involvement of nitric oxide in the head regeneration of *Hydra vulgaris. Nitric Oxide, 21*, 164–170.

Colasanti, M., Venturini, G., Merante, A., Musci, G., & Lauro, M. G., (1997). Nitric oxide involvement in *Hydra vulgaris* very primitive olfactory-like system. *Journal of Neuroscience, 17*, 493–499.

Collier, M., Ji, G., Wang, Y. X., & Kotlikoff, M., (2000). Calcium-induced calcium release in smooth muscle: Loose coupling between the action potential and calcium release. *The Journal of general Physiology, 115*, 653–662.

Costas-Insua, C., Merino-Gracia, J., Aicart-Ramos, C., & Rodríguez-Crespo, I., (2018). Chapter 5 – subcellular targeting of nitric oxide synthases mediated by their n-terminal motifs. In: Donev, R., (ed.), *Advances in Protein Chemistry and Structural Biology* (pp. 165–195). Academic Press.

Cottrell, G., Green, K., & Davies, N., (1990). The neuropeptide Phe-Met-Arg-Phe-NH_2 (FMRFamide) can activate a ligand-gated ion channel in *Helix* neurons. *Pflügers Archiv, 416*, 612–614.

Croset, V., et al., (2010). Ancient protostome origin of chemosensory ionotropic glutamate receptors and the evolution of insect taste and olfaction. *PLoS Genetics, 6*, e1001064.

Cymes, G. D., & Grosman, C., (2016). Identifying the elusive link between amino acid sequence and charge selectivity in pentameric ligand-gated ion channels. *Proceedings of the National Academy of Sciences, 113*, E7106–E7115.

Daff, S., Sagami, I., & Shimizu, T., (1999). The 42-amino acid insert in the FMN domain of neuronal nitric-oxide synthase exerts control over Ca^{2+}/calmodulin-dependent electron transfer. *Journal of Biological Chemistry, 274*, 30589–30595.

Das, G. R. R., Scheurer, L., Pelczar, P., Wildner, H., & Zeilhofer, U. H., (2021). Neuron-specific spinal cord translatomes reveal a neuropeptide code for mouse dorsal horn excitatory neurons. *Scientific Reports, 11*, 5232.

Davey, F., Harris, S. J., & Cottrell, A. G., (2001). Histochemical localisation of FMRFamide-gated Na+ channels in *Helisoma trivolvis* and *Helix aspersa* neurones. *Journal of Neurocytology, 30*, 877–884.

De Hoog, E., Lukewich, M. K., & Spencer, E. G., (2019). Retinoid receptor-based signaling plays a role in voltage-dependent inhibition of invertebrate voltage-gated Ca^{2+} channels. *Journal of Biological Chemistry, 294*, 10076–10093.

De Mendoza, A., Sebé-Pedrós, A., & Ruiz-Trillo, I., (2014). The evolution of the GPCR signaling system in eukaryotes: Modularity, conservation, and the transition to metazoan multicellularity. *Genome Biology and Evolution, 6*, 606–619.

De Oliveira, A. L., Calcino, A., & Wanninger, A., (2019). Ancient origins of arthropod moulting pathway components. *eLife, 8*, e46113.

Delfosse, V., Maire, A. L., Balaguer, P., & Bourguet, W., (2015). A structural perspective on nuclear receptors as targets of environmental compounds. *Acta Pharmacol. Sin., 36*, 88–101.

Dellaporta, S. L., et al., (2006). Mitochondrial genome of *Trichoplax adhaerens* supports Placozoa as the basal lower metazoan phylum. *Proceedings of the National Academy of Sciences, 103*, 8751–8756.

Devine, E. L., Oprian, D. D., & Theobald, L. D., (2013). Relocating the active-site lysine in rhodopsin and implications for evolution of retinylidene proteins. *Proceedings of the National Academy of Sciences of the United States of America, 110*, 13351–13355.

Dmetrichuk, J. M., Carlone, R. L., Jones, T. R. B., Vesprini, N. D., & Spencer, E. G., (2008). Detection of endogenous retinoids in the molluscan CNS and characterization of the trophic and tropic actions of 9-cis retinoic acid on isolated neurons. *The Journal of Neuroscience, 28*, 13014.

Donald, J. A., Forgan, L. G., & Cameron, S. M., (2015). The evolution of nitric oxide signaling in vertebrate blood vessels. *Journal of Comparative Physiology B, 185*, 153–171.

Dong, N., Lee, D. W., Sun, H. S., & Feng, Z. P., (2018). Dopamine-mediated calcium channel regulation in synaptic suppression in *L. stagnalis* interneurons. *Channels, 12*, 153–173.

Dozin, B., Cahnmann, H. J., & Nikodem, M. V., (1985). Identification of thyroid hormone receptors in rat liver nuclei by photoaffinity labeling with L-thyroxine and triiodo-L-thyronine. *Biochemistry, 24*, 5197–5202.

Drillon, G., Carbone, A., & Fischer, G., (2014). SynChro: A fast and easy tool to reconstruct and visualize synteny blocks along eukaryotic chromosomes. *PloS One, 9*, e92621.

Driscoll, T., Gillespie, J. J., Nordberg, E. K., Azad, A. F., & Sobral, W. B., (2013). Bacterial DNA sifted from the *Trichoplax adhaerens* (Animalia: Placozoa) genome project reveals a putative rickettsial endosymbiont. *Genome Biology and Evolution, 5*, 621–645.

Driss-Ecole, D., Lefranc, A., & Perbal, G., (2003). A polarized cell: The root statocyte. *Physiologia Plantarum, 118*, 305–312.

DuBuc, T. Q., Ryan, J. F., & Martindale, Q. M., (2019). "Dorsal–ventral" genes are part of an ancient axial patterning system: Evidence from *Trichoplax adhaerens* (Placozoa). *Molecular Biology and Evolution, 36*, 966–973.

Eakin, R. M., & Kuda, A., (1970). Ultrastructure of sensory receptors in ascidian tadpoles. *Zeitschrift für Zellforschung und Mikroskopische Anatomie, 112*, 287–312.

Eckert, R., & Brehm, P., (1979). Ionic mechanisms of excitation in paramecium. *Annual Review of Biophysics and Bioengineering, 8*, 353–383.

Edgar, R. C., (2004). MUSCLE: Multiple sequence alignment with high accuracy and high throughput. *Nucleic Acids Research, 32*, 1792–1797.

Ehlers, U., (1991). Comparative morphology of statocysts in the plathelminthes and the xenoturbellida. *Hydrobiologia, 227*, 263–271.

Eipper, B. A., Stoffers, D. A., & Mains, E. R., (1992). The biosynthesis of neuropeptides: Peptide alpha-amidation. *Annual Review of Neuroscience, 15*, 57–85.

Eitel, M., & Schierwater, B., (2010). The phylogeography of the Placozoa suggests a taxon-rich phylum in tropical and subtropical waters. *Molecular Ecology, 19*, 2315–2327.

Eitel, M., et al., (2018). Comparative genomics and the nature of placozoan species. *Plos Biology, 16*, e2005359.

Eitel, M., Guidi, L., Hadrys, H., Balsamo, M., & Schierwater, B., (2011). New insights into placozoan sexual reproduction and development. *PLOS One, 6*, e19639.

Eitel, M., Osigus, H. J., DeSalle, R., & Schierwater, B., (2013). Global diversity of the Placozoa. *PloS One, 8*, e57131–e57131.

Elkhatib, W., Smith, C. L., & Senatore, A., (2019). A Na^+ leak channel cloned from *Trichoplax adhaerens* extends extracellular pH and Ca^{2+} sensing for the DEG/ENaC family close to the base of Metazoa. *Journal of Biological Chemistry, 294*, 16320–16336.

Emery, P., So, W. V., Kaneko, M., Hall, J. C., & Rosbash, M., (1998). CRY, a drosophila clock and light-regulated cryptochrome, is a major contributor to circadian rhythm resetting and photosensitivity. *Cell, 95*, 669–679.

Ender, A., & Schierwater, B., (2003). Placozoa are not derived cnidarians: Evidence from molecular morphology. *Molecular Biology and Evolution, 20*, 130–134.

Ertel, E. A., et al., (2000). Nomenclature of voltage-gated calcium channels. *Neuron, 25*, 533–535.

Escriva, H., Langlois, M, C., Mendonca, R. L., Pierce, R., & Laudet, V., (1998). Evolution and diversification of the nuclear receptor superfamily A. *Annals of the New York Academy of Sciences, 839*, 143–146.

Evans, R. M., & Mangelsdorf, J. D., (2014). Nuclear receptors, RXR, and the big bang. *Cell, 157*, 255–266.

Fain, G. L., Hardie, R., & Laughlin, B. S., (2010). Phototransduction and the evolution of photoreceptors. *Current Biology, 20*, R114–R124.

Farah, C., Michel, L. Y., & Balligand, J. L., (2018). Nitric oxide signaling in cardiovascular health and disease. *Nature Reviews Cardiology, 15*, 292–316.

Fei, F., et al., (2019). Transgenic strategies to generate heterogeneous hepatic cancer models in zebrafish. *Journal of Molecular Cell Biology, 11*, 1021–1023.

Feske, S., Wulff, H., & Skolnik, Y. E., (2015). Ion channels in innate and adaptive immunity. *Annual Review of Immunology, 33*, 291–353.

Feuda, R., Hamilton, S. C., McInerney, J. O., & Pisani, D., (2012). Metazoan opsin evolution reveals a simple route to animal vision. *Proceedings of the National Academy of Sciences, 109*, 18868–18872.

Feuda, R., Rota-Stabelli, O., Oakley, T. H., & Pisani, D., (2014). The comb jelly opsins and the origins of animal phototransduction. *Genome Biology and Evolution, 6*, 1964–1971.

Finch, E. A., Turner, T. J., & Goldin, M. S., (1991). Calcium as a coagonist of inositol 1, 4, 5-trisphosphate-induced calcium release. *Science, 252*, 443–446.

Fleming, J. F., Feuda, R., Roberts, N. W., & Pisani, D., (2020). A novel approach to investigate the effect of tree reconstruction artifacts in single-gene analysis clarifies opsin evolution in nonbilaterian metazoans. *Genome Biology and Evolution, 12*, 3906–3916.

Fortunato, A., & Aktipis, A., (2019). Social feeding behavior of *Trichoplax adhaerens. Frontiers in Ecology and Evolution, 7.*

Franzini-Armstrong, C., & Protasi, F., (1997). Ryanodine receptors of striated muscles: A complex channel capable of multiple interactions. *Physiological Reviews, 77*, 699–729.

Fritze, O., Filipek, S., Kuksa, V., Palczewski, K., Hofmann, K. P., & Ernst, P. O., (2003). Role of the conserved NPxxY(x)5,6F motif in the rhodopsin ground state and during activation. *Proceedings of the National Academy of Sciences of the United States of America, 100*, 2290–2295.

Fuxe, K., Borroto-Escuela, D. O., Romero-Fernandez, W., Zhang, W. B., & Agnati, F. L., (2013). Volume transmission and its different forms in the central nervous system. *Chinese Journal of Integrative Medicine, 19*, 323–329.

Gamper, N., & Shapiro, S. M., (2007). Target-specific PIP2 signaling: How might it work? *The Journal of Physiology, 582*, 967–975.

Gan, Q., Salussolia, C. L., & Wollmuth, P. L., (2015). Assembly of AMPA receptors: Mechanisms and regulation. *The Journal of Physiology, 593*, 39–48.

Gao, T., et al., (1997). cAMP-dependent regulation of cardiac L-type Ca^{2+} channels requires membrane targeting of PKA and phosphorylation of channel subunits. *Neuron, 19*, 185–196.

Gauberg, J., et al., (2020). Conserved biophysical features of the Ca_V2 presynaptic Ca^{2+} channel homologue from the early-diverging animal *Trichoplax adhaerens*. *Journal of Biological Chemistry, 295*, 18553–18578.

Gauberg, J., W. Elkhatib, C. L. Smith, A. Singh, and A. Senatore., A. (2022). Divergent Ca^{2+}/calmodulin feedback regulation of Ca_V1 and Ca_V2 voltage-gated calcium channels evolved in the common ancestor of Placozoa and Bilateria. *The Journal of Biological Chemistry*, 298, 101741.

Gegear, R. J., Casselman, A., Waddell, S., & Reppert, M. S., (2008). Cryptochrome mediates light-dependent magnetosensitivity in drosophila. *Nature, 454*, 1014–1018.

Gilbert, G., et al., (2017). T-type voltage gated calcium channels are involved in endothelium-dependent relaxation of mice pulmonary artery. *Biochemical* pharmacology, *138*, 61–72.

Glanzman, D. L., (2010). Common mechanisms of synaptic plasticity in vertebrates and invertebrates. *Current Biology, 20*, R31–R36.

Golubovic, A., et al., (2007). A peptide-gated ion channel from the freshwater polyp *Hydra*. *Journal of Biological Chemistry, 282*, 35098–35103.

Gosselin-Badaroudine, P., et al., (2016). Biophysical characterization of the honeybee DSC1 orthologue reveals a novel voltage-dependent Ca^{2+} channel subfamily: CaV4. *Journal of General Physiology, 148*, 133–145.

Green, K., Falconer, S., & Cottrell, G., (1994). The neuropeptide Phe-Met-Arg-Phe-NH_2 (FMRFamide) directly gates two ion channels in an identified *Helix* neurone. *Pflügers Archiv, 428*, 232–240.

Grell, K. G., & Benwitz, G., (1981). Ergänzende untersuchungen zur ultrastruktur von *Trichoplax adhaerens* FE Schulze (Placozoa). *Zoomorphology, 98*, 47–67.

Grell, K. G., (1972). Eibildung und furchung von *Trichoplax adhaerens* FE Schulze (*Placozoa*). *Zeitschrift für Morphologie der Tiere, 73,* 297–314.

Grell K. G., (1980). Einführung, Protozoa, Placozoa, Porifera, Stamm Placozoa. *Lehrbuch der Speziellen Zoologie Wirbellose Tiere* (pp. 247–250). Jena: Gustav Fischer Verlag.

Grell, K., & Benwitz, G., (1971). Die ultrastruktur von *Trichoplax adhaerens* FE schulze. *Cytobiologie, 4*, 216–240.

Grell, K., & Benwitz, G., (1974). Spezifische verbindungsstrukturen der faserzellen von *Trichoplax adhaerens* FE Schulze/Special connecting structures between fiber cells of *Trichoplax adhaerens* FE Schulze. *Zeitschrift für Naturforschung C, 29*, 790–790b.

Grell, K., & Ruthmann, A., (1991). *Placozoa* In: Harrison, F. W., & Westfall, J. A., (eds.), *Microscopic Anatomy of Invertebrates*, *Placozoa*, *Porifera*, *Cnidaria*, *and Ctenophora*. New York: Wiley-Liss.

Grell, K., (1971). Embryonalentwicklung bei *Trichoplax adhaerens* FE Schulze. *Naturwissenschaften, 58*, 570–570.

Grell, K., Ruthmann, A., Harrison, F., & Ruppert, E., (1991). In: Harrison, F. W., & Westfall, J. A., (eds.), *Microscopic Anatomy of Invertebrates, Placozoa, Porifera, Cnidaria and Ctenophora* (pp. 13–28).

Gruber-Vodicka, H. R., et al., (2019). Two intracellular and cell type-specific bacterial symbionts in the placozoan *Trichoplax* H2. *Nature Microbiology, 4*, 1465–1474.

Gründer, S., & Assmann, M., (2015). Peptide-gated ion channels and the simple nervous system of *Hydra*. *Journal of Experimental Biology, 218*, 551–561.

Guidi, L., Eitel, M., Cesarini, E., Schierwater, B., & Balsamo, M., (2011). Ultrastructural analyses support different morphological lineages in the phylum Placozoa Grell, 1971. *Journal of Morphology, 272,* 371–378.

Häder, D. P., Braun, M., Grimm, D., & Hemmersbach, R., (2017). Gravireceptors in eukaryotes—A comparison of case studies on the cellular level. *NPJ Microgravity, 3*, 13.

Hamasaki, T., Barkalow, K., Richmond, J., & Satir, P., (1991). cAMP-stimulated phosphorylation of an axonemal polypeptide that copurifies with the 22S dynein arm regulates microtubule translocation velocity and swimming speed in *Paramecium*. *Proceedings of the National Academy of Sciences, 88*, 7918–7922.

Han, G., Chang, B., Connor, M. J., & Sidell, N., (1995). Enhanced potency of 9-cis versus all-trans-retinoic acid to induce the differentiation of human neuroblastoma cells. *Differentiation, 59*, 61–69.

Han, M., Smith, S. O., & Sakmar, P. T., (1998). Constitutive activation of opsin by mutation of methionine 257 on transmembrane helix 6. *Biochemistry, 37*, 8253–8261.

Hanmer, S. B., (2014). *Comparative Molecular Physiology of Novel P2X Receptors: Identification, Cloning and Functional Characterization*. Cardiff University.

Hejnol, A., et al., (2009). Assessing the root of bilaterian animals with scalable phylogenomic methods. *Proceedings of the Royal Society B: Biological Sciences, 276*, 4261–4270.

Hemmersbach, R., & Braun, M., (2006). Gravity-sensing and gravity-related signaling pathways in unicellular model systems of protists and plants. *Signal Transduction, 6*, 432–442.

Hemmersbach, R., Volkmann, D., & Häder, & D, P., (1999). Graviorientation in protists and plants. *Journal of Plant Physiology, 154*, 1–15.

Heyland, A., Croll, R., Goodall, S., Kranyak, J., & Wyeth, R., (2014). *Trichoplax adhaerens*, an enigmatic basal metazoan with potential. *Developmental Biology of the Sea Urchin and Other Marine Invertebrates* (pp. 45–61.). Springer.

Hillier, B. J., Christopherson, K. S., Prehoda, K. E., Bredt, D. S., & Lim, A. W., (1999). Unexpected modes of PDZ domain scaffolding revealed by structure of nNOS-syntrophin complex. *Science, 284*, 812–815.

Hitomi, K., et al., (2009). Functional motifs in the (6-4) photolyase crystal structure make a comparative framework for DNA repair photolyases and clock cryptochromes. *Proceedings of the National Academy of Sciences, 106*, 6962–6967.

Hochstrasser, M., Carlson, G. L., Walczak, C. E., & Nelson, L. D., (1996). Paramecium has two regulatory subunits of cyclic AMP-dependent protein kinase, one unique to cilia. *Journal of Eukaryotic Microbiology, 43*, 356–362.

Horridge, G. A., (1965). Intracellular action potentials associated with the beating of the cilia in ctenophore comb plate cells. *Nature, 205*, 602–602.

Huang, X., Senatore, A., Dawson, T. F., Quan, Q., & Spafford, D. J., (2010). G-proteins modulate invertebrate synaptic calcium channel (LCav2) differently from the classical voltage-dependent regulation of mammalian Cav2. 1 and Cav2. 2 channels. *Journal of Experimental Biology, 213*, 2094–2103.

Iacobucci, G. J., & Popescu, K. G., (2017). NMDA receptors: Linking physiological output to biophysical operation. *Nature Reviews Neuroscience, 18*, 236–249.

Jacklet, J. W., (1997). Nitric oxide signaling in invertebrates. *Invertebrate Neuroscience, 3*, 1–14.

Jackson, A. M., & Buss, W. L., (2009). Shiny spheres of placozoans (*Trichoplax*) function in anti-predator defense. *Invertebrate Biology, 128*, 205–212.

Jakob, W., Sagasser, S., Dellaporta, S., Holland, P., Kuhn, K., & Schierwater, B., (2004). The trox-2 Hox/ParaHox gene of *Trichoplax* (Placozoa) marks an epithelial boundary. *Development Genes and Evolution, 214*, 170–175.

Jékely, G., (2009). Evolution of phototaxis. *Philosophical Transactions of the Royal Society B: Biological Sciences, 364*, 2795–2808.

Jékely, G., (2011). Origin and early evolution of neural circuits for the control of ciliary locomotion. *Proceedings of the Royal Society B: Biological Sciences, 278*, 914–922.

Jékely, G., (2013). Global view of the evolution and diversity of metazoan neuropeptide signaling. *Proceedings of the National Academy of Sciences, 110*, 8702–8707.

Jékely, G., (2021). The chemical brain hypothesis for the origin of nervous systems. *Philosophical Transactions of the Royal Society B: Biological Sciences, 376*, 20190761.

Jékely, G., Godfrey-Smith, P., & Keijzer, F., (2021). Reafference and the origin of the self in early nervous system evolution. *Philosophical Transactions of the Royal Society B: Biological Sciences, 376*, 20190764.

Jin, L., et al., (2016). A designed tryptophan-and lysine/arginine-rich antimicrobial peptide with therapeutic potential for clinical antibiotic-resistant *Candida albicans* vaginitis. *Journal of Medicinal Chemistry, 59*, 1791–1799.

Jonz, M. G., Buck, L. T., Perry, S. F., Schwerte, T., & Zaccone, G., (2016). Sensing and surviving hypoxia in vertebrates. *Annals of the New York Academy of Sciences, 1365*, 43–58.

Kaelin, W. G., & Ratcliffe, J. P., (2008). Oxygen sensing by metazoans: The central role of the HIF hydroxylase pathway. *Molecular Cell, 30*, 393–402.

Kamm, K., Osigus, H. J., Stadler, P. F., DeSalle, R., & Schierwater, B., (2018). *Trichoplax* genomes reveal profound admixture and suggest stable wild populations without bisexual reproduction. *Scientific Reports, 8*, 11168.

Kamm, K., Osigus, H. J., Stadler, P. F., DeSalle, R., & Schierwater, B., (2019). Genome analyses of a placozoan rickettsial endosymbiont show a combination of mutualistic and parasitic traits. *Scientific Reports, 9*, 1–12.

Kapli, P., & Telford, J. M., (2020). Topology-dependent asymmetry in systematic errors affects phylogenetic placement of Ctenophora and Xenacoelomorpha. *Science Advances, 6*, eabc5162.

Katz, P. S., & Lillvis, L. J., (2014). Reconciling the deep homology of neuromodulation with the evolution of behavior. *Current Opinion in Neurobiology, 29*, 39–47.

Kellenberger, S., & Schild, L., (2002). Epithelial sodium channel/degenerin family of ion channels: A variety of functions for a shared structure. *Physiological Reviews, 82*, 735–767.

Kenny, N. J., et al., (2020). Tracing animal genomic evolution with the chromosomal-level assembly of the freshwater sponge *Ephydatia muelleri*. *Nature Communications, 11*, 3676.

Kim, J. M., Altenbach, C., Kono, M., Oprian, D. D., Hubbell, W. L., & Khorana, G. H., (2004). Structural origins of constitutive activation in rhodopsin: Role of the K296/E113 salt bridge. *Proceedings of the National Academy of Sciences, 101*, 12508–12513.

Knudsen, G. M., Nishida, C. R., Mooney, S. D., & De Montellano, P. R. O., (2003). Nitric-oxide synthase (NOS) reductase domain models suggest a new control element in endothelial NOS that attenuates calmodulin-dependent activity. *Journal of Biological Chemistry, 278*, 31814–31824.

Kojima, K., Yamashita, T., Imamoto, Y., Kusakabe, T. G., Tsuda, M., & Shichida, Y., (2017). Evolutionary steps involving counterion displacement in a tunicate opsin. *Proceedings of the National Academy of Sciences, 114*, 6028.

Kostrouch, Z., Kostrouchova, M., Love, W., Jannini, E., Piatigorsky, J., & Rall, E. J., (1998). Retinoic acid X receptor in the diploblast, *Tripedalia cystophora*. *Proceedings of the National Academy of Sciences, 95*, 13442–13447.

Kozmik, Z., (2008). The role of Pax genes in eye evolution. *Brain Research Bulletin, 75*, 335–339.

Krishnan, A., Almén, M. S., Fredriksson, R., & Schiöth, B. H., (2012). The origin of GPCRs: Identification of mammalian like rhodopsin, adhesion, glutamate and frizzled GPCRs in fungi. *PloS One, 7*, e29817.

Krumbach, T., (1907). *Trichoplax*, die umgewandelte planula einer hydromeduse. *Zool. Anz., 31*, 450–454.

Kuhl, W., & Kuhl, G., (1963). Bewegungsphysiologische untersuchungen an *Trichoplax adhaerens* EF Schulze (Kurzvortrag mit film). *Zool. Anz., 26*, 460–469.

Kuhl, W., & Kuhl, G., (1966). Untersuchungen über das bewegungsverhalten von *Trichoplax adhaerens* FE Schulze (Zeittransformation: Zeitraffung). *Zeitschrift für Morphologie und Ökologie der Tiere, 56*, 417–435.

Kumar, D., Blaby-Haas, C. E., Merchant, S. S., Mains, R. E., King, S. M., & Eipper, A. B., (2016). Early eukaryotic origins for cilia-associated bioactive peptide-amidating activity. *Journal of Cell Science, 129,* 943–956.

Kumar, D., et al., (2017). A bioactive peptide amidating enzyme is required for ciliogenesis. *eLife, 6*, e25728.

Kuo, I. Y., & Ehrlich, E. B., (2015). Signaling in muscle contraction. *Cold Spring Harbor Perspectives in Biology, 7*, a006023.

Lattin, J., Zidar, D. A., Schroder, K., Kellie, S., Hume, D. A., & Sweet, J. M., (2007). G-protein-coupled receptor expression, function, and signaling in macrophages. *Journal of leukocyte Biology, 82*, 16–32.

Laudet, V., (1997). Evolution of the nuclear receptor superfamily: Early diversification from an ancestral orphan receptor. *Journal of Molecular Endocrinology, 19*, 207–226.

Lefebvre, P., Benomar, Y., & Staels, B., (2010). Retinoid X receptors: Common heterodimerization partners with distinct functions. *Trends in Endocrinology & Metabolism, 21*, 676–683.

Levina, N. N., Lew, R. R., Hyde, G. J., & Heath, B. I., (1995). The roles of Ca^{2+} and plasma membrane ion channels in hyphal tip growth of *Neurospora crassa*. *Journal of Cell Science, 108*, 3405–3417.

Leys, S. P., (2003). The significance of syncytial tissues for the position of the *Hexactinellida* in the metazoa. *Integrative and Comparative Biology, 43*, 19–27.

Leys, S., & Meech, R., (2006). Physiology of coordination in sponges. *Canadian Journal of Zoology, 84*, 288–306.

Li, X., et al., (2015). Major diversification of voltage-gated K+ channels occurred in ancestral parahoxozoans. *Proceedings of the National Academy of Sciences, 112*, E1010–E1019.

Liaudet, L., Soriano, F. G., & Szabó, C., (2000). Biology of nitric oxide signaling. *Critical Care Medicine, 28*, N37–N52.

Liebeskind, B. J., Hillis, D. M., & Zakon, H. H., (2011). Evolution of sodium channels predates the origin of nervous systems in animals. *Proceedings of the National Academy of Sciences, 108*, 9154–9159.

Liebeskind, B. J., D. M. Hillis, and H. H. Zakon., (2015). Convergence of ion channel genome content in early animal evolution. *Proceedings of the National Academy of Sciences, 112*, E846–E851.

Lin, C., & Shalitin, D., (2003). Cryptochrome structure and signal transduction. *Annual Review of Plant Biology, 54*, 469–496.

Lin, C., (2000). Plant blue-light receptors. *Trends in Plant Science, 5*, 337–342.

Lin, C., Yang, H., Guo, H., Mockler, T., Chen, J., & Cashmore, R. A., (1998). Enhancement of blue-light sensitivity of *Arabidopsis* seedlings by a blue light receptor cryptochrome. *Proceedings of the National Academy of Sciences, 95,* 2686.

Lin, H., Mann, K. J., Starostina, E., Kinser, R. D., & Pikielny, W. C., (2005). A *Drosophila* DEG/ENaC channel subunit is required for male response to female pheromones. *Proceedings of the National Academy of Sciences of the United States of America, 102*, 12831–12836.

Lingueglia, E., Champigny, G., Lazdunski, M., & Barbry, P., (1995). Cloning of the amiloride-sensitive FMRFamide peptide-gated sodium channel. *Nature, 378*, 730–733.

Liu, L., et al., (2003). Contribution of *Drosophila* DEG/ENaC genes to salt taste. *Neuron, 39*, 133–146.

Loenarz, C., et al., (2011). The hypoxia-inducible transcription factor pathway regulates oxygen sensing in the simplest animal, *Trichoplax adhaerens*. *EMBO Reports, 12*, 63–70.

Lüscher, C., & Slesinger, A. P., (2010). Emerging roles for G protein-gated inwardly rectifying potassium (GIRK) channels in health and disease. *Nature Reviews Neuroscience, 11*, 301–315.

Luxmi, R., Kumar, D., Mains, R. E., King, S. M., & Eipper, A. B., (2019). Cilia-based peptidergic signaling. *PLOS Biology, 17*, e3000566.

Lynagh, T., Mikhaleva, Y., Colding, J. M., Glover, J. C., & Pless, A. S., (2018). Acid-sensing ion channels emerged over 600 Mya and are conserved throughout the deuterostomes. *Proceedings of the National Academy of Sciences, 115*, 8430–8435.

Mackie, G., & Singla, C., (1983). Studies on hexactinellid sponges. I. Histology of *Rhabdocalyptus dawsoni* (Lambe, 1873). *Philosophical Transactions of the Royal Society of London. B, Biological Sciences, 301*, 365–400.

Mackie, G., (1990). The elementary nervous system revisited. *American Zoologist, 30*, 907–920.

Maden, M., & Hind, M., (2003). Retinoic acid, a regeneration-inducing molecule. *Developmental Dynamics: An Official Publication of the American Association of Anatomists, 226*, 237–244.

Maden, M., Sonneveld, E., Van, D. S. P. T., & Gale, E., (1998). The distribution of endogenous retinoic acid in the chick embryo: Implications for developmental mechanisms. *Development, 125*, 4133–4144.

Maeng, S., Kim, G. J., Choi, E. J., Yang, H. O., Lee, D. S., & Sohn, C. Y., (2012). 9-Cis-retinoic acid induces growth inhibition in retinoid-sensitive breast cancer and sea urchin embryonic cells via retinoid X receptor α and replication factor C3. *Molecular Endocrinology, 26*, 1821–1835.

Malacombe, M., Bader, M. F., & Gasman, S., (2006). Exocytosis in neuroendocrine cells: New tasks for actin. *Biochimica et Biophysica Acta (BBA)-Molecular Cell Research, 1763*, 1175–1183.

Mansvelder, H. D., & Kits, S. K., (2000). Calcium channels and the release of large dense core vesicles from neuroendocrine cells: Spatial organization and functional coupling. *Progress in Neurobiology, 62*, 427–441.

Marder, E., (2012). Neuromodulation of neuronal circuits: Back to the future. *Neuron, 76*, 1–11.

Martindale, M. Q., Pang, K., & Finnerty, R. J., (2004). Investigating the Origins of Triploblasty: Mesodermal' Gene Expression in a Diploblastic Animal, the Sea Anemone Nematostella Vectensis (Phylum, Cnidaria; Class, Anthozoa). *Development, 131*, 2463–2474.

Martinelli, C., & Spring, J., (2003). Distinct expression patterns of the two T-box homologues brachyury and Tbx2/3 in the placozoan *Trichoplax adhaerens*. *Development Genes and Evolution, 213*, 492–499.

Mayer, M. L., & Jegla, T., (2018). Glutamate receptors: Family matters. *Elife, 7*, e43815.

Mayer, M. L., (2021). Glutamate receptors from diverse animal species exhibit unexpected structural and functional diversity. *The Journal of Physiology, 10*, 2605–2613.

Mayorova, T. D., et al., (2018). Cells containing aragonite crystals mediate responses to gravity in *Trichoplax adhaerens* (Placozoa), an animal lacking neurons and synapses. *PLoS One, 13*, e0190905.

Mayorova, T. D., et al., (2021). Placozoan fiber cells: Mediators of innate immunity and participants in wound healing. *Scientific Reports, 11*, 1–18.

Mayorova, T. D., Hammar, K., Winters, C. A., Reese, T. S., & Smith, L. C., (2019). The ventral epithelium of *Trichoplax adhaerens* deploys in distinct patterns cells that secrete digestive enzymes, mucus or diverse neuropeptides. *Biol. Open, 8*, bio045674.

Mayr, E., (1942). *Systematics and the Origin of Species.* Columbia Univ. Press, New York.

Mazaira, G. I., et al., (2019). *Nuclear Receptors: A Historical Perspective* (pp. 1–5). Nuclear Receptors. Springer.

McEwan, I. J., Escriva, H., Bertrand, S., & Laudet, V., (2004). The evolution of the nuclear receptor superfamily. *Essays in Biochemistry, 40*, 11–26.

Milletti, F., (2012). Cell-penetrating peptides: Classes, origin, and current landscape. *Drug Discovery Today, 17*, 850–860.

Mills, D. B., et al., (2018). The last common ancestor of animals lacked the HIF pathway and respired in low-oxygen environments. *eLife, 7*, e31176.

Minamishima, Y. A., Moslehi, J., Padera, R. F., Bronson, R. T., Liao, R., & Kaelin, Jr. W. G., (2009). A feedback loop involving the Phd3 prolyl hydroxylase tunes the mammalian hypoxic response *in vivo*. *Molecular and Cellular Biology, 29*, 5729–5741.

Mirabeau, O., & Joly, J. S., (2013). Molecular evolution of peptidergic signaling systems in bilaterians. *Proceedings of the National Academy of Sciences, 110*, E2028–E2037.

Miyazawa, H., & Nakano, H., (2018). Multiple surveys employing a new sample-processing protocol reveal the genetic diversity of placozoans in Japan. *Ecology and Evolution, 8*, 2407–2417.

Miyazawa, H., Osigus, H. J., Rolfes, S., Kamm, K., Schierwater, B., & Nakano, H., (2020). Mitochondrial genome evolution of placozoans: Gene rearrangements and repeat expansions. *Genome Biology and Evolution, 13*.

Miyazawa, H., Yoshida, M. A., Tsuneki, K., & Furuya, H., (2012). Mitochondrial genome of a Japanese placozoan. *Zoological Science, 29*, 223–228.

Mockler, T., et al., (2003). Regulation of photoperiodic flowering by *Arabidopsis* photoreceptors. *Proceedings of the National Academy of Sciences of the United States of America, 100*, 2140–2145.

Monticelli, F., (1893). Treptoplax reptans ng, n. sp. Atti dell'Academia dei Lincei, Rendiconti (5) II: 39–40.

Moran, Y., & Zakon, H. H., (2014). The evolution of the four subunits of voltage-gated calcium channels: Ancient roots, increasing complexity, and multiple losses. *Genome Biology and Evolution, 6*, 2210–2217.

Moran, Y., Barzilai, M. G., Liebeskind, B. J., & Zakon, H. H., (2015). Evolution of voltage-gated ion channels at the emergence of Metazoa. *Journal of Experimental Biology, 218*, 515–525.

Moroz, L. L., & Kohn, B. A., (2015). Unbiased view of synaptic and neuronal gene complement in ctenophores: Are there pan-neuronal and pan-synaptic genes across metazoa? *Integrative and Comparative Biology*, icv104.

Moroz, L. L., (2009). On the independent origins of complex brains and neurons. *Brain, Behavior and Evolution, 74*, 177–190.

Moroz, L. L., et al., (2014a). The ctenophore genome and the evolutionary origins of neural systems. *Nature, 510*, 109–114.

Moroz, L. L., et al., (2014b). The ctenophore genome and the evolutionary origins of neural systems. *Nature, 510*, 109–114.

Moroz, L. L., et al., (2020a). The diversification and lineage-specific expansion of nitric oxide signaling in Placozoa: Insights in the evolution of gaseous transmission. *Scientific Reports, 10,* 13020.

Moroz, L. L., Meech, R. W., Sweedler, J. V., & Mackie, O. G., (2004). Nitric oxide regulates swimming in the jellyfish *Aglantha digitale*. *Journal of Comparative Neurology, 471*, 26–36.

Moroz, L. L., Romanova, D. Y., & Kohn, B. A., (2021). Neural versus alternative integrative systems: Molecular insights into origins of neurotransmitters. *Philosophical Transactions of the Royal Society B: Biological Sciences, 376*, 20190762.

Moroz, L. L., Sohn, D., Romanova, D. Y., & Kohn, A. B., (2020b). Microchemical identification of enantiomers in early-branching animals: Lineage-specific diversification in the usage of D-glutamate and D-aspartate. *Biochemical and Biophysical Research Communications, 527*, 947–952.

Moss, A. G., & Tamm, L. S., (1986). Electrophysiological control of ciliary motor responses in the ctenophore *Pleurobrachia*. *Journal of Comparative Physiology A, 158*, 311–330.

Murad, F., Lehman, D. C., Bennett, B. M., Molina, C., & Waldman, A. S., (1987). Regulation of guanylate cyclase by atrial natriuretic factor and the role of cyclic GMP in vasodilation. *The American Journal of the Medical Sciences, 294*, 139–143.

Murad, F., Mittal, C., Arnold, W., Katsuki, S., & Kimura, H., (1978). Guanylate cyclase: Activation by azide, nitro compounds, nitric oxide, and hydroxyl radical and inhibition by hemoglobin and myoglobin. *Advances in Cyclic Nucleotide Research, 9*, 145–158.

Naitoh, Y., & Kaneko, H., (1972). Reactivated triton-extracted models of paramecium: Modification of ciliary movement by calcium ions. *Science, 176*, 523–524.

Nakano, H., (2014). Survey of the Japanese coast reveals abundant placozoan populations in the Northern Pacific Ocean. *Scientific Reports, 4*, 1–5.

Nathan, C., & Xie, Q. W., (1994). Nitric oxide synthases: Roles, tolls, and controls. *Cell, 78*, 915–918.

Nathans, J., (1990). Determinants of visual pigment absorbance: Identification of the retinylidene Schiff's base counterion in bovine rhodopsin. *Biochemistry, 29*, 9746–9752.

Neff, E. P., (2018). What is a lab animal? *Lab Animal, 47*, 223–227.

Nguyen, L. T., Schmidt, H. A., Von, H. A., & Minh, Q. B., (2015). IQ-TREE: A fast and effective stochastic algorithm for estimating maximum-likelihood phylogenies. *Molecular Biology and Evolution, 32,* 268–274.

Ni, L., (2020). The structure and function of ionotropic receptors in drosophila. *Frontiers in Molecular Neuroscience, 13*, 266.

Nikitin, M., (2015). Bioinformatic prediction of *Trichoplax adhaerens* regulatory peptides. *General and Comparative Endocrinology, 212,* 145–155.

Nilsson, D. E., (2009). The evolution of eyes and visually guided behavior. *Philosophical Transactions of the Royal Society B: Biological Sciences, 364,* 2833–2847.

Nishida, C. R., & De Montellano, P. R. O., (1999). Autoinhibition of endothelial nitric-oxide synthase: Identification of an electron transfer control element. *Journal of Biological Chemistry, 274*, 14692–14698.

Nishida, Clinton R. et al., (2001). Control of electron transfer in nitric-oxide synthases: Swapping of autoinhibitory elements among nitric-oxide synthase isoforms. *Journal of Biological Chemistry, 276*, 20116–20124.

Nordström, K. J. V., Lagerström, M. C., Wallér, L. M. J., Fredriksson, R., & Schiöth, B. H., (2009). The secretin GPCRs descended from the family of adhesion GPCRs. *Molecular Biology and Evolution, 26*, 71–84.

Novotný, J. P., et al., (2017). *Trichoplax adhaerens* reveals a network of nuclear receptors sensitive to 9-cis-retinoic acid at the base of metazoan evolution. *PeerJ, 5*, e3789.

O'Brien, J. A., & Lummis, C. S., (2011). Nano-biolistics: A method of biolistic transfection of cells and tissues using a gene gun with novel nanometer-sized projectiles. *BMC Biotechnology, 11*, 1–6.

O'Hagan, R., Chalfie, M., & Goodman, B. M., (2005). The MEC-4 DEG/ENaC channel of *Caenorhabditis* elegans touch receptor neurons transduces mechanical signals. *Nature Neuroscience, 8*, 43–50.

Osigus, H. J., Rolfes, S., Herzog, R., Kamm, K., & Schierwater, B., (2019). *Polyplacotoma mediterranea* is a new ramified placozoan species. *Current Biology, 29*, R148, R149.

Ozturk, N., (2017). Phylogenetic and functional classification of the photolyase/cryptochrome family. *Photochemistry and Photobiology, 93*, 104–111.

Ozturk, N., et al., (2008). Purification and characterization of a type III photolyase from *Caulobacter crescentus. Biochemistry, 47*, 10255–10261.

Panchin, Y. V., Arshavsky, Y. I., Deliagina, T. G., Popova, L. B., & Orlovsky, N. G., (1995). Control of locomotion in marine mollusk *Clione limacina*. IX. Neuronal mechanisms of spatial orientation. *Journal of Neurophysiology, 73*, 1924–1937.

Paoletti, P., Bellone, C., & Zhou, Q., (2013). NMDA receptor subunit diversity: Impact on receptor properties, synaptic plasticity and disease. *Nature Reviews Neuroscience, 14*, 383–400.

Paps, J., & Holland, W. P., (2018). Reconstruction of the ancestral metazoan genome reveals an increase in genomic novelty. *Nature Communications, 9*, 1–8.

Pearse, V. B., & Voigt, O., (2007). Field biology of placozoans (*Trichoplax*): Distribution, diversity, biotic interactions. *Integrative and Comparative Biology, 47*, 677–692.

Pearse, V. B., (1989). Growth and Behavior of *Trichoplax adhaerens*: First Record of the Phylum Placozoa in Hawaii. *Pac Sci, 13*, 117–121.

Peirson, S. N., Halford, S., & Foster, G. R., (2009). The evolution of irradiance detection: Melanopsin and the non-visual opsins. *Philosophical Transactions of the Royal Society B: Biological Sciences, 364*, 2849–2865.

Peng, G., Shi, X., & Kadowaki, T., (2015). Evolution of TRP channels inferred by their classification in diverse animal species. *Molecular Phylogenetics and Evolution, 84*, 145–157.

Perez-Reyes, E., (2003). Molecular physiology of low-voltage-activated t-type calcium channels. *Physiological Reviews, 83*, 117–161.

Perez-Reyes E., (2010). G protein-mediated inhibition of Cav3.2 T-type channels revisited. *Molecular Pharmacology, 77*, 136–138.

Petty, K. J., Desvergne, B., Mitsuhashi, T., & Nikodem, M. V., (1990). Identification of a thyroid hormone response element in the malic enzyme gene. *Journal of Biological Chemistry, 265*, 7395–7400.

Petty, K. J., Morioka, H., Mitsuhashi, T., & Nikodem, M. V., (1989). Thyroid hormone regulation of transcription factors involved in malic enzyme gene expression. *Journal of Biological Chemistry, 264*, 11483–11490.

Piekut, T., et al., (2020). Early metazoan origin and multiple losses of a novel clade of RIM presynaptic calcium channel scaffolding protein homologs. *Genome Biology and Evolution, 12*, 1217–1239.

Pisani, D., et al., (2015). Genomic data do not support comb jellies as the sister group to all other animals. *Proceedings of the National Academy of Sciences, 112*, 15402–15407.

Pozdnyakov, I., Matantseva, O., & Skarlato, S., (2018). Diversity and evolution of four-domain voltage-gated cation channels of eukaryotes and their ancestral functional determinants. *Scientific Reports, 8*.

Price, M. B., Jelesko, J., & Okumoto, S., (2012). Glutamate receptor homologs in plants: Functions and evolutionary origins. *Frontiers in Plant Science, 3*, 235.

Prindle, A., Liu, J., Asally, M., Garcia-Ojalvo, J., & Süel, M. G., (2015). Ion channels enable electrical communication in bacterial communities. *Nature, 527*, 59–63.

Prole, D. L., & Taylor, W. C., (2019). Structure and function of IP3 receptors. *Cold Spring Harbor Perspectives in Biology, 11*, a035063.

Putnam, N. H., et al., (2007). Sea anemone genome reveals ancestral eumetazoan gene repertoire and genomic organization. *Science, 317*, 86–94.

Ramirez, M. D., et al., (2016). The last common ancestor of most bilaterian animals possessed at least nine opsins. *Genome Biology and Evolution, 8*, 3640–3652.

Ramos-Vicente, D., et al., (2018). Metazoan evolution of glutamate receptors reveals unreported phylogenetic groups and divergent lineage-specific events. *eLife, 7*, e35774.

Ramos-Vicente, D., Grant, S. G. N., & Bayés, À., (2021). Metazoan evolution and diversity of glutamate receptors and their auxiliary subunits. *Neuropharmacology, 195*, 108640.

Rassat, J., & Ruthmann, A., (1979). *Trichoplax adhaerens* FE Schulze (Placozoa) in the scanning electron microscope. *Zoomorphologie, 93*, 59–72.

Reardon, S., (2019). CRISPR gene-editing creates wave of exotic model organisms. *Nature, 568*, 441–443.

Reitzel, A. M., & Tarrant, M. A., (2009). Nuclear receptor complement of the cnidarian *Nematostella vectensis*: Phylogenetic relationships and developmental expression patterns. *BMC Evolutionary Biology, 9*, 1–13.

Reitzel, A. M., Behrendt, L., & Tarrant, M. A., (2010). Light entrained rhythmic gene expression in the sea anemone *Nematostella vectensis*: The evolution of the animal circadian clock. *PloS One, 5*, e12805–e12805.

Reitzel, A. M., et al., (2011). Nuclear receptors from the ctenophore *Mnemiopsis leidyi* lack a zinc-finger DNA-binding domain: Lineage-specific loss or ancestral condition in the emergence of the nuclear receptor superfamily? *Evodevo, 2*, 1–12.

Reitzel, A. M., Macrander, J., Mane-Padros, D., Fang, B., Sladek, F. M., & Tarrant, M. A., (2018). Conservation of DNA and ligand binding properties of retinoid X receptor from the placozoan *Trichoplax adhaerens* to human. *The Journal of Steroid Biochemistry and Molecular Biology, 184*, 3–10.

Reppert, S. M., (1997). Melatonin receptors: Molecular biology of a new family of G protein-coupled receptors. *Journal of Biological Rhythms, 12*, 528–531.

Rhinn, M., & Dollé, P., (2012). Retinoic acid signaling during development. *Development, 139*, 843–858.

Riedel, B., et al., (2014). Effect of hypoxia and anoxia on invertebrate behavior: Ecological perspectives from species to community level. *Biogeosciences, 11*, 1491–1518.

Ringrose, J. H., et al., (2013). Deep proteome profiling of *Trichoplax adhaerens* reveals remarkable features at the origin of metazoan multicellularity. *Nature Communications, 4*, 1408.

Rivera, A. S., et al., (2012). Blue-light-receptive cryptochrome is expressed in a sponge eye lacking neurons and opsin. *The Journal of Experimental Biology, 215*, 1278–1286.

Rizzuto, R., et al., (1998). Close contacts with the endoplasmic reticulum as determinants of mitochondrial Ca^{2+} responses. *Science, 280*, 1763–1766.

Rohde, K., Watson, N., & Faubel, A., (1993). Ultrastructure of the statocyst in an undescribed species of Iuridae (*Platyhelminthes, Rhabdocoela, Luridae*). *Australian Journal of Zoology, 41*, 215–224.

Roller, L., et al., (2008). The unique evolution of neuropeptide genes in the silkworm *Bombyx mori. Insect Biochemistry and Molecular Biology, 38*, 1147–1157.

Romanova, D. Y., et al., (2020a). Glycine as a signaling molecule and chemoattractant in *Trichoplax* (Placozoa): Insights into the early evolution of neurotransmitters. *NeuroReport, 31.*

Romanova, D. Y., et al., (2020b). Sodium action potentials in Placozoa: Insights into behavioral integration and evolution of nerveless animals. *Biochemical and Biophysical Research Communications, 532*, 120–126.

Romanova, D. Y., et al., (2021). Hidden cell diversity in Placozoa: Ultrastructural insights from *Hoilungia hongkongensis*. *Cell and Tissue Research.*

Rosenblum, E. B., Parent, C. E., & Brandt, E. E., (2014). The molecular basis of phenotypic convergence. *Annual Review of Ecology, Evolution, and Systematics, 45*, 203–226.

Rothwell, C. M., De Hoog, E., & Spencer, E. G., (2017). The role of retinoic acid in the formation and modulation of invertebrate central synapses. *Journal of Neurophysiology, 117*, 692–704.

Rovati, G. E., Capra, V., & Neubig, R. R., (2007). The highly conserved DRY motif of class AG protein-coupled receptors: Beyond the ground state. *Molecular Pharmacology, 71*, 959–964.

Ruthmann, A., & Terwelp, U., (1979) Disaggregation and reaggregation of cells of the primitive metazoon *Trichoplax adhaerens*. *Differentiation, 13*, 185–198.

Ruthmann, A., Behrendt, G., & Wahl, R., (1986). The ventral epithelium of *Trichoplax adhaerens* (Placozoa): Cytoskeletal structures, cell contacts and endocytosis. *Zoomorphology, 106*, 115–122.

Ruthmann, A., Grell, K., & Benwitz, G., (1981). DNA content and fragmentation of the egg nucleus of *Trichoplax adhaerens*. *Zeitschrift für Naturforschung C, 36*, 564–567.

Ryan, J. F., et al., (2013). The genome of the ctenophore *Mnemiopsis leidyi* and its implications for cell type evolution. *Science, 342*, 1242592.

Ryan, J. F., Pang, K., Mullikin, J. C., Martindale, M. Q., & Baxevanis, D. A., (2010). The homeodomain complement of the ctenophore *Mnemiopsis leidyi* suggests that Ctenophora and Porifera diverged prior to the ParaHoxozoa. *Evodevo, 1*, 1–18.

Sakmar, T. P., Franke, R. R., & Khorana, G. H., (1989). Glutamic acid-113 serves as the retinylidene Schiff base counterion in bovine rhodopsin. *Proceedings of the National Academy of Sciences, 86*, 8309–8313.

Salathe, M., (2007). Regulation of mammalian ciliary beating. *Annu. Rev. Physiol. 69*, 401–422.

Salerno, J. C., et al., (1997). An autoinhibitory control element defines calcium-regulated isoforms of nitric oxide synthase. *Journal of Biological Chemistry, 272*, 29769–29777.

Sancar, A., (2003). Structure and function of DNA photolyase and cryptochrome blue-light photoreceptors. *Chemical Reviews, 103*, 2203–2238.
Sassone-Corsi, P., (2012). The cyclic AMP pathway. *Cold Spring Harbor Perspectives in Biology, 4*, a011148.
Sather, W. A., & McCleskey, W. E., (2003). Permeation and selectivity in calcium channels. *Annual Review of Physiology, 65*, 133–159.
Say, T. E., & Degnan, M. S., (2020). Molecular and behavioral evidence that interdependent photo-and chemosensory systems regulate larval settlement in a marine sponge. *Molecular Ecology, 29*, 247–261.
Scheitlin, C. G., Julian, J. A., Shanmughapriya, S., Madesh, M., Tsoukias, N. M., & Alevriadou, R. B., (2016). Endothelial mitochondria regulate the intracellular Ca^{2+} response to fluid shear stress. *American Journal of Physiology-Cell Physiology, 310*, C479–C490.
Schierwater, B., (2005). My favorite animal, *Trichoplax adhaerens*. *BioEssays, 27*, 1294–1302.
Schierwater, B., De Jong, D., & DeSalle, R., (2009). Placozoa and the evolution of Metazoa and intrasomatic cell differentiation. *The International Journal of Biochemistry & Cell Biology, 41*, 370–379.
Schiöth, H. B., & Fredriksson, R., (2005). The GRAFS classification system of G-protein coupled receptors in comparative perspective. *General and Comparative Endocrinology, 142*, 94–101.
Schmidt, A., Bauknecht, P., Williams, E. A., Augustinowski, K., Gründer, S., & Jékely, G., (2018). Dual signaling of Wamide myoinhibitory peptides through a peptide-gated channel and a GPCR in *Platynereis*. *The FASEB Journal, 32*, 5338–5349.
Schnitzler, C. E., et al., (2012). Genomic organization, evolution, and expression of photoprotein and opsin genes in *Mnemiopsis leidyi*: A new view of ctenophore photocytes. *BMC Biology, 10*, 1–26.
Scholz, C. B., & Technau, U., (2003). The ancestral role of brachyury: Expression of NemBra1 in the basal cnidarian *Nematostella vectensis* (Anthozoa). *Development Genes and Evolution, 212*, 563–570.
Schoofs, L., Loof, A. D., & Hiel, M. B. V., (2017). Neuropeptides as regulators of behavior in insects. *Annual Review of Entomology, 62*, 35–52.
Schubert, M., & Gibert, Y., (2020). *Retinoids in Embryonic Development* (p. 1278). Multidisciplinary Digital Publishing Institute.
Schuchert, P., (1993). *Trichoplax adhaerens* (Phylum Placozoa) has cells that react with antibodies against the neuropeptide RFamide. *Acta Zoologica, 74*, 115–117.
Schulze, F. E., (1883). *Trichoplax adhaerens*, Nov. gen., Nov. spec. *Zool. Anz., 6*, 92–97.
Schulze, F. E., (1892). *Über Trichoplax Adhaerens*. Verlag der königlichen akademie der wissenschaften, in Commission bei G. Reimer.
Schwartz, V., (1984). Das radialpolare differenzierungsmuster bei *Trichoplax adhaerens* FE Schulze (*Placozoa*)/the radial polar pattern of differentiation in *Trichoplax adhaerens* FE Schulze (*Placozoa*). *Zeitschrift für Naturforschung, C, 39*, 818–832.
Sebé-Pedrós, A., et al., (2018). Early metazoan cell type diversity and the evolution of multicellular gene regulation. *Nature Ecology & Evolution, 2*, 1176–1188.
Senatore, A., & Spafford, D. J., (2010). Transient and big are key features of an invertebrate T-type channel (LCav3) from the central nervous system of *Lymnaea stagnalis*. *Journal of Biological Chemistry, 285*, 7447–7458.

Senatore, A., Guan, W., Boone, A. N., & Spafford, D. J., (2014). T-type channels become highly permeable to sodium ions using an alternative extracellular turret region (S5-P) outside the selectivity filter. *Journal of Biological Chemistry, 289*, 11952–11969.

Senatore, A., Raiss, H., & Le, P., (2016). Physiology and evolution of voltage-gated calcium channels in early diverging animal phyla: Cnidaria, Placozoa, Porifera and Ctenophora. *Frontiers in Physiology, 7*, 481.

Senatore, A., Reese, T. S., & Smith, L. C., (2017). Neuropeptidergic integration of behavior in *Trichoplax adhaerens*, an animal without synapses. *The Journal of Experimental Biology, 220*, 3381–3390.

Senatore, A., Zhorov, B. S., & Spafford, D. J., (2012). Cav3 T-type calcium channels. *Wiley Interdisciplinary Reviews: Membrane Transport and Signaling, 1*, 467–491.

Sever, R., & Glass, K. C., (2013). Signaling by nuclear receptors. *Cold Spring Harbor Perspectives in Biology, 5*, a016709.

Shah, A. S., Ben-Shahar, Y., Moninger, T. O., Kline, J. N., & Welsh, J. M., (2009). Motile cilia of human airway epithelia are chemosensory. *Science, 325*, 1131.

Shearman, M. S., Sekiguchi, K., & Nishizuka, Y., (1989). Modulation of ion channel activity: A key function of the protein kinase C enzyme family. *Pharmacological Reviews, 41*, 211–237.

Shichida, Y., & Matsuyama, T., (2009). Evolution of opsins and phototransduction. *Philosophical Transactions of the Royal Society B: Biological Sciences, 364*, 2881–2895.

Sigg, M. A., et al., (2017). Evolutionary proteomics uncovers ancient associations of cilia with signaling pathways. *Developmental Cell, 43*, 744–762. e711.

Signorovitch, A. Y., Buss, L. W., & Dellaporta, L. S., (2007). Comparative genomics of large mitochondria in placozoans. *PLoS Genetics, 3*, e13.

Signorovitch, A. Y., Dellaporta, S. L., & Buss, W. L., (2005). Molecular signatures for sex in the Placozoa. *Proceedings of the National Academy of Sciences, 102*, 15518–15522.

Simão, F. A., Waterhouse, R. M., Ioannidis, P., Kriventseva, E. V., & Zdobnov, M. E., (2015). BUSCO: Assessing genome assembly and annotation completeness with single-copy orthologs. *Bioinformatics, 31*, 3210 3212.

Simion, P., et al., (2017). A large and consistent phylogenomic dataset supports sponges as the sister group to all other animals. *Current Biology, 27*, 958–967.

Singla, C. L., (1975). Statocysts of hydromedusae. *Cell and Tissue Research, 158*, 391–407.

Smith, C. L., & Mayorova, D. T., (2019). Insights into the evolution of digestive systems from studies of *Trichoplax adhaerens*. *Cell and Tissue Research, 377*, 353–367.

Smith, C. L., & Reese, S. T., (2016). Adherens junctions modulate diffusion between epithelial cells in *Trichoplax adhaerens*. *The Biological Bulletin, 231*, 216–224.

Smith, C. L., et al., (2014). Novel cell types, neurosecretory cells, and body plan of the early-diverging metazoan *Trichoplax adhaerens*. *Current Biology, 24*, 1565–1572.

Smith, C. L., et al., (2017). Evolutionary insights into T-type Ca^{2+} channel structure, function, and ion selectivity from the *Trichoplax adhaerens* homologue. *Journal of General Physiology, 149*, 483–510.

Smith, C. L., Pivovarova, N., & Reese, S. T., (2015). Coordinated feeding behavior in *Trichoplax*, an animal without synapses. *PloS One, 10*, e0136098.

Smith, C. L., Reese, T. S., Govezensky, T., & Barrio, R. A., (2019a). Coherent directed movement toward food modeled in *Trichoplax*, a ciliated animal lacking a nervous system. *Proceedings of the National Academy of Sciences, 116*, 8901–8908.

Smith, S. J., et al., (2019b). Single-cell transcriptomic evidence for dense intracortical neuropeptide networks. *eLife, 8*, e47889.
Spafford, J. D., Chen, L., Feng, Z. P., Smit, A. B., & Zamponi, G. W., (2003a). Expression and modulation of an invertebrate presynaptic calcium channel α1 subunit homolog. *Journal of Biological Chemistry, 278*, 21178–21187.
Spafford, J. D., Dunn, T., Smit, A. B., Syed, N. I., & Zamponi, W. G., (2006). *In vitro* characterization of L-type calcium channels and their contribution to firing behavior in invertebrate respiratory neurons. *Journal of Neurophysiology, 95*, 42–52.
Spafford, J. D., et al., (2003b). Calcium channel structural determinants of synaptic transmission between identified invertebrate neurons. *Journal of Biological Chemistry, 278*, 4258–4267.
Srivastava, M., et al., (2008). The *Trichoplax* genome and the nature of placozoans. *Nature, 454*, 955–960.
Srivastava, M., et al., (2010). The *Amphimedon queenslandica* genome and the evolution of animal complexity. *Nature, 466*, 720–726.
Steinmetz, P. R., et al., (2012). Independent evolution of striated muscles in cnidarians and bilaterians. *Nature, 487*, 231–234.
Südhof, T. C., (2012). The presynaptic active zone. *Neuron, 75*, 11–25.
Suter, B., Livingstone-Zatchej, M., & Thoma, F., (1997). Chromatin structure modulates DNA repair by photolyase *in vivo*. *The EMBO Journal, 16*, 2150–2160.
Syed, T., & Schierwater, B., (2002a). The evolution of the Placozoa: A new morphological model. *Senckenbergiana Lethaea, 82*, 315–324.
Syed, T., & Schierwater, B., (2002b). *Trichoplax adhaerens*: Discovered as a missing link forgotten as a hydrozoan, re-discovered as a key to metazoan evolution. *Vie et Milieu/Life & Environment*, 177–187.
Szabo, C., (1995). Alterations in nitric oxide production in various forms of circulatory shock. *New Horizons (Baltimore, Md.), 3*, 2–32.
Taiakina, V., et al., (2013). The calmodulin-binding, short linear motif, NSCaTE is conserved in L-type channel ancestors of vertebrate Cav1. 2 and Cav1. 3 channels. *PloS One, 8*, e61765.
Takashima, S., Gold, D., & Hartenstein, V., (2013). Stem cells and lineages of the intestine: A developmental and evolutionary perspective. *Development Genes and Evolution, 223*, 85–102.
Talavera, K., & Nilius, B., (2006). Biophysics and structure–function relationship of T-type Ca^{2+} channels. *Cell Calcium, 40*, 97–114.
Tamm, S. L., (2014). Cilia and the life of ctenophores. *Invertebrate Biology, 133*, 1–46.
Tarade, D., Lee, J. E., & Ohh, M., (2019). Evolution of metazoan oxygen-sensing involved a conserved divergence of VHL affinity for HIF1α and HIF2α. *Nature Communications, 10*, 3293.
Tedford, H. W., & Zamponi, W. G., (2006). Direct G protein modulation of Cav2 calcium channels. *Pharmacological Reviews, 58*, 837–862.
Terakita, A., Koyanagi, M., Tsukamoto, H., Yamashita, T., Miyata, T., & Shichida, Y., (2004). Counterion displacement in the molecular evolution of the rhodopsin family. *Nature structural & Molecular Biology, 11*, 284–289.
Thiemann, M., & Ruthmann, A., (1988). *Trichoplax adhaerens* F. E. Schulze (Placozoa): The formation of swarmers. *Zeitschrift für Naturforschung, C, 43*, 955–957.

Thiemann M, & Ruthmann A., (1989). Microfilaments and microtubules in isolated fiber cells of *Trichoplax adhaerens* (Placozoa). *Zoomorphology, 109*, 89–96.

Thiemann M, & Ruthmann A., (1990). Spherical forms of *Trichoplax adhaerens* (Placozoa). *Zoomorphology, 110*, 37–45.

Thorn, P., Zorec, R., Rettig, J., & Keating, J. D., (2016). Exocytosis in non-neuronal cells. *Journal of Neurochemistry, 137*, 849–859.

Tichelaar, W., Safferling, M., Keinänen, K., Stark, H., & Madden, R. D., (2004). The three-dimensional structure of an ionotropic glutamate receptor reveals a dimer-of-dimers assembly. *Journal of Molecular Biology, 344*, 435–442.

Tochio, H., Mok, Y. K., Zhang, Q., Kan, H. M., Bredt, D. S., & Zhang, M., (2000). Formation of nNOS/PSD-95 PDZ dimer requires a preformed β-finger structure from the nNOS PDZ domain. *Journal of Molecular Biology, 303*, 359–370.

Todo, T., (1999). Functional diversity of the DNA photolyase/blue light receptor family. Mutation research. *DNA Repair, 434*, 89–97.

Tonini, R., et al., (1999). Expression of Ras-GRF in the SK-N-BE neuroblastoma accelerates retinoic-acid-induced neuronal differentiation and increases the functional expression of the IRK1 potassium channel. *European Journal of Neuroscience, 11*, 959–966.

Ueda, T., Koya, S., & Maruyama, K. Y., (1999). Dynamic patterns in the locomotion and feeding behaviors by the placozoan *Trichoplax adhaerence*. *Biosystems, 54*, 65–70.

Van, G. L., & Garrity, A. P., (2017). More than meets the IR: The expanding roles of variant ionotropic glutamate receptors in sensing odor, taste, temperature and moisture. *F1000Research, 6*.

Varoqueaux, F., et al., (2018). High cell diversity and complex peptidergic signaling underlie placozoan behavior. *Current Biology, 28*, 3495–3501. e3492.

Verkhratsky, A., & Burnstock, G., (2014). Biology of purinergic signaling: Its ancient evolutionary roots, its omnipresence and its multiple functional significance. *BioEssays, 36*, 697–705.

Verkhratsky, A., (2021). Early evolutionary history (from bacteria to hemichordata) of the omnipresent purinergic signaling: A tribute to Geoff Burnstock inquisitive mind. *Biochemical Pharmacology, 187*, 114261.

Vesprini, N. D., Dawson, T. F., Yuan, Y., Bruce, D., & Spencer, E. G., (2014). Retinoic acid affects calcium signaling in adult molluscan neurons. *Journal of Neurophysiology, 113*, 172–181.

Vick, J. S., & Askwith, C. C., (2015). ASICs and neuropeptides. *Neuropharmacology, 94*, 36–41.

Vinnikov, Y. A., (1974). Gravity reception. In: Vinnikov, Y. A., (ed.), *Sensory Reception: Cytology, Molecular Mechanisms and Evolution* (pp. 272–321). Springer Berlin Heidelberg.

Wagner, M., Han, B., & Jessell, T., (1992). Regional differences in retinoid release from embryonic neural tissue detected by an *in vitro* reporter assay. *Development, 116*, 55–66.

Wang, X., Iyer, A., Lyons, A. B., Körner, H., & Wei, W., (2019). Emerging roles for G-protein coupled receptors in development and activation of macrophages. *Frontiers in Immunology, 10*, 2031.

Watras, J., Bezprozvanny, I., & Ehrlich, B., (1991). Inositol 1, 4, 5-trisphosphate-gated channels in cerebellum: Presence of multiple conductance states. *Journal of Neuroscience, 11*, 3239–3245.

Weir, K., Dupre, C., Van, G. L., Lee, A. S., & Bellono, W. N., (2020). A molecular filter for the cnidarian stinging response. *Elife, 9*, e57578.

Weiss, N., & Zamponi, W. G., (2013). Control of low-threshold exocytosis by T-type calcium channels. *Biochimica et Biophysica Acta (BBA)-Biomembranes, 1828*, 1579–1586.

Wellendorph, P., & Bräuner-Osborne, H., (2009). Molecular basis for amino acid sensing by family CG-protein-coupled receptors. *British Journal of Pharmacology, 156*, 869–884.

Wenderoth, H., (1986). Transepithelial cytophagy by *Trichoplax adhaerens* F. E. Schulze (Placozoa) feeding on yeast. *Zeitschrift für Naturforschung C. 41*, 343–347.

Whalan, S., Webster, N. S., & Negri, P. A., (2012). Crustose coralline algae and a cnidarian neuropeptide trigger larval settlement in two coral reef sponges. *PLOS One, 7*, e30386.

Whelan, N. V., et al., (2017). Ctenophore relationships and their placement as the sister group to all other animals. *Nature Ecology & Evolution, 1*, 1737–1746.

Wickman, K., & Clapham, E. D., (1995). Ion channel regulation by G proteins. *Physiological reviews, 75*, 865–885.

Williams, E. A., et al., (2017). Synaptic and peptidergic connectome of a neurosecretory center in the annelid brain. *eLife, 6*, e26349.

Wolgemuth, C. W., (2005). Lamellipodial contractions during crawling and spreading. *Biophysical Journal, 89,* 1643–1649.

Wong, L. B., Park, C. L., & Yeates, B. D., (1998). Neuropeptide Y inhibits ciliary beat frequency in human ciliated cells via nPKC, independently of PKA. *American Journal of Physiology-Cell Physiology, 275*, C440–C448.

Wong, Y. Y., Le, P., Elkhatib, W., Piekut, T., & Senatore, A., (2019). Transcriptome Profiling of *Trichoplax adhaerens* Highlights its Digestive Epithelium and a Rich Set of Genes for Fast Electrogenic and Slow Neuromodulatory Cellular Signaling. *Research Square.*

Wuarin, L., Sidell, N., & De Vellis, J., (1990). Retinoids increase perinatal spinal cord neuronal survival and astroglial differentiation. *International Journal of Developmental Neuroscience, 8*, 317–326.

Yang, Q., Bermingham, N. A., Finegold, M. J., & Zoghbi, Y. H., (2001). Requirement of Math1 for secretory cell lineage commitment in the mouse intestine. *Science, 294*, 2155–2158.

Yañez-Guerra, L. A., D. Thiel, and G. Jékely. 2022. Premetazoan origin of neuropeptide signaling. Molecular Biology and Evolution 39. doi:10.1093/molbev/msac051

Young, J. Z., (1960). The statocysts of *Octopus vulgaris*. *Proceedings of the Royal Society of London. Series B. Biological Sciences, 152*, 3–29.

Yu, A., et al., (2016). Molecular lock regulates binding of glycine to a primitive NMDA receptor. *Proceedings of the National Academy of Sciences, 113*, E6786–E6795.

Yuan, Q., Metterville, D., Briscoe, A. D., & Reppert, M. S., (2007). Insect cryptochromes: Gene duplication and loss define diverse ways to construct insect circadian clocks. *Molecular Biology and Evolution, 24*, 948–955.

Zapata, F., et al., (2015). Phylogenomic analyses support traditional relationships within Cnidaria. *PLOS One, 10*, e0139068.

Zhang, T., Liu, Z., Song, W., Du, Y., & Dong, K., (2011). Molecular characterization and functional expression of the DSC1 channel. *Insect. Biochem. Mol. Biol., 41*, 451–458.

Zhou, L., & Zhu, & D, Y., (2009). Neuronal nitric oxide synthase: Structure, subcellular localization, regulation, and clinical implications. *Nitric Oxide, 20*, 223–230.

Zhou, W., Chung, I., Liu, Z., Goldin, A. L., & Dong, K., (2004). A voltage-gated calcium-selective channel encoded by a sodium channel-like gene. *Neuron, 42*, 101–112.

Index

B

H

I

T